International Series in Operations Research & Management Science

Volume 150

AF785749

For further volumes:
http://www.springer.com/series/6161

Gerd Infanger
Editor

Stochastic Programming

The State of the Art
In Honor of George B. Dantzig

Springer

Editor
Gerd Infanger
Department of Management
Science and Engineering
Stanford University
Stanford, CA 94305, USA
infanger@stanford.edu

Infanger Investment Technology, LLC
2685 Marine Way, Suite 1315
Mountain View, CA 94043, USA

ISSN 0884-8289
ISBN 978-1-4614-2762-9 ISBN 978-1-4419-1642-6 (eBook)
DOI 10.1007/978-1-4419-1642-6
Springer New York Dordrecht Heidelberg London

Printed on acid-free paper

Springer is part of Springer Science+Business Media (www.springer.com)

To Alex

Preface

The preparation of this book started in 2004, when George B. Dantzig and I, following a long-standing invitation by Frederick S. Hillier to contribute a volume to his International Series in Operations Research and Management Science, decided finally to go ahead with editing a volume on stochastic programming. The field of stochastic programming (also referred to as optimization under uncertainty or planning under uncertainty) had advanced significantly in the last two decades, both theoretically and in practice. George Dantzig and I felt that it would be valuable to showcase some of these advances and to present what one might call the state of the art of the field to a broader audience. We invited researchers whom we considered to be leading experts in various specialties of the field, including a few representatives of promising developments in the making, to write a chapter for this volume.

Unfortunately, to the great loss of all of us, George Dantzig passed away on May 13, 2005. Encouraged by many colleagues, I decided to continue with the book and edit it as a volume dedicated to George Dantzig.

Management Science published in 2005 a special volume featuring the "ten most influential papers of the first 50 years of management science." George Dantzig's original 1955 stochastic programming paper, "Linear Programming Under Uncertainty," was featured among these ten. Hearing about this, George Dantzig suggested that his 1955 paper be the first chapter of this book. The vision expressed in that paper gives an important scientific and historical perspective to the book.

An important approach for solving practical stochastic programs, especially when the number of stochastic parameters is large, involves Monte Carlo sampling. Sampling may be employed in two ways: within a decomposition scheme to estimate coefficients and right-hand sides of cutting planes or to generate an approximate (sampled) problem to be solved by a suitable deterministic algorithm. In each case, one is interested in the quality of the obtained solution, i.e., how close the obtained objective is to the (unknown) true optimal objective of the problem. Chapter 2 (by Dantzig and Infanger) derives a theory for probabilistic upper and lower bounds for the former approach. Chapter 3 (by Bayraksan, Morton, and Partani) discusses in detail various probabilistic bounds for the latter case. For the former case of using sampling, Chapter 4 (by Sen, Zhou, and Huang) reviews stochastic decomposition, a method introduced by Higle and Sen in the early 1990s, and extensions thereof, including regularized decomposition.

Exact bounds on the optimal objective of a stochastic program are typically based on Jensen's inequality for the lower bound and on refinements of the inequalities of Edmundson and Madansky for the upper bound. Tight bounds and accurate solutions require partitioning the sample space using a suitable geometry. The approach based on partitioning has been shown to be efficient especially when the number of (independent) stochastic parameters in a problem is small. Chapter 5 (by Frauendorfer, Kuhn, and Schürle) discusses an approach called barycentric approximation, wherein the upper bounds and the partitioning are based on simplices. The approach was successfully applied to financial and electric power planning problems. Chapter 6 (by Edirisinghe) gives an overview of deterministic approximation techniques based on limited moment information and discusses an application in the area of asset allocation.

Scenario generation is an important part of stochastic programming, in particular when, as part of a solution approach, it is necessary to approximate a given (e.g., continuous) distribution (or process) of random parameters with a discrete distribution (or process) over a manageable number of realizations. Chapter 7 (by Heitsch and Römisch) discusses (conditions for) the stability of the optimal objective with respect to perturbations of the underlying probability distribution for multistage stochastic programs. It presents a scenario tree generation approach based on recursive branching and deletion of branches while controlling the approximation error. The approach is applied to problems in electric power planning.

Chapter 8 (by Schultz) reviews risk aversion in the context of integer stochastic programming. It discusses two-stage mean–risk models with selected risk measures, such as excess probability and conditional value at risk, stability (with respect to change in measures), and also solution algorithms.

The starting point for any portfolio analysis is Harry Markowitz's mean–variance model. In contrast to traditional finance, the stochastic programming approach to portfolio analysis and construction represents (or approximates) the distribution (or process) of asset returns or prices directly through discrete scenarios. Considering explicitly higher moments of the joint distribution of asset returns or prices (e.g., skewness), considering explicitly the multiperiod nature of many investment problems, framing the problem as expected utility maximization, and/or applying various other one-sided risk measures typically result in different and often dynamic portfolio strategies. Chapters 9 through 13 address various aspects of portfolio optimization. Chapter 9 (by Dentcheva and Ruszczyński) presents portfolio construction based on stochastic dominance and discusses its relationship to expected utility maximization and to the dual (or rank-dependent) utility theory. Chapter 10 (by Markowitz and van Dijk) shows that mean–variance efficient portfolios provide a close approximation to the optimal solution of the more general dynamic (multiperiod) utility maximization problem. The chapter provides an extension of the mean–variance analysis that laid the foundation of modern finance—for which Harry Markowitz received the 1990 Nobel Prize in economics. Chapter 11 (by Konno) reviews the mean–absolute deviation portfolio optimization model and discusses its close relationship to the mean–variance portfolio optimization model of Markowitz. The chapter further discusses computational advantages of the mean–absolute deviation model for large-scale portfolio optimization. Chapter 12 (by Mulvey and Kim) discusses, in the context of multiperiod portfolio optimization, the integration of multistage stochastic programming with policy simulation and optimization. The chapter demonstrates that the combined approach results in practical portfolio strategies for asset-only and asset-liability situations. Chapter 13 (by MacLean, Zhao, and Ziemba) discusses the fundamental investment problem of allocating between risky and riskless investments over time in the framework of capital growth versus security, where (instead of traditional expected utility) stochastic dominance is used as a measure of investor risk preference.

Chapter 14 (by Higle and Kempf) develops a multistage stochastic programming model for production planning, where both supplies and demands are considered uncertain and driven by stochastic processes represented by Markov chains.

Chapter 15 on analyzing climate change under uncertainty was contributed by Alan Manne in August 2005. Alan Manne passed away on September 27, 2005, leaving a legacy of mathematical economic research behind. Chapter 15 was his final paper.

A related concept to multistage stochastic programming is stochastic dynamic programming, and one may interpret the former as an approximation to the latter. Another approximation is approximate dynamic programming, where the recourse (or value-to-go) function of a dynamic program is approximated via a linear combination of basis functions. Value function approximations have shown much promise when the state space is reasonably small. Chapter 16 (by Han and Van Roy) discusses advances in approximate stochastic dynamic programming, where, in particular, a linear program is solved to obtain the weights of the linear combination, and, instead of modeling discrete time decision processes, an approximation is constructed for diffusion processes directly.

I like to thank all authors for their great contributions to the book.

All chapters of this book were reviewed. The following referees contributed greatly in helping improve the chapters: Chanaka Edirisinghe (University of Tennessee), Karl Frauendorfer (University of St. Gallen), Peter Glynn (Stanford University), Julia Higle (Ohio State University), Daniel Kuhn (Imperial College of Science, Technology and Medicine), Leonard C. MacLean (Dalhousie University), David Morton (University of Texas, Austin), John Mulvey (Princeton University), Andrzej Ruszczyński (Rutgers University), Rene Schaub (formerly of Stanford University), Suvrajeet Sen (Ohio State University), John Stone (Infanger Investment Technology, LLC, and formerly of Stanford University), Peter Wöhrmann (Swiss Institute of Banking), Wei Wu (Stanford University), and two anonymous referees. I want to thank the reviewers for their valuable contributions to this book. Christopher Maes, a doctoral student at Stanford University, assisted in the finer issues of formatting the book. I greatly appreciate his assistance.

The book would have not happened without the initial insistence of Fred Hillier on preparing a book on stochastic programming. My thanks are extended to Fred for this.

I hope that researchers, graduate students, and practitioners alike find ideas expressed in the various chapters of this book useful for their research and practice and that the book not only presents a state of the art in an exciting field but helps spawn new developments in theory and application.

Stanford, California Gerd Infanger

George B. Dantzig and Optimization Under Uncertainty

George Bernard Dantzig (November 8, 1914–May 13, 2005) is considered by many as one of the great mathematicians of the twentieth century and was an icon in the fields of operations research and the management sciences. He was the inventor of the simplex method, an algorithm for solving the linear programming problem, for which efficient solution methodology led to its widespread application in the military, government, and industry. Today, despite the development of modern interior point methods, the simplex method remains a fast and reliable workhorse for solving large linear programming problems on a regular basis for many applications in a multitude of industries.

For his contribution, George B. Dantzig was honored considerably during his lifetime. In 1975 he was awarded by President Ford the National Medal of Science. He was the first recipient of the John von Neumann Prize in 1975, an honor established by the Institute for Operations Research and the Management Sciences that is awarded annually to an individual who has made fundamental and sustained contributions to theory in operations research and the management sciences. The Mathematical Programming Society and the Society for Industrial and Applied Mathematics established in 1979 the George B. Dantzig Prize, awarded for original research having a major impact on the field of mathematical programming. He was a member of the National Academy of Sciences, the National Academy of Engineering, and the American Academy of Arts and Sciences. In addition, he received numerous honorary doctorates from universities all over the world.

While he was honored and is commonly known mostly for his work in linear programming, his contributions are far wider and include large-scale linear systems, non-linear programming, integer linear programming, complementary programming, stochastic programming, mathematical statistics, and economics. An excellent overview of George B. Dantzig's important work is given in the book edited by Richard Cottle *The Basic George Dantzig* (Cottle 2003). Also the reminiscenses, Cottle (2005, 2006), Cottle et al. (2007), and Gill et al. (2008), demonstrate his lifetime breadth of scientific contributions.

Within the wide range of his work, one theme seemed to persist throughout his scientific career. His quest to "automatize (or mechanize) the planning process" grew out of an assignment at the Pentagon (where during the Second World War he was in charge of planning military supplies) and persisted during his years at the RAND Corporation (1952–1960), as a professor in the Department of Industrial Engineering at the University of California at Berkeley (1960–1966), where he founded and directed the Operations Research Center, and as the C. A. Criley Professor of Transportation Sciences and Professor of Operations Research and Computer Science at Stanford University (1966–1996), where he founded and directed the Systems Optimization Laboratory in the Department of Operations Research, and into forced retirement. His "basic" interest was in the theory, solution, and application of time-staged linear systems, where the treatment of uncertainty in the data became more and more his favorite interest.

Early on, and not too long after having proposed the simplex method in 1948 (Dantzig 1949), George Dantzig recognized that the "real problem" concerned programming under uncertainty. In 1955 he wrote "Linear Programming Under Uncertainty" (Dantzig 1955), reprinted in this book as the first chapter, a seminal paper in which he introduced uncertain parameters as coefficients and right-hand sides of the linear programming problem. He outlined by way of examples when using expected values for the distributions suffices (noting that "it may not be desirable to minimize the expected value of the costs if the decision has too great a variation in the actual total costs" and referencing Harry Markowitz's work on mean–variance analysis (Markowitz 1953) that enables an investor to sacrifice some of his expectation to control his risks), the simple recourse model, the two-stage linear program with recourse, and the multistage linear program with recourse, now all important archetypes in the field of stochastic programming. Dantzig's vision in his 1955 paper was extraordinary, and the paper has been reprinted in 2005 by *Management Science* as one of the ten most influential papers of the first 50 years of management science.

Independently, as Dantzig said, E. M. L. Beale, at almost the same time, proposed ways to solve stochastic programs (Beale 1955). Beale's paper "On minimizing a convex function subject to linear inequalities," included in its Section 5 titled "Linear programming with random coefficients" a generalization of linear programming in which the coefficients in the constraints are random, showed this to involve the minimization of a convex function. Beale's paper gives a footnote added at proof that "similar ideas have been put forward by Dantzig (1955) in Report-P596 of the Rand Corporation, Santa Monica, California, entitled "Linear Programming Under Uncertainty." Both papers are the earliest and thus founding expositions of stochastic programming, putting forward the recourse problem. Another early, but later treatment of stochastic programming is Charnes and Cooper (1959), establishing the different methodology of stochastic programs with chance constraints.

Dantzig's early work on stochastic programming was motivated by work with A.R. Ferguson on the assignment of aircraft to routes (Ferguson and Dantzig 1954), where Ferguson and Dantzig (1956) considered uncertainty in the demands. His paper "On the Solution of Two-Stage Linear Programs Under Uncertainty" (Dantzig and Madansky 1961) applied the Dantzig–Wolfe decomposition principle (Dantzig and Wolfe 1960) to solving the two-stage stochastic linear problem. Dantzig–Wolfe decomposition preceded Benders decomposition (Benders 1962). The latter, originally put forward for solving mixed-integer linear programs, in its linear version is dual (or identical, depending on the viewpoint) to Dantzig–Wolfe decomposition, and it is routinely used for solving stochastic programs with recourse. Van Slyke and Wets, both students of Dantzig, by applying Benders decomposition, originated the well-known L-shaped method for solving two-stage stochastic linear programs (Van Slyke and Wets 1969).

The Dantzig and Madansky (1961) paper presented supporting hyperplanes to the recourse function of a stochastic program noting it to be an application of Kelley's cutting plane algorithm (Kelley 1960). The paper concludes with the remark, "An interesting area of future consideration is the effect of sampling from the distribution [of the right-hand side]" to statistically estimate cut gradients and intercepts. Approaches based on sampling were three decades later pioneered by Dantzig, Glynn, and Infanger (Dantzig and Glynn 1990, Infanger 1992), and Higle and Sen (1994). As of this writing, using sampling is the most powerful approach for solving general types of stochastic programming problems.

Linear programming under uncertainty is also represented in Dantzig's book *Linear Programming and Extensions* (Dantzig 1963), detailing further early important cases and applications in Chapter 25, titled "Uncertainty," and in Chapter 28, titled "The Allocation of Aircraft to Routes Under Uncertainty," the latter being noted as based on the work with Ferguson. Dantzig considered his work on "Time-Staged Methods in Linear Programs" (Dantzig 1982) and generally his work on large-scale methods as all being useful for eventually solving large-scale stochastic programs.

A new approach for solving large-scale stochastic programs that combines decomposition, Monte Carlo sampling, and parallel processors was explored, starting in 1989, by Dantzig and Glynn (1990) and Infanger (1992). The approach used importance sampling to estimate expected second-stage costs, as well as coefficients and right-hand sides of Benders cuts. Success with the approach spawned a close collaboration over several years between Dantzig and Infanger on stochastic programming and its applications. Dantzig and Infanger (1992) give numerical results for practical large-scale electric power planning and financial problems, showing that a significant variance reduction in the estimation of cuts resulted in small confidence intervals around the optimal objective of problems. Such problems were hitherto considered intractable due to their size. Dantzig and Infanger (1993) discussed multistage portfolio optimization problems, Dantzig and Infanger (1997) outlined stochastic multistage control problems, and Dantzig and Infanger (1995) provided the theory for probabilistic bounds on the objective of two-stage stochastic programs, when solved by sampling.

George Dantzig considered stochastic programming as one of the most promising areas of management science and saw "planning under uncertainty," as he liked to call it, as "an unresolved problem of finding an optimal allocation of scarce resources that is fundamental to operations research, control theory, and economics." He was a relentless promoter of the subject throughout the last 30 years and inspired many of his students and collaborators to become leaders in the field. During the 1990s he pushed to establish an "Institute for Planning Under Uncertainty," with the goal to promote the exchange of ideas within the field.

In a special session at the Tenth International Conference On Stochastic Programming (Tucson, Arizona, USA, October 9–15, 2004), George B. Dantzig was honored as a Pioneer of Stochastic Programming (along with ten others for their long-standing contributions). At that event, a poster (put together by David Morton) and a presentation (by Gerd Infanger) on Dantzig's contributions to stochastic programming recognized Dantzig, in addition to being well known as the father of linear programming, also as the father of stochastic programming.

George Dantzig's discoveries were generally motivated by applications. As he tells the story, the decomposition principle was discovered with Philip Wolfe on a plane flight from Texas to Santa Monica, after visiting an oil company and initially underestimating the size of their linear program. Dantzig later put it in the following perspective: "You have to first promise the world and then work hard to make it all work."

Let me add some personal observations from working closely with George Dantzig between 1989 and 1996 and more sporadically thereafter. When a student walked into his office to present an idea, George would somehow mentally clear his cluttered (with piles of papers and books) table and lend his full attention to the student, as if there were only this one thought and there was no history to it. He would then make a comment, giving a unique insight, a different perspective, or a thought extending the scope. Then he would say, "This is really interesting; you should keep working on this." The student would walk away, feeling that he/she truly could do something really important, and would be even more motivated to make the best out of his/her idea. George must have heard one or the other idea before, but he always gave his full attention to the thought. There might be something new to it.

His great way of motivating others is represented in his many successful students with important contributions of their own right.

When I came to Stanford at George's invitation in January of 1989, I had planned to work on large-scale systems and to study using decomposition on a special class of such systems. I was aware of optimization under uncertainty, but thought to stay initially away from this area, especially the algorithmic side of it. After George and I had sat together and I had explained what I wanted to do, George asked me if I would like to participate in his uncertainty research. Without hesitation I simply said yes, changing on the spot part of the plans I had made earlier. That very moment surely has changed the course of my life. After a few meetings that followed, I also felt that I could contribute something to this area and was highly motivated to work in the field of optimization under uncertainty.

Just as George's influence changed the course of my life, George touched many lives and by his pioneering work provided the basis for many successful careers.

After having initial success in solving large stochastic programs using an approach of combining decomposition and Monte Carlo sampling techniques, George and I were very focused on trying to establish bounds on the optimal objective of a stochastic program, when sampled cuts were used for its solution. This research issue occupied us over many years with uncountable interesting discussions and fights (as George called it) of different ideas. At times, the interest in the subject was so great that almost no day passed without a discussion about it. The problem seemed to be constantly on George's mind. I remember well, when we started discussing the subject at Gail Stein's (former Operations Research department administrator and George's administrative assistant) holiday party on the evening of December 24, 1989. Despite being among many good friends at a great party, George said "Hey Gerd, we need to get back to our favorite subject. Let's not waste time here, let's go to my study and work on our theory." And we did—we scribbled on the blackboard until about 11 at night, when George's wife Anne brought us a wonderful meal. We continued the discussion until well past midnight. It was a mathematically intense yet sort of festive evening.

George Dantzig was a true believer in what he was doing. It is well known that he once had his wife Anne cook according to his linear programming results on the diet problem. But he would go even further. He timed his (heart) valve replacement surgery according to a stochastic decision model. He had decided that the physicians did not have the right objective, at least for his goals, and he implemented a solution that maximized his expected *useful* life expectancy. He planned then for a few more productive years that, happily, in the end were exceeded by far.

One day George and I walked from the faculty club toward the office. We could see a huge crane being used for constructing the Earth Sciences building. George said "Whoever invented the crane must not have been a construction engineer!" and commented further that the crane represents such an out-of-the-box approach that only an outsider could have had the boldness to consider it. George was correct. The tower crane was patented by a contractor, not a studied engineer, an entrepreneur subsequently building a crane manufacturing company that now operates worldwide. George always tended to have the right hunch.

George B. Dantzig was ever in pursuit of big ideas. In his quest he never felt constrained by perceived limits of what could be done. He was a visionary and true believer in the application of mathematics to real-life decision making. George and I shared the vision that stochastic programming will one day be widely used in practical decision making. Experience shows the vision is well under way.

George Dantzig was one of the most generous persons I have known and one of the nicest persons around. To me he was a mentor, colleague, and great friend. With warm feelings and great respect I wish to dedicate this book to the memory of George B. Dantzig.

Stanford, California Gerd Infanger

References

Beale, E.M.L.: On minimizing a convex function subject to linear inequalities. J. R. Stat. Soc. **17b**, 173–184 (1955)

Benders, J.F.: Partitioning procedures for solving mixed-variable programming problems. Numerische Math. **4**, 238–252 (1962)

Charnes, A., Cooper, A.A.: Chance constrained programming. Manage. Sci. **6**, 73–79 (1959)

Cottle, R.W.: The Basic George Dantzig. Stanford University Press, Stanford, CA (2003)
Cottle, R.W.: George B. Dantzig: Operations research icon. Oper. Res. **53**(6), 892–898 (2005)
Cottle, R.W.: George B. Dantzig: A legendary life in mathematical programming. Math. Program. **105**(1), 1–8 (2006)
Cottle, R.W., Johnson, E., Wets, R.: George B. Dantzig (1914–2005). Notices. Am. Math. Soc. **54**(3), 344–362 (2007)
Dantzig, G.B.: Programming in a linear structure. Econometrica **17**, 73–74 (1949)
Dantzig, G.B.: Linear programming under uncertainty. Manage. Sci. **1**, 197–206 (1955)
Dantzig, G.B.: Linear Programming and Extensions. Princeton University Press, Princeton, NJ (1963)
Dantzig, G.B.: Time-staged methods in linear programs. In: Haims, Y.Y. (ed.) Large-Scale Systems. Studies in Management Science, Vol. 7. North-Holland, Amsterdam Holland, (1982)
Dantzig, G.B., Glynn, P.W.: Parallel processors for planning under uncertainty. Ann. Oper. Res. **22**, 1–21 (1990)
Dantzig, G.B., Infanger, G.: Large-scale stochastic linear programs—importance sampling and Benders decomposition. In: Brezinski, C., Kulisch, U. (eds.) Computational and Applied Mathematics, I (Dublin, 1991), pp. 111–120. North-Holland, Amsterdam, Holland (1992)
Dantzig, G.B., Infanger, G.: Multi-stage stochastic linear programs for portfolio optimization. Ann. Oper. Res. **45**, 59–76 (1993)
Dantzig, G.B., Infanger, G.: A probabilistic lower bound for two-stage stochastic programs, Department of Operations Research, Stanford University, Stanford, CA, (1995)
Dantzig, G.B., Infanger, G.: Stochastic intelligent control and optimization under uncertainty with application to hydro-power systems. Eur. J. Oper. Res. **97**(2), 396–407 (1997)
Dantzig, G.B., Madansky, M.: On the solution of two-staged linear programs under uncertainty. In Neyman, J. (ed.) Proceeding of 4th Berkeley Symp on Mathematical Statistics and Probability I, pp. 165–176. University of California Press, Berkeley, CA (1961)
Dantzig, G.B., Wolfe, P.: The decomposition principle for linear programs. Oper. Res. **8**, 110–111 (1960)
Ferguson, A.R., Dantzig, G.B.: Notes on linear programming: Part XVI—the problem of routing aircraft. Aeronaut. Eng. Rev. **14**(4), 51–55 (1954)
Ferguson, A.R., Dantzig, G.B.: The allocation of aircraft to routes: An example of linear programming under uncertain demand. Manage. Sci. **3**, 45–73 (1956)
Gill, P.E., Murray, W., Saunders, M.A., Tomlin, J.A., Wright, M.H.: George B. Dantzig and systems optimization. Discrete Optim. **5**(2), 151–158 (2008) In Memory of Dantzig, G.B.
Higle, J.L., Sen, S.: Finite master programs in regularized stochastic decomposition. Math. Program. **67**(2, Ser. A), 143–168 (1994)
Infanger, G.: Monte Carlo (importance) sampling within a Benders decomposition algorithm for stochastic linear programs. Ann. Oper. Res. **39**(1–4), 69–95 (1992)
Kelley, J.E.: The cutting-plane method for solving convex programs. J. Soc. Ind. Appl. Math. **8**(4), 703–712 (1960)
Markowitz, H.: Portfolio Selection. PhD thesis, The University of Chicago, Chicago, IL (1953)
Van Slyke, R.M., Wets, R.J.: L-shaped linear programs with applications to optimal control and stochastic programming. SIAM J. Appl. Math. **17**, 638–663 (1969)

Contents

1 **Linear Programming Under Uncertainty** 1
George B. Dantzig

2 **A Probabilistic Lower Bound for Two-Stage Stochastic Programs** ... 13
George B. Dantzig and Gerd Infanger

3 **Simulation-Based Optimality Tests for Stochastic Programs** 37
Güzin Bayraksan, David P. Morton, and Amit Partani

4 **Stochastic Decomposition and Extensions** 57
Suvrajeet Sen, Zhihong Zhou, and Kai Huang

5 **Barycentric Bounds in Stochastic Programming: Theory and Application** 67
Karl Frauendorfer, Daniel Kuhn, and Michael Schürle

6 **Stochastic Programming Approximations Using Limited Moment Information, with Application to Asset Allocation** 97
N. Chanaka P. Edirisinghe

7 **Stability and Scenario Trees for Multistage Stochastic Programs** ... 139
Holger Heitsch and Werner Römisch

8 **Risk Aversion in Two-Stage Stochastic Integer Programming** 165
Rüdiger Schultz

9 **Portfolio Optimization with Risk Control by Stochastic Dominance Constraints** 189
Darinka Dentcheva and Andrzej Ruszczyński

10 Single-Period Mean–Variance Analysis in a Changing World 213
Harry M. Markowitz and Erik L. van Dijk

11 Mean–Absolute Deviation Model 239
Hiroshi Konno

12 Multistage Financial Planning Models: Integrating Stochastic Programs and Policy Simulators 257
John M. Mulvey and Woo Chang Kim

13 Growth–Security Models and Stochastic Dominance 277
Leonard C. MacLean, Yonggan Zhao, and William T. Ziemba

14 Production Planning Under Supply and Demand Uncertainty: A Stochastic Programming Approach 297
Julia L. Higle and Karl G. Kempf

15 Global Climate Decisions Under Uncertainty 317
Alan S. Manne

16 Control of Diffusions via Linear Programming 329
Jiarui Han and Benjamin Van Roy

Index 355

List of Figures

2.1 APL1P: lower bound distributions 30
2.2 STORM: lower bound distributions 32
2.3 Scatter diagrams of $\tilde{\mu}^k$ versus $\tilde{\delta}^k$ (problem PGP2, 1000 observations) 33
4.1 Solution times for regularized SD: with and without resampling 64
4.2 Comparison of solution quality: SD vs SDR 65
5.1 Coverages for the uncorrelated and correlated case with barycenters 91
5.2 Possible refinements of a scenario tree and simplices. (a) No refinements; (b) Refinement in A; (c) Refinement in B; (d) Split of $\mathcal{E}_t^{\mathrm{o}}$ or $\mathcal{E}_t^{\mathrm{r}}$; (e) Alternative edges; (f) Alternative points 93
6.1 First-moment upper bound 107
6.2 Mean–covariance lower/upper bounds 115
6.3 Mean–covariance lower bound for convex case 116
6.4 Nonlinearity measure $\Delta(i, j) = \min\{AB, CD\}$ 124
6.5 Conditional mean partitioning of a simplicial cell 125
6.6 Cell-redefining procedure for a scenario-based simplex in two dimensions 126
6.7 Embarrassment penalty by wealth target shortfall 127
6.8 SSS performance on AAM with 500,000 scenarios (0% TC) 129
6.9 Risk–return trade-off in AAM with 500,000 scenarios (0% TC) 129
6.10 CPU time comparison with risk tolerance parametric analysis (0% TC) 132
6.11 Efficient frontiers under transaction costs 132
6.12 CPU time sensitivity with transaction costs 133
6.13 CPU time comparison with Bender's decomposition 134
6.14 Optimal allocation sensitivity with sample size 134
7.1 Illustration of the tree construction for an example with $T = 5$ time periods 157
7.2 Time plot of load profile for 1 year 159
7.3 Time plot of spot price profile for 1 year 160
7.4 Yearly demand–price scenario trees obtained by Algorithm 7.1. 162
9.1 First-degree stochastic dominance: $R(x) \succeq_{(1)} R(y)$ 194

9.2 Second-order dominance: $R(x) \succeq_{(2)} R(y)$ . . . 194
9.3 First-degree stochastic dominance: $R(x) \succeq_{(1)} R(y)$ in the inverse form . . . 197
9.4 Second-order dominance $R(x) \succeq_{(2)} R(y)$ in the inverse form . . . 197
9.5 Implied utility functions corresponding to dominance constraints for four benchmark portfolios . . . 209
10.1 Expected discounted utility for various strategies, unending game, cost = 0.005 . . . 229
10.2 Expected discounted utility for various strategies, unending game, cost = 0.02 . . . 230
11.1 Variance and absolute deviation . . . 242
11.2 MAD efficient frontier . . . 244
11.3 Market portfolio . . . 245
11.4 Value of stock . . . 247
11.5 β and θ of a typical stock . . . 250
11.6 Transaction cost . . . 251
12.1 Illustration of the dual strategy . . . 258
12.2 Log-prices of S&P 500 index and S&P EWI during July 2003–December 2006 . . . 262
12.3 Decomposition of MLM index returns for different time periods . . . 263
12.4 Efficient frontiers of the portfolios with/without alternative assets . . . 264
12.5 Performance of the dynamic diversification portfolio with leverage . . 266
12.6 Improvements by employing overlay strategies for an illustrative auto company . . . 269
12.7 Performance of dynamic diversification portfolio . . . 272
13.1 Dynamic investment process . . . 280
14.1 Typical throughput distribution . . . 305
14.2 A Markov model for throughput times . . . 306
14.3 Progression of due dates . . . 308
14.4 Due date distribution . . . 308
14.5 Markov chain model for the change in order quantity . . . 310
14.6 Distribution of changes in order quantity . . . 311
15.1 Global carbon emissions . . . 322
15.2 Carbon price . . . 322
15.3 Global and regional population . . . 323
15.4 Projections of GDP per capita at market exchange rates . . . 323
15.5 Total primary energy . . . 324
15.6 Global carbon emissions with alternative climate sensitivities . . . 326
15.7 Global carbon price with alternative climate sensitivities . . . 326
15.8 Increase in mean global temperature with alternative climate sensitivities . . . 326
15.9 Global carbon emissions . . . 327

Contributors

Güzin Bayraksan Systems and Industrial Engineering, University of Arizona, Tucson, AZ 85721, USA, guzinb@sie.arizona.edu

George B. Dantzig Department of Operations Research, Stanford University, Stanford, CA 94305, USA; The RAND Corporation, Santa Monica, CA 90401, USA, klasspiano@gmail.com

Darinka Dentcheva Department of Mathematical Sciences, Stevens Institute of Technology, Hoboken, NJ 07030, USA, darinka.dentcheva@stevens.edu

N. Chanaka P. Edirisinghe Department of Statistics, Operations, and Management Science, College of Business Administration, University of Tennessee, Knoxville, TN 37996, USA, chanaka@utk.edu

Karl Frauendorfer University of St. Gallen, St. Gallen, Switzerland, karl.frauendorfer@unisg.ch

Jiarui Han Department of Management Science and Engineering, Stanford University, Stanford, CA 94305, USA, hanjiarui@gmail.com

Holger Heitsch Institute of Mathematics, Humboldt-University Berlin, Berlin, Germany, heitsch@math.hu-berlin.de

Julia L. Higle Department of Integrated Systems Engineering, The Ohio State University, Columbus, OH 43210, USA, higle.1@osu.edu

Kai Huang Area of Operations Management, DeGroote School of Business McMaster University, Hamilton, Ontario L8S4M4, Canada, khuang@mcmaster.ca

Gerd Infanger Department of Management Science and Engineering, Stanford University, Stanford, CA 94305, USA, infanger@stanford.edu

Karl G. Kempf Decision Technologies Group, Intel Corporation, Chandler, AZ 85226, USA, karl.g.kempf@intel.com

Woo Chang Kim Department of Industrial and Systems Engineering, Korea Advanced Institute of Science and Technology, Daejeon, South Korea, woochang.kim@gmail.com

Hiroshi Konno Department of Industrial and Systems Engineering, Chuo University, Tokyo, Japan, konno@indsys.chuo-u.ac.jp

Daniel Kuhn Imperial College of Science, Technology and Medicine, London, UK, dkuhn@doc.ic.ac.uk

Leonard C. MacLean Herbert Lamb Chair, School of Business Administration, Dalhousie University, Halifax, NS, Canada, B3H 3J5, l.c.maclean@dal.ca

Alan S. Manne Department of Management Science and Engineering, Stanford University, Stanford, CA 94305, USA, jackie.manne@gmail.com

Harry M. Markowitz Harry Markowitz Company, San Diego, CA 92109, USA, harryhmm@aol.com

David P. Morton Graduate Program in Operations Research, The University of Texas at Austin, Austin, TX 78712, USA, morton@mail.utexas.edu

John M. Mulvey Department of Operations Research and Financial Engineering, Princeton University, Princeton, NJ 08544, USA, mulvey@princeton.edu

Amit Partani Graduate Program in Operations Research, The University of Texas at Austin, Austin, TX 78712, USA, amitpartani@mail.utexas.edum

Werner Römisch Institute of Mathematics, Humboldt-University Berlin, Berlin, Germany, romisch@math.hu-berlin.de

Andrzej Ruszczyński Department of Management Science and Information Systems, Rutgers University, Piscataway, NJ 08854, USA, rusz@business.rutgers.edu

Michael Schürle University of St. Gallen, St. Gallen, Switzerland, michael.schuerle@unisg.ch

Rüdiger Schultz Department of Mathematics, University of Duisburg-Essen, Campus Duisburg, Duisburg, Germany, schultz@math.uni-duisburg.de

Suvrajeet Sen The Data Driven Decisions Lab, Industrial, Welding, and Systems Engineering, The Ohio State University, Columbus, OH 43210, USA; Department of Systems and Industrial Engineering, The MORE Institute, The University of Arizona, Tucson, AZ, 85721, USA, sen.22@osu.edu

Erik L. van Dijk LMG Emerge Ltd., 3704 HK Zeist, The Netherlands, erikvandijk@lmg-emerge.nl

Benjamin Van Roy Department of Management Science and Engineering, Stanford University, Stanford, CA 94305, USA, bvr@stanford.edu

Yonggan Zhao Canada Research Chair (Tier II), School of Business Administration, Dalhousie University, Halifax, NS, Canada, B3H 3J5, yonggan.zhao@dal.ca

Zhihong Zhou Department of Systems and Industrial Engineering, The MORE Institute, The University of Arizona, Tucson, AZ, 85721, USA, zhzhou@email.arizona.edu

William T. Ziemba Sauder School of Business, University of British Columbia, Vancouver, BC, Canada V6T 1Z2; Mathematical Institute, University of Oxford, Oxford, England; ICMA Center, University of Reading, Reading, UK, wtzimi@mac.com

Chapter 1
Linear Programming Under Uncertainty

George B. Dantzig

Example 1. *Minimum Expected Cost Diet*

A nutrition expert wishes to advise his followers on a minimum cost diet without prior knowledge of the prices (Stigler 1945). Since prices of food (except for general inflationary trends) are likely to show variability due to weather conditions, supply, etc., he wishes to assume a distribution of possible prices rather than a fixed price for each food and determine a diet that meets specified nutritional requirements and minimizes expected costs. Let x_j be the quantity of jth food purchased in pounds, p_j its price, a_{ij} the quantity of the ith nutrient (e.g., vitamin A) contained in a unit quantity of the jth food, and b_i the minimum quantity required by an individual for good health. Then the x_i must be chosen so that

$$\sum_{j=1}^{n} a_{ij}x_j \geq b_j \quad x_j \geq 0\ (i = 1, \ldots, m) \tag{1.1}$$

and the cost of the diet will be

$$C = \sum_{j=1}^{n} p_j x_j. \tag{1.2}$$

G.B. Dantzig (✉)
The RAND Corporation, Santa Monica, CA 90401, USA
e-mail: klasspiano@gmail.com

This chapter originally appeared in *Management Science*, April–July 1955, Vol. 1, Nos. 3 and 4, pp. 197–206, published by The Institute of Management Sciences. Copyright is held by the Institute for Operations Research and the Management Sciences (INFORMS), Linthicum, Maryland.

This chapter was also reprinted in a special issue of *Management Science*, edited by Wallace Hopp, featuring the "Ten Most Influential Papers of Management Science's First Fifty Years," Vol. 50, No. 12, Dec., 2004, pp. 1764–1769. For this special issue George B. Dantzig provided the following commentary:

"I am very pleased that my first paper on planning under uncertainty is being republished after all these years. It is a fundamental paper in a growing field."
— George B. Dantzig

G. Infanger (ed.), *Stochastic Programming*, International Series in Operations Research & Management Science 150, DOI 10.1007/978-1-4419-1642-6_1,

The x_j are chosen before the prices are known so that the expected costs of such a diet are clearly

$$\text{Exp}\, C = \sum_{j=1}^{n} \bar{p}_j x_j, \tag{1.3}$$

where $\bar{p}_j$ is its expected price. Since the $\bar{p}_j$ are known in advance, the best choices of x_j are those which satisfy (1.1) and minimize (1.3). Hence in this case expected prices may be used in place of the distribution of prices and the usual linear programming problem solved.[1]

Example 2. *Shipping to an Outlet to Meet an Uncertain Demand*

Let us consider a simple two-stage case: A factory has 100 items on hand which may be shipped to an outlet at the cost of \$1 apiece to meet an uncertain demand d_2. In the event that the demand should exceed the supply, it is necessary to meet the unsatisfied demand by purchases on the local market at \$2 apiece. The equations that the system must satisfy are

$$\begin{array}{rcrcrcrl} 100 & = & x_{11} & + & x_{12} & & & \\ d_2 & = & x_{11} & + & x_{21} & - & x_{22} & (x_{ij} \geq 0) \\ C & = & x_{11} & + & 2x_{21}, & & & \end{array} \tag{1.4}$$

where

x_{11} = number shipped from the factory;
x_{12} = number stored at factory;
x_{21} = number purchased on open market;
x_{22} = excess of supply over demand;
d_2 = unknown demand uniformly distributed between 70 and 80;
C = total costs.

It is clear that whatever be the amount shipped and whatever be the demand d_2, it is possible to choose x_{21} and x_{22} consistent with the second equation. The unused stocks $x_{12} + x_{22}$ are assumed to have no value or are written off at some reduced value (like last year's model automobiles when the new production comes in). To illustrate some of the concepts of this chapter, a solution will be presented later.

---

[1] In some applications, however, it may not be desirable to minimize the expected value of the costs if the decision has too great a variation in the actual total costs. Markowitz (1953) in his analysis of investment portfolios develops a technique for computing for each possible expected value the minimum variance. This enables the investor to sacrifice some of his expectation to control his risks.

Example 3. *A Three-Stage Case*

For this purpose it is easy to construct an extension of the previous example by allowing the surpluses x_{12} and x_{22} to be carried over to a third stage, i.e.,

$$\begin{array}{lrllllllll}
\text{First stage} & 100 & = x_{11} & + x_{12} & & & & & & \\
\hline
\text{Second stage} & d_2 & = x_{11} & & + x_{21} & - x_{22} & & & & \\
 & 70 & = & - x_{12} & & & + x_{23} & + x_{24} & & \\
\hline
\text{Third stage} & d_3 & = & & & x_{22} & + x_{23} & & + x_{31} & - x_{32} \\
 & C & = x_{11} & & + 2x_{21} & & + x_{23} & & + 2x_{31}, & \\
\hline
\end{array} \tag{1.5}$$

where

x_{23} = number shipped from factory in second stage;
x_{24} = number stored at factory in second stage;
70 = number produced second stage;
d_3 = unknown demand in third stage uniformly distributed between 70 and 80;
x_{31} = number purchased on the open market in third stage;
x_{32} = excess of supply over demand in third stage.[2]

It will be noted that the distribution of d_3 is independent of d_2. However, the approach which we shall use will apply even if the distribution of d_3 depends on d_2. This is important in problems where there may be some postponement of the timing of demand. For example, it may be anticipated that the potential refrigerator buyers will buy in November or December. However, those buyers who failed to purchase in November will affect the demand distribution for December.

Example 4. *A Class of Two-Stage Problems*

In the Ferguson problem and in many supply problems the total costs may be divided into two parts: first, the costs of assigning various resources to several destinations j, and second, the costs (or lost revenues) incurred because of the failure of the total amounts $u_1, u_2, \ldots, u_n$ assigned to meet demands at various destinations in unknown amounts $d_1, d_2, \ldots, d_n$, respectively.

The special class of two-stage programming problems we are considering has the following structure.[3] For the first stage

---

[2] No solution for this example will be given in this chapter. For this case perhaps the simplest approach is through the techniques of dynamic programming; see Bellman (1953).

[3] The remarks of this section apply if (1.6) and (1.7) are replaced more generally by $AX = a$, $BX = U$, where X is the vector of activity levels in the first stage, A and B are given matrices, a is a given initial status vector, and $U = (u_1, u_2, \ldots, u_n)$.

$$\sum_{j=1}^{n} x_{ij} = a_i \quad (x_{ij} \geq 0), \tag{1.6}$$

$$\sum_{i=1}^{m} b_{ij} x_{ij} = u_j, \tag{1.7}$$

where x_{ij} represents the amount of ith resource assigned to the jth destination and b_{ij} represents the number of units of demand at destination j that can be satisfied by one unit of resource i. For the second stage

$$d_j = u_j + v_j - s_j \quad (j = 1, 2, \ldots, n), \tag{1.8}$$

where v_j is the shortage[4] of supply and s_j is the excess of supply.

The total cost function is assumed to be of the form

$$C = \sum_{i=1}^{m} \sum_{j=1}^{n} c_{ij} x_{ij} + \sum_{j=1}^{n} a_j v_j, \tag{1.9}$$

i.e., depends linearly on the choice x_{ij} and on the shortages v_j (which depend on assignments u_j and the demands d_j).

Our objective will be to minimize total expected costs.[5] Let $\phi_j(u_j|d_j)$ be the minimum costs at a destination if the supply is u_j and the demand is d_j. It is clear that

$$\phi_j(u_j|d_j) = \begin{cases} \alpha_j(d_j - u_j) & \text{if} \quad d_j \geq u_j \\ 0 & \text{if} \quad d_j < u_j, \end{cases} \tag{1.10}$$

where α_j is the coefficient of proportionality. We shall now give a result due to H. Scarf.

Theorem *The expected value of $\phi_j(u_j|d_j)$, denoted by $\phi_j(u_j)$, is a convex function of u_j.*

Proof Let $p(d_j)$ be the probability density of d_j, then

[4] Equation (1.8) should be viewed more generally than simply as a statement about the shortage and excess of supply. In fact, given any u_j and d_j, there is an infinite range of possible values of v_j and s_j satisfying (1.8). For example, v_j might be interpreted as the amount obtained from some new source (perhaps at some premium price) and s_j the amount not used. When the cost form is as in (1.9), it becomes clear that in order for C to be a minimum the values of v_j and s_j will have the more restrictive meaning above.

[5] Markowitz in his analysis of portfolios considers the interrelation of the variance with the expected value. See Markowitz (1953).

$$\phi_j(u_j) = \alpha_j \int_{x=u_j}^{+\infty} (x - u_j)p(x)dx$$
$$= \alpha_j \int_{x=u_j}^{+\infty} xp(x)dx - \alpha_j u_j \int_{x=u_j}^{+\infty} p(x)dx, \qquad (1.11)$$

whence differentiating $\phi(u_j)$

$$\phi_j'(u_j) = -\alpha_j \int_{x=u_j}^{+\infty} p(x)dx. \qquad (1.12)$$

It is clear that $\phi_j'(u_j)$ is a non-decreasing function of u_j with $\phi_j''(u_j) \geq 0$ and that $\phi_j(u_j)$ is convex. An alternative proof (due also to Scarf) is obtained by applying a lemma which we shall use later on. □

Lemma *If $\phi(x_1, x_2, \ldots, x_n|\theta)$ is a convex function over a fixed region Ω for every value of θ, then any positive linear combination of such functions is also convex in Ω.*

In particular if θ is a random variable with probability density $p(\theta)$, then the expected value of ϕ

$$\phi(x_1, x_2, \ldots, x_n) = \int_{-\infty}^{+\infty} \phi(x_1, x_2, \ldots, x_n)|\theta)p(\theta)d\theta \qquad (1.13)$$

is convex. For example, from (1.10), $\phi(u_j|d_j)$, plotted below, is convex.

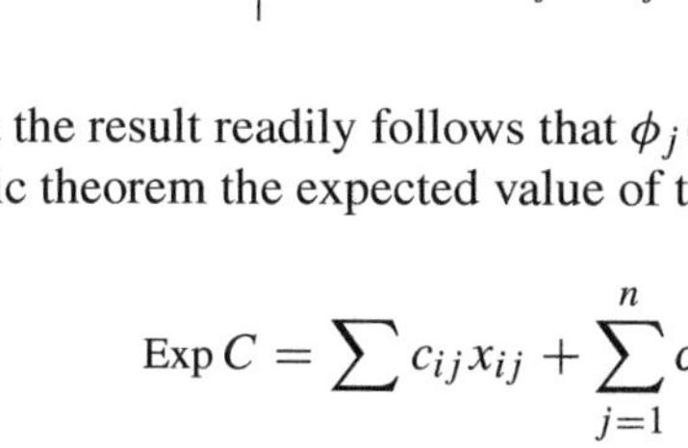

(1.14)

From the lemma the result readily follows that $\phi_j(u_j)$ is convex.

From the basic theorem the expected value of the objective function is

$$\text{Exp}\, C = \sum c_{ij}x_{ij} + \sum_{j=1}^{n} \alpha_j \phi_j(u_j), \qquad (1.15)$$

where $\phi_j(u_j)$ are *convex functions*. Thus the original problem has been reduced to minimizing (1.15) subject to (1.6) and (1.7).

This permits application of a well-known device for *approximating* such a problem by a standard linear programming problem in the case the objective function can be represented by a sum of convex functions. See, for example, Dantzig (1954) or Charnes and Lemke (1954). To do this one approximates the derivative of $\phi(u)$ in some sufficiently large range $0 \leq u \leq u_0$ by a step function

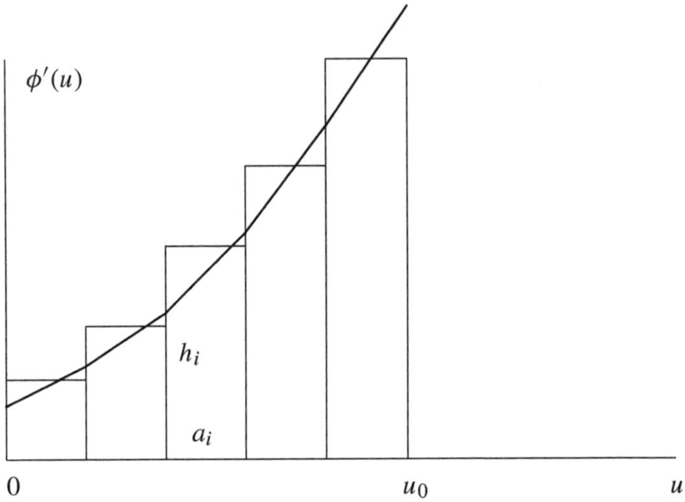

(1.16)

involving k steps where the size of the ith base is a_i and its height is h_i; where $h_1 \leq h_2 \leq \cdots \leq h_k$ because ϕ is convex. An approximation for $\phi(u)$ is given by

$$\phi(u) \doteq \phi(0) + \text{Min} \sum_1^k h_i \Delta_i \tag{1.17}$$

subject to

$$u = \sum_{i=1}^{k} \Delta_i, \quad 0 \leq \Delta_i \leq a_i. \tag{1.18}$$

Indeed, it is fairly obvious that the approximation achieves its minimum by choosing $\Delta_1 = a_1, \Delta_2 = a_2, \ldots$ until the cumulative sum of the Δ_i exceeds u for some $i = r$; Δ_r is then chosen as the value of the residual with all remaining $\Delta_{r+i} = 0$. In other words, we have approximated an integral by the sum of rectangular areas under the curve up to u, i.e.,

$$\phi(u) = \phi(0) + \int_0^u \phi'(x)dx \doteq \sum_{i=1}^{r} h_i a_i + h_r \Delta_r. \tag{1.19}$$

The next step is to replace $\phi(u)$ by $\sum_1^k h_i \Delta_i$ and u by $\sum_1^k \Delta_i$ in the programming problem and add the restrictions $0 \leq \Delta_i \leq a_i$. If the objective is minimization of total costs, it will, of necessity, for whatever value of $u = \sum_1^n \Delta_i$ and $0 \leq \Delta_i \leq a_i$, minimize $\sum_1^k h_i \Delta_i$. Thus, this class of two-stage linear programming problems involving uncertainty can be reduced to a standard linear programming-type problem. In addition, simplifying computational methods exist when variables have upper bounds such as $\Delta_i \leq a_i$; see Dantzig (1954).

Example 5. *The Two-Stage Problem with General Linear Structure*

We shall prove a general theorem on convexity for the two-stage problem that forms the inductive step for the multistage problem. We shall say a few words about the significance of this convexity later on. The assumed structure of the general[6] two-stage model is

$$\begin{array}{rclcl} b_1 & = & A_{11}X_1 & & \\ b_2 & = & A_{21}X_1 & + & A_{22}X_2 \\ C & = & \phi(X_1, X_2 | E_2), & & \end{array} \tag{1.20}$$

where A_{ij} are known matrices and b_1 is a known vector of initial inventories. For example,

$$\begin{array}{ll} a_i = \sum_{j=1}^m x_{ij} & \text{here } b_1 = (a_1, a_2, \ldots, a_m) \\ & \text{here } X_1 = (x_{11}, \ldots, x_{1n}, x_{21}, \ldots, x_{2n}, \ldots, x_{mn}) \\ d_j = \sum_{i=1}^m b_{ij} x_{ij} + v_j - s_j & \text{here } b_2 = (d_1, d_2, \ldots, d_n) \\ C = \sum \sum c_{ij} x_{ij} + \sum a_j v_j & \text{here } X_2 = (v_1, v_2, \ldots, v_n, s_1, s_2, \ldots, s_n) \end{array}$$

b_2 an unknown vector whose components are determined by a chance mechanism.[7] (Mathematically, E_2 is a sample point drawn from a multidimensional sample space with known probability distribution); X_1 is the vector of nonnegative activity levels to be determined in the first stage, while X_2 is the vector of nonnegative activity levels for the second stage. It is assumed that whatever be the choice of X_1 satisfying the first-stage equations and whatever be the particular values of b_2 determined by chance, there exists at least one vector X_2 satisfying the second-stage equations. The total costs C of the program are assumed to depend on the choice of X_1, X_2, and parametrically on E_2. The basic problem is to choose X_1 and later X_2 in the second stage such that the expected value of C is a minimum.

Theorem *If $\phi(X_1, X_2|E_2)$ is a convex function in X_1, X_2 whatever be X_1 in Ω_1, i.e., satisfying the first-stage restrictions and whatever be X_2 in $\Omega_2 = \Omega_2(X_1|b_2)$, i.e., satisfying the second-stage restrictions given b_2 and X_1, then there exists a convex function $\phi_0(X_1)$ such that the optimal choice of X_1 subject to $b_1 = A_{11}X_1$ is found by minimizing $\phi_0(X_1)$ where*

---

[6] A special case of the general model given in (1.20) is found in Example 4.

[7] The chance mechanism may be the "market," the "weather."

$$\begin{aligned}\phi_0(X_1) &= \text{Exp}_{E_2}[\text{Inf}_{X_2\in\Omega_2}\phi(X_1, X_2|E_2)],\\ \text{Exp}\, C &= \text{Inf}_{X_1\in\Omega_1}\phi_0(X_1);\end{aligned} \tag{1.21}$$

the expectation (Exp) is taken with respect to the distribution of E_2 and the greatest lower bound (Inf)[8] is taken with respect to all $X_2 = \Omega_2(X_1|E_2)$.

Proof [9] In order to minimize the $\text{Exp}\,\phi(X_1, X_2|E_2)$, it is clear that once X_1 has been selected, E_2 determined by chance, that X_2 must be selected so that $\phi(X_1, X_2|E_2)$ is minimized for fixed X_1 and E_2. Thus, the costs for *given* X_1 and E_2 are given by

$$\phi_1(X_1|E_2) = \text{Inf}_{X_2\in\Omega_2}\phi(X_1, X_2|E_2). \tag{1.22}$$

The expected costs for a *given* X_1 are then simply the expected value of $\phi(X_1|E_2)$ and this we denote by $\phi_0(X_1)$. The optimal choice of X_1 to minimize expected costs C is thus reduced to choosing X_1 so as to minimize $\phi_0(X_1)$. There remains only to establish the convexity property. We shall show first that $\phi_1(X_1|E_2)$ for bounded ϕ_1 is convex for X_1 in Ω_1. If true, then applying the lemma, the result that $\phi_0(X_1)$ is convex readily follows. Let us suppose that $\phi_1(X_1|E_2)$ is not convex, then there exist three points in Ω_1, X_1', X_1'', $X_1''' = \lambda X_1' + \mu X_1''$ ($\lambda + \mu = 1,\ 0 \le \lambda \le 1$), that violate the condition for convexity, i.e.,

$$\lambda\phi_1(X_1'|E_2) + \mu\phi_1(X_1''|E_2) < \phi_1(X_1'''|E_2) \tag{1.23}$$

or

$$\lambda\phi_1(X_1'|E_2) + \mu\phi_1(X_1''|E_2) = \phi_1(X_1'''|E_2) - \epsilon_0 \quad \epsilon_0 > 0. \tag{1.24}$$

For any $\epsilon_0 > 0$, however, there exists X_2' and X_2'' such that

$$\begin{aligned}\phi_1(X_1'|E_2) &= \phi(X_1', X_2'|E_2) - \epsilon_1 \quad 0 \le \epsilon_1 < \epsilon_0\\ \phi_1(X_1''|E_2) &= \phi(X_1'', X_2''|E_2) - \epsilon_2 \quad 0 \le \epsilon_2 < \epsilon_0.\end{aligned} \tag{1.25}$$

Setting $X_2''' = \lambda X_2' + \mu X_2''$ we note because of the assumed linearity of the model (1.20) that $(\lambda X_2' + \mu X_2'') \in \Omega_2(\lambda X_1' + \mu X_1''|E_2)$ and hence by convexity of ϕ

$$\lambda\phi(X_1', X_2'|E_2) + \mu\phi(X_1'', X_2''|E_2) \ge \phi(X_1''', X_2'''|E_2) \tag{1.26}$$

whence by (1.25)

---

[8] The greatest lower bound instead of minimum is used to avoid the possibility that the minimum value is not attained for any admissible point $X_2 \in \Omega_2$ or $X_1 \in \Omega_1$. In case where the latter occurs, it should be understood that while there exists no X_i where the minimum is attained, there exists X_i for which values as close to minimum as desired are attained.

[9] This proof is along lines suggested by I. Glicksberg.

$$\lambda\phi_1(X_1'|E_2) + \mu\phi_1(X_1''|E_2) \geq \phi(X_1''', X_2'''|E_2) - \lambda\epsilon_1 - \mu\epsilon_2 \qquad (1.27)$$

and by (1.24)

$$\phi_1(X_1'''|E_2) \geq \phi(X_1''', X_2'''|E_2) - \lambda\epsilon_1 - \mu\epsilon_2 + \epsilon_0 \quad (0 \geq \lambda\epsilon_1 + \mu\epsilon_2 < \epsilon_0), \quad (1.28)$$

which contradicts the assumption that $\phi_1(X_1'''|E_2) = \text{Inf}\,\phi(X_1''', X_2|E_2)$. The proof for unbounded ϕ is omitted. □

Example 6. *The Multistage Problem with General Linear Structure*

The structure assumed is

$$\begin{array}{lllllllll} b_1 & = & A_{11}X_1 & & & & & & \\ b_2 & = & A_{21}X_1 & + & A_{22}X_2 & & & & \\ b_3 & = & A_{31}X_1 & + & A_{32}X_2 & + & A_{33}X_3 & & \\ b_4 & = & A_{41}X_1 & + & A_{42}X_2 & + & A_{43}X_3 & + & A_{44}X_4 \\ \cdots & & \cdots & & \cdots & & \cdots & & \cdots \\ b_m & = & A_{m1}X_1 & + & A_{m2}X_2 & + & A_{m3}X_3 & + \cdots & + \; A_{mm}X_m \\ C & = & \phi(X_1, X_2, \ldots, X_m|E_2, E_3, \ldots, E_m), & & & & & & \end{array} \qquad (1.29)$$

where b_1 is a known vector; b_i is a chance vector ($i = 2, \ldots, m$) whose components are functions of a point E_i drawn from a known multidimensional distribution; A_{ij} are known matrices. The sequence of decisions is as follows: X_1, the vector of nonnegative activity levels in the first stage, is chosen so as to satisfy the first-stage restrictions $b_1 = A_{11}X_1$; the values of components of b_2 are chosen by chance by determining E_2; X_2 is chosen to satisfy the second-stage restrictions $b_2 = A_{21}X_1 + A_{22}X_2$, etc., iteratively for the third and higher stages. It is further assumed that

(1) The components of X_j are nonnegative;
(2) There exists at least one X_j satisfying the jth-stage restraints, whatever be the choice of $X_1, X_2, \ldots, X_{j-1}$ satisfying the earlier restraints or the outcomes $b_l, b_2, \ldots, b_m$.
(3) The total cost C is a convex function in $X_1, \ldots, X_m$ which depends on the values of the sample points $E_2, E_3, \ldots, E_m$.

Theorem *An equivalent* $(m-1)$ *stage programming problem with a convex pay-off function can be obtained by dropping the mth-stage restrictions and replacing the convex cost function* ϕ *by*

$$\begin{aligned} &\phi_{m-1}(X_1, X_2, \ldots, X_{m-1}|E_2, \ldots, E_{m-1}) \\ &\quad = \text{Exp}_{E_m}\text{Inf}_{X_m \in \Omega_m}\phi(X_1, X_2, \ldots, X_m|E_2, \ldots, E_m), \end{aligned} \qquad (1.30)$$

where Ω_m *is the set of possible* X_m *that satisfy the mth-stage restrictions.*

Since the proof of the above theorem is identical to the two-stage case no details will be given. The fact that a cost function for the $(m-1)$ stage can be obtained from the mth stage is simply a consequence that optimal behavior for the mth stage is well defined, i.e., given any state, e.g., $(X_1, X_2, \ldots, X_{m-1})$, at the beginning of this stage, the best possible actions can be determined and the minimum expected cost evaluated. This is a standard technique in "dynamic programming." For the reader interested in methods built around this approach refer Bellman's book on *Dynamic Programming* (Bellman 1953).

While the existence of convex functions has been demonstrated that permit reduction of an m-stage problem to equivalent $m-1, m-2, \ldots, 1$-stage problems, it appears hopeless that such functions can be computed except in very simple cases. The convexity theorem was demonstrated not as a solution to an m-stage problem but only in the hope that it will aid in the development of an efficient computational theory for such models. It should be remembered that any procedure that yields a local optimum will be a true optimum if the function is convex. This is important because multidimensional problems in which nonconvex functions are defined over nonconvex domains lead as a rule to local optimum and an almost hopeless task, computationally, of exploring other parts of the domain for the other extremes.

Solution for Example 2. *Shipping to an Outlet to Meet an Uncertain Demand*

Let us consider the two-stage case given earlier (1.4). It is clear that if supply exceeds demand $(x_{11} \geq d_2)$ $x_{21} = 0$ gives minimum costs and, if $x_{11} \leq d_2$, $x_{21} = d_2 - x_{11}$ gives minimum costs. Hence

$$\text{Min}_{x_{21}} \phi = \begin{cases} x_{11} & \text{if} \quad x_{11} > d_2 \\ x_{11} + 2(d_2 - x_{11}) & \text{if} \quad x_{11} \leq d_2. \end{cases} \tag{1.31}$$

Since d_2 is assumed to be uniformly distributed between 70 and 80

$$\text{Exp}_{\,d_2}[\text{Min}_{x_{21}} \phi] = \begin{cases} -x_{11} + 150 & \text{if} \quad x_{11} \leq 70 \\ 77.5 + \frac{1}{10}(75 - x_{11})^2 & \text{if} \quad 70 < x_{11} \leq 80 \\ x_{11} & \text{if} \quad 80 \leq x_{11}. \end{cases} \tag{1.32}$$

This function is clearly convex and attains its minimum 77.5, which is the expected cost, at $x_{11} = 75$. Since $x_{11} = 75$ is in the range of possible values of x_{11} as determined by $100 = x_{11} + x_{12}$ this is clearly the optimal shipment. In this case it pays to ship $x_{11} = \bar{d}_2 = 75$, the expected demand.

It can be shown by simple examples that one cannot replace, in general, the chance vectors b_i by $\bar{b}_i$, the vector of expected values of the components of b_i. Nevertheless, this procedure, which is quite common, probably provides an excellent starting solution for any improvement technique that might be devised. For example, in the problem of Ferguson (application of Example 4), using as a start the solution based on expected values of demand, it was an easy matter to improve the solution to an optimal one whose expected costs were 15% less.

Solution for Example 5. *The General Two-Stage Case*

When the number of possibilities for the chance vector b_2 is $b_2^{(1)}, b_2^{(2)}, \ldots, b_2^{(k)}$ with probabilities $p_1, p_2, \ldots, p_k$, $\left(\sum p_i = 1\right)$, it is not difficult to obtain a direct linear programming solution for small k, say $k = 3$. Since this type of structure is very special, it appears likely that techniques can be developed to handle large k. For $k = 3$, the problem is equivalent to determining vectors X_1 and vectors $X_2^{(1)}, X_2^{(2)}, X_2^{(3)}$ such that

$$
\begin{array}{lcllll}
b_1 & = & A_{11}X_1 & & & \\
b_2^{(1)} & = & A_{21}X_1 & + \; A_{22}X_2^{(1)} & & \\
b_2^{(2)} & = & A_{21}X_1 & & + \; A_{22}X_2^{(2)} & \\
b_2^{(3)} & = & A_{21}X_1 & & & + \; A_{22}X_2^{(3)} \\
\mathrm{Exp}\, C & = & \gamma_1 X_1 & + \; p_1\gamma_2 X_2^{(1)} & + \; p_2\gamma_2 X_2^{(2)} & + \; p_3\gamma_2 X_2^{(3)} = \mathrm{Min},
\end{array}
\tag{1.33}
$$

where for simplicity we have assumed a linear objective function.

References

Bellman, R.: An introduction to the theory of dynamic programming. Report R-245, The RAND Corporation, June (1953)
Charnes, A., Lemke, C.E.: Minimization of non-linear separable functions, Graduate School of Industrial Administration, Carnegie Institute of Technology, May (1954)
Dantzig, G.B.: Notes on linear programming: Parts VIII, XVI,X—upper bounds, secondary constraints, and block triangularity in linear programming. Research Memorandum RM-1367, The RAND Corporation, October, 4 (1954)
Ferguson, A.R., Dantzig, G.B.: Notes on linear programming: Part XVI—the problem of routing aircraft—a mathematical solution. Research Memorandum RM-1369, The RAND Corporation, September, 1 (1954)
Markowitz, H.: Portfolio Selection. PhD thesis, The University of Chicago, Chicago, IL (1953)
Stigler, G.F.: The cost of subsistence. J. Farm Econ., **27**, 303–314 May (1945)

Chapter 2
A Probabilistic Lower Bound for Two-Stage Stochastic Programs

George B. Dantzig and Gerd Infanger

2.1 Introduction

It has long been recognized that traditional deterministic models do not reflect the true dynamic behavior of real-world applications because they fail to take into account uncertainty. Since Dantzig (1955) and Beale (1955) independently proposed a stochastic model formulation, these models have been studied intensively. In the literature, a number of different algorithmic approaches have been proposed that we can broadly categorize as deterministic methods, approximation schemes, sampling-based algorithms, and others.

Deterministic methods attempt to solve the deterministic equivalent problem, either directly or by exploiting structure. Prominent among these are the L-shaped method of Van Slyke and Wets (1969), its multi-cut variant by Birge and Louveaux (1988), the progressive hedging algorithm of Rockafellar and Wets (1989), the use of interior point methods by Lustig et al. (1991), the regularized decomposition method by Ruszczyński (1986), and other large-scale techniques implemented on serial and parallel computers. Clearly, even the most sophisticated deterministic techniques can only solve problems with a limited number of scenarios. So far, problems with up to several million scenarios have been solved using deterministic techniques.

Approximation schemes calculate deterministic lower and upper bounds on the optimal objective of the problem via the inequalities of Jensen (1906) (lower bound) and Edmundson (1956) and Madansky (1959) (upper bound) and successively improve these bounds. Refinements of these bounds have been proposed by many authors, e.g., Kall (1974), Huang et al. (1977), Frauendorfer and Kall (1988), Frauendorfer (1988, 1992), Birge and Wallace (1988), Birge and Wets (1987, 1989), Prékopa (1990), and others. Approximation techniques work very well for problems with a small number of stochastic parameters, but seem to encounter difficulties when the number of stochastic parameters is large.

G. Infanger (✉)
Department of Management Science and Engineering, Stanford University, Stanford, CA 94305, USA
e-mail: infanger@stanford.edu

G. Infanger (ed.), *Stochastic Programming*, International Series in Operations Research & Management Science 150, DOI 10.1007/978-1-4419-1642-6_2,

Sampling-based algorithms can be further categorized into methods that presample a number of scenarios to create a tractable deterministic equivalent problem, which is then solved by a suitable deterministic method, and methods that use sampling within the algorithm. For the former category, confidence intervals on the optimal objective for the class of sampling-based procedures based on "sample average approximation" have been proposed by Shapiro and Homem-de-Mello (1998) and by Mak et al. (1999) that are asymptotically valid as the sample size used for generating the approximate problem approaches infinity. In the latter category fall stochastic quasigradient methods (Ermoliev (1983) and Gaivoronski (1988)) that select sequentially random search directions based on a limited number of observations of the random function in each iteration. Others are based on modifications of deterministic decomposition techniques to allow for sampling. The stochastic decomposition method by Higle and Sen (1991) relies on taking only one observation or a very small number of observations per iteration, while Pereira et al. (1989) used control variables as a variance reduction technique in Monte Carlo sampling in a modified Benders decomposition framework.

Sampling seems to be the best method for practical problems with a large number of stochastic parameters. The approach by Dantzig and Glynn (1990) and Infanger (1992) combines Benders decomposition and Monte Carlo importance sampling for solving stochastic linear programs. Importance sampling serves as a variance reduction technique and in practice often results in accurate estimates being obtained with only small sample sizes. Infanger (1992) and Dantzig and Infanger (1992) report the solution on personal computers of large-scale problems that seemed to be intractable in the past, even on large mainframes.

In this chapter we present a theory for obtaining a probabilistic lower bound for the true optimal objective value when using Benders (1962) decomposition and Monte Carlo sampling for estimating coefficients and right-hand sides of cuts to solve two-stage stochastic linear programs. In Section 2.2, we state the problem. In Section 2.3 we review the original Benders decomposition algorithm for two-stage stochastic linear programs (Van Slyke and Wets 1969). We then derive in Section 2.4 the theory of a probabilistic lower bound. Finally, in Section 2.5, we discuss the numerical results obtained from testing the theory on a number of practical problems.

This chapter is based on Dantzig and Infanger (1995), issued as a technical report, with minor modifications and an appendix added. A complete and rigorous discussion of the confidence bounds introduced in Dantzig and Infanger (1995) has been put forward by Glynn and Infanger (2008).

2.2 The Problem

We consider the following two-stage stochastic linear program:

$$\begin{array}{rrcrcl} \min z = & cx & + & E^{\omega} f^{\omega} y^{\omega} & & \\ \text{s/t} & Ax & & & = & b \\ & -B^{\omega} x & + & D^{\omega} y^{\omega} & = & d^{\omega}, \quad \omega \in \Omega \\ & x, & & y^{\omega} & \geq & 0. \end{array} \tag{2.1}$$

The matrix of constraint coefficients A, the right-hand side vector b, and the objective function coefficients c of the first stage are assumed to be known with certainty. In the second stage, the transition matrix B, the technology (or recourse) matrix D, the right-hand side vector d, and the objective function coefficients f are not known with certainty—only their joint distribution of values is assumed to be known.

We denote particular outcomes by $B = B^\omega$, $D = D^\omega$, $d = d^\omega$, $f = f^\omega$, $\omega = 1, \ldots, W$ and the known probability of realization ω occurring by $p(\omega)$. The set of all possible realizations of ω is denoted by $\Omega = \{1, \ldots, W\}$ or by $\omega \in \Omega$. Since in many practical applications, D and f do not depend on ω, we have omitted their ω superscript in the rest of the chapter to simplify the presentation; the analysis is the same if D and f are replaced by D^ω and f^ω.

The equivalent deterministic form of (2.1) is (2.2):

$$
\begin{array}{rllllll}
\min z = & cx + & \theta & & & & \\
-1: & & -\theta + p^1 f y^1 + \cdots & + p^\omega f y^\omega + \cdots & + p^W f y^W & = 0 \\
\rho: & Ax & & & & = b \\
p^1\pi^1: & -B^1 x & + \quad Dy^1 & & & = d^1 \\
& \vdots & \qquad \ddots & & & \vdots \\
p^\omega \pi^\omega: & -B^\omega x & & + \quad Dy^\omega & & = d^\omega \\
& \vdots & & \qquad \ddots & & \vdots \\
p^W \pi^W: & -B^W x & & & + \quad Dy^W & = d^W \\
& x, & y^1, \ldots & y^\omega, \ldots, & y^W & \geq 0,
\end{array}
\tag{2.2}
$$

where θ denotes the expected second-stage costs.

The dual variables corresponding to the primal constraints of (2.2) are displayed in the column to the left of the equations. In particular, ρ is the vector of dual variables associated with the first-stage constraints $Ax = b$ and $p^\omega \pi^\omega$ is the vector of dual variables associated with the second-stage constraint $-B^\omega x + Dy^\omega = d^\omega$, for each $\omega \in \Omega$.

In practical applications the number of possible second-stage realizations W can be very large (e.g., 10^{20} or even 10^{100}) and it is impossible to express system (2.2) explicitly. The different realizations are implicitly generated as needed by a combination of an underlying small set of h independent random parameters $\phi_1, \phi_2, \ldots, \phi_h$, where h might be 100 or so. Sampling is required to solve such problems. The ϕ structure underlying the set Ω makes it possible to use variance reduction techniques, such as *importance sampling*, to reduce the size of the sample required. Since the theory for the lower bound estimate is the same with variance reduction, we omit the discussion of the latter to simplify the presentation.

The stochastic algorithm follows the same steps as the original Benders algorithm, except that the necessary conditions (the *true cuts*) are approximated by *pseudo-cuts* obtained by summing over a random sample of ω instead of all ω. After a preassigned number of iterations K the algorithm terminates with a proposed first-stage decision $x = \xi^l$, which yields the lowest approximate expected first-stage and second-stage cost found so far. We review the original algorithm first and then use estimated cuts based on random samples to determine the first-stage decision and estimate bounds on how close its objective is to the true minimum of z.

2.3 Benders Decomposition

Assumption 1

The problem $\min cx,\ Ax = b,\ x \geq 0$ has a finite optimal solution $x = \xi^1$. It is used to initiate iteration 1 of the Benders algorithm for solving (2.2).

Assumption 2

For any feasible first-stage decision $x = \xi$, each second-stage subproblem ω,

$$\begin{array}{lcrcl} \min\ \theta_\xi^\omega & = & fy^\omega & & \\ \text{s/t}\ \pi^\omega: & & Dy^\omega & = & d^\omega + B^\omega \xi \\ & & y^\omega & \geq & 0,\ \omega \in \Omega, \end{array} \tag{2.3}$$

has optimal primal and dual feasible solutions $y^\omega = y_\xi^\omega$, $\pi^\omega = \pi_\xi^\omega$ that depend on ξ. By the duality theorem, these satisfy

$$f - \pi_\xi^\omega D \geq 0, \quad Dy_\xi^\omega = d^\omega + B^\omega \xi, \quad fy_\xi^\omega = \pi_\xi^\omega (d^\omega + B^\omega \xi), \quad (f - \pi_\xi^\omega D) y_\xi^\omega = 0. \tag{2.4}$$

Assumptions 1 and 2 imply that the convex set C of feasible solutions to (2.2) is nonempty and that there exists an optimal solution to (2.2): z^*, x^*, θ^*, y_*^ω, for $\omega \in \Omega$.

Generating True Cuts

Let $x, \theta, y^1, \ldots, y^W$ be any feasible solution to (2.2). It satisfies

$$-B^\omega x + Dy^\omega = d^\omega, \quad \text{for each } \omega \in \Omega. \tag{2.5}$$

For any given first-stage solution ξ and any dual feasible π_ξ^ω, multiplying by π_ξ^ω and then adding and subtracting $p^\omega f y^\omega$ we obtain

$$-\pi_\xi^\omega B^\omega x + (\pi_\xi^\omega D y^\omega - fy^\omega) + fy^\omega = \pi_\xi^\omega d^\omega. \tag{2.6}$$

Note that

$$\pi_\xi^\omega D y^\omega - fy^\omega \leq 0 \tag{2.7}$$

because $f - \pi_\xi^\omega D \geq 0$ by (2.4) and $y^\omega \geq 0$. Dropping this term in parentheses in (2.6) we obtain

$$-\pi_\xi^\omega B^\omega x + f y^\omega \geq \pi_\xi^\omega d^\omega, \quad \omega \in \Omega. \tag{2.8}$$

Multiplying by $p^\omega \geq 0$ and taking the sum over all $\omega \in \Omega$ yields

$$-\sum_{\omega\in\Omega} p^\omega \pi_\xi^\omega B^\omega x + \sum_{\omega\in\Omega} p^\omega f y^\omega \geq \sum_{\omega\in\Omega} p^\omega \pi_\xi^\omega d^\omega. \tag{2.9}$$

In iteration k the value of ξ is denoted by $\xi = \xi^k$, and we denote

$$G^k = \sum_{\omega\in\Omega} p^\omega \pi_\xi^\omega B^\omega, \quad g^k = \sum_{\omega\in\Omega} p^\omega \pi_\xi^\omega d^\omega, \quad \xi = \xi^k. \tag{2.10}$$

Substituting these into (2.9), and noting from (2.2), the second term is θ by definition, we obtain the

True Cut

$$-G^k x + \theta \geq g^k, \tag{2.11}$$

which is valid for all feasible solutions to (2.2).

Starting with the first-stage solution $x = \xi^1$ (see Assumption 1), cut (1) is generated and used to form master program (1). The optimal solution x of master problem (1) generates the first-stage solution $x = \xi^2$. After k iterations of the Benders decomposition algorithm, we arrive at

Master Problem (k)

$$\begin{array}{rcrcccl} \min v & = & cx & + & \theta & & \\ & & Ax & & & = & b \\ & & -G^1 x & + & \theta & \geq & g^1 \\ & & \vdots & & \vdots & & \vdots \\ & & -G^k x & + & \theta & \geq & g^k \\ & & x, & & & \geq & 0. \end{array} \tag{2.12}$$

Master problem (k) is optimized and its optimum solution is denoted $x = \xi^{k+1}$, $\theta = \theta^{k+1}$, $v = v^{k+1}$. ξ^{k+1} is used to initialize iteration $k+1$. Cut ($k+1$) is then determined by (2.10) and adjoined to master (k) to generate master ($k+1$), and so on.

Lower Bound

In particular, an optimal solution to (2.2), x^*, θ^*, y_*^ω, $\omega \in \Omega$, satisfies (2.12). Therefore

$$z^* = cx^* + \theta^* \tag{2.13}$$

and

$$-G^k x^* + \theta^* \geq g^k, \quad k = 1, 2, \ldots. \tag{2.14}$$

Since master problem (k) is a set of necessary conditions derived from (2.2), clearly its optimum objective $v = v^{k+1}$ is a lower bound for z^*, i.e.,

$$v^{k+1} \leq z^*, \tag{2.15}$$

because master (k) may be viewed as a relaxation of its conditions evaluated at $x = x^*, \theta = \theta^*$ as shown here:

$$\begin{array}{rcrcl} cx^* & + & \theta^* & = & z^* \\ Ax^* & & & = & b \\ -G^1 x^* & + & \theta^* & \geq & g^1 \\ \vdots & & \vdots & & \vdots \\ -G^k x^* & + & \theta^* & \geq & g^k \\ x^* & & & \geq & 0. \end{array} \tag{2.16}$$

Upper Bound

Given the first-stage solution $x = \xi^{k+1}$, the corresponding minimum second-stage cost is

$$\theta_\xi^{k+1} = \sum_{\omega \in \Omega} p^\omega f y_\xi^\omega, \quad \xi = \xi^{k+1}. \tag{2.17}$$

The minimum expected cost, given the first-stage decision $\xi = \xi^{k+1}$, is $z_\xi^{k+1} = c\xi^{k+1} + \theta_\xi^{k+1}$. Therefore,

$$z^* \leq z_\xi^{k+1}. \tag{2.18}$$

Termination

If $v^{k+1} = z_\xi^l$ for some $l \leq k+1$, then $x = \xi^l$ is an optimal first-stage decision and $z_\xi^l = z^*$. The condition $v^{k+1} = z_\xi^l$ will be reached in a finite number of iterations, see Benders (1962). In practice, however, the iterative process is terminated if

$$z_{\xi}^{l} - v^{k+1} \leq TOL, \quad l \leq k+1, \tag{2.19}$$

where *TOL* is a preassigned "close enough" criterion, implying

$$z_{\xi}^{l} - z^{*} \leq TOL. \tag{2.20}$$

The objective value of the first-stage decision $x = \xi^{l}$ is then deemed to be "close enough" to z^{*}. Otherwise, the iterative process continues with increasing k until k reaches a preassigned maximum number K and the solution $x = \xi^{l}$, where $l = \arg\min z_{\xi}^{k}$, $k = 1, \ldots, K$, is then chosen as the first-stage decision.

2.4 Probabilistic Lower Bound

2.4.1 Cut Generation Using Sampling

The need for sampling arises if W, the number of possible realizations, is large. We then use Monte Carlo sampling to estimate the cutting planes. If a cut is estimated by a sample of ω, we will call it a *pseudo-cut*. Given a first-stage solution $x = \xi$, to generate a pseudo-cut, a random sample S of size $|S|$ of the necessary conditions (2.5) with replacement is taken according to the distribution $p^{1}, p^{2}, \ldots, p^{W}$ and averaged. Thus, summing (2.8) for $\omega \in S$ and averaging, we obtain a true cut, called a *stochastic cut*. Adding $\theta = \sum_{\omega\in\Omega} p^{\omega} f y^{\omega}$ to both sides and rearranging terms, we obtain

$$-\frac{1}{|S|}\sum_{\omega\in S}\pi_{\xi}^{\omega}B^{\omega}x + \theta \geq \frac{1}{|S|}\sum_{\omega\in S}\pi_{\xi}^{\omega}d^{\omega} - \left(\frac{1}{|S|}\sum_{\omega\in S} f y^{\omega} - \theta\right). \tag{2.21}$$

On iteration k, we denote ξ and S by ξ^{k} and S^{k}, respectively, and we denote the various terms of (2.21) by

$$\tilde{G}^{k} = \frac{1}{|S^{k}|}\sum_{\omega\in S^{k}}\pi_{\xi}^{\omega}B^{\omega}, \quad \tilde{g}^{k} = \frac{1}{|S^{k}|}\sum_{\omega\in S^{k}}\pi_{\xi}^{\omega}d^{\omega}, \quad \xi = \xi^{k}, \tag{2.22}$$

and define

$$\tilde{\epsilon}^{k} = \frac{1}{|S^{k}|}\sum_{\omega\in S^{k}} f y^{\omega} - \theta \tag{2.23}$$

as the *sampling error* of estimating $\theta = \sum_{\omega\in\Omega} p^{\omega} f y^{\omega}$ by a random sample S^{k} of size $|S^{k}|$ from the distribution of $f y^{\omega}$ with density p^{ω}. Substituting $\tilde{G}^{k}$, $\tilde{g}^{k}$, and $\tilde{\epsilon}^{k}$ for the expressions in (2.21) we obtain the

Stochastic Cut (k)

$$-\tilde{G}^k x + \theta \geq \tilde{g}^k - \tilde{\epsilon}^k, \quad \xi = \xi^k, \quad S = S^k, \tag{2.24}$$

which we note is a true statement for any feasible solution of (2.2), x, θ, y^ω, for all $\omega \in \Omega$. The corresponding *pseudo-cut* (k) is obtained by dropping the error term:

Pseudo-cut (k)

$$-\tilde{G}^k x + \theta \geq \tilde{g}^k, \quad \xi = \xi^k, \quad S = S^k. \tag{2.25}$$

2.4.2 Stochastic Benders Algorithm

Starting with any feasible solution $x = \xi^1$ to the first-stage problem (such as the optimal solution of $\min cx$, $Ax = b$, $x \geq 0$) each subsequent iteration k is initialized using as its feasible first-stage solution $x = \xi^k$. A random sample S^k of size $|S^k|$ is drawn and the parameters $\tilde{G}^k$ and $\tilde{g}^k$ of the pseudo-cut (k) are computed according to (2.22). The pseudo-cut (k) is then adjoined to pseudo-master ($k-1$) to generate the pseudo-master problem (k), see (2.26).

Pseudo-master (k)

$$\begin{array}{rcrcrcl}
\min \tilde{v} & = & cx & + & \theta & & \\
\tilde{\rho}: & & Ax & & & = & b \\
\tilde{\lambda}^1: & & -\tilde{G}^1 x & + & \theta & \geq & \tilde{g}^1 \\
\tilde{\lambda}^2: & & -\tilde{G}^2 x & + & \theta & \geq & \tilde{g}^2 \\
 & & \vdots & & \vdots & & \vdots \\
\tilde{\lambda}^k: & & -\tilde{G}^k x & + & \theta & \geq & \tilde{g}^k \\
 & & x & & & \geq & 0.
\end{array} \tag{2.26}$$

Pseudo-cuts are iteratively adjoined in this manner until the maximum iteration $k = K$ is reached.

2.4.3 First-Stage Decision $x = \xi^l$, Upper Confidence Interval for z^*

An unbiased estimate of the *minimum expected second-stage costs* θ_ξ^k, calculated on iteration k, is

$$\tilde{\theta}_{\xi}^{k} = \frac{1}{|S^k|} \sum_{\omega \in S^k} f y_{\xi}^{\omega}, \quad \xi = \xi^k, \tag{2.27}$$

where y_{ξ}^{ω} minimizes subproblem ω given $x = \xi^k$, see (2.3). An unbiased estimate of the variance of $\tilde{\theta}_{\xi}^{k}$ about its true mean

$$\theta_{\xi}^{k} = \sum_{\omega \in \Omega} p^{\omega} f y_{\xi}^{\omega} \tag{2.28}$$

is also calculated on iteration k:

$$\tilde{\sigma}_k^2 = \frac{1}{|S^k|(|S^k|-1)} \sum_{\omega \in S^k} (f y_{\xi}^{\omega} - \tilde{\theta}^k)^2, \quad \xi = \xi^k. \tag{2.29}$$

Therefore an unbiased estimate of the first-stage and second-stage costs (given $\xi = \xi^k$) is

$$\tilde{z}_{\xi}^{k} = c\xi^k + \tilde{\theta}_{\xi}^{k}, \quad \xi = \xi^k, \tag{2.30}$$

and $\tilde{\sigma}_k^2$ is an unbiased estimate of the variance of $\tilde{z}_{\xi}^{k}$.

At the termination of the stochastic Benders algorithm, we choose $x = \xi^l$ as the first-stage decision, where $l = \arg\min \tilde{z}^k$. However, the minimum of several minima $\tilde{z}^k$ is no longer an unbiased estimate of z^l. To obtain an unbiased estimate of z^l, we re-estimate $\tilde{\theta}_{\xi}^{l}$, $\tilde{z}_{\xi}^{k}$, and $\tilde{\sigma}_l^2$ using a new independently drawn sample $\hat{S}^l$. In our applications, sample sizes are $|S| = 100$ or more, so it is reasonable (by the central limit theorem) to assume $\tilde{z}_{\xi}^{l}$ is normally distributed. Therefore an α upper confidence interval for z^* is

$$\text{Prob}(\tilde{z}_{\xi}^{l} + t\sigma_l \geq z^*) \geq \alpha, \tag{2.31}$$

where t is defined as

$$\frac{1}{\sqrt{2\pi}} \int_{-\infty}^{t} e^{-\frac{t^2}{2}} dt = \alpha. \tag{2.32}$$

2.4.4 Lower Confidence Interval for z^*

In order to know how close the objective associated with the first-stage decision $x = \xi^l$ is to the true $\min z = z^*$, we estimate a lower confidence level for z^*. All feasible solutions to (2.2) satisfy the stochastic cuts (2.24), $-\tilde{G}^k x + \theta \geq \tilde{g}^k - \tilde{\epsilon}^k$, where $\tilde{\epsilon}^k = \frac{1}{|S^k|} \sum_{\omega \in S^k} f y^{\omega} - \theta$.

In particular, the optimum $z^*, x^*, \theta^*, \epsilon_*^k$, $k = 1, \ldots, K$, satisfies the following

System of Inequalities

$$
\begin{array}{rcrcl}
cx^* & + & \theta^* & = & z^* \\
Ax^* & & & = & b \\
-\tilde{G}^1 x^* & + & \theta^* & \geq & \tilde{g}^1 - \tilde{\epsilon}_*^1 \\
-\tilde{G}^2 x^* & + & \theta^* & \geq & \tilde{g}^2 - \tilde{\epsilon}_*^2 \\
\vdots & & \vdots & & \vdots \\
-\tilde{G}^K x^* & + & \theta^* & \geq & \tilde{g}^K - \tilde{\epsilon}_*^K \\
x^* & & & \geq & 0,
\end{array}
\tag{2.33}
$$

where for each k,

$$
\tilde{\epsilon}_*^k = \frac{1}{|S^k|} \sum_{\omega \in S^k} f y_*^\omega - \theta^* \tag{2.34}
$$

is a sample average about the true mean θ^* drawn from the same distribution of fy_*^ω. We now show how the optimal dual multipliers of the pseudo-master (K), depicted in (2.35), can be applied to (2.33) to derive a lower bound for the optimal solution z^*.

Pseudo-master Problem (K)

$$
\begin{array}{rcrcrcl}
\min \tilde{v} & = & cx & + & \theta & & \\
\tilde{\rho}: & & Ax & & & = & b \\
\tilde{\lambda}^1: & & -\tilde{G}^1 x & + & \theta & \geq & \tilde{g}^1 \\
\tilde{\lambda}^2: & & -\tilde{G}^2 x & + & \theta & \geq & \tilde{g}^2 \\
& & \vdots & & \vdots & & \vdots \\
\tilde{\lambda}^K: & & -\tilde{G}^K x & + & \theta & \geq & \tilde{g}^K \\
& & x & & & \geq & 0.
\end{array}
\tag{2.35}
$$

Let $\tilde{\rho}, \tilde{\lambda}^1, \ldots, \tilde{\lambda}^K$ be the optimal dual variables of (2.35) and let $\tilde{v}^* = \min \tilde{v}$. These satisfy

$$\tilde{v}^* = \tilde{\rho} b + \sum_{k=1}^{K} \tilde{\lambda}^k \tilde{g}^k, \tag{2.36}$$

$$\tilde{\rho} A - \sum_{k=1}^{K} \tilde{\lambda}^k \tilde{G}^k + \gamma = c, \quad \gamma \geq 0, \tag{2.37}$$

$$\sum_{k=1}^{K} \tilde{\lambda}^k = 1, \tag{2.38}$$

$$\tilde{\lambda}^k \geq 0, \quad k = 1, \ldots, K. \tag{2.39}$$

(Note that $\gamma \geq 0$ may be interpreted as the optimal dual multiplier associated with $x \geq 0$.) Applying these same multipliers to the corresponding relations in (2.33) and subtracting from the first relation of (2.33), we obtain

$$0 \leq \gamma x^* \leq z^* - \left(\tilde{\rho} b + \sum_{k=1}^{K} \tilde{\lambda}^k \tilde{g}^k \right) + \sum_{k=1}^{K} \tilde{\lambda}^k \tilde{\epsilon}_*^k. \tag{2.40}$$

Dropping $\gamma x^* \geq 0$, substituting $\tilde{v}^*$ from (2.36) for the middle term in parentheses, and rearranging terms, we obtain a lower bound for the optimal objective z^* of (2.2):

$$\tilde{v}^* - \Delta \leq z^*, \quad \Delta = \sum_{k=1}^{K} \tilde{\lambda}^k \tilde{\epsilon}_*^k, \tag{2.41}$$

where $\tilde{\lambda}^k$ are optimal dual multipliers of the pseudo-master (K), $\tilde{\lambda}^k \geq 0$, $\sum_{k=1}^{K} \tilde{\lambda}^k = 1$, and $\tilde{\epsilon}_*^k$ for $k = 1, \ldots, K$ are the deviations from their true mean θ^* of the sample means of samples S^k of size $|S^k|$ of terms $f y_*^{\omega}$.

2.4.5 The Distribution of the Error Term Δ

Our goal is to derive an approximation of the probability distribution for the random variable $\Delta = \sum_{k=1}^{K} \tilde{\lambda}^k \tilde{\epsilon}_*^k$. In fact we develop two such distributions upper bounding Δ; the first we call "worst-case bound" and the second we call "conservative bound." For this analysis we need the distributions of the error terms $\tilde{\epsilon}_*^k$.

By definition, σ_*^2, the variance of the population of the optimal second-stage costs $f y_*^{\omega}$, $\omega \in \Omega$, corresponding to the optimal first-stage solution x^* is given by

$$\sigma_*^2 = \sum_{\omega \in \Omega} p^{\omega} (f y_*^{\omega} - \theta^*)^2. \tag{2.42}$$

We assume that all sample sizes used in the various iterations are equal to N, i.e.,

$$|S^k| = N, \quad k = 1, \ldots, K. \tag{2.43}$$

Since we have fixed $x = x^*$ at the optimum, the random samples S^k of size N for computing the means $\frac{1}{|S^k|}\sum_{\omega \in S^k} f y_*^{\omega}$ are *all drawn from the same distribution of values* $f y_*^{\omega}$. Therefore each variance of $\tilde{\epsilon}_*^k$, the expected value of $\left(\frac{1}{|S^k|}\sum_{\omega \in S^k} f y_*^{\omega} - \theta^*\right)^2$, is equal to

$$\mathrm{var}(\tilde{\epsilon}_*^k) = \frac{\sigma_*^2}{N}, \quad \text{for } k = 1, \ldots, K. \tag{2.44}$$

Note that this is a theoretical result, since we do not know x^* and hence we cannot generate sampled values of $f y_*^{\omega}$ to estimate σ_*^2. We estimate σ_*^2 by setting it equal to the estimated variance of $f y_{\xi}^{\omega}$, where $\xi = \xi^l$:

$$\text{estimated } \sigma_*^2 = \tilde{\sigma}_l = \frac{1}{N-1}\sum_{\omega \in S}\left(f y_{\xi}^{\omega} - \frac{1}{N} f y_{\xi}^{\omega}\right)^2, \quad \xi = \xi^l. \tag{2.45}$$

Each error term $\tilde{\epsilon}_*^k$ is the difference of the average of N independently drawn observations with replacements *from the same distribution* of $f y_*^{\omega}$ minus its true mean θ^*. Because sample sizes in our applications typically satisfy $N \geq 100$ and are often several hundreds, it is reasonable (by the central limit theorem) to assume $\tilde{\epsilon}_*^k$ are normally distributed:

$$\tilde{\epsilon}_*^k \sim N\left(0, \frac{\sigma_*}{\sqrt{N}}\right) = \frac{\sigma_*}{\sqrt{N}} N(0, 1). \tag{2.46}$$

Our goal now is to determine an upper bound for $\Delta = \sum_{k=1}^{K} \tilde{\lambda}^k \tilde{\epsilon}_*^k$, say Δ^{α}, such that $\mathrm{Prob}(\Delta \leq \Delta^{\alpha}) \geq \alpha$. Since we do not know the distribution of Δ, we cannot compute Δ^{α} directly. Instead, we determine two distributions, a worst-case distribution Δ_{W} and a "conservative" distribution Δ_{C}, each of which dominates the Δ distribution, and find an α-point for each distribution.

2.4.6 Worst-Case Lower Bound for z^*

Since $\tilde{\lambda}^k \geq 0$ and $\sum_{k=1}^{K} \tilde{\lambda}^k = 1$, the worst-case upper bound for Δ is Δ_{W},

$$\Delta_{\mathrm{W}} = \max_k \tilde{\epsilon}_*^k \geq \sum_{k=1}^{K} \tilde{\lambda}^k \tilde{\epsilon}_*^k = \Delta. \tag{2.47}$$

We do not know what particular realization of $\max_k \tilde{\epsilon}_*^k$ gave rise to the worst-case lower bound (2.47), but we know from (2.46) that $\Delta_{\mathrm{W}} = \max \tilde{\epsilon}_*^k$ is a random variable distributed as the maximum of K independent normal $(\sigma_*/\sqrt{N})N(0,1)$ variates. We determine the point $\Delta_{\mathrm{W}}^{\alpha}$ on this distribution such that

$$\mathrm{Prob}(\Delta_{\mathrm{W}} \leq \Delta_{\mathrm{W}}^{\alpha}) \geq \alpha, \tag{2.48}$$

where

$$\Delta_{\mathrm{W}}^{\alpha} = \tilde{v}^* - t_K \frac{\sigma_*}{\sqrt{N}}, \tag{2.49}$$

and t_K is defined as the Kth order statistic of a unit normal random variable:

$$\frac{1}{\sqrt{2\pi}} \int_{-\infty}^{t_K} e^{-\frac{t^2}{2}} dt = \alpha^{\frac{1}{K}}. \tag{2.50}$$

Since $\Delta \leq \Delta_{\mathrm{W}}$ by (2.47),

$$\mathrm{Prob}(\Delta_{\mathrm{W}} \leq \Delta_{\mathrm{W}}^{\alpha}) \geq \alpha, \tag{2.51}$$

and

$$\mathrm{Prob}(\tilde{v}^* - \Delta \geq \tilde{v}^* - \Delta_{\mathrm{W}}^{\alpha}) \geq \alpha. \tag{2.52}$$

Since $z^* \geq \tilde{v}^* - \Delta$ by (2.41),

$$\mathrm{Prob}(\tilde{v}^* - \Delta_{\mathrm{W}}^{\alpha} \leq z^*) \geq \alpha. \tag{2.53}$$

We call the lower bound $(\tilde{v}^* - \Delta_{\mathrm{W}}^{\alpha})$ the probabilistic worst-case α lower bound for z^*.

2.4.7 Conservative Lower Bound for z^*

The conservative lower bound is based on the observation that the $\tilde{\lambda}^k$ and $\tilde{\epsilon}_*^k$ in the expression $\Delta = \sum_{k=1}^{K} \tilde{\lambda}^k \tilde{\epsilon}_*^k$ tend to be positively correlated. This can be seen intuitively. It is reasonable to expect, in forming the pseudo-master problem, that dropping the term $\tilde{\epsilon}_*^k$ when it is large and positive will give rise to a tighter constraint k and dropping it when negative will slacken the constraint. This positive correlation of $\tilde{\lambda}^k$ with $\tilde{\epsilon}_*^k$ is very evident in practical problems, see the empirical evidence presented at the end of this chapter, Fig. 2.3 and Table 2.5.

In order to obtain Δ_{C} (the conservative upper bound for Δ) and hence $\tilde{v}^* - \Delta_{\mathrm{C}}$ (the conservative lower bound for z^*), we reorder the $\tilde{\lambda}^k$ from high to low and relabel them as

$$\tilde{\mu}^1 \geq \tilde{\mu}^2 \geq \cdots \geq \tilde{\mu}^K \geq 0. \tag{2.54}$$

We also order the $\tilde{\epsilon}_*^k$ from high to low and relabel them as

$$\tilde{\delta}^1 \geq \tilde{\delta}^2 \geq \cdots \geq \tilde{\delta}^K. \tag{2.55}$$

It is then obvious (and trivial to prove) that

$$\Delta = \sum_{k=1}^{K} \tilde{\lambda}^k \tilde{\epsilon}_*^k \leq \sum_{k=1}^{K} \tilde{\mu}^k \tilde{\delta}^k. \tag{2.56}$$

The difference $\sum_{k=1}^{K} \tilde{\mu}^k \tilde{\delta}^k - \Delta$ depends on how correlated $\tilde{\lambda}^k$ and $\tilde{\epsilon}_*^k$ are. The higher they are correlated, the smaller will be the difference. We have observed empirically that after $\tilde{\lambda}^k$ and $\tilde{\epsilon}_*^k$ are reordered, $\tilde{\mu}^k$ and $\tilde{\delta}^k$ are no longer correlated; accordingly, for the development of the conservative lower bound, we make the assumption that $\tilde{\mu}^k$ and $\tilde{\delta}^k$ are *independent.*

Notice that the $\tilde{\delta}^k$s are the distribution of instances of K normal $(\sigma_*/\sqrt{N})N(0, 1)$ variates ordered from high to low. We do not know which particular instance of the K ordered normal deviates $\tilde{\delta}^k$ formed the products with the ordered $\tilde{\mu}^k$ and were then summed. We can, however, view

$$\Delta_C = \sum_{k=1}^{K} \tilde{\mu}^k \tilde{\delta}^k, \quad \text{for given } \tilde{\mu}^k, \tag{2.57}$$

as a random variable, generated by all possible instances of ordered sets of K ordered normal deviates $\tilde{\delta}^k$. The distribution of Δ_C can be determined either by numerical integration or by Monte Carlo approximation. We approximate the distribution of Δ_C by generating many K independent $(\sigma_*/\sqrt{N})N(0, 1)$ normal deviates and reordering them into many ordered instances of $\tilde{\delta}^1, \ldots, \tilde{\delta}^K$. Substituting each of these ordered instances into (2.57) we obtain an approximate distribution that we call the *probabilistic* upper bound distribution for Δ_C. We determine Δ_C^α, the α percent point of this distribution, for which

$$\text{Prob}(\Delta_c \leq \Delta_c^\alpha) \geq \alpha. \tag{2.58}$$

Since $\Delta \leq \sum_{k=1}^{K} \tilde{\mu}^k \tilde{\delta}^k = \Delta_C$ by (2.56) and (2.57),

$$\text{Prob}(\Delta \leq \Delta_C^\alpha) \geq \alpha, \tag{2.59}$$

and

$$\text{Prob}(\tilde{v}^* - \Delta \geq \tilde{v}^* - \Delta_C^\alpha) \geq \alpha. \tag{2.60}$$

Since $z^* \geq \tilde{v}^* - \Delta$ by (2.41),

$$\text{Prob}(\tilde{v}^* - \Delta_C^\alpha \leq z^*) \geq \alpha. \tag{2.61}$$

We call the lower bound $(\tilde{v}^* - \Delta_C^\alpha)$ the probabilistic conservative α lower bound for z^*.

To summarize, the main steps are as follows:

- Obtain $\tilde{\rho}$, $\tilde{\lambda}^k$, $k = 1, \ldots, K$, and $\tilde{v}^* = \tilde{\rho}b + \sum_{k=1}^{K} \tilde{\lambda}^k \tilde{g}^k$ by solving the *pseudo-master problem* (2.35) for $\tilde{v}^* = \min \tilde{v}$.
- Order $\tilde{\lambda}^k$ from high to low to obtain $\tilde{\mu}^k$, $k = 1, \ldots, K$.
- Obtain the distribution Δ_C by generating many sets of K independent observations $\tilde{\epsilon}_*^k$ of the distribution $(\sigma_*/\sqrt{N})N(0, 1)$, reordering them from high to low to form the $\tilde{\delta}^1, \ldots, \tilde{\delta}^K$ sets and substituting into $\Delta_C = \sum_{k=1}^{K} \tilde{\mu}^k \tilde{\delta}^k$ to obtain instances of Δ_C.
- Determine the Δ_C^α point of the distribution Δ_C by Monte Carlo and construct the confidence interval $\text{Prob}(\tilde{v}^* - \Delta_c^\alpha \le z^*) \ge \alpha$, where the estimate of the lower bound for z^* is $\tilde{v}^* - \Delta_C^\alpha$.

Comment

When generating the distribution Δ_C for approximating $\sum_{k=1}^{K} \tilde{\mu}^k \tilde{\delta}^k$, we are assuming that $\tilde{\mu}^k$ and $\tilde{\delta}^k$, the kth-order variate of the normal distribution, are uncorrelated. Empirically, once we reorder, that seems to be the case (see, e.g., the evidence presented in the appendix of this chapter). Even if there is some correlation remaining it cannot have much effect because the kth-order variate of the normal distribution is confined to a narrow range of values.

What deserves more study, and is a goal for future research, is how best to estimate σ_*. If ξ^l is a nearly optimal solution, $\tilde{\sigma}_l$ should be a good approximation to the true value of σ_*. Alternatively, $\tilde{\sigma}_{K+1}$, the estimated variance of $\tilde{z}_\xi^{K+1}$ corresponding to $\xi = xi^{K+1}$, the optimal solution of the pseudo-master problem (K), could be used as a good estimate for σ_*. As a third way of estimating σ_*, we suggest computing the solution

$$\hat{\xi} = \sum_{k=1}^{K} \tilde{\lambda}^k \xi^k \tag{2.62}$$

and estimating the variance of $\tilde{\theta}_\xi = \frac{1}{|S|} \sum_{\omega \in S} f y_\xi^\omega$, $\xi = \hat{\xi}$, around its true mean $\theta_\xi = \sum_{\omega \in \Omega} f p^\omega y_\xi^\omega$, $\xi = \hat{\xi}$, as

$$\tilde{\sigma}_\xi^2 = \frac{1}{|S|(|S| - 1)} \sum_{\omega \in S} (f y_\xi^\omega - \tilde{\theta}_\xi)^2, \quad \xi = \hat{\xi}, \tag{2.63}$$

using an independent sample $S = \hat{S}$. We recommend in practice using $\tilde{\sigma}_l$ as an estimator for σ_*. Nevertheless, to be on the safe side in practice, it may be a good idea to inflate the estimated σ_*, say by 5%.

2.5 Test

In order to test the theory, we used a number of problems discussed in the literature (see, e.g., Holmes (1994)) and also problems we designed in-house. For each of these, the number W of scenarios ω was small enough to allow us to compute the optimal solution exactly. For the test we solved the universe problem to obtain the optimal universe solution z^*, x^*, y_*^ω, and the true population variance σ_*^2. We also recorded the sequence of solutions $x = \xi^k$, $k = 1, \ldots, K$, that led to the optimal solution. The solutions ξ^k, $k = 1, \ldots, K$, then represented the solutions corresponding to the cutting planes stored in the master problem when it terminated at the optimal solution.

To carry out the test we used the recorded solutions ξ^k, $k = 1, \ldots, K$, to generate the stochastic cuts. For each of these, we computed the true correction terms $\tilde{\epsilon}_*^k$. Based on the true correction terms we computed the true value of the lower bound $\tilde{v}^* - \Delta$, where $\tilde{v}^* = \tilde{\rho} b + \sum_{k=1}^K \tilde{\lambda}^k \tilde{g}^k$ is the optimal objective of the pseudo-master problem (K), and we computed the less tight but true lower bound $\tilde{v}^* - \Delta_C$, based on the reordering of the $\tilde{\lambda}^k$ and $\tilde{\epsilon}_*^k$.

We then computed empirically an upper bound distribution Δ_C by repeating R times the sampling of K observations from the normal distribution $(\sigma_*/\sqrt{N})N(0, 1)$ and ordering the resulting values to obtain R observations of $\Delta_C = \sum_{k=1}^K \tilde{\mu}^k \tilde{\delta}^k$. We used this empirical distribution of Δ_C to calculate Δ_C^α and used Δ_C^α to calculate $\tilde{v}^* - \Delta_C^\alpha$, a one-sided α lower confidence interval for z^*.

For illustration purposes we describe in detail the results of applying the methodology to the test problem APL1P, which is a small electric power expansion planning problem with uncertainty in three demands and in the availability of two generators, see Infanger (1992). The master problem has three rows and three columns and each second-stage scenario has six rows and nine columns. The total number of scenarios is $W = 1280$. The optimal values of x^* and y_*^ω resulted in $z^* = 24642.3$, $\theta^* = 13513.7$, and $\sigma_* = 4808.8$. For the experiment we estimated $K = 20$ cuts using a sample size of $N = 100$. The optimal dual multipliers for the pseudo-master problem $\tilde{\lambda}^k$, and the values of $\tilde{\epsilon}_*^k$, as well as the ordered values $\tilde{\mu}^k$ and $\tilde{\delta}^k$ for $k = 1, \ldots, 20$, are displayed in Table 2.1. The lower bound for z^* before reordering the $\tilde{\lambda}^k$ and $\tilde{\epsilon}_*^k$ is

$$z^* \geq \tilde{v}^* - \sum_{k=1}^{K} \tilde{\lambda}^k \tilde{\epsilon}^k = 25188.6 - 546.3 = 24642.3, \tag{2.64}$$

which verifies that $\tilde{v}^* - \sum_{k=1}^K \tilde{\lambda}^k \tilde{\epsilon}^k$ is, as stated earlier, a true lower bound (24642.3) for z^*. It happened in this case that the 20 corrected cuts were sufficient to determine the true minimum $z^* = 24642.3$. We then reordered the $\tilde{\lambda}^k$ to obtain $\tilde{\mu}^k$ and reordered the $\tilde{\epsilon}_*^k$ to obtain $\tilde{\delta}^k$. Clearly,

Table 2.1 Test results for problem APL1P

Cut(k)	$\tilde{\lambda}^k$	$\tilde{\epsilon}_*^k$	$\tilde{\mu}^k$	$\tilde{\delta}^k$
1	0.558	636.757	0.558	636.757
2	0.000	−656.181	0.276	504.965
3	0.276	504.965	0.166	388.265
4	0.000	−500.738	0.000	339.834
5	0.000	−71.361	0.000	316.619
6	0.000	−519.535	0.000	311.268
7	0.000	−457.835	0.000	114.471
8	0.000	−181.275	0.000	96.202
9	0.000	388.265	0.000	−48.166
10	0.000	96.202	0.000	−71.361
11	0.000	−362.272	0.000	−117.495
12	0.000	−117.495	0.000	−181.275
13	0.000	−929.552	0.000	−331.046
14	0.000	339.834	0.000	−362.272
15	0.000	114.471	0.000	−457.835
16	0.166	311.268	0.000	−500.738
17	0.000	−601.078	0.000	−519.535
18	0.000	−48.166	0.000	−601.078
19	0.000	316.619	0.000	−656.181
20	0.000	−331.046	0.000	−929.552
	$\sum \tilde{\lambda}^k = 1$	$\sum \tilde{\lambda}^k \tilde{\epsilon}_*^k = 546.309$	$\sum \tilde{\mu}^k = 1$	$\sum \tilde{\mu}^k \tilde{\delta}^k = 559.086$

$$z^* \geq \tilde{v}^* - \sum_{k=1}^{K} \tilde{\mu}^k \tilde{\delta}^k = 25188.6 - 559.1 = 24629.5, \tag{2.65}$$

which verifies that $\tilde{v}^* - \sum_{k=1}^{K} \tilde{\mu}^k \tilde{\delta}^k$ is a smaller true lower bound for z^*. Using the largest observed value $\max_k \tilde{\epsilon}_*^k$, the worst-case lower bound for z^* is

$$z^* \geq \tilde{v}^* - \max_k \tilde{\epsilon}_*^k = 25188.6 - 636.6 = 24552.0. \tag{2.66}$$

So far, all the evaluations for the example were exact; there were no estimates. We now discuss how we estimated the term $\sum_{k=1}^{K} \tilde{\mu}^k \tilde{\delta}^k$. We know that the $\tilde{\delta}^k$ are a reordering of $\tilde{\epsilon}_*^k$, and that these δ^k, $k = 1, \ldots, K$, are randomly drawn (approximately) $(\sigma_*/\sqrt{N})N(0, 1)$ deviates. What we did was draw $R = 500$ samples of $K = 20$ independent normally distributed $(\sigma_*/\sqrt{N})N(0, 1)$ deviates. We ordered the ϵ_*^k in each of the sets of 20 from high to low, and for each reordered set i of the 20 δ_k we computed $\sum_{k=1}^{20} \tilde{\mu}^k \delta^k = \Delta_C^i$. Next we generated the distribution of Δ_C^i, $i = 1, \ldots, 500$, cases of Δ_C and computed the point $\Delta_C^{0.95}$ as the upper bound estimate of Δ_C, where $\Delta_C^{0.95}$ is the value of Δ_C that was exceeded in 95% of the cases. Hence

$$z^* \geq \tilde{v}^* - \Delta_C^{0.95} = 25188.6 - 1143.7 = 24044.9, \tag{2.67}$$

where $\sigma_* = 4808.90$. In practice, of course, we do not have the true value of σ_*. Using $\tilde{\sigma}_{K-2} = 4449.8$ (to simulate a situation where the optimum x^* has not been reached yet) as an estimate of σ_*, we obtain

$$z^* \geq \tilde{v}^* - \Delta_C^{0.95}|_{\tilde{\sigma}_{K-2}} = 25188.6 - 1058.3 = 24130.3, \tag{2.68}$$

which under-bounds the true value of $z^* = 24642.3$ by 2.1%. We calculated the worst-case lower bound using the true value of σ^* as

$$z^* \geq \tilde{v}^* - \Delta_W^{0.95} = 25188.6 - 1345.9 = 23842.7 \tag{2.69}$$

and using the estimate $\tilde{\sigma}_{K-2} = 4449.8$ as

$$z^* \geq \tilde{v}^* - \Delta_W^{0.95}|_{\tilde{\sigma}_{K-2}} = 25188.6 - 1245.4 = 23943.2. \tag{2.70}$$

To test the coverage of the computed lower bound, we repeatedly (say 100 times) ran the test described above with different seeds. That is, for each replication we used the sequence of solutions ξ^k, $k = 1, \ldots, K$, to compute cutting planes using independent samples S^k and solved the corresponding pseudo-master problem (K) to obtain the optimal $\tilde{\rho}$ and $\tilde{\lambda}^1, \ldots, \tilde{\lambda}^K$. We computed for each of these replications the minimum objective function value of the pseudo-master problem $\tilde{v}^* = \tilde{\rho}b + \sum_{k=1}^K \tilde{\lambda}^k \tilde{g}^k$; a true lower bound $\tilde{v}^* - \sum_{k=1}^K \tilde{\mu}^k \tilde{\delta}^k$, based on the true ordered values of the correction terms $\tilde{\epsilon}_*^k$; and the 95% point $\tilde{v}^* - \Delta_C^{0.95}$, based on the reordering procedure. We also computed and recorded $\tilde{v}^* - \Delta_W^{0.95}$, the 95% point of the worst-case lower bound distribution $\tilde{v}^* - \Delta_W$. Based on the 100 replications of the experiment we estimated the coverage of the probabilistic lower bounds by computing the percent of the 100 cases in which the probabilistic lower bound was actually less than or equal to the true optimal objective value z^*.

The results for the test problem APL1P are represented in Fig. 2.1, which displays the values of $\tilde{v}^*$, $\tilde{v}^* - \sum_{k=1}^K \tilde{\mu}^k \tilde{\delta}^k$, $\tilde{v}^* - \Delta_C^{0.95}$, and $\tilde{v}^* - \Delta_W^{0.95}$, respectively, as a histogram of the 100 values from the replications. Instead of the actual values,

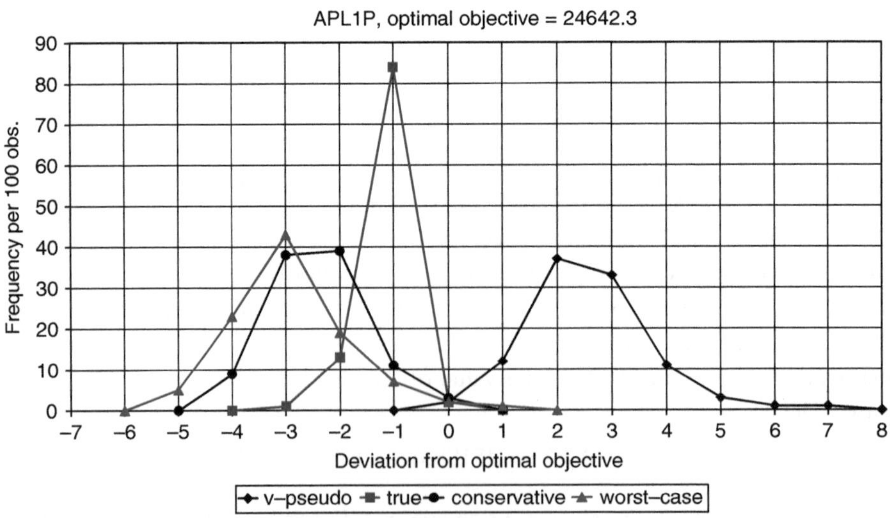

Fig. 2.1 APL1P: lower bound distributions

we report the quantities as percent deviation from the true objective function value z^*. The true objective function value in the graph is labeled as 0.

The curve labeled "v-pseudo" represents the histogram of $\tilde{v}^*$, the minimum objective of the pseudo-master problem. One can see that most of the 100 replications had $\tilde{v}^*$ values larger than the true optimal objective z^*. It clearly reveals the bias of the optimal objective of the pseudo-master $\tilde{v}^*$ as an underestimator for z^*. The curve labeled "true" represents the histogram of the true lower bound $\tilde{v}^* - \sum_{k=1}^{K} \tilde{\mu}^k \tilde{\delta}_*^k$ based on the reordering of $\tilde{\lambda}^k$ and $\tilde{\epsilon}_*^k$. There was (as predicted by the theory) no instance in which this true lower bound exceeded the true optimal objective. The two observations at the zero point of the curve are two observations in the interval between -1 and 0. The curve labeled "conservative" represents the histogram (constructed from the 100 replications) of $\tilde{v}^* - \Delta_{\mathrm{C}}^{0.95}$, the 95% point of the $\tilde{v}^* - \Delta_{\mathrm{C}}^{0.95}$ distribution. The curve shows that $\Delta_{\mathrm{C}}^{0.95}$ is a conservative estimate of Δ. The coverage of the $\tilde{v}^* - \Delta_{\mathrm{C}}^{0.95}$ turned out to be 96%. Finally, the curve labeled "worst case" represents the histogram of $\tilde{v}^* - \Delta_{\mathrm{W}}^{0.95}$, the 95% point of the probabilistic worst-case lower bound. As expected, the $\tilde{v}^* - \Delta_{\mathrm{W}}^{0.95}$ values turned out to be smaller than the $\tilde{v}^* - \Delta_{\mathrm{C}}^{0.95}$ values, which makes the probabilistic worst-case lower bound $\tilde{v}^* - \Delta_{\mathrm{W}}^{0.95}$ an even smaller but nevertheless tight lower bound for z^*. Its coverage turned out to be 96%.

With the other test problems, we obtained very similar results. As a representative example, we show the results for the test problem STORM described in Mulvey and Ruszczyński (1992). The problem is a freight-scheduling problem with uncertainty in demands. The version we used had a total of 40 universe scenarios. The size of the master problem was 126 rows and 289 columns, and each of the 40 subproblems had 347 rows and 769 columns. The optimal objective of the universe problem was $z^* = 15.569 \times 10^6$ and the variance σ_* was 89159.5. For the experiment we used $K = 30$ cutting planes estimated with a sample size of $N = 20$. Figure 2.2 gives the results. The figure looks very similar to the one for APL1P, except that the pseudo-cutting planes are better estimates of the true cutting planes. Looking at the distribution labeled "v-pseudo" one can see the bias of the optimal solution of the pseudo-master problem $\tilde{v}^*$ as an estimator for z^*. There was no instance where the true conservative lower bound $\tilde{v}^* - \sum_{k=1}^{K} \tilde{\mu}^k \tilde{\delta}^k$ (labeled "true") exceeded the true objective z^*. The 95% point $\tilde{v}^* - \Delta_{\mathrm{C}}^{0.95}$ of the estimated conservative lower bound distribution (labeled "conservative") gave a conservative estimate but nevertheless an excellent lower bound for z^*. The coverage of $\tilde{v}^* - \Delta_{\mathrm{C}}^{0.95}$ turned out to be 100%. The point $\Delta_{\mathrm{W}}^{0.95}$ of the worst-case lower bound distribution (labeled "worst case") gave a smaller but nevertheless tight lower bound for z^*. The coverage of the worst-case lower bound $\tilde{v}^* - \Delta_{\mathrm{W}}^{0.95}$ proved to be 100%.

We tested further with the following problems: PGP2, CEP1, and SCTAP1, all described in Holmes (1994). PGP2 (Louveaux and Smeers 1988) is a small electric power capacity expansion planning test problem (master: 2 rows and 4 columns, sub: 7 rows and 16 columns) with uncertain parameters in three demands. The number of universe scenarios was $W = 576$. CEP1 (Higle and Sen 1994) is a small machine capacity expansion planning problem (master: 9 rows and 8 columns, sub: 7 rows and 15 columns) with uncertain parameters in the right-hand side. The

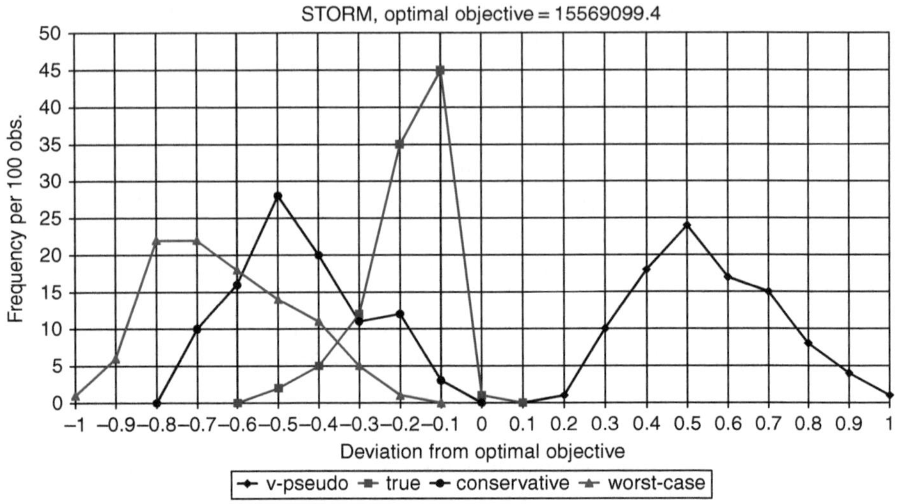

Fig. 2.2 STORM: lower bound distributions

Table 2.2 Test data

Problem	z^*	σ_*	N	K
APL1P	24642.3	4808.8	100	20
PGP2	447.3	77.60	100	20
CEP1	355159.5	420458.4	100	6
STORM	$15.569{\cdot}10^6$	89159.5	20	30
SCTAP1	248.5	24.01	100	3

Table 2.3 Coverage test results (lower bounds calculated using σ_*)

Problem	$\tilde{v}^* - \Delta_{\mathrm{C}}^{0.95}$ (%)	Coverage (%)	$\tilde{v}^* - \Delta_{\mathrm{W}}^{0.95}$ (%)	Coverage (%)
APL1P	−1.78	95	−2.39	94
PGP2	−2.94	97	−3.51	98
CEP1	−12.12	95	−12.27	94
STORM	−0.40	100	−0.58	100
SCTAP1	−0.9	99	−0.91	99

number of universe scenarios was $W = 1000$. SCTAP (Ho 1980) is a traffic assignment problem (master: 30 rows and 48 columns, sub: 60 rows and 96 columns) with stochastic right-hand sides. The number of universe scenarios was $W = 864$. The values of z^* and σ_* for the various test problems, as well as the sample sizes N and the number of cutting planes K used for the experiments, are given in Table 2.2. The coverage results of the test problems are summarized in Table 2.3 based on using the true value of σ_* and in Table 2.4 based on using the estimate $\tilde{\sigma}_{K-2}$ (in order to simulate a situation where the optimum x^* has not been reached yet) as approximation for σ_*.

Table 2.4 Coverage test results (lower bounds calculated using $\tilde{\sigma}_{K-2}$)

Problem	$\tilde{v}^* - \Delta_C^{0.95}$ (%)	Coverage (%)	$\tilde{v}^* - \Delta_W^{0.95}$ (%)	Coverage (%)
APL1P	−1.80	96	−2.41	96
PGP2	−2.35	91	−2.84	94
CEP1	−12.07	92	−12.23	92
STORM	−0.40	98	−0.58	99
SCTAP1	−0.9	98	−0.91	98

Appendix

We present empirical observations with respect to the correlation between $\tilde{\mu}^k$ and $\tilde{\delta}^k$, the reordered values of $\tilde{\lambda}^k$ and $\tilde{\epsilon}_*^k$, respectively. The results presented in Fig. 2.3 were obtained by solving the test problem PGP2 1000 times with different seeds. The figure presents four scatter diagrams of values for $\tilde{\mu}^k$ versus $\tilde{\delta}^k$. The top left diagram represents all values (i.e., $\tilde{\lambda}^k$ and $\tilde{\epsilon}_*^k$). One can clearly see the positive cor-

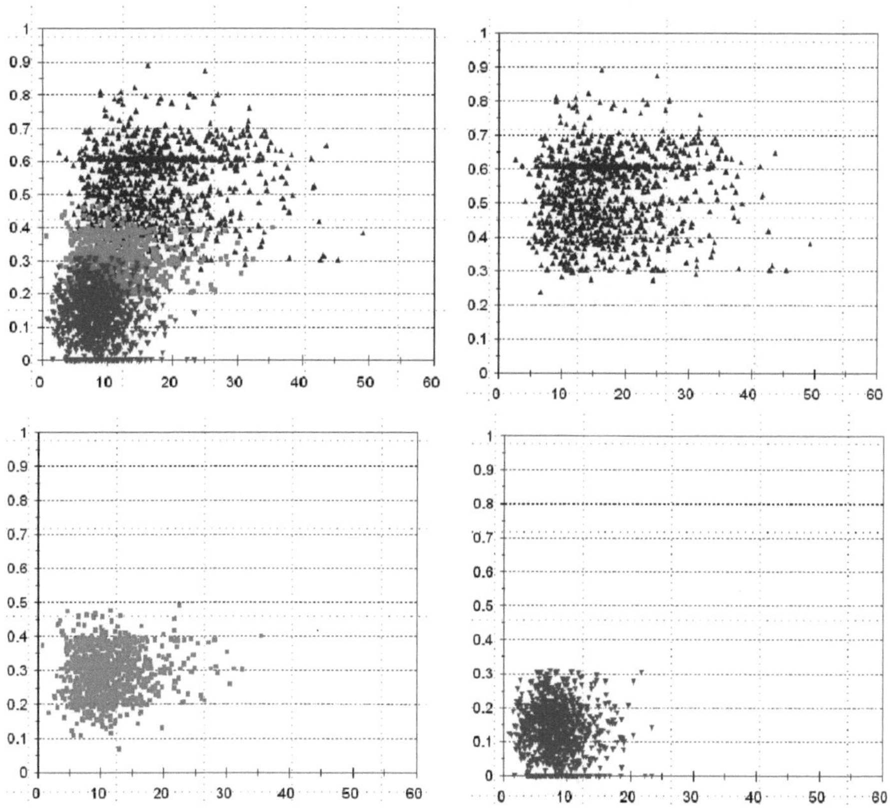

Fig. 2.3 Scatter diagrams of $\tilde{\mu}^k$ versus $\tilde{\delta}^k$ (problem PGP2, 1000 observations). All values (*top left*), first-ordered values (*top right*), second-ordered values (*bottom left*), and third-ordered values (*bottom right*)

relation between the values of $\tilde{\lambda}^k$ and $\tilde{\epsilon}_*^k$. The top right diagram represents the first-ordered values, the bottom left diagram the second-ordered values, and the bottom right diagram the third-ordered values. Each ordered group, graphed by itself, does not exhibit correlation between the values for $\tilde{\mu}^k$ and $\tilde{\delta}^k$. Ordering maximizes the correlation between values. After accounting for the maximum correlation through the ordering procedure in Section 2.4.7, the values within each ordering are very close to uncorrelated.

Table 2.5 gives the estimated correlation coefficients between each ordering of $\tilde{\mu}^k$ and $\tilde{\delta}^k$ (based on 1000 observations) for the test problems APL1P, PGP2, and SCSD8. The first-ordered values of $\tilde{\mu}^k$ and $\tilde{\delta}^k$ for problem APL1P exhibit a small (but statistically significant) correlation coefficient. All other correlation coefficients of the ordered values $\tilde{\mu}^k$ and $\tilde{\delta}^k$ are small and statistically insignificant.

Table 2.5 Correlation coefficients between $\tilde{\mu}^k$ and $\tilde{\delta}^k$ of first-, second- and third-ordered values

Problem	First ordered	Second ordered	Third ordered
APL1P	0.15	−0.0006	−0.05
PGP2	0.06	0.09	−0.06
SCSD8	0.05	0.03	0.0

References

Beale, E.M.L.: On minimizing a convex function subject to linear inequalities. J. R. Stat. Soc. **17b**, 173–184 (1955)

Benders, J.F.: Partitioning procedures for solving mixed-variable programming problems. Numerische Math. **4**, 238–252 (1962)

Birge, J.R., Louveaux, F.V.: A multicut algorithm for two-stage stochastic linear programs. Eur. J. Oper. Res. **34**(3), 384–392 (1988) ISSN 0377-2217

Birge, J.R., Wallace, S.W.: A separable piecewise linear upper bound for stochastic linear programs. SIAM J. Control. Optim. **26**(3) (1988)

Birge, J.R., Wets, R.J.: Computing bounds for stochastic programming problems by means of a generalized moment problem. Math. Oper. Res. **12**, 149–162 (1987)

Birge, J.R., Wets, R.J.: Sublinear upper bounds for stochastic programs with recourse. Math. Program. **43**, 131–149 (1989)

Dantzig, G.B.: Linear programming under uncertainty. Manage. Sci. 1, 197–206 (1955)

Dantzig, G.B., Glynn, P.W.: Parallel processors for planning under uncertainty. Ann. Oper. Res. **22**, 1–21 (1990)

Dantzig, G.B., Infanger, G.: Large-scale stochastic linear programs—importance sampling and Benders decomposition. In: Brezinski, C. Kulisch, U. (eds.) Computational and Applied Mathematics, I (Dublin, 1991), pp. 111–120. North-Holland, Amsterdam (1992)

Dantzig, G.B., Infanger, G.: A probabilistic lower bound for two-stage stochastic programs, Department of Operations Research, Stanford University, Stanford, CA (1995)

Edmundson, H.P.: Bounds on the expectation of a convex function of a random variable. Paper 982, Rand Corporation Santa Monica, California (1956)

Ermoliev, Y.: Stochastic quasi-gradient methods and their applications to systems optimization. Stochastics **9**, 1–36 (1983)

Frauendorfer, K.: Solving slp recourse problems with arbitrary multivariate distributions—the dependent case. Math. Oper. Res. **13**(3), 377–394 (1988)

Frauendorfer, K.: Stochastic Two-Stage Programming. Lecture Notes in Economics and Mathematical Systems, vol. 392. Springer, Berlin (1992)
Frauendorfer, K., Kall, P.: Solving slp recourse problems with arbitrary multivariate distributions—the independent case. Probl. Control. Inf. Theory. **17**(4), 177–205 (1988)
Gaivoronski, A.: Stochastic quasigradient methods and their implementation. In: Ermoliev, Y. Wets, R.J-B. (eds.) Numerical Techniques for Stochastic Optimization. Springer Series is Computational Mathematics, vol. 10, pp. 313–351. Springer, Berlin (1988)
Glynn, P.W., Infanger, G.: Technical Report SOL-2008-1, Systems Optimization Laboratory, Department of Management Science and Engineering, Stanford University, Stanford, CA (2008)
Higle, J.L., Sen, S.: Stochastic decomposition: An algorithm for two-stage linear programs with recourse. Math. Oper. Res. **16**(3), 650–669 (1991)
Higle, J.L., Sen, S.: Finite master programs in regularized stochastic decomposition. Math. Program. **67**(2), 143–168 (1994)
Ho, J.K.: A successive linear optimization approach to the dynamic traffic assignment problem. Transport. Sci. **14**(4), 295–305 (1980)
Holmes, D.: A collection of stochastic programming problems. Technical Report 94–11, Department of Industrial and Operations Engineering, The University of Michigan, Ann Arbor, MI, pp. 48109–2117 (1994)
Huang, C.C., Ziemba, W.T., Ben-Tal, A.: Bounds on the expectation of a convex function with a random variable with applications to stochastic programming. Oper. Res. **25**(2), 315–325 (1977)
Infanger, G.: Monte Carlo (importance) sampling within a Benders decomposition algorithm for stochastic linear programs. Ann. Oper. Res. **39**(1–4), 69–95 (1992)
Jensen, J.L.: Sur les fonctions convexes et les inegalites entres les valeurs moyennes. Acta Math. **30**, 175–193 (1906)
Kall, P.: Approximations to stochastic programs with complete fixed recourse. Numerische Math. **22**, 333–339 (1974)
Louveaux, F., Smeers, Y.: Optimal investments for electricity generation, a stochastic model and a test problem. In: Ermoliev, Y., Wets, R.J-B. (eds.) Numerical Techniques for Stochastic Optimization, pp. 33–64. Springer, Berlin (1988)
Lustig, I.J., Mulvey, J.M., Carpenter, T.J.: Formulating two-stage stochastic programs for interior point methods. Oper. Res. **39**, 757–770 (1991)
Madansky, A.: Bounds on the expectation of a convex function of a multivariate random variable. Ann. Math. Stat. **30**, 743–746 (1959)
Mak, W.K., Morton, D.P., Wood, R.K.: Monte Carlo bounding techniques for determining solution quality in stochastic programs. Oper. Res. Lett. **24**, 47–56 (1999)
Mulvey, J.M., Ruszczyński, A.: A new scenario decomposition method for large-scale stochastic optimization. Technical Report SOR-91-19, Department of Civil Engineering and Operations Research, Princeton University, Princeton, NJ (1992)
Pereira, M.V., Pinto, L.M.V.G., Oliveira, G.C., Cunha, S.H.F.: A technique for solving lp-problems with stochastic right-hand sides, CEPEL, Centro Del Pesquisas De Energia Electria, Rio De Janeiro, Brazil (1989)
Prékopa, A.: Sharp bounds on probabilities using linear programming. Oper. Res. **38**(2) 227–239 (1990)
Rockafellar, R.T., Wets, R.J.: Scenario and policy aggregation in optimization under uncertainty. Math. Oper. Res. **16**, 119–147 (1989)
Ruszczyński, A.: A regularized decomposition method for minimizing a sum of polyhedral functions. Math. Program. **35**, 309–333 (1986)
Shapiro, A., Homem-de-Mello, T.: A simulation-based approach to two-stage stochastic programming with recourse. Math. Program. **81**, 301–325 (1998)
Van Slyke, R.M., Wets, R.J.: L-shaped linear programs with applications to optimal control and stochastic programming. SIAM J. Appl. Math. **17**, 638–663 (1969)

Chapter 3
Simulation-Based Optimality Tests for Stochastic Programs

Güzin Bayraksan, David P. Morton, and Amit Partani

3.1 Introduction

We consider the following stochastic optimization model:

$$z^* = \min_{x \in X} Ef(x, \xi). \qquad \text{(SP)}$$

Here, f is a real-valued function measuring the "cost" of a system of interest, x is a decision vector constrained to obey physical and policy rules represented by the set $X \subset \Re^{d_x}$, ξ is a d_ξ-dimensional random vector defined on probability space $(\Xi, \mathcal{B}, \mathcal{P})$, and E is the associated expectation operator. We denote an optimal solution and the optimal value of (SP) as x^* and z^*, respectively, recognizing that x^* may not be unique.

It is widely recognized that we cannot solve model (SP) exactly. Approximations are required. When we obtain a candidate solution, say $\hat{x} \in X$, via an approximation we seek to understand the quality of $\hat{x}$. The purpose of this chapter is to describe procedures for assessing whether $\hat{x}$ is of high quality. We measure quality using the optimality gap $Ef(x, \xi) - z^*$ as opposed to some metric in the space of x.

The two-stage stochastic linear program introduced by Dantzig (1955) and Beale (1955) is an example of model (SP). In such a model, X is a polyhedron, f is the optimal value of a linear program, and the parameters of that linear program are random. Stochastic programming applications emerge from a wide variety of areas involving optimization under uncertainty, including electric power systems, homeland security, finance, production supply chains, communications networks, transportation systems, and water resources management, see, e.g., the volume by Wallace and Ziemba (2005).

One approach for approximately optimizing (SP) is to sample n independent and identically distributed (i.i.d.) observations $\xi^1, \xi^2, \ldots, \xi^n$ from the distribution of ξ and then solve the approximating problem

G. Bayraksan (✉)
Systems and Industrial Engineering, University of Arizona, Tucson, AZ 85721, USA
e-mail: guzinb@sie.arizona.edu

G. Infanger (ed.), *Stochastic Programming*, International Series in Operations Research & Management Science 150, DOI 10.1007/978-1-4419-1642-6_3,

$$z_n^* = \min_{x \in X} \frac{1}{n} \sum_{j=1}^{n} f(x, \xi^j). \tag{SP$_n$}$$

Non-i.i.d. sampling schemes are also possible, and we return to one such scheme in Section 3.6.

We justify solving (SP$_n$) in place of (SP) by consistency results that establish conditions on f, X, ξ and the sampling procedure under which solutions (x_n^*, z_n^*) to (SP$_n$) are optimal to (SP) as the sample size n grows to infinity, see, for example, Attouch and Wets (1981), Dupačová and Wets (1988), Kall (1987), King and Wets (1991), Korf and Wets (2001), Robinson (1996), Robinson and Wets (1987), Rubinstein and Shapiro (1993), Shapiro (2003), and Shapiro et al. (2009). As an alternative to solving (SP$_n$), we can instead employ a sampling-based algorithm to solve (SP) such as a stochastic approximation algorithm, e.g., Lan et al. (2008), or a sampling-based cutting plane method, Higle and Sen (1999), Dantzig and Glynn (1990), and Infanger (1994). Such an algorithm can produce a sequence of iterates that has limit points that solve (SP), with probability one (w.p.1). Such consistency and convergence results are clearly needed, but they are insufficient because we must terminate finitely. So, in this chapter we describe a class of procedures that can be used to assess solution quality using approximate confidence interval statements. After additional background and motivation in the next section, we provide our core procedure for assessing solution quality in Section 3.3. Then, the three sections that follow present enhancements that can be employed if the core procedure yields inadequate results for a given computational budget.

The procedures we describe use both a candidate solution and a sample size, which are fixed. This is analogous to so-called fixed sample size procedures in the simulation literature. When attempting to control the size of errors associated with an interval estimator in simulation, one can employ a two-stage procedure or a sequential procedure, see, e.g., Law (2007). In a two-stage procedure, the first stage involves a pilot study used to estimate variance, and the second stage is carried out with a sample size that achieves the desired level of error, assuming that variance estimate to be correct. Sequential procedures iterate until the error estimate is sufficiently small. A two-stage procedure can be employed in our context, if the goal is to control sampling error. However, if the candidate solution is not optimal, our interval estimator will not shrink to zero with growing sample size. Hence, when employing sampling in stochastic optimization, sequential procedures are arguably most natural. We discuss this in Section 3.7 and refer to Bayraksan and Morton (2009a,b) for more detail. Other methods for estimating required sample sizes have also been put forward in various settings, e.g., Campi and Garatti (2007), Shapiro et al. (2002), Shapiro and Homem-de-Mello (1998). That said, these largely neglect the sequential aspects that we regard as inherent.

The material of Section 3.3 is based on Mak et al. (1999) and that of Section 3.4 on Bayraksan and Morton (2006). The two main theorems of Section 3.5 appear in Partani et al. (2006), where a variant of the bias reduction procedure in Section 3.5 is applied to an asset allocation model. A version of this chapter appeared in

Bayraksan and Morton (2009b), with Sections 3.2, 3.3, 3.4, and 3.6 largely taken from Bayraksan and Morton (2009b). Our approach to bias reduction in Section 3.5 differs from that of Bayraksan and Morton (2009b), and the manner in which they are distinguished is discussed further in Section 3.7. Moreover, as we indicate in the previous paragraph we address sequential issues in Bayraksan and Morton (2009b) in greater detail than we do here.

3.2 Background and Motivation

Our primary goal is to have efficient methods for assessing the quality of a feasible candidate solution $\hat{x} \in X$. This candidate solution might be obtained by solving (SP$_n$) or possibly through other means. The optimal value, z_n^*, to (SP$_n$) plays an important role in accomplishing this goal. So, before proceeding, we give a simple example that illustrates a number of properties of z_n^*.

Example 3.1 Define (SP) through $X = [-1, 1]$, $\xi \sim N(0, 1)$, and $f(x, \xi) = \xi x$. Clearly, $z^* = 0$ and every feasible solution is an optimal solution to this instance of (SP). Even though this problem is trivial we can form the approximating problem

$$z_n^* = \min_{-1 \le x \le 1} \left(\frac{1}{n} \sum_{j=1}^{n} \xi^j \right) x,$$

and we find $x_n^* = 1$ if $\sum_{j=1}^n \xi^j \le 0$ and $x_n^* = -1$ if $\sum_{j=1}^n \xi^j > 0$. The objective coefficient is an average of n i.i.d. standard normals and so $z_n^* = -|N(0, 1/n)|$. In this example, z_n^* has the following properties:

1. $Ez_n^* \le z^*, \forall n$ negative bias
2. $Ez_n^* \le Ez_{n+1}^*$ monotonically shrinking bias
3. $z_n^* \to z^*$, w.p.1 strong consistency
4. $\sqrt{n}(z_n^* - z^*) = -|N(0, 1)|$ nonnormal errors
5. $b(z_n^*) \equiv Ez_n^* - z^* = a_1/\sqrt{n}$ $O(n^{-1/2})$ bias. □

These first two properties on z_n^* hold much more generally (Mak et al. 1999), as does the third, under mild conditions. The fourth property is not in a form that holds for more general instances of (SP). Instead of "=" we usually have "⇒," i.e., convergence in distribution, and the limiting distribution differs. That said, the result is representative of the more general case with respect to both the $n^{-1/2}$ rate of convergence and the "folded" or "projected" normal random variables that arise (Dupačová 1991, King and Rockafellar 1993, Rubinstein and Shapiro 1993, Shapiro 1989). The fifth property concerns the rate at which z_n^*'s bias shrinks to zero. When (SP) has multiple optimal solutions, as in this example, $O(n^{-1/2})$ bias arises in a very general setting. This is important because for certain classes of mathematical programs (including stochastic programs), having multiple optimal solutions or having a "large" set of ϵ-optimal solutions is the norm rather than the exception.

The canonical rate at which sampling error in Monte Carlo methods shrinks to zero is $O(n^{-1/2})$ as illustrated here in the fourth property, and the fifth property suggests that rate at which bias shrinks to zero can be as slow as that of sampling error.

These results contrast sharply with those that arise when the optimization operator "$\min_{x \in X}$" is not present in (SP) and (SP$_n$). If $x \in X$ is fixed, we have an unbiased estimator that is strongly consistent with normally distributed errors (assuming, e.g., finite second moments). Optimization leads to biased estimators with nonnormal errors. Even consistency can be lost as can be seen by replacing $X = [-1, 1]$ with $X = \Re$ in Example 3.1. Finally, that the optimization operator can yield an $O(n^{-1/2})$ bias result is also atypical relative to what one usually obtains when other "smoother" types of operators are applied to sample means. (We return to the nature of bias that can arise in (SP$_n$) in Section 3.5.) For example, if an estimator is a smooth nonlinear function of a sample mean, then using a second-order Taylor expansion we find that the estimator has bias that shrinks to zero with $O(n^{-1})$. Thus, the optimization operator qualitatively changes the nature of results arising when employing Monte Carlo sampling. This suggests that we may need to exercise care when developing point and interval estimators that are rooted in (SP$_n$).

A key observation is that z_n^* gives a lower bound, in expectation, on the optimal solution value z^*. The intuition is as follows. In solving the original problem (SP), we seek a decision, x, that hedges against *all* realizations of ξ. When solving (SP$_n$), we optimize with respect to a subset of ξ's support, selected according to the distribution of ξ. Because of this "inside information," we over-optimize and, on average, obtain an optimistic objective value. Based on this same intuition, we expect the value of the bound to grow as n increases (property 2 in Example 3.1).

As indicated above, the z_n^* bound is analogous to lower bounds that arise from optimizing relaxations in other areas of optimization, except that it is statistical in nature. Therefore, it yields a different type of optimality statement compared to deterministic problems as discussed in Section 3.1. Now, suppose we take as a candidate solution the solution $\hat{x} = x_n^*$ of (SP$_n$) along with its optimal value z_n^*. The following question then arises: In what sense should we seek to characterize x_n^* and/or z_n^* as being "good" estimators of their population counterparts x^* and z^* from (SP)? We could pursue a number of different goals regarding such inference. For example, we could try to prove that

1. $x_n^* \to x^*$, w.p.1 and $\sqrt{n}(x_n^* - x^*) \Rightarrow Y_x$ (for some limiting random variable Y_x),
2. $z_n^* \to z^*$, w.p.1 and $\sqrt{n}(z_n^* - z^*) \Rightarrow Y_z$ (for some limiting random variable Y_z),
3. $Ef(x_n^*, \xi) \to z^*$, w.p.1, or
4. $\lim_{n \to \infty} P\left(Ef(x_n^*, \xi) \le z^* + \epsilon_n\right) = 1 - \alpha$, where the random width satisfies $\epsilon_n \to 0$, w.p.1.

Goal 1 would provide a route to forming an approximate CI on the optimal solution x^* using the limiting distribution Y_x. Similarly, we could attempt to use goal 2 to form an approximate confidence interval on z^*. If (SP$_n$) were a maximum likelihood estimation problem, then goal 1 would be appropriate, and if (SP) were a model to price a financial option then goal 2 would be appropriate. However, when

(SP) is a decision-making model then we believe that goal 1 is more than we need and goal 2 is of secondary interest. In this case, goals 3 and 4 are arguably what we should try to achieve. Note that the expectation in goal 3 is with respect to ξ and is conditional on x_n^*, i.e., $Ef(x_n^*, \xi)$ is a random variable. The probability in goal 4 is with respect to the random width ϵ_n and is also conditional on x_n^*. We further note that we cannot expect $\{x_n^*\}_{n=1}^{\infty}$ to converge when (SP) has multiple optimal solutions. In this case, we weaken the consistency result to "limit points of $\{x_n^*\}_{n=1}^{\infty}$ solve (SP)." If we achieve this "limit points result" when X is compact and $Ef(x, \xi)$ is continuous, then we obtain goal 3. As we have seen above, the limiting distributions Y_x and Y_z may not be normal. Finally, as noted via the references given above, these goals—particularly in the form of 1–3—have been pursued and established under various conditions on f, ξ, and X.

Finally, with respect to background, we recall some basic notions associated with confidence intervals (see, e.g., Casella and Berger 1990 Section 9.1). An *interval estimator*, I_n, of a real-valued parameter of interest is a random set based on sample size n. For example, for finite n with $\hat{x} = x_n^*$, $I_n = [0, \epsilon_n]$ in goal 4 is an interval estimator for the *optimality gap*, $\mu_{\hat{x}} = Ef(\hat{x}, \xi) - z^*$. The *coverage probability*, or simply the *coverage*, of an interval estimator is the probability that the random interval I_n contains the parameter of interest, e.g., $P(\mu_{\hat{x}} \in I_n)$. Typically, asymptotic results such as that in goal 4 provide theoretical justification for a specific procedure to form the interval estimator I_n, provided n is sufficiently large. As indicated in Section 3.1, practical interpretation of what is meant by sufficiently large is usually guided by empirical testing, specifically, by empirically assessing coverage probabilities.

3.3 Multiple Replications Procedure

In the spirit of goal 4 of the previous section, we measure the quality of a candidate solution $\hat{x}$, e.g., $\hat{x} = x_n^*$, by the optimality gap, $\mu_{\hat{x}} = Ef(\hat{x}, \xi) - z^*$. If the gap is sufficiently small then $\hat{x}$ is of high quality. Unfortunately, in our setting this gap cannot be computed exactly. In Mak et al. (1999), we circumvented the issue of nonnormal errors for z_n^* via a replications procedure to construct a confidence interval (CI) of the form

$$P\left(Ef(\hat{x}, \xi) - z^* \leq \epsilon\right) \approx 1 - \alpha. \tag{3.1}$$

Here, $\hat{x} \in X$ is a candidate solution, $Ef(\hat{x}, \xi)$ is its "true" and unknown expected performance measure, ϵ is the (random) CI width, and $1 - \alpha$ is the confidence level, e.g., 0.95. Let $t_{n,\alpha}$ be the $1 - \alpha$ quantile of the t distribution with n degrees of freedom. We summarize below our multiple replications procedure for constructing (3.1) from Mak et al. (1999).

MRP:

Input: Value $\alpha \in (0, 1)$ (e.g., $\alpha = 0.05$), sample size n, replication size n_g, and a candidate solution $\hat{x} \in X$.
Output: $(1-\alpha)$-level confidence interval on $\mu_{\hat{x}}$.

1. For $k = 1, 2, \ldots, n_g$,
   1.1. Sample i.i.d. observations $\xi^{k1}, \xi^{k2}, \ldots, \xi^{kn}$ from the distribution of ξ.
   1.2. Solve (SP$_n^k$) using $\xi^{k1}, \xi^{k2}, \ldots, \xi^{kn}$ to obtain x_n^{k*}.
   1.3. Calculate $G_n^k(\hat{x}) = \frac{1}{n}\sum_{j=1}^n \left(f(\hat{x}, \xi^{kj}) - f(x_n^{k*}, \xi^{kj})\right)$.
2. Calculate gap estimate and sample variance by

$$\bar{G}_n(n_g) = \frac{1}{n_g}\sum_{k=1}^{n_g} G_n^k(\hat{x}) \quad \text{and} \quad s_G^2(n_g) = \frac{1}{n_g - 1}\sum_{k=1}^{n_g}\left(G_n^k(\hat{x}) - \bar{G}_n(n_g)\right)^2.$$

3. Let $\epsilon_g = t_{n_g-1,\alpha} s_G(n_g)/\sqrt{n_g}$, and output one-sided CI on $\mu_{\hat{x}}$,

$$\left[0, \bar{G}_n(n_g) + \epsilon_g\right].$$

An upper bound on the optimality gap for $\hat{x}$ is given by $Ef(\hat{x}, \xi) - Ez_n^*$, since $Ez_n^* \leq z^*$. We estimate this quantity by

$$G_n(\hat{x}) = \frac{1}{n}\sum_{j=1}^n f(\hat{x}, \xi^j) - \min_{x \in X} \frac{1}{n}\sum_{j=1}^n f(x, \xi^j), \tag{3.2}$$

where $\xi^1, \xi^2, \ldots, \xi^n$ are i.i.d. as ξ. We know $G_n(\hat{x})$ can be nonnormal and so we produce n_g i.i.d. replicates in step 1 and form the resulting sample mean $\bar{G}_n(n_g)$ and sample variance $s_G^2(n_g)$ in step 2. Note that like $G_n(\hat{x})$, the sample mean $\bar{G}_n(n_g)$ depends on $\hat{x}$ but we suppress that dependence to keep the notation from becoming cumbersome. The CI $[0, \bar{G}_n(n_g) + \epsilon_g]$ on $\mu_{\hat{x}} = Ef(\hat{x}, \xi) - z^*$ is inferred from the central limit theorem (CLT)

$$\sqrt{n_g}\left[\bar{G}_n(n_g) - EG_n(\hat{x})\right] \Rightarrow N(0, \sigma_g^2) \ \text{ as } n_g \to \infty \ \text{ where } \sigma_g^2 = \operatorname{var} G_n(\hat{x}),$$

and the fact that $s_G^2(n_g)$ is a consistent estimator of σ_g^2.

The CLT holds by requiring that the batches $\xi^{k1}, \xi^{k2}, \ldots, \xi^{kn}$, $k = 1, \ldots, n_g$, be i.i.d. We have more freedom with respect to the random vectors within a batch. Specifically, for fixed values of x, we only require that the random vectors within a batch produce unbiased estimators, i.e., they satisfy $E\frac{1}{n}\sum_{j=1}^n f(x, \xi^{kj}) = Ef(x, \xi)$, $k = 1, 2, \ldots, n_g$. This condition implies that the first term on the right-hand side of (3.2) is an unbiased estimator of $Ef(\hat{x}, \xi)$. And, it implies that the second term of (3.2) satisfies $Ez_n^* = E\min_{x\in X}\frac{1}{n}\sum_{j=1}^{n} f(x,\xi^j) \leq \min_{x\in X} E\frac{1}{n}\sum_{j=1}^{n} f(x,\xi^j) = \min_{x\in X} Ef(x,\xi) = z^*$. That is, we have an estimator whose expectation provides an upper bound on the optimality gap, $E\bar{G}_n(n_g) = EG_n(\hat{x}) \geq Ef(\hat{x},\xi) - z^*$. The freedom to employ non-i.i.d. sampling within a batch is exploited in Section 3.6, where we employ Latin hypercube sampling. The random confidence interval width ϵ in (3.1) is simply $\epsilon = \bar{G}_n(n_g)+\epsilon_g$ in the interval estimator produced by the MRP.

Norkin et al. (1998) used the statistical lower bound z_n^* in global optimization of stochastic programs within a branch-and-bound method. Other algorithmic work that uses Monte Carlo simulation-based bounds and multiple replications includes Ahmed and Shapiro (2002) and Kleywegt et al. (2001). MRP has been applied to different kinds of problems in the literature including financial portfolio models (Bertocchi et al. 2000, Morton et al. 2006), stochastic vehicle routing problems (Kenyon and Morton 2003, Verweij et al. 2003), supply chain network design (Santoso et al. 2005), a stochastic water resources model to contain groundwater contaminants (Watkins, Jr. et al. 2005), an employee and production scheduling problem (Morton and Popova 2003), and a stochastic integer knapsack problem (Morton and Wood 1998).

There is other related work on assessing solution quality in stochastic programs via Monte Carlo methods, some being in the context of specific algorithms. Higle and Sen (1991b) derive a bound on the optimality gap for two-stage stochastic linear programs that is motivated by the Karush–Kuhn–Tucker (KKT) optimality conditions, see also Shapiro and Homem-de-Mello (1998). Higle and Sen (1996b) have also proposed a statistical lower bound that is rooted in duality. Dantzig, Glynn and Infanger (Dantzig and Glynn 1990, Dantzig and Infanger 1995, Infanger 1992, 1994) and Higle and Sen (1991a, 1999) use Monte Carlo versions of lower bounds obtained in sampling-based adaptations of deterministic cutting plane algorithms. And, Lan et al. (2008) use such lower bounds in a stochastic approximation algorithm.

The MRP has a number of important advantages. Foremost is its wide applicability. The same approach is valid regardless of whether (SP) is a two-stage stochastic linear or nonlinear program, X includes integer restrictions on x and/or second-stage variables are integer constrained, or $Ef(x,\xi)$ is a "nonstandard" objective function, e.g., the probability we exceed a threshold. Of course, the techniques employed to solve (SP_n) in step 1.2 in these cases will vary, as will the computational effort required to do so. That said, the validity of the CI on the optimality gap produced by the MRP is not an issue. To carry out the method we must be able to generate i.i.d. observations of ξ, and in order for the CLT to apply we require that $f(x,\xi)$ has finite second moments. The MRP is not tied to any particular algorithm so that (SP_n) can be solved by any number of means, and the effort to implement the MRP can be modest. For example, the MRP can be implemented with relative ease in a modeling language such as GAMS (Brooke et al. 2006). Another property we have observed is that the MRP tends to be conservative in its coverage properties.

For many problems, one can carry out the MRP and obtain satisfactory results with modest computational effort. The remainder of this chapter is devoted to what one can do when this is *not* the case. Three factors contribute to the CI width: (a) the bias of z_n^* can be large, (b) the sampling error, ϵ_g, can be large, and (c) the candidate solution $\hat{x}$ can be far from optimal to (SP). Any of these three can result in a CI width that is too large to be useful. A fourth shortcoming is (d) the computational cost of performing multiple replications, i.e., having to solve (say) $n_g = 30$ instances of (SP_n), can be prohibitive.

In this chapter we do not pursue issue (c). There is a large literature on algorithms and approximation schemes for stochastic programs. We assume that we have a solution, $\hat{x}$, believed to be of reasonably high quality, and we focus on the issue of quantitatively assessing that quality. Said another way, if we cannot obtain what is believed to be a high-quality solution then establishing its quality is not a high priority. We now briefly discuss (d) and then (a) and (b), in turn.

Single and Two Replication Procedures: The technical reason for performing n_g replications in MRP is that z_n^* can have nonnormal errors. Despite this, in Section 3.4 we describe an approach that enables us to assess the solution quality of $\hat{x}$ by solving either one or two instances of (SP_n). We may wish to pursue this option when the output of the MRP is acceptable, but the computational cost of achieving that result is excessive.

Bias Reduction: In integer or nonlinear programming, sometimes we have an optimal or near-optimal $\hat{x}$, but it takes a long time to prove optimality because considerable computational effort is required to tighten weak lower bounds. Issue (a) is an analogous dilemma in stochastic programming. As we have discussed $Ez_n^* \le z^*$, and sometimes this bias is the largest contributor to the CI width. In Section 3.5, we use a jackknife estimator that is designed to tighten weak lower bounds.

Variance Reduction: The error due to sampling, ϵ_g, can dominate the CI width and make it impossible to recognize a high-quality solution. Decreasing ϵ_g by increasing n_g or n can be prohibitively expensive because this shrinks the error at rate $n_g^{-1/2}$ or $n^{-1/2}$. In Section 3.6, we consider employing a randomized quasi-Monte Carlo procedure in order to reduce ϵ_g. (We note that if bias is small but ϵ_g is large and we seek to reduce ϵ_g *within* the MRP then we should increase n_g instead of n. For example, solving $n_g = 50$ instances of (SP_n) with $n = 100$ is usually computationally cheaper than solving $n_g = 25$ instances with $n = 200$.)

Our single and two replication procedures, the use of a jackknife estimator, and application of quasi-Monte Carlo methods are not tied to specific solution algorithms and hence have broad applicability. We emphasize that the three techniques we discuss are not designed to be used in concert. Rather, they are a set of tools that can be employed to improve the efficiency of the MRP depending on the nature of the computational bottleneck, i.e., issue (a), (b), or (d), for the specific problem under consideration. (While we do not explore the issue here, this assessment can usually be made via a two-stage procedure commonly used in simulation, e.g.,

Law 2007.) Indeed, the sections are written in a modular manner so that the reader can skip directly to the recommended remedy.

3.4 Single and Two Replication Procedures

When applying the MRP, the replication size, n_g, is taken to be, say, 20 or 30, in an attempt to have a valid statistical inference. This section is based on Bayraksan and Morton (2006) and begins with a single replication procedure to make a valid statistical inference on the quality of a candidate solution.

As before, let the candidate solution $\hat{x} \in X$ be given. We use the following additional notation. For a feasible solution, $x \in X$, let $\bar{f}_n(x) = \frac{1}{n}\sum_{j=1}^{n} f(x, \xi^j)$, $\sigma_{\hat{x}}^2(x) = \mathrm{var}[f(\hat{x}, \xi) - f(x, \xi)]$, and $s_n^2(x) = \frac{1}{n-1}\sum_{j=1}^{n}[(f(\hat{x}, \xi^j) - f(x, \xi^j)) - (\bar{f}_n(\hat{x}) - \bar{f}_n(x))]^2$. Note that $G_n(\hat{x})$ given in (3.2) can be written as $\bar{f}_n(\hat{x}) - z_n^*$, with the understanding that the same n observations $\xi^1, \xi^2, \ldots, \xi^n$ are used in $\bar{f}_n(\hat{x})$ and z_n^*. The single replication procedure follows:

SRP:

Input: Value $\alpha \in (0, 1)$, sample size n, and a candidate solution $\hat{x} \in X$.
Output: $(1 - \alpha)$-level confidence interval on $\mu_{\hat{x}}$.

1. Sample i.i.d. observations $\xi^1, \xi^2, \ldots, \xi^n$ from the distribution of ξ.
2. Solve (SP$_n$) to obtain x_n^*.
3. Calculate $G_n(\hat{x})$ as given in (3.2) and

$$s_n^2(x_n^*) = \frac{1}{n-1}\sum_{j=1}^{n}\left[(f(\hat{x}, \xi^j) - f(x_n^*, \xi^j)) - (\bar{f}_n(\hat{x}) - \bar{f}_n(x_n^*))\right]^2.$$

4. Let $\epsilon_g = t_{n-1,\alpha} s_n(x_n^*)/\sqrt{n}$, and output one-sided CI on $\mu_{\hat{x}}$,

$$\left[0,\ G_n(\hat{x}) + \epsilon_g\right].$$

The SRP and the MRP differ in how the sample variance is calculated. In the MRP, n_g i.i.d. observations of $G_n(\hat{x})$ are calculated, and the sample variance of these gap estimates is used to form the CI. In contrast, only one value of $G_n(\hat{x})$ is calculated in the SRP, and the individual observations, $f(\hat{x}, \xi^j) - f(x_n^*, \xi^j)$ for $j = 1, \ldots, n$, are used to calculate the sample variance. In fact, $G_n(\hat{x})$ is the sample mean of these individual observations and $s_n^2(x_n^*)$ is the corresponding sample variance. The following theorem provides asymptotic validity of the SRP.

Theorem 3.1 *Assume $X \neq \emptyset$ and is compact, that $Ef(\cdot, \xi)$ is lower semicontinuous on X, and $E\sup_{x\in X} f^2(x, \xi) < \infty$. Let $\hat{x} \in X$, and $\xi^1, \xi^2, \ldots, \xi^n$ be i.i.d. as ξ. Let $X^* = \arg\min_{x\in X} Ef(x, \xi)$, $x_{\min}^* \in \arg\min_{x\in X^*} var[f(\hat{x}, \xi) - f(x, \xi)]$, $x_{\max}^* \in \arg\max_{x\in X^*} var[f(\hat{x}, \xi) - f(x, \xi)]$, and assume*

$$\sigma_{\hat{x}}^2(x_{\min}^*) \leq \liminf_{n\to\infty} s_n^2(x_n^*) \leq \limsup_{n\to\infty} s_n^2(x_n^*) \leq \sigma_{\hat{x}}^2(x_{\max}^*), \textit{ w.p.1}. \tag{3.3}$$

Then, given $0 < \alpha < 1$ *in the SRP,*

$$\liminf_{n\to\infty} P\left(\mu_{\hat{x}} \leq G_n(\hat{x}) + \frac{t_{n-1,\alpha} s_n(x_n^*)}{\sqrt{n}}\right) \geq 1 - \alpha.$$

Theorem 3.1 justifies construction of the approximate $(1-\alpha)$-level one-sided confidence interval for $\mu_{\hat{x}} = Ef(\hat{x}, \xi) - z^*$ given in the SRP without requiring $G_n(\hat{x}) = \bar{f}_n(\hat{x}) - z_n^*$ to be asymptotically normal. The hypotheses of Theorem 3.1 ensure $\bar{f}_n(x_n^*) = z_n^*$ converges to z^* and hence $G_n(\hat{x})$ converges to $\mu_{\hat{x}}$, w.p.1. Even if (SP) has multiple optimal solutions, $Ef(x, \xi)$ takes value z^* on X^*. However, $f(x, \xi)$'s second moment is not constant on X^* and hence we cannot expect $s_n^2(x_n^*)$ to converge. Given that $\{x_n^*\}_{n=1}^{\infty}$ has all of its limit points in X^*, w.p.1, hypothesis (3.3) is a natural notion of consistency for $s_n^2(x_n^*)$. Sufficient conditions to ensure (3.3) are provided in Bayraksan and Morton (2006). We note that only the lower bound in (3.3) is needed when $\alpha < \frac{1}{2}$, as is typically the case.

The MRP is a procedure in which we use $n_g \geq$ 20–30 replications and the SRP is a procedure with just one replication, $n_g = 1$. As we show in Bayraksan and Morton (2006), empirical coverage results for the SRP do not always inherit the relatively conservative nature of those of the MRP. For this reason, in practice we recommend the following two replication variant of the SRP.

2RP:

Recall the definition of the MRP and fix $n_g = 2$. Replace steps 1.3, 2, and 3 by

1.3′. Calculate $G_n^k(\hat{x})$ and $s_n^2(x_n^{k*})$.

2′. Calculate the estimates by taking the average

$$G_n'(\hat{x}) = \frac{1}{2}\left(G_n^1(\hat{x}) + G_n^2(\hat{x})\right) \quad \text{and} \quad (s_n')^2 = \frac{1}{2}\left(s_n^2(x_n^{1*}) + s_n^2(x_n^{2*})\right).$$

3′. Output one-sided CI on $\mu_{\hat{x}}$

$$\left[0,\ G_n'(\hat{x}) + \frac{t_{2n-1,\alpha} s_n'}{\sqrt{2n}}\right].$$

The sample variance, $s_n^2(x_n^{k*})$, for each sample $k = 1, 2$, in the 2RP is calculated as in the single replication procedure and these are averaged to obtain the variance estimator of the 2RP (in step 2′). We provide in Bayraksan and Morton (2006) an asymptotic justification for the 2RP that is analogous to Theorem 3.1. When concern for conservative coverage results is paramount but the MRP is too computationally expensive to employ, we recommend the 2RP. In Bayraksan and Morton (2006) we further note computational and coverage benefits that arise from suboptimizing the (SP_n) used in the SRP and 2RP. For many problems the cost of performing multiple replications, i.e., issue (d), is the primary computational bottleneck in efficiently assessing solution quality. In such cases, the 2RP can yield considerable computational savings.

3.5 Bias Reduction

When lower bounds are weak, we may fail to recognize an optimal, or near-optimal, candidate solution, i.e., the bias of z_n^* can significantly degrade our ability to assess the quality of a candidate solution. Therefore, we develop techniques to reduce this bias. Our approach is rooted in jackknife estimators. The jackknife is a bias reduction tool first used by Quenouille (1949a,b) and has become an important tool in simulation and data analysis. The *standard* jackknife estimator is used to eliminate $O(n^{-1})$ bias. Let g_n be an estimator of g_μ that is based on n i.i.d. observations and suppose that

$$b(g_n) = Eg_n - g_\mu = a_1/n + a_2/n^2 + \cdots . \tag{3.4}$$

Now, $nEg_n = ng_\mu + a_1 + a_2/n + \cdots$ and $(n-1)Eg_{n-1} = (n-1)g_\mu + a_1 + a_2/(n-1) + \cdots$. So, $E[ng_n - (n-1)g_{n-1}] = g_\mu + O(n^{-2})$, i.e., $J_n = ng_n - (n-1)g_{n-1}$ has lower order bias than g_n. Forming J_n involves computing g_n with n samples and then computing g_{n-1} after *deleting* the nth observation. The n underlying observations are i.i.d. so we can delete any single observation and achieve the same effect. Deleting each observation in turn we obtain *pseudo-values*, $J_{n\backslash i}, i = 1, \ldots, n$, and the jackknife estimator is given by $\frac{1}{n}\sum_{i=1}^n J_{n\backslash i}$. One advantage of the jackknife is that the procedure can be applied when the coefficients, $a_1, a_2, \ldots$, are unknown.

We have already seen in Example 3.1 that we can have $b(z_n^*) = Ez_n^* - z^* = O(n^{-1/2})$, so that, in general, the standard jackknife estimator is not appropriate in our setting. The following example shows that we can have $b(z_n^*) = O(n^{-p})$ for any $p \in [1/2, \infty)$.

Example 3.2 Define (SP) through $X = \Re$, $\xi \sim N(0,1)$, and $f(x,\xi) = \xi x + |x|^\beta$, where $\beta > 1$. Clearly, $z^* = 0$ and $x^* = 0$. The approximating problem

$$z_n^* = \min_x \left(\frac{1}{n} \sum_{j=1}^n \xi^j \right) x + |x|^\beta$$

has optimal value given by

$$z_n^* = -\beta^{-\frac{\beta}{\beta-1}} (\beta - 1) \, |N(0,1)| \, n^{-p},$$

where $p = \frac{\beta}{2(\beta-1)}$. Taking expectations we obtain $b(z_n^*) = Ez_n^* = -an^{-p}$, where $a > 0$ is a constant independent of n. As $\beta \to \infty$, $p \to 1/2$ and as $\beta \to 1$, $p \to \infty$. □

Example 3.2 compels us to consider the following form of bias:

$$b(z_n^*) = Ez_n^* - z^* = a_1/n^p + o(n^{-p}). \tag{3.5}$$

Mimicking the derivation of the standard jackknife estimator leads to the following definition of pseudo-values for the generalized jackknife estimator as developed in Gray and Schucany (1972):

$$J_{n\backslash i}^p = \frac{n^p z_n^* - (n-1)^p z_{n\backslash i}^*}{n^p - (n-1)^p}, \quad i = 1, \ldots, n,$$

where z_n^* is the optimal value to (SP_n) and $z_{n\backslash i}^*$ is the optimal value of (SP_n) except that the ith sample is removed, i.e.,

$$z_{n\backslash i}^* = \min_{x \in X} \frac{1}{n-1} \sum_{j=1, j \neq i}^{n} f(x, \xi^j), \tag{$\mathrm{SP}_{n\backslash i}$}$$

$i = 1, \ldots, n$. The generalized jackknife estimator is then

$$z_n^J(p) = \frac{1}{n} \sum_{i=1}^{n} J_{n\backslash i}^p. \tag{3.6}$$

Unfortunately, we are unlikely to know the true order of the bias, p. In what follows we form the generalized jackknife estimator using parameter q and reserve notation p to denote the true (and unknown) order of the bias. We return to the issue of how to choose q below.

The following simple but useful result rests only on the monotonicity property $Ez_n^* \geq Ez_{n-1}^*$.

Theorem 3.2 *Let* $0 < q_1 < q_2$. *Then* $Ez_n^J(q_1) \geq Ez_n^J(q_2) \geq Ez_n^*$ *and* $\lim_{q \to \infty} Ez_n^J(q) = Ez_n^*$.

As $q \to \infty$ for fixed n, $z_n^J(q) \to z_n^*$, w.p.1, i.e., large values of q effectively correspond to our original lower bound estimator and smaller values of q are more aggressive with respect to removing bias, with the associated risk that the "lower bound" estimator exceeds z^*.

Of course, in our MRP of Section 3.3 our estimation is centered on the optimality gap, $\mu_{\hat{x}}$, as opposed to z^*. That said, $b(G_n(\hat{x})) = -b(z_n^*)$ and so the bias considerations are captured in the lower bound estimator. The following multiple replication procedure with generalized jackknife estimators uses the above ideas except that the jackknife estimator is expressed in terms of the underlying gap estimator.

MRP-GJ(q):

Input: Value $\alpha \in (0, 1)$, sample size n, replication size n_g, jackknife parameter q, and a candidate solution $\hat{x} \in X$.
Output: $(1-\alpha)$-level confidence interval on $\mu_{\hat{x}}$.

1. For $k = 1, 2, \ldots, n_g$,

   1.1. Sample i.i.d. observations $\xi^{k1}, \xi^{k2}, \ldots, \xi^{kn}$ from the distribution of ξ.
   1.2. Let

$$G_n^k(\hat{x}) = \frac{1}{n}\sum_{j=1}^{n} f(\hat{x}, \xi^{kj}) - \min_{x \in X} \frac{1}{n}\sum_{j=1}^{n} f(x, \xi^{kj}).$$

   1.3. For $i = 1, \ldots, n$ let

$$G_{n\backslash i}^k = \frac{1}{n-1}\sum_{j=1, j\neq i}^{n} f(\hat{x}, \xi^{kj}) - \min_{x \in X} \frac{1}{n-1}\sum_{j=1, j\neq i}^{n} f(x, \xi^{kj}).$$

   1.4. Form

$$J_n^k = \frac{n^q G_n^k(\hat{x}) - (n-1)^q \frac{1}{n}\sum_{i=1}^{n} G_{n\backslash i}^k(\hat{x})}{n^q - (n-1)^q}.$$

2. Calculate gap estimate and sample variance by

$$\bar{G}_n^q(n_g) = \frac{1}{n_g}\sum_{k=1}^{n_g} J_n^k \quad \text{and} \quad s_J^2(n_g) = \frac{1}{n_g - 1}\sum_{k=1}^{n_g} \left(J_n^k - \bar{G}_n^q(n_g)\right)^2.$$

3. Output one-sided CI on $\mu_{\hat{x}}$

$$\left[0, \bar{G}_n^q(n_g) + \frac{t_{n_g-1,\alpha} s_J(n_g)}{\sqrt{n_g}}\right].$$

The following quantity, ρ, compares the asymptotic decrease in bias of the generalized jackknife estimator relative to the bias of the original estimator, $\bar{G}_n(n_g)$, from the MRP of Section 3.3.

$$\rho = \left| \lim_{n\to\infty} \frac{E\bar{G}_n^q(n_g) - \mu_{\hat{x}}}{E\bar{G}_n(n_g) - \mu_{\hat{x}}} \right|.$$

The next result is adapted to our setting from Gray and Schucany (1972).

Theorem 3.3 *Consider the gap estimator of the MRP-GJ(q). Assume* $b(z_n^*) = a_1 n^{-p_1} + o(n^{-p_1})$, *where* $a_1 \neq 0$. *Then,*

(i) if $q = p_1$ *then* $\rho = 0$;
(ii) if $q > \frac{p_1}{2}$ *and* $q \neq p_1$ *then* $0 < \rho < 1$;
(iii) if $q = \frac{p_1}{2}$ *then* $\rho = 1$; *and*
(iv) if $0 < q < \frac{p_1}{2}$ *then* $1 < \rho < \infty$.

Typically, when attempting to reduce bias via the generalized jackknife, one obtains an asymptotic expansion (e.g., (3.4) or (3.5) or otherwise) and then employs the order-p jackknife estimator if the series is $O(n^{-p})$. However, we seek the one-sided CI (3.1) on $\mu_{\hat{x}} = Ef(\hat{x}, \xi) - z^*$, and in reducing bias in z_n^*, the estimator of z^*, we prefer to be conservative and err on the "low" side. Theorem 3.2 provides guidance in this regard. For example, if $b(z_n^*) = O(n^{-p})$ then we recommend using MRP-GJ(q) with $q > p$. Theorem 3.3 ensures that we achieve (asymptotic) bias reduction by using these conservative values of q. Of course, this is complicated by the fact that p is unknown in our setting. In Section 3.7 we point to ongoing work on estimating p and using that estimate, or rather a conservative estimate as described above, as the q value in the MRP-GJ(q).

3.6 Variance Reduction

Uniform random variables (or pseudo-random variates) on the unit interval form the basis for numerical procedures for generating i.i.d. observations of a nonuniform random variable ξ. Using an appropriate transformation ϕ we can express $f(\xi) = f(\phi(U)) \equiv h(U)$, where U is a uniform random vector on the unit cube, $[0, 1]^d$, and where, for simplicity, we temporarily suppress x. Here, we would have $d = d_\xi$ if, for example, the components of ξ are independent and each is generated via inversion. So, computing $Ef(\xi)$ may be viewed as numerical integration of h over $[0, 1]^d$. In their simplest form, numerical quadrature rules for computing $Eh(U)$ (e.g., d-dimensional variants of Simpson's rule) use a *uniform grid* of points over $[0, 1]^d$. However, the computational effort needed to achieve a specific level of error with quadrature grows exponentially in the dimension d, and in the absence of special structures, such techniques are not usually viable for d larger than 3 or 4. Monte Carlo (MC) methods instead draw i.i.d. observations of U over $[0, 1]^d$. Each observation is generated independently and so the resulting collection of n points is not as "evenly spread" across $[0, 1]^d$ as it could be. Loosely speaking, quasi-Monte Carlo (QMC) methods (e.g., Niederreiter 1992) lie somewhere between these two approaches. To avoid the dimensional effect of quadrature, QMC does not use a uniform grid. But, it spreads points more evenly than MC to improve on MC's $O(n^{-1/2})$ convergence rate and hence improve on its efficiency.

A number of authors have used variance reduction techniques for stochastic programming. Importance sampling has been employed by Dantzig and Glynn (1990) and Infanger (1992). Antithetic variates, control variates, importance sampling, and stratified sampling are explored in Higle (1998). The use of common random numbers in Mak et al. (1999) reduces variance and this technique is also used in Higle and Sen (1996a). Latin hypercube sampling is used in Bailey et al. (1999), Diwekar and Kalagnanam (1997), Freimer et al. (2005), and Linderoth et al. (2006). Application of QMC in stochastic programming is examined in Diwekar and Kalagnanam (1997) and Pennanen and Koivu (2005).

QMC methods use *low-discrepancy* sequences of points that can achieve an $O(n^{-1}\log^d n)$ rate of convergence. For sufficiently large n, $n^{-1}\log^d n \le n^{-1/2}$. However, when $d = 14$ (which is small for stochastic programs) this requires $n \ge 8.3 \times 10^{59}$. Based on this worst-case bound, we have little hope for QMC methods working well on moderate- to high-dimensional problems. Yet, there is considerable computational evidence in the literature that QMC can perform quite well relative to this bound and, more importantly, relative to MC. Work on the notion of an *effective dimension* of the integrand has been put forward to explain this behavior (Caflisch et al. 1997). Methods for constructing low-discrepancy sequences include (t, m, s)-nets; sequences due to Faure, Halton, Hammersley, Niederreiter, and others; and lattices (L'Ecuyer and Lemieux 2000, Sloan and Joe 1994). These schemes produce infinite (deterministic) sequences of vectors in $[0, 1]^d$ that have partial sequences with low discrepancy.

One shortcoming of QMC is the difficulty associated with obtaining good error statements. (In contrast, MC methods infer such error estimates from CLTs.) Another difficulty with employing QMC, particularly in our setting, is that the point estimates are not unbiased estimators—they are not even random variables. These two issues make it difficult to simply "plug in" a QMC scheme in our MRP in a sensible way.

Randomized QMC (RQMC) methods (e.g., Fox 1999) circumvent both of these difficulties. Let the n points generated via a QMC scheme be denoted $\mathcal{U}_n = \{u^1, u^2, \ldots, u^n\}$. RQMC performs a random shift of these points. In particular, let v be a *random* vector from the uniform distribution on $[0, 1]^d$. We redefine each element of $\mathcal{U}_n$ as $\tilde{u}^j = [u^j + v]$ where $[\cdot]$ returns the fractional part of its argument and then we form $\tilde{h} = \frac{1}{n}\sum_{j=1}^n h(\tilde{u}^j)$. Performing n_g i.i.d. shifts of $\mathcal{U}_n$ in this way yields n_g observations $\tilde{h}^k$, $k = 1, \ldots, n_g$, and we define $\bar{h}_{n_g} = \frac{1}{n_g}\sum_{k=1}^{n_g} \tilde{h}^k$. Each $\tilde{u}^i$ is uniform on $[0, 1]^d$ and the n_g random shifts are done independently so $\tilde{h}^k$, $k = 1, \ldots, n_g$, are i.i.d., $\bar{h}_{n_g}$ is an unbiased estimator of $Eh(U)$, the standard sample variance is an unbiased and strongly consistent estimator of $\mathrm{var}(\tilde{h}^1)$, and the standard CLT for i.i.d. random variables applies (Fox 1999, L'Ecuyer and Lemieux 2000, Sloan and Joe 1994). As an immediate consequence we obtain the following theorem.

Theorem 3.4 *Consider the variant of the MRP from Section 3.1 in which we use RQMC to generate the i.i.d. batches in step 1.1. Then,*

(i) $EG_n(\hat{x}) \ge Ef(\hat{x}, \xi) - z^*$,

(ii) $\sqrt{n_g}\left[\bar{G}_n(n_g) - EG_n(\hat{x})\right] \Rightarrow N(0, \sigma_g^2)$ *as* $n_g \to \infty$ *where* $\sigma_g^2 = \mathrm{var}\, G_n(\hat{x})$, *and*

(iii) $Es_G^2(n_g) = \sigma_g^2$ *and* $s_G^2(n_g) \to \sigma_g^2$, *w.p.1.*

From Theorem 3.4 we can infer the output of a RQMC variant of the MRP satisfies (asymptotically) the CI (3.1) on the optimality gap. We have observed significant decreases in sampling error, ϵ_g, by using the RQMC variant of the MRP. Moreover, while our initial motivation for exploring RQMC is to reduce variance, we have found empirically that bias is also reduced relative to MC. It does follow from the $O(n^{-1} \log^d n)$ rate of convergence that the bias shrinks more quickly than the worst-case $O(n^{-1/2})$ for MC but the caveats on the size of n still hold. The intuitive explanation of the nature of the bias from Section 3.2 (i.e., having to hedge against only n realizations and not all of ξ's support) suggests that it would be more difficult to "over-optimize" n evenly spread scenarios than n random scenarios. Similar results are obtained in Freimer et al. (2005) for Latin hypercube sampling.

3.7 Conclusions

Stochastic programs can be challenging to solve, and much attention in the literature has been devoted to approximation schemes. One such approach involves solving (SP_{n_x}), i.e., (SP_n) with $n = n_x$, and obtaining solution $\hat{x} = x^*_{n_x}$. Conditions under which $x^*_{n_x}$ is optimal to (SP) as n_x grows large have received significant attention in the literature. Ways to assess whether $\hat{x} = x^*_{n_x}$ is near optimal for finite n_x have received much less attention. This has been the focus of this chapter.

We have presented in Section 3.3 a simple simulation-based procedure called the MRP for forming a confidence interval on the optimality gap, $\mu_{\hat{x}} = Ef(\hat{x}, \xi) - z^*$. This approach involves solving n_g instances of (SP_n). At first, solving (say) $n_g = 20$ instances of (SP_n) in order to assess the quality of a solution $\hat{x}$ obtained by solving a single instance (SP_{n_x}) may seem excessive. However, for many problems we can obtain sufficiently tight confidence intervals using $n \ll n_x$ (e.g., Freimer et al. 2005, Mak et al. 1999) so that the computational effort required to assess $\hat{x}$'s quality is less than the effort required to compute $\hat{x}$. Sections 3.4, 3.5, and 3.6 are devoted to remedies that we recommend when this is not the case, i.e., when the computational effort required to form a sufficiently tight confidence interval on $\mu_{\hat{x}}$ via the MRP is excessive.

Section 3.4 allows for forming a confidence interval on $\mu_{\hat{x}}$ using a single replication. In some "nonsmooth" cases our experience is that this procedure can be risky with respect to coverage properties and so we recommend a more conservative two replication procedure.

When the bias of the lower bound estimator z^*_n is large, the confidence interval is loose. So, Section 3.5 introduces a generalized jackknife estimator to reduce bias within the context of a multiple replications procedure. The procedure requires as input an estimate of the rate at which the bias shrinks to zero. An alternative procedure (Bayraksan and Morton 2009b, Partani 2007) is called an adaptive jackknife procedure that estimates this rate as part of the procedure. Specifically, the jackknife estimator of Section 3.5 uses two equations to estimate the desired unknown parameter (e.g., z^* or $\mu_{\hat{x}}$) and the leading term in the bias expansion. Our adaptive scheme uses a third equation to simultaneously estimate these two values as well as the value of p in the order of the bias $O(n^{-p})$.

The jackknife approach is just one way to reduce bias, even within the context of the approaches considered in this chapter. Specifically, suppose with a fixed computational budget we can carry out (i) the MRP of Section 3.3 with n_g instances of (SP_n) with sample size n formed using i.i.d. sampling, (ii) the SRP (or 2RP) of Section 3.4 with a sample size $n' > n$, (iii) the MRP-GJ(q) jackknife procedure of Section 3.5 with n_g replications of $(\text{SP}_{n''})$ with $n'' < n$, or (iv) the MRP of Section 3.6 with n_g instances of (SP_n) with sample size n formed using randomized quasi-Monte Carlo sampling. To date our results indicate that option (iii) can significantly outperform option (i) but this issue deserves further investigation, particularly with respect to options (ii) and (iv).

Finally, the SRP and 2RP approaches we have described take as input sample sizes n and the MRP and MRP-GJ(q) procedures also require replication sizes n_g. The user does not have direct control over the width of the resulting confidence interval, ϵ. An attractive alternative is a sequential procedure (Bayraksan and Morton 2009a,b) that takes the desired confidence interval width as input and adaptively determine the required sample size.

Acknowledgments The authors thank Georg Pflug for valuable discussions, particularly with respect to Example 3.2. This research was supported by the National Science Foundation under grants CMMI-0653916 and EFRI-0835930.

References

Ahmed, S., Shapiro, A.: The sample average approximation method for stochastic programs with integer recourse. Optim. Online, http://www.optimization-online.org, 2002, accessed September 26, (2002)

Attouch, H., Wets, R.J.-B.: Approximation and convergence in nonlinear optimization. In: Mangasarian, O., Meyer, R., Robinson, S., (eds.) Nonlinear Programming 4, pp. 367–394. Academic, New York, NY (1981)

Bailey, T.G., Jensen, P., Morton, D.P.: Response surface analysis of two-stage stochastic linear programming with recourse. Naval Res. Logistics, **46**, 753–778 (1999)

Bayraksan, G., Morton, D.P.: Assessing solution quality in stochastic programs. Math. Program. **108**, pp. 495–514 (2006)

Bayraksan, G., Morton, D.P.: A sequential sampling procedure for stochastic programming Oper. Res. (2010)

Bayraksan, G., Morton, D.P.: Assessing solution quality in stochastic programs via sampling. In Oskoorouchi, M. R. (ed.) Tutorials in Operations Research, pp. 102–122. INFORMS, Hanover, MD (2009b)

Beale, E.M.L.: On minimizing a convex function subject to linear inequalities. J. R. Stat. Soc. **17B**, 173–184 (1955)

Bertocchi, M., Dupačová, J., Moriggia, V.: Sensitivity of bond portfolio's behavior with respect to random movements in yield curve: A simulation study. Ann. Oper. Res. **99**, 267–286 (2000)

Brooke, A., Kendrick, D., Meeraus, A., Raman, R.: GAMS, A User's Guide, 2006. GAMS Development Corporation, Washington, DC, http://www.gams.com/, accessed September 26 (2006)

Caflisch, R.E., Morokoff, W.J., Owen, A.B.: Valuation of mortgage backed securities using Brownian bridges to reduce effective dimension. J. Comput. Finance **1**, 27–46 (1997)

Campi, M.C., Garatti, S.: The exact feasibility of randomized solutions of robust convex programs. Optim. Online. http://www.optimization-online.org, accessed September 26 (2007)
Casella, G., Berger, R.L.: Statistical Inference. Duxbury Press, Belmont, CA (1990)
Dantzig, G.B.: Linear programming under uncertainty. Manage. Sci. **1**, 197–206 (1955)
Dantzig, G.B., Glynn, P.W.: Parallel processors for planning under uncertainty. Ann. Oper. Res. **22**, 1–21 (1990)
Dantzig, G.B., Infanger, G.: A probabilistic lower bound for two-stage stochastic programs, Department of Operations Research, Stanford University, Stanford, CA, November (1995)
Diwekar, U.M., Kalagnanam, J.R.: An efficient sampling technique for optimization under uncertainty. Am. Inst. Chem. Eng. J. **43**, 440 (1997)
Dupačová, J.: On non-normal asymptotic behavior of optimal solutions for stochastic programming problems and on related problems of mathematical statistics. Kybernetika. **27**, 38–52 (1991)
Dupačová, J., Wets, R.J.-B.: Asymptotic behavior of statistical estimators and of optimal solutions of stochastic optimization problems. Ann. Stat. **16**, 1517–1549 (1988)
Fox, B.L.: Strategies for Quasi-Monte Carlo. Kluwer, Boston, MA (1999)
Freimer, M., Thomas, D., Linderoth, J.T.: Reducing bias in stochastic linear programming with sampling methods, Technical Report 05T-002, Industrial and Systems Engineering, Lehigh University (2005)
Gray, H.L., Schucany, W.R.: The Generalized Jackknife Statistic. Marcel Dekker, New York, NY (1972)
Higle, J.L.: Variance reduction and objective function evaluation in stochastic linear programs. INFORMS J. Comput. **10**, 236–247 (1998)
Higle, J.L., Sen, S.: Statistical verification of optimality conditions for stochastic programs with recourse. Ann. Oper. Res. **30**, 215–240 (1991a)
Higle, J.L., Sen, S.: Stochastic decomposition: An algorithm for two-stage linear programs with recourse. Math. Oper. Res. **16**, 650–669 (1991b)
Higle, J.L., Sen, S.: Stochastic Decomposition: A Statistical Method for Large Scale Stochastic Linear Programming. Kluwer, Dordrecht (1996a)
Higle, J.L., Sen, S.: Duality and statistical tests of optimality for two stage stochastic programs. Math. Program. **75**, 257–275 (1996b)
Higle, J.L., Sen, S.: Statistical approximations for stochastic linear programming problems. Ann. Oper. Res. **85**, 173–192 (1999)
Infanger, G.: Monte Carlo (importance) sampling within a Benders decomposition algorithm for stochastic linear programs. Ann. Oper. Res. **39**, 69–95 (1992)
Infanger, G.: Planning Under Uncertainty: Solving Large-Scale Stochastic Linear Programs. The Scientific Press Series, Boyd & Fraser, Danvers, MA (1994)
Kall, P.: On approximations and stability in stochastic programming. In: Guddat, J., Jongen, H.Th., Kummer, B., Nožička, F. (eds.) Parametric Optimization and Related Topics, pp. 387–407. Akademie-Verlag, Berlin (1987)
Kenyon, A.S., Morton, D.P.: Stochastic vehicle routing with random travel times. Transport. Sci. **37**, 69–82 (2003)
King, A.J., Rockafellar, R.T.: Asymptotic theory for solutions in statistical estimation and stochastic programming. Math. Oper. Res. **18**, 148–162 (1993)
King, A.J., Wets, R.J.-B.: Epi-consistency of convex stochastic programs. Stochastics and Stochastics Reports **34**, 83–92 (1991)
Kleywegt, A.J., Shapiro, A., Homem-de-Mello, T.: The sample average approximation method for stochastic discrete optimization. SIAM J. Optim. **12**, 479–502 (2001)
Korf, L.A., Wets, R.J.-B.: Random lsc functions: an ergodic theorem. Math. Oper. Res. **26**, 421–445 (2001)
Lan, G., Nemirovski, A., Shapiro, A.: Validation analysis of robust stochastic approximation method. Optim. Online. http://www.optimization-online.org (2008)
Law, A.M.: Simulation Modeling and Analysis, 4th edn. McGraw-Hill, Boston, MA (2007)
L'Ecuyer, P., Lemieux, C.: Variance reduction via lattice rules. Manage. Sci. **46**, 1214–1235 (2000)

Linderoth, J.T., Shapiro, A., Wright, S.: The empirical behavior of sampling methods for stochastic programming. Ann. Oper. Res. **142**, 215–241 (2006)

Mak, W.K., Morton, D.P., Wood, R.K.: Monte Carlo bounding techniques for determining solution quality in stochastic programs. Oper. Res. Lett. **24**, 47–56 (1999)

Morton, D.P., Popova, E.: A Bayesian stochastic programming approach to an employee scheduling problem. IIE Trans. Oper. Eng. **36**, 155–167 (2003)

Morton, D.P., Wood, R.K.: On a stochastic knapsack problem and generalizations. In: Woodruff, D.L. (ed.) Advances in Computational and Stochastic Optimization, Logic Programming, and Heuristic Search: Interfaces in Computer Science and Operations Research, pp. 149–168. Kluwer, Boston, MA (1998)

Morton, D.P., Popova, E., Popova, I.: Efficient fund of hedge funds construction under downside risk measures. J. Bank. Finance **30**, 503–518 (2006)

Niederreiter, H.: Random Number Generation and Quasi-Monte Carlo Methods. CBMS-NSF Regional Conference Series in Applied Mathematics, SIAM (1992)

Norkin, V.I., Pflug, G.Ch., Ruszczyński, A.: A branch and bound method for stochastic global optimization. Math. Program. **83**, 425–450 (1998)

Partani, A.: Adaptive Jackknife Estimators for Stochastic Programming. PhD thesis, The University of Texas at Austin (2007)

Partani, A., Morton, D. P., Popova, I.: Jackknife estimators for reducing bias in asset allocation. In: Proceedings of the Winter Simulation Conference Monterey, California (2006)

Pennanen, T., Koivu, M.: Epi-convergent discretizations of stochastic programs via integration quadratures. Numerische Math. **100**, 141–163 (2005)

Quenouille, M.H.: Approximate tests of correlation in time-series. J. R. Stat. Soc. Series B **11**, 68–84 (1949a)

Quenouille, M.H.: Problems in plane sampling. Ann. Math. Stat. **20**, 355–375 (1949b)

Robinson, S.M.: Analysis of sample-path optimization. Math. Oper. Res. **21**, 513–528 (1996)

Robinson, S.M., Wets, R.J.-B.: Stability in two-stage stochastic programming. SIAM J. Control Optim. **25**, 1409–1416 (1987)

Rubinstein, R.Y., Shapiro, A.: Discrete Event Systems: Sensitivity and Stochastic Optimization by the Score Function Method. Wiley, Chichester (1993)

Santoso, T., Ahmed, S., Goetschalckx, M., Shapiro, A.: A stochastic programming approach for supply chain network design under uncertainty. Eur. J. Oper. Res. **167**, 96–115 (2005)

Shapiro, A.: Asymptotic properties of statistical estimators in stochastic programming. Ann. Stat. **17**, 841–858 (1989)

Shapiro, A.: Monte Carlo sampling methods. In: Ruszczyński, A., Shapiro, A. (eds.) Stochastic Programming, Handbooks in Operations Research and Management Science. Elsevier, Amsterdam (2003)

Shapiro, A., Homem-de-Mello, T.: A simulation-based approach to two-stage stochastic programming with recourse. Math. Program. **81**, 301–325 (1998)

Shapiro, A., Homem-de-Mello, T., Kim, J.: Conditioning of convex piecewise linear stochastic programs. Math. Program. **94**, 1–19 (2002)

Shapiro, A., Dentcheva, D., Ruszczyński, A.: Lectures on Stochastic Programming: Modeling and Theory. MPS-SIAM Series on Optimization, Philadelphia, PA (2009)

Sloan, I.H., Joe, S.: Lattice Methods for Multiple Integration. Clarendon Press, Oxford (1994)

Verweij, B., Ahmed, S., Kleywegt, A., Nemhauser, G., Shapiro, A.: The sample average approximation method applied to stochastic vehicle routing problems: A computational study. Comput. Appl. Optim. **24**, 289–333 (2003)

Wallace, S.W., Ziemba, W.T. (eds.) Applications of Stochastic Programming. MPS-SIAM Series on Optimization, Philadelphia, PA (2005)

Watkins, D.W., Jr., McKinney, D.C., Morton, D.P.: Groundwater pollution control. In: Wallace, S.W., Ziemba, W.T. (eds.) Applications of Stochastic Programming, pp. 409–424. MPS-SIAM Series on Optimization, Philadelphia, PA (2005)

Chapter 4
Stochastic Decomposition and Extensions

Suvrajeet Sen, Zhihong Zhou, and Kai Huang

4.1 Introduction

The title of this chapter is intended to evoke memories of one of the classics in the field of optimization: *Linear Programming and Extensions* by George B. Dantzig. This was so much more than a textbook; it was a veritable fountain of ideas including dynamic linear programs, generalized linear programming, and, of course, stochastic programming.

From his pulpit as the "Father of Linear Programming," George Dantzig inspired an entire community of researchers to study the "uncertainty problem," which is how he referred to stochastic programming. This area is now a vibrant part of mathematical programming, covering research ranging from statistical approaches to robust optimization, risk measures, and, of course, stochastic programming. Among his many lasting legacies, George Dantzig was instrumental in promoting the use of sampling in stochastic programming (Dantzig and Infanger 1991). His intuitive explanation that "one does not need to know the height of every American to estimate the average height of Americans," was intended to convince the mathematical programming community that in order to develop scalable methods for stochastic programming, one needed to move away from deterministic algorithms.

Algorithms referred to as "Stochastic Decomposition" (SD) constitute a class of decomposition methods that combines traditional deterministic decomposition (e.g., Benders' decomposition) with statistical sampling methods (Higle and Sen 1991, 1994, 1999). It can be shown that such methods provide asymptotically optimal solutions. More importantly, SD algorithms have been successful in providing solutions using workstation class computers where other methods have resorted to high-performance computing to solve the same instances (e.g., Sen et al. 1994). Nevertheless, there are issues that call for further investigation.

S. Sen (✉)
The Data Driven Decisions Lab, Industrial, Welding, and Systems Engineering, The Ohio State University, Columbus, OH 43210, USA; Department of Systems and Industrial Engineering, The MORE Institute, The University of Arizona, Tucson, AZ 85721, USA
e-mail: sen.22@osu.edu

G. Infanger (ed.), *Stochastic Programming*, International Series in Operations Research & Management Science 150, DOI 10.1007/978-1-4419-1642-6_4,

The specific extensions that we study in this chapter cover both convergence theory and algorithmic considerations. In the area of convergence theory, we study conditions under which the regularized version of SD provides a unique limit. On the algorithmic side, we address the issue of scalability. We show that without loss of asymptotic properties, it is possible to let a user speed up calculations in any iteration by avoiding the need to use every previously generated outcome in the "argmax" process of cut generation. The results associated with these extensions are provided in Section 4.3. Prior to that, however, a summary of SD methods is given in Section 4.2.

4.2 Stochastic Decomposition

In general, the two-stage stochastic linear programming model has the following form:

$$\min_{x \in \Re^{n_1}} \; f(x) := c^\top x + E[h(x, \tilde{\omega})] \tag{4.1a}$$

$$\text{s.t.} \quad Ax \le b, \tag{4.1b}$$

where A is $m_1 \times n_1$, b is $m_1 \times 1$, $\tilde{\omega}$ is a random variable defined on the probability space $(\Omega, \mathcal{A}, \mathcal{F})$, and

$$h(x, \omega) = \min_{y \in \Re^{n_2}} \; g^\top y \tag{4.2a}$$

$$\text{s.t.} \quad Wy = r(\omega) - T(\omega)x, \tag{4.2b}$$

$$y \ge 0. \tag{4.2c}$$

In this model, we have restricted (4.2) to a fixed recourse version in which random variables do not appear in the data elements g and W.

In essence, SD is a stochastic version of Benders' decomposition (Benders 1962), which has been recognized as the L-shaped method in the SP literature (Van Slyke and Wets 1969). The main idea of these decomposition methods is to construct a piecewise linear approximation to the recourse function and update the approximation during each iteration of the algorithm. The major difference between SD and deterministic decomposition methods (Benders' decomposition, L-shaped method) is the computational effort to create a new cut in the approximation. In traditional deterministic approximations, the recourse function needs to be evaluated for every outcome of the random variable, which requires the solution of as many second-stage problems (4.2) as there are realizations ω of random variables $\tilde{\omega}$. Consequently, deterministic decomposition algorithms are computationally feasible for only those instances in which a few outcomes are sufficient to model randomness. Moreover, deterministic decomposition methods are restricted to only those instances in which the random variables are discrete. In contrast, SD successfully overcomes these limitations by combining sampling with sequential approximations in such a manner as to reduce the computational effort in generating a new cut (a new piecewise linear approximation) in each iteration.

In SD, a new sampled outcome (ω^k) is obtained independently of all previous outcomes, $\{\omega^1, \ldots, \omega^{k-1}\}$, at iteration k. Hence, this corresponds to the i.i.d. samples. In fact, the SD algorithm constructs a *lower bounding linear approximation* of the sample mean function

$$H_k(x) = \frac{1}{k}\sum_{t=1}^{k} h(x, \omega^t). \tag{4.3}$$

In order to approximate the above function by a lower bound, one might adopt the ideas underlying the Benders' cut. However, such a strategy would require us to solve k linear programs to obtain a subgradient for each outcome $h(x, \omega^t)$, $t = 1, \ldots, k$ (at the point x^k). As a matter of fact, SD theory suggests that asymptotic convergence can be achieved even without solving all these LPs. Instead, it is sufficient to solve only one second-stage LP (4.2) in any iteration. Furthermore, we incorporate previously obtained data (on the optimal dual solutions for (4.2)) to define a lower bounding approximation of $h(x, \omega^t)$, for $t < k$. The details are as follows.

For the most recent outcome denoted ω^k, we evaluate $h(x^k, \omega^k)$, which includes solving (4.2) with (x^k, ω^k) as the inputs. Let π_k^k denote the dual optimum to this problem. Then, the data collection process within the SD algorithm accumulates this optimal dual vector into a set of previously discovered optimal dual vectors denoted as V_{k-1}. As a result, the updated set of the optimal dual vectors at iteration k is denoted as V_k (i.e., $V_k = V_{k-1} \cup \pi_k^k$). Thus the set V_k is a collection of the optimal dual vector discovered during the construction of $h(x^t, \omega^t)$, $t = 1, \ldots, k$, in each iteration. Next, a lower bounding function for the kth sample mean function (4.3) is generated by assigning a dual feasible solution for each previously observed outcome $\{\omega^t\}$, $t < k$. To see this, note that LP duality ensures that for $\pi \in V_k$,

$$\pi^\top[r(\omega^t) - T(\omega^t)x] \le h(x, \omega^t) \; \forall x. \tag{4.4}$$

Thus, in iteration k, the dual vector in V_k that provides the best lower bounding approximation at $\{h(x^k, \omega^t)\}$, for $t < k$, is given by the dual vector for which

$$\pi_t^k(x) \in \operatorname{argmax}\{\pi^\top[r(\omega^t) - T(\omega^t)x] \mid \pi \in V_k\}. \tag{4.5}$$

When the above operation is undertaken with appropriate data structures (see Higle and Sen 1996), it is computationally faster than solving a linear program from scratch. In any event, it follows that

$$H_k(x) \ge \frac{1}{k}\sum_{t=1}^{k}(\pi_t^k)^\top[r(\omega^t) - T(\omega^t)x]. \tag{4.6}$$

Using the lower bound in (4.6), SD applies a procedure similar to Benders' decomposition, in that the right hand side of (4.6) is added as a new cut of the piecewise linear approximation of $E[h(x, \tilde{\omega})]$.

Finally, we note one more distinction between deterministic and stochastic decomposition. Since the number of outcomes increases as the number of iterations grows, different number of sample sizes is used in generating the approximate cuts. Therefore, in iteration t of SD, one uses a cut that approximates $H_t(x)$, not $H_k(x)$, for $t < k$. As a result, the former cuts need to be readjusted to be guaranteed to be the lower bounds for $H_k(x)$. Without loss of generality, we can assume that $h(x, \omega) \geq 0$ almost surely. With this assumption, it is clear that $H_k(x) \geq \frac{t}{k} H_t(x)$. Hence, by multiplying each previously generated cut ($t = 1, \ldots, k-1$) by the multiplier t/k, all previously generated cuts provide a lower bound for the sample mean approximation $H_k(x)$. In any event, the approximation for the first-stage objective function at iteration k is then given by

$$f_k(x) := c^\top x + \max_{t=1,\ldots,k} \left\{ \frac{t}{k} \times \frac{1}{t} \sum_{j=1}^{t} (\pi_j^t)^\top [r(\omega^j) - T(\omega^j)x] \right\}.$$

Since these approximations are generated in a recursive manner, it is best to consult Higle and Sen (1996) regarding the data structures and updates to be used for efficient implementations.

The most basic version of SD uses the sequence $\{x^k\}$ such that

$$x^{k+1} \in \operatorname{argmin}\{f_k(x) \mid x \in X\},$$

where $X = \{x \mid Ax \leq b\}$ denotes the first-stage feasible region. However, to improve the algorithmic properties of SD, we recommend the use of a regularized approximations as defined in Higle and Sen (1994). Denoting an incumbent at iteration k as $\bar{x}^k$, we recommend the following:

$$x^{k+1} \in \operatorname{argmin}\left\{ f_k(x) + \frac{1}{2}\|x - \bar{x}^k\|^2 \mid Ax \leq b \right\}. \tag{4.7}$$

In the following, we will use $\rho_k(x) = \frac{1}{2}\|x - \bar{x}^k\|^2$. The condition to update the incumbent is whether the (sample mean) point estimate of the objective value at x^{k+1} is better than the point estimate of the objective value of the incumbent $\bar{x}^k$. If it is the case, then $\bar{x}^{k+1} = x^{k+1}$; else, $\bar{x}^{k+1} = \bar{x}^k$. This procedure is referred to as the regularized stochastic decomposition (RSD) method which is used for our computational study.

In the last part of this summary, we should point out one aspect in computing implementations. We note that as k changes, so does $H_k(x)$. However, all but one of the observations used in defining $H_k(x)$ are used in $H_{k-1}(x)$. Hence, as k becomes large, the use of common random numbers will reduce variance.

4.3 Extensions of Regularized SD

In this section, we discuss two extensions for regularized stochastic decomposition. First, we examine conditions under which the regularized SD algorithm produces a sequence of incumbent solutions that converge to a unique limit. Next, we discuss a computational enhancement that allows the formations of cuts using resampling.

4.3.1 A Sufficient Condition for a Unique Limit

Higle and Sen (1996) proved that the regularized SD algorithm converges to an optimal solution with probability one. This theorem is summarized below.

Theorem 4.1 *Let the sequences of incumbent and candidate solutions be* $\{\bar{x}^k\}_{k=0}^{\infty}$ *and* $\{x^k\}_{k=1}^{\infty}$*, obtained by regularized stochastic decomposition algorithm. Then with probability one, there exists a subsequence of iterations,* K^* *such that* $\lim_{k\in K^*} f_k(x^{k+1}) + \rho_k(x^{k+1}) - f_k(\bar{x}^k) = 0$ *and each accumulation point of* $\{\bar{x}^k\}_{k\in K^*}$ *is optimal to* (4.1).

Proof See Higle and Sen (1996). □

Higle and Sen (1999) show how one may test for a convergence of subsequence by combining the definition of K^* with stopping rules for regularized SD. However, these results do not guarantee that the sequence has a unique limit. The following theorem provides this results:

Theorem 4.2 *Let* $\{\bar{x}^k\}_K \rightarrow x^*$ *denote a convergent subsequence generated by the regularized SD algorithm. Let* f *be defined as in* (4.1) *and assume that* f *is continuously differentiable at* x^* *and the row vectors corresponding to the tight constraints at* x^* *(i.e.,* $a_i^T x^* = b_i$*) are linearly independent. Then* $\{\bar{x}^{k+1}\}_K \rightarrow x^*$*, implying that* x^* *is a unique limit.*

Proof Because of the epi-consistency of the SD approximations (Higle and Sen 1992) and the continuity of ∇f in a neighborhood of x^*, it follows that there are subgradients $\beta^k \in \partial H_k(\bar{x}^k)$ such that $\{c + \beta^k\}_{k\in K} \rightarrow \nabla f(x^*)$. Let λ^k denote the optimal dual multiplier to (4.7). Since regularized SD provides asymptotically optimal primal and dual solutions, it follows that $\{\lambda^k\}_{k\in K} \rightarrow \lambda^*$, an optimal dual vector. Hence $\{||(c + \beta^k - \nabla f(x^*)) + A^\top(\lambda^k - \lambda^*)||\}_{k\in K} \rightarrow 0$. Using (2.1e) of Chapter 4 in Higle and Sen (1996), we have

$$A^\top \lambda^k + c + \beta^k = -(x^{k+1} - \bar{x}^k).$$

It follows that $\{||x^{k+1} - \bar{x}^k||\}_{k\in K} \rightarrow 0$. As a matter of fact, we have

$$\bar{x}^{k+1} = \begin{cases} \bar{x}^k \\ x^{k+1} \end{cases}$$

in SD. If $\bar{x}^{k+1} = x^{k+1}$, then

$$||\bar{x}^{k+1} - x^*|| = ||x^{k+1} - x^*|| \leq ||x^{k+1} - \bar{x}^k|| + ||\bar{x}^k - x^*||.$$

Moreover, the same inequality holds if $\bar{x}^{k+1} = \bar{x}^k$. Since the right-hand side of the above inequality goes to zero, it follows that $\{\bar{x}^{k+1}\}_K \to x^*$ and the result follows. □

4.3.2 SD With Resampling

In this section, we discuss the potential reduction of computational effort by using resampling, which is also known as bootstrapping (Efron 1979). Generally speaking, in stochastic decomposition, there are two sources of errors; one is the statistical approximation error, and the other error results from linearization using cuts. Although the regularized SD algorithm can achieve asymptotic optimality with finite master programs, asymptotic convergence (with probability one) requires that the number of observations in any cut grow indefinitely. The issue can be explained as follows: in order to generate a new cut, the stochastic decomposition algorithm requires all previously generated samples. As the number of iterations increase, the computational effort in cut generation also grows. However, it may not be necessary to make full use of all the generated samples to construct a new cut. In addition, some of the samples may not be very useful in forming the new cut, which motivates us to apply the concept of resampling in stochastic decomposition.

The idea of resampling is straightforward: at iteration k, instead of using all the generated samples (k) to construct a new cut, we only use a relatively small number of samples (L_k and $L_k < k$) to form a new cut. When k is large, we can reduce the computational effort significantly by choosing $L_k \ll k$. With this modification, one only needs to carry out the comparison in (4.5) for L_k elements. Moreover, the convergence results of SD carry over into the modified SD with resampling. Before providing the convergence results of SD with resampling, we briefly summarize the bootstrap method as follows.

Let $\{\omega^1, \ldots, \omega^k\}$ be a random i.i.d. sample of size k with distribution F and let F_k be the empirical distribution of $\{\omega^1, \ldots, \omega^k\}$. Define a random variable $T(\omega^1, \ldots, \omega^k; F)$, which depends upon distribution F. The bootstrap method is to approximate the distribution $T(\omega^1, \ldots, \omega^k; F)$ by $T(\theta^1, \ldots, \theta^k; F_k)$ under F_k, where $\{\theta^1, \ldots, \theta^k\}$ denotes a random sample of size k under distribution F_k. Next, we summarize an important theorem by Singh (1981).

Lemma 4.3 *Let* $\mu = E_F[\omega]$, $\bar{\omega}^k = 1/k \sum_{i=1}^k \omega^i$, $\bar{\theta}^k = 1/k \sum_{i=1}^k \theta^i$ *and assume* $E[\omega^2] < \infty$. *Let* P *and* P_k *denote the probabilities under* F *and* F_k, *respectively. Then*

$$\lim_{k\to\infty} |P\{k^{1/2}(\bar{\omega}^k - \mu) \leq s\} - P_k\{k^{1/2}(\bar{\theta}^k - \bar{\omega}^k) \leq s\}| = 0 \quad a.s. \tag{4.8}$$

Proof See Singh (1981). □

Basically, Lemma 4.3 studies the convergence to zero of the discrepancy between distribution $k^{1/2}(\bar{\omega}^k - \mu)$ and the bootstrap approximation of it. However, in stochastic decomposition with resampling, the asymptotic properties of approximations of the sample mean function $H_k(x^k)$ are of interest. Therefore, we provide the following theorem.

Theorem 4.4 *(a) Let* $\mu_k = E_F[h(x^k, \tilde{\omega})]$, $\hat{h}_k(x^k) = (1/\sum_{i=1}^k I_i) \sum_{i=1}^k I_i h(x^k, \omega^i)$, *where*

$$I_i = \begin{cases} 1 & \textit{with prob} \quad p \\ 0 & \textit{otherwise} \end{cases}$$

and assume $E[h(x^k, \tilde{\omega})]^2 < \infty$. *Let* P *and* P_k *denote the probabilities under* F *and* F_k, *respectively. Then*

$$\lim_{k \to \infty} |P\{k^{1/2}(H_k(x^k) - \mu_k) \leq s\} - P_k\{k^{1/2}(\hat{h}_k(x^k) - H_k(x^k)) \leq s\}| = 0 \quad a.s. \tag{4.9}$$

(b) Let the coefficient of cut $f_k(x)$ *be*

$$\Theta(x^k, \omega^k) = c^\top + \max_{t=1,\ldots,k} \left\{ \tfrac{t}{k} \times \tfrac{1}{t} \textstyle\sum_{j=1}^t (\pi_j^t)^\top [-T(\omega^j)] \right\},$$

let $\mu_{\Theta(x^k)} = E_F[\Theta(x^k, \tilde{\omega})]$, $\bar{\Theta}_k(x^k) = 1/k \sum_{i=1}^k \Theta(x^k, \omega^i)$, $\hat{\Theta}_k(x^k) = (1/\sum_{i=1}^k I_i) \sum_{i=1}^k I_i \Theta(x^k, \omega^i)$, *where*

$$I_i = \begin{cases} 1 & \textit{with prob} \quad p \\ 0 & \textit{otherwise} \end{cases},$$

and assume $E[||\Theta(x^k, \tilde{\omega})||^2] < \infty$. *Let* P *and* P_k *denote the probabilities under* F *and* F_k, *respectively. Then*

$$\lim_{k \to \infty} |P\{k^{1/2}(\bar{\Theta}_k(x^k) - \mu_{\Theta(x^k)}) \leq s\} - P_k\{k^{1/2}(\hat{\Theta}_k(x^k) \tag{4.10}$$

$$- \bar{\Theta}_k(x^k)) \leq s\}| = 0 \quad a.s. \tag{4.11}$$

Proof (a) and (b): See details in Sen et al. (2009). □

For readers familiar with implementing SD, we observe that one can implement this resampling version without changing the code a great deal. One relatively straightforward way to include resampling within the cut generation process is to accept/reject an outcome within a sample, based on a Bernoulli random variable. Thus, if p is the acceptance (success) probability, then we generate a uniform random variate for each previously generated outcome, and use only those outcomes

$(\omega^t, t < k)$ for which the random number is less than p. Clearly, as p increases, cuts tend to use more outcomes in the approximation.

4.3.3 Preliminary Computational Results

In this section, we provide preliminary computational results comparing similar versions of Regularized SD, with and without resampling. We demonstrate the effectiveness of resampling by solving an instance referred to as 20Term which arose in freight scheduling (Infanger 1999 "personal communication"). For this instance, we fix the maximum number of iterations at 800 for both versions of Regularized SD (with and without resampling) and perform 20 replicated runs. In the resampled version, we start the resampling process after 300 iterations and choose the acceptance probability p as 0.7. Finally, the LP solver in our computation uses the ILOG CPLEX callable library, version 10.0, and all programs were compiled and run in a Unix environment running on a Sun workstation (Sun Fire V440).

Figure 4.1 reports the solution times for the two versions we are comparing. We record the CPU time (in seconds) every 100 iterations. As illustrated in Fig. 4.1, there is no difference between the two versions for the first 300 iterations because resampling was started only after 300 iterations. However, after 300 iterations, the resampled version is faster as iterations proceed. Moreover, at iteration 800, the resampled version takes 62 seconds compared to 84.5 s for Regularized SD. Thus the resampled version results in a reduction of 26.7% in computational time.

Figure 4.2 demonstrates the solution quality obtained by the two versions. As expected, there is no difference between the two versions SD for the first 300 iterations. At iterations 400, there is a jump in objective function value due to resampling. These values were obtained by running an out-of-sample evaluator that samples the

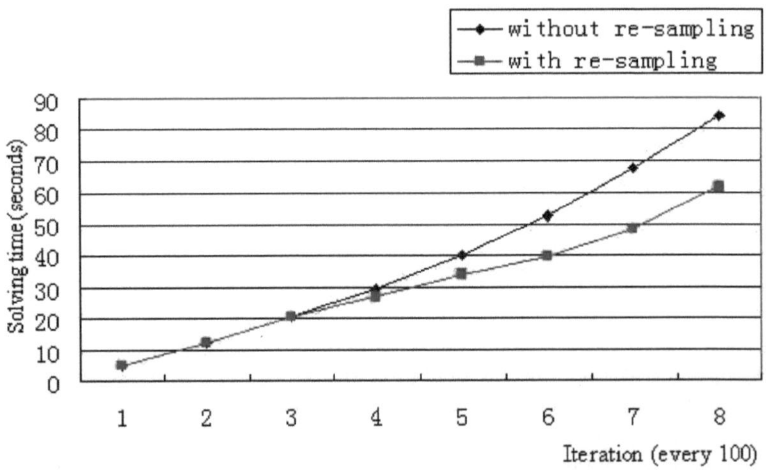

Fig. 4.1 Solution times for regularized SD: with and without resampling

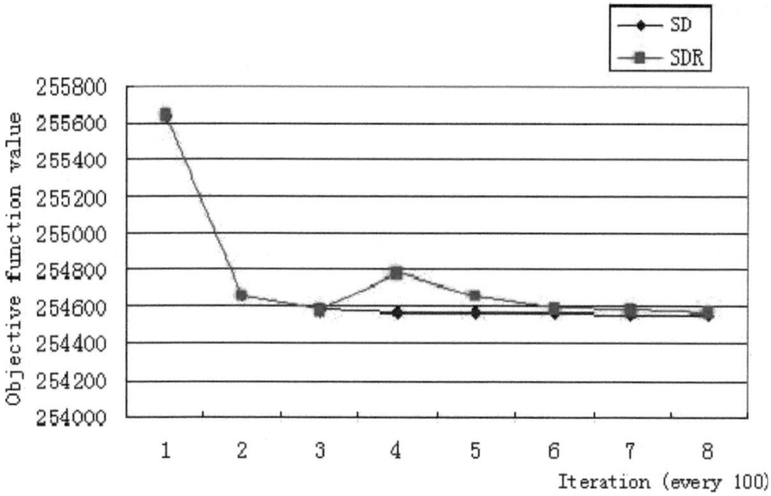

Fig. 4.2 Comparison of solution quality: SD vs SDR

objective function, given a first-stage solution used in a particular iteration. It is interesting to observe that although the objective values obtained by the resampled version are not as good in the early iterations of resampling, the two versions begin to converge to the same value as iterations proceed. As a matter of fact, at iteration 800, one observes scant differences in objective function value between the two versions, with the original Regularized SD version yielding 254561.8310 and the sampled version providing a slightly higher value at 254572.1976.

4.4 Conclusion

The specific extensions that we have investigated cover both convergence theory and algorithmic considerations for SD. In the area of convergence theory, we show that differentiability of the recourse function at any accumulation point is sufficient for a unique limit. While this cannot be verified a priori, it is possible to estimate the likelihood of uniqueness at termination. On the algorithmic side, we address the issue of scalability. We show that without loss of asymptotic properties, it is possible to control the number of outcomes used for the "argmax" process. This modification helps SD algorithms overcome the need to use every previously generated outcome in the process.

Acknowledgments We are grateful to the referee for a thorough report. This research was funded, in part, by NSF grant CMMI - 0900070, and AFOSR grant FA9550-08-1-0117.

References

Benders, J.F.: Partitioning procedures for solving mixed variables programming problems. Numerische Math. **4**, 238–252 (1962)

Dantzig, G.B., Infanger, G.: Large-scale stochastic linear programs: importance sampling and Benders' decomposition. In Brezinski, C.: Kulisch, U. (eds.) Computational and applied mathematics I–algorithms and theory, pp. 111–120. Proceedings of the 13th IMACS World Congress, Dublin, Ireland (1991)

Efron, B.: Bootstrap methods: Another look at the jackknife. Ann. Stat. **7**, 1–26 (1979)

Higle, J.L., Sen, S.: Stochastic decomposition: An algorithm for two-stage linear programs with recourse. Math. Oper. Res. **16**, 650–669 (1991)

Higle, J.L., Sen, S.: On the convergence of algorithms with implications for stochastic and nondifferentiable optimization. Math. Oper. Res. **17**, 112–131 (1992)

Higle, J.L., Sen, S.: Finite master programs in regularized stochastic decomposition. Math. Program. **67**, 143–168 (1994)

Higle, J.L., Sen, S.: Stochastic decomposition: A statistical method for large scale stochastic linear programming. Nonconvex optimization and its applications, vol. 8. Kluwer, Dordrecht The Netherlands Jun (1996)

Higle, J.L., Sen, S.: Statistical approximations for stochastic linear programming problems. Ann. Oper. Res. **85**, 173–192 (1999)

Sen, S., Doverspike, R. D., Cosares, S.: Network planning with random demand. Telecommun. Syst. **3**, 11–30 (1994)

Sen, S., Zhou, Z., Huang, K.: Enhancements of two-stage stochastic decomposition. Comput. Oper. Res. **36**(8), 2434–2439 (2009)

Singh, K.: On the asymptotic accuracy of Efron's bootstrap. Ann. Stat. **9**(6), 1187–1195 (1981)

Van Slyke, R.M., Wets, R.J.-B: L-shaped linear programs with applications to optimal control and stochastic programming. SIAM J. App. Math. **17**, 638–663 (1969)

Chapter 5
Barycentric Bounds in Stochastic Programming: Theory and Application

Karl Frauendorfer, Daniel Kuhn, and Michael Schürle

5.1 Introduction

Multistage stochastic programs arise whenever sequential decisions have to be taken under incomplete information about some decision-relevant parameters. The general setup is as follows: a decision maker observes some random parameter and selects a decision based on this first observation. Next, a new random parameter is observed, in response to which a recourse decision is taken. The second decision may depend on both previous observations (and on the first decision, which adds no extra information, however, since it is merely a function of the first observation). This basic scheme of alternating observations and decisions is continued until the end of the planning horizon. Typically, decisions cannot be chosen freely but are subject to constraints of physical or regulatory nature. Moreover, rationality requires decision makers to select utility-maximizing actions. When formulating a stochastic optimization model, therefore, a suitable objective criterion must be specified. In order to simplify prose, we will henceforth assume that the objective is to maximize expected profit.

A stochastic optimization problem is said to be linear if the objective and constraint functions are linear affine in the decision variables, which will always be assumed in the remainder of this chapter. Linear two- and multistage stochastic programs were first studied by Dantzig (1955) and Beale (1955). Their immense popularity originates from a rich variety of applications in the fields of management, engineering, finance, logistics, etc. If the underlying random variables are finitely supported, the extensive form (Birge and Louveaux 1997) of a given stochastic program represents a large-scale linear program, which can in principle be solved by using Dantzig's simplex algorithm (Dantzig and Thapa 1997). However, if the supports of the underlying random parameters are very large or (un)countably infinite, a stochastic program will in most cases allow only for an approximate solution.

K. Frauendorfer (✉)
University of St. Gallen, St. Gallen, Switzerland
e-mail: karl.frauendorfer@unisg.ch

G. Infanger (ed.), *Stochastic Programming*, International Series in Operations Research & Management Science 150, DOI 10.1007/978-1-4419-1642-6_5,

The present chapter addresses an approximation scheme based on discretization. Concretely speaking, we will establish two auxiliary stochastic programs that emerge from the original problem by replacing the true distribution of the random parameters by two finitely supported approximate distributions. After a suitable transformation offsetting possible nonconvexities in the random parameters, the optimal values of the auxiliary stochastic programs can be shown to provide bounds for the optimal value of the original problem. Moreover, repeated solution of the auxiliary problems under different parameter settings will allow us to construct compact bounding sets for the true optimal decisions. Further emphasis will be put on convergence issues.

Our approximation scheme belongs to the class of bounding methods, which have a long tradition in stochastic optimization, see, e.g., Birge and Louveaux (1997, Chapter 9) for a textbook introduction. In developing so-called bounding probability measures, one typically exploits structural properties of the recourse functions (profit-to-go functions) associated with the underlying stochastic program. Frequently, bounding measures may be interpreted as solutions of generalized moment problems (Birge and Wets 1987, Edirisinghe and Ziemba (1994a,b), Frauendorfer 1992, Kall 1988). This idea is credited to Dupačová (1966) and has fruitful applications also in models where the underlying probability distribution is only known in limited manner (Dupačová 1976, 1987, 1994).

When the recourse functions are convex in the random parameters, a lower bound is found via Jensen's inequality (Jensen 1906), while an upper bound arises from the Edmundson–Madansky inequality.[1] Edmundson (1956) treats the univariate case, whereas Madansky (1959, 1960) and Frauendorfer (1988) consider the multivariate setting, given that the components of the random parameters are independent and dependent, respectively. Elaborate extensions are due to Gassmann and Ziemba (1986) and Birge and Wets (1986, 1987). Note that these bounds are based only on first-order moment information. One way of tightening them is by inclusion of higher order information in the construction of bounding measures. Edirisinghe (1996) was the first to develop second-order lower bounds, which were later extended by Dokov and Morton (2005). Birge and Dulá (1991), Dupačová (1966), and Kall (1991) propose second-order upper bounds. Higher order upper bounds are suggested in Dokov and Morton (2002). For more information on higher order bounds, see also the Chapter 6 in this book by Edirisinghe. All bounds discussed so far can be improved to an arbitrary degree of precision by applying them on increasingly small subsets of the underlying domain. This so-called *partitioning* technique is documented in Birge and Wets (1986), Frauendorfer and Kall (1988), and Huang et al. (1977).

Similar bounding methods are also available for problems whose recourse functions are convex–concave saddle functions. First-order bounds closely related to those of Jensen and Edmundson–Madansky are proposed by Frauendorfer (1992) and Edirisinghe and Ziemba (1994a,b). Second-order bounds on saddle functions

---

[1] In the presence of concave recourse functions upper and lower bounds switch roles.

are due to Edirisinghe (1996) and generalized bounds for specific problems with nonconvex recourse functions are developed by Kuhn (2004). Furthermore, suitable partitioning schemes are discussed in Edirisinghe and You (1996), Frauendorfer (1992), and Kuhn (2004).

The present chapter reviews some cornerstone results related to bounding techniques in stochastic linear programming and presents an illustrative application to finance. Our theoretical exposition follows the lines of Frauendorfer (1992, 1994, 1996) and Kuhn (2004). Section 5.2 formally introduces the class of optimization problems to be studied, while Section 5.3 elaborates a set of regularity conditions ensuring well-definedness of our mathematical models. In Section 5.4 we sketch the construction of discrete approximate distributions for the random parameters; these so-called *barycentric measures* facilitate the calculation of bounds on the expectations of subdifferentiable saddle functions. The theoretical main results are presented in Section 5.5. Most importantly, we will demonstrate the construction of tight bounds on the recourse functions and establish compact bounding sets for the optimal decisions of linear stochastic programs. Furthermore, we will report on convergence results. In order to illustrate the practical implementation of the theoretical concepts, Section 5.6 discusses the problem of reinvesting money from nonmaturing accounts (e.g., savings products), which is a major concern for financial institutions. This particular decision problem is formulated as a linear multistage stochastic program and can be analyzed with our bounding methods. Numerical experiments show that the bounds on the optimal objective value can be made tight with affordable computational effort. In a series of test calculations, we investigate the convergence of the bounds for different families of scenario trees.

5.2 Problem Formulation

We consider a class of constrained profit maximization problems under uncertainty. We assume that decisions may be selected at several time points $t = 1, \dots, T$. At the outset, we must establish a probabilistic model for the underlying uncertainty. All random objects are defined on an abstract probability space (Ω, Σ, P). Adopting the standard terminology of probability theory, we will refer to Ω as the *sample space*. Furthermore, we use the following definition of a stochastic process.

Definition 5.2.1 (Stochastic Process) We say that $\boldsymbol{\xi}$ is a stochastic process with state space Ξ if $\boldsymbol{\xi} = (\boldsymbol{\xi}_1, \dots, \boldsymbol{\xi}_T)$ and $\Xi = \times_{t=1}^{T} \Xi_t$ such that each random vector $\boldsymbol{\xi}_t$ maps (Ω, Σ) to the Borel space $(\Xi_t, \mathcal{B}(\Xi_t))$ and each Ξ_t is a convex closed subset of some finite-dimensional Euclidean space. Moreover, we define combined random vectors $\boldsymbol{\xi}^t := (\boldsymbol{\xi}_1, \dots, \boldsymbol{\xi}_t)$ valued in $\Xi^t := \times_{\tau=1}^{t} \Xi_\tau$ for all $t = 1, \dots, T$.[2]

As a notational convention, throughout this chapter, random objects will be represented in boldface, while their realizations will be denoted by the same symbols in

[2] Sometimes, notation is simplified by further introducing a deterministic dummy random variable $\boldsymbol{\xi}^0$ of the past.

normal face. We will frequently encounter stochastic processes with compact state spaces. Then, the corresponding random vectors are bounded, thus having finite moments of all orders.

Let $\boldsymbol{\eta}$ and $\boldsymbol{\xi}$ be two stochastic processes in the sense of Definition 5.2.1 with state spaces Θ and Ξ, respectively. These processes are assumed to describe the random data which is revealed sequentially in the decision-making process. Thus, the information $\mathcal{F}^t$ available at stage t amounts to the σ-algebra generated by the random variables observed in stages 1 through t, i.e., $\mathcal{F}^t = \sigma(\boldsymbol{\eta}^t, \boldsymbol{\xi}^t)$. With these conventions, the space of non-anticipative *decision processes*[3] is defined as

$$\begin{aligned}\mathcal{N} = \{&\boldsymbol{x} \text{ is a stochastic process with state space } X \text{ such} \\ &\text{that } \boldsymbol{x}_t \in \mathcal{L}^\infty(\Omega, \mathcal{F}^t, P; X_t) \text{ for each } t = 1, \ldots, T\}.\end{aligned}$$

We will assume that $X = \mathbb{R}^n$ and $X_t = \mathbb{R}^{n_t}$. Then, consistency requires that $n = n_1 + \cdots + n_T$. In order to simplify notation, we introduce $\boldsymbol{x}_0$ as a fictitious fixed decision of the past. The *static version* of a linear multistage stochastic program can now be formulated as

$$\begin{aligned}&\underset{\boldsymbol{x}\in\mathcal{N}}{\text{maximize}}\; E\left[\sum_{t=1}^{T}\langle c_t^*(\boldsymbol{\eta}^t), \boldsymbol{x}_t\rangle\right] \qquad (5.1)\\ &\text{s.t.}\quad W_t(\boldsymbol{\xi}^t)\boldsymbol{x}_t + T_t(\boldsymbol{\xi}^t)\boldsymbol{x}_{t-1} \sim_t h_t(\boldsymbol{\xi}^t) \quad P\text{-a.s.} \quad t = 1, \ldots, T.\end{aligned}$$

The objective and constraint functions are linear in the decision variables. However, linearity in the stochastic parameters is not required. Note that the process $\boldsymbol{\eta}$ exclusively appears in the objective function, whereas $\boldsymbol{\xi}$ only affects the dynamic constraints (but both $\boldsymbol{\eta}$ and $\boldsymbol{\xi}$ influence the set $\mathcal{N}$ accounting for the so-called non-anticipativity constraints). As argued in Frauendorfer (1992, Section 5.5), a distinction between $\boldsymbol{\eta}$ and $\boldsymbol{\xi}$ is always possible. However, some components of $\boldsymbol{\eta}$ and $\boldsymbol{\xi}$ may represent the same random parameter and must therefore be modeled as perfectly correlated random variables.[4]

We choose to work explicitly with inequality (less-or-equal, greater-or-equal) and equality constraints. Thus, for every stage we introduce an r_t-dimensional "vector" $\sim_t$ of binary relations, each of whose entries is either "$\leq$" or "$\geq$" for inequalities or "$=$" for equalities. Of course, the stochastic program (5.1) can be brought to the standard form with only "less-or-equal" constraints if the equality constraints are replaced by two opposing inequality constraints, and the "greater-or-equal" constraints are multiplied by -1. Notice that the constraints in (5.1) only couple neighboring decision stages. This can always be enforced, i.e., dependencies across more than one decision stage can systematically be eliminated by introducing additional

[3] Decision processes will also be referred to as *strategies* or *policies*.

[4] Random variables appearing in both the objective function and the constraints of the stochastic program must be duplicated. The first copy of such a random variable is appended to $\boldsymbol{\eta}$, while the second copy is appended to $\boldsymbol{\xi}$. Note that the two copies will be discretized differently.

decision variables. For every $t = 1, \ldots, T$, the $r_t \times n_t$ matrix W_t is termed *recourse matrix* and may generally depend on ξ^t. The $r_t \times n_{t-1}$ matrix T_t is referred to as *technology matrix* in literature. Obviously, the technology matrix determines the intertemporal coupling and may also depend on ξ^t. Moreover, the *right-hand side (rhs) vector* h_t and the vector of *objective function coefficients* c_t^* may depend on ξ^t and η^t, respectively.

For both theoretical and practical purposes it is sometimes more comfortable to work with the *dynamic version* of a given multistage stochastic program which relies on a backward recursion scheme. Setting $\Phi_{T+1} \equiv 0$, the optimal value functions or *recourse functions* of stages $t = T, \ldots, 1$ are defined as

$$\begin{aligned} \Phi_t(x^{t-1}, \eta^t, \xi^t) = \max_{x_t \in \mathbb{R}^{n_t}} \; & \langle c_t^*(\eta^t), x_t \rangle + (E_t \Phi_{t+1})(x^t, \eta^t, \xi^t) \\ \text{s.t.} \quad & W_t(\xi^t)x_t + T_t(\xi^t)x_{t-1} \sim_t h_t(\xi^t). \end{aligned} \tag{5.2}$$

Thereby, the expected recourse functions or *expectation functionals* are given by

$$(E_t \Phi_{t+1})(x^t, \eta^t, \xi^t) = \int \Phi_{t+1}(x^t, \eta^{t+1}, \xi^{t+1})\, dP_{t+1}(\eta_{t+1}, \xi_{t+1} | \eta^t, \xi^t), \tag{5.3}$$

and P_{t+1} denotes a regular conditional probability distribution of the random parameters observed at stage $t + 1$ given the realizations of the random parameters observed at stages 1 through t. In real-life applications it is usually assumed that $\boldsymbol{\eta}_1$ and $\boldsymbol{\xi}_1$ are deterministic, i.e., they represent degenerate random vectors with a Dirac distribution (as was assumed for the fictitious past decision $\boldsymbol{x}_0$). Then, the first-stage recourse function Φ_1 needs to be evaluated only at one point, where it coincides with the optimal value of the underlying stochastic program.

To facilitate the formulation of precise statements about dynamic stochastic programs, it proves useful to introduce specific subsets of parameter space. For each $t = 1, \ldots, T$ we define

$$\begin{aligned} Z^t = \{(x^{t-1}, \eta^t, \xi^t) \in X^{t-1} \times \Theta^t \times \Xi^t \mid \forall s = 1, \ldots, t-1 : \\ W_s(\xi^s)x_s + T_s(\xi^s)x_{s-1} \sim_s h_s(\xi^s) \}. \end{aligned} \tag{5.4}$$

Points in the complement of Z^t correspond to impossible realizations of the random parameters or to forbidden decisions. Thus, only the restriction of Φ_t to Z^t has physical meaning, and we will refer to Z^t as the *natural domain* of the recourse function of stage t.

Without suitable regularity conditions, it is not clear whether the static and dynamic versions of a given multistage stochastic program are well defined and solvable (as is suggested by the use of the "max"-operators). In particular, measurability of the integrands in (5.1) and (5.3) must be ensured. To this end, the next section will elaborate and discuss a set of suitable regularity conditions which guarantee well-definedness, solvability, and equivalence of the static and dynamic versions of a linear multistage stochastic program.

5.3 Regularity Conditions

In order to establish a set of concise regularity conditions which ensure that the linear stochastic program given by representation (5.1) or (5.2) is well behaved, we have to introduce some terminology. A major concern is usually the functional dependence of the objective and rhs vectors on the stochastic parameters. Modeling c_t^* and h_t as differences of convex functions turns out to be particularly advantageous. Let us therefore state the following formal definition.

Definition 5.3.1 Let Ξ be a convex subset of a finite-dimensional Euclidean space. A function $f : \Xi \to \mathbb{R}^r$ is called d.c. (abbreviation for difference of convex functions) if there are two componentwise convex functions κ^+ and κ^- such that

$$f(\xi) = \kappa^+(\xi) - \kappa^-(\xi) \quad \forall \xi \in \Xi.$$

Notice that the decomposition of a d.c. function is never unique. In fact, by adding the same convex mapping to both κ^+ and κ^-, one obtains a valid alternative decomposition. Recently, d.c. functions have experienced considerable attention in the field of global optimization (Floudas 2000, Horst and Tuy 1996, Pardalos et al. 2000). A survey of their properties is provided in Hiriart-Urruty (1985). Here, we recall only a few properties relevant in our context. First, the class of twice continuously differentiable functions is a linear subspace of the space of d.c. functions. This can easily be proved if Ξ is compact. However, the statement remains true for Ξ open or unbounded, as pointed out by Hartman (1959). Moreover, if Ξ is compact, the d.c. functions are dense in the space of continuous mappings from Ξ to $\mathbb{R}^r$ endowed with the topology of uniform convergence with respect to the Euclidean norm in $\mathbb{R}^r$. This follows directly from the Stone–Weierstrass theorem and the fact that all vector-valued polynomials are d.c. functions.

Well-definedness of the integrands in (5.1) and (5.3) depends not only on the properties of the objective and rhs vectors (and the constraint matrices), but also essentially on the properties of the stochastic data process.

Definition 5.3.2 We say that the random data $(\boldsymbol{\eta}, \boldsymbol{\xi})$ follows a block-diagonal autoregressive process if it is driven by two serially independent stochastic processes $\boldsymbol{\varepsilon}^{\mathrm{o}}$ and $\boldsymbol{\varepsilon}^{\mathrm{r}}$ with state spaces $\mathcal{E}^{\mathrm{o}}$ and $\mathcal{E}^{\mathrm{r}}$, respectively. Moreover, for each t there are two matrices H_t^{o} and H_t^{r} with appropriate dimensions such that

$$\begin{bmatrix} \boldsymbol{\eta}_t \\ \boldsymbol{\xi}_t \end{bmatrix} = \begin{bmatrix} H_t^{\mathrm{o}} & 0 \\ 0 & H_t^{\mathrm{r}} \end{bmatrix} \begin{bmatrix} \boldsymbol{\eta}^{t-1} \\ \boldsymbol{\xi}^{t-1} \end{bmatrix} + \begin{bmatrix} \boldsymbol{\varepsilon}_t^{\mathrm{o}} \\ \boldsymbol{\varepsilon}_t^{\mathrm{r}} \end{bmatrix}, \quad t = 1, \ldots, T.$$

Block-diagonal autoregressive processes exhibit a linear dependence on history. Moreover, the AR coefficient matrices of all stages are block-diagonal. Note that Definition 5.3.2 allows the noise processes $\boldsymbol{\varepsilon}^{\mathrm{o}}$ and $\boldsymbol{\varepsilon}^{\mathrm{r}}$ to be correlated. In the extreme case where the same random processes affect the objective and the constraint functions, we are obliged to set $\boldsymbol{\varepsilon}^{\mathrm{o}} \equiv \boldsymbol{\varepsilon}^{\mathrm{r}}$ (thereby doubling the dimension of the relevant state space, which is computationally expensive but not necessarily prohibitive).

In the remainder of this chapter we will study linear multistage stochastic programs of the form (5.1) which satisfy the following regularity conditions:

(C1) the marginal spaces Θ_t and Ξ_t are compact regular simplices, $t = 1, \ldots, T$;
(C2) the vector of objective function coefficients c_t^* is continuous and d.c. on a convex neighborhood of Θ^t, $t = 1, \ldots, T$;
(C3) the rhs vector h_t is continuous and d.c. on a convex neighborhood of Ξ^t, the matrices W_t and T_t are independent of the random parameters, and the recession cone $\{x_t | W_t x_t \sim_t 0\}$ is given by $\{0\}$, $t = 1, \ldots, T$;
(C4) the random data $(\boldsymbol{\eta}, \boldsymbol{\xi})$ follows a block-diagonal autoregressive process;
(C5) at any reference point in Z^t the parametric maximization problem (5.2) has a feasible point where the gradients of the active constraint functions are linearly independent, $t = 1, \ldots, T$.

The first condition requires the state spaces of the random parameters to be compact, which implies the data processes to have finite moments of all orders. Consequently, (C1) constitutes a restrictive condition. Assumption (C2), however, is nonrestrictive since the set of d.c. functions is dense in the set of continuous functions. Therefore, we have much flexibility in modeling the functional form of c_t^*. Analogously, condition (C3) offers considerable flexibility in choosing the rhs vector h_t, but it is restrictive in that it requires the recourse and technology matrices to be nonrandom and the feasible sets of the parametric maximization problems (5.2) to be compact. Assumption (C4) is certainly restrictive as it only allows for linear dependencies between random parameters of different stages. But this is not a serious deficiency. In fact, it is frequently possible to absorb all nonlinearities in the definition of the functions c_t^* and h_t. However, requiring these functions to be linear affine, as is frequently done, and, at the same time, assuming the random parameters to follow a block-diagonal autoregressive process, would severely limit the scope of our methodology. Assumption (C5) has the character of a generalized Slater condition and implies non-anticipativity of the constraint multifunction in the sense of Rockafellar and Wets (1976a). This condition is nonrestrictive, since any constraint multifunction can be made non-anticipative by explicitly introducing the so-called *induced constraints*, see, e.g., Rockafellar and Wets (1976b) and Wets (1972). Notice that, as opposed to popular strict feasibility conditions, our assumption (C5) is not in conflict with the presence of equality constraints.

Some basic consequences of the above regularity conditions are summarized in the following theorem.

Theorem 5.1 *Under the assumptions* (C1)–(C5) *the static and dynamic versions of a linear multistage stochastic program are both well defined and solvable and the optimal values coincide. Moreover, the recourse functions are finite and continuous on their natural domains.*

Proof The assumptions (C1)–(C5) imply the weaker conditions used by Rockafellar and Wets (1976a). Thus, the claim follows from Rockafellar and Wets (1976a, Theorem 1). Alternatively, see Kuhn (2004, Proposition 2.5 and Theorem 2.6). □

5.4 Barycentric Probability Measures

The solution of stochastic programs poses severe difficulties, especially in the multistage case. If the underlying probability measure is absolutely continuous with respect to Lebesgue measure, the static version of a stochastic program represents an optimization problem over an infinite-dimensional function space. Then, analytical solutions are available only for simple models of questionable practical relevance. Analytical treatment of the dynamic version of a stochastic program is no less challenging. Instead of a single optimization problem over a function space one faces a sequence of nested optimization problems over finite-dimensional Euclidean spaces. Evaluation of the expectation functionals is particularly involved: it requires multivariate integration of a function which is only known implicitly as the result of a subordinate parametric optimization problem.

Numerical solutions are usually based on discretization of the underlying probability space. The standard approach is to solve the stochastic program with respect to a finitely supported auxiliary measure instead of the original measure. Thereby, one effectively approximates the original stochastic program by an auxiliary optimization problem over a finite-dimensional space, which is numerically tractable. The auxiliary probability measure should approximate the original measure in a specific sense, i.e., it should be designed so as to guarantee that the optimal value and the solution set of the auxiliary problem are close to the optimal value and the solution set of the original stochastic program, respectively. Thereby, distance of optimal values is measured with respect to the Euclidean metric on the real line, while distance of the solution sets is measured, e.g., with respect to the Pompeiu–Hausdorff metric. In this sense, proximity of the auxiliary and the original probability measures depends on the underlying stochastic program.

The selection of an appropriate discrete probability measure is referred to as *scenario tree construction* and represents a primary challenge in the field of stochastic programming. Our scenario tree construction method is based on the following procedure, which stems from Frauendorfer (1996, Section 3). As usual, we let P_t be a regular conditional probability distribution of the stochastic parameters observed at stage t given the history of the stochastic parameters observed at stages 1 through $t-1$. Suppose that for each history of realizations $(\eta^{t-1}, \xi^{t-1}) \in \Theta^{t-1} \times \Xi^{t-1}$ a discrete measure $P_t^d(\cdot|\eta^{t-1}, \xi^{t-1})$ approximates the true conditional probability measure $P_t(\cdot|\eta^{t-1}, \xi^{t-1})$ on $\mathcal{B}(\Theta_t \times \Xi_t)$. Furthermore, assume that the assignment

$$P_t^d : \begin{cases} \mathcal{B}(\Theta_t \times \Xi_t) \times \Theta^{t-1} \times \Xi^{t-1} \to [0, 1] \\ (B, \eta^{t-1}, \xi^{t-1}) \mapsto P_t^d(B|\eta^{t-1}, \xi^{t-1}) \end{cases}$$

characterizes a *transition probability*, i.e., it satisfies the following conditions:

(i) $P_t^d(\cdot|\eta^{t-1}, \xi^{t-1})$ is a probability measure on $\mathcal{B}(\Theta_t \times \Xi_t)$ for any fixed outcome history $(\eta^{t-1}, \xi^{t-1}) \in \Theta^{t-1} \times \Xi^{t-1}$;
(ii) $P_t^d(B|\cdot)$ is a Borel measurable function on $\Theta^{t-1} \times \Xi^{t-1}$ for every fixed Borel subset B of $\Theta_t \times \Xi_t$.

By the product measure theorem (Ash 1972, Section 2.6), the transition probabilities P_t^d of all stages can be nested to form a unique probability measure P^d on the measurable space $(\Theta \times \Xi, \mathcal{B}(\Theta \times \Xi))$. Under suitable conditions, the discrete measure P^d then approximates the original measure P with respect to the underlying stochastic optimization problem. The above reasoning implies that we should first focus on the discretization of the conditional distributions $P_t(\cdot|\eta^{t-1}, \xi^{t-1})$. In a second step, the transition probabilities must be combined to form a discrete scenario tree.

In any efficient scenario tree construction method, the choice of discrete transition probabilities should account for the structural properties of the underlying stochastic program. Our approach exploits distinct convexity properties of the recourse functions (5.2), which follow from the prevailing regularity conditions. In fact, the assumptions (C1)–(C5) imply that the recourse functions are concave in the decisions and d.c. in the stochastic variables on a neighborhood of their natural domains (Kuhn 2004, Chapter 5). By adding suitable correction terms, which may depend on the random parameters but not on the decision variables, the recourse functions can be transformed to saddle functions being convex in η and jointly concave in x and ξ.[5] Thus, evaluation of the expectation functionals is intimately related to calculating the expected value of a saddle function on a compact domain.

Approximating the conditional distributions of the random parameters in a multistage stochastic program is basically equivalent to approximating the unconditional distribution of the random parameters in a one-stage problem. Therefore, we may temporarily omit time indices and suppress any dependencies on the outcome and decision history. Our task now reduces to approximating the joint distribution P of two random vectors $\boldsymbol{\eta}$ and $\boldsymbol{\xi}$ by a discrete probability measure P^d. Recall also that $\boldsymbol{\eta}$ and $\boldsymbol{\xi}$ are valued in compact simplices Θ and Ξ, respectively. Having in mind the remarks of the previous paragraph, we try to establish a definite relation between the expectations of some (a priori unknown) saddle function $\Phi(\eta, \xi)$ with respect to the complementary measures P and P^d, assuming only that Φ is convex in η, concave in ξ, and subdifferentiable. This problem has been extensively studied in the monograph (Frauendorfer 1992); see also the related work in Edirisinghe (1996), Edirisinghe and You (1996), and Edirisinghe and Ziemba (1994a,b). A promising approach is via moment problems, as will be outlined below. Let $\mathcal{P}(m)$ be the set of all Borel probability measures on $\Theta \times \Xi$ which have the same first-order and second-order cross-moments m as the original measure P, i.e.,

$$m = \int_{\Theta \times \Xi} \begin{bmatrix} 1 \\ \eta \end{bmatrix} \begin{bmatrix} 1 \\ \xi \end{bmatrix}^\top dP(\eta, \xi).$$

[5] If the objective function coefficients and the rhs vectors are linear affine functions of the stochastic parameters, then the recourse functions exhibit a saddle shape themselves and no correction terms are needed. This situation is investigated in Frauendorfer (1994, 1996).

Next, choose a measure P^l from the minimizer set of

$$\inf_{Q \in \mathcal{P}(m)} \int_{\Theta \times \Xi} \Phi(\eta, \xi)\, dQ(\eta, \xi). \tag{5.5}$$

Note that P^l exists since Φ is subdifferentiable and saddle shaped (Frauendorfer 1992, Chapter 3). Furthermore, choose a measure P^u from the maximizer set of the symmetric problem

$$\sup_{Q \in \mathcal{P}(m)} \int_{\Theta \times \Xi} \Phi(\eta, \xi)\, dQ(\eta, \xi), \tag{5.6}$$

which is solvable for the same reasons as (5.5). By construction, we find the following chain of inequalities for the expected values of the saddle function Φ with respect to the three measures under consideration:

$$\int_{\Theta \times \Xi} \Phi(\eta, \xi)\, dP^l(\eta, \xi) \le \int_{\Theta \times \Xi} \Phi(\eta, \xi)\, dP(\eta, \xi) \le \int_{\Theta \times \Xi} \Phi(\eta, \xi)\, dP^u(\eta, \xi). \tag{5.7}$$

An elegant duality argument shows that P^l and P^u can be chosen to be discrete, while depending solely on the matrix m of cross-moments and the geometry of the state space $\Theta \times \Xi$.[6] This implies that for fixed distributions P, P^l, and P^u, the estimate (5.7) is universally valid for all subdifferentiable saddle functions and not just for the specific function Φ. Universality is a very useful feature, since the recourse functions of a given multistage stochastic program are a priori unknown. Note that universality may be lost if we naively attempt to match higher order moments in the semi-infinite linear programs (5.5) and (5.6). However, universality can be restored if one changes the shape of the underlying domain of the random variables, see Edirisinghe (1996) and Edirisinghe and You (1996).

Loosely speaking, the discrete measure P^l concentrates probability mass at the barycenter of Θ and at the extreme points of Ξ, while P^u concentrates probability mass at the barycenter of Ξ and at the extreme points of Θ. Therefore, we will refer to P^l and P^u as lower and upper *barycentric probability measures*. Although we have established a definite relation between the original and the barycentric probability measures, it is not clear whether the inequalities in (5.7) are tight. If not, the probability measure P can be partitioned into smaller pieces, i.e., it may be represented as a convex combination of specific probability measures with smaller supports. Then, the barycentric measures are constructed for each component separately and their convex combinations provide improved estimates in (5.7). By successively increasing the number of components in a partition, one can construct two sequences of refined barycentric probability measures $\{P^l_J\}_{J \in \mathbb{N}_0}$ and $\{P^u_J\}_{J \in \mathbb{N}_0}$,

---

[6] Analytical formulae for the masses and coordinates of the atoms of P^l and P^u are provided in Frauendorfer (1992); cf. also Frauendorfer (1994, 1996) and Kuhn (2004).

where the integer J indices the current partition and will be called the *refinement parameter*. It can be shown that both sequences converge weakly to the original measure P if the supports of all components become uniformly small for large J. This will always be assumed in the remainder of this chapter, for details see Frauendorfer (1996) and Kuhn (2004).

Let us now return to the multistage case. The recipe of the previous paragraph can be used to approximate the conditional probability measure $P_t(\cdot|\eta^{t-1}, \xi^{t-1})$ by lower and upper barycentric measures $P^l_{J,t}(\cdot|\eta^{t-1}, \xi^{t-1})$ and $P^u_{J,t}(\cdot|\eta^{t-1}, \xi^{t-1})$, which may depend on the refinement parameter, the stage index, and the outcome history. We assume that all lower (upper) barycentric measures have the same number of atoms for fixed values of J and t. This is no major restriction since any atom can be viewed as a group of collapsed single atoms. If the probability masses and the coordinates of the discretization points are measurable functions of the outcome history (η^{t-1}, ξ^{t-1}), then $P^l_{J,t}$ and $P^u_{J,t}$ satisfy the defining conditions of a transition probability. By the product measure theorem (Ash 1972, Section 2.6), the barycentric transition probabilities $\{P^l_{J,t}\}_{t=1}^T$ ($\{P^u_{J,t}\}_{t=1}^T$) can be combined to form a unique barycentric probability measure P^l_J (P^u_J) on the joint state space of the stochastic processes $\boldsymbol{\eta}$ and $\boldsymbol{\xi}$. It can be shown that the inequalities (5.7) still hold in the multistage case for any subdifferentiable saddle function Φ being convex in $\eta = (\eta_1, \ldots, \eta_T)$ and concave in $\xi = (\xi_1, \ldots, \xi_T)$. Moreover, the overall barycentric measures converge weakly to the true probability measure P as the refinement parameter J tends to infinity (Kuhn 2004, Proposition 4.4).

For each fixed $J \in \mathbb{N}_0$, the barycentric transition probabilities can be used to establish two sequences of *auxiliary recourse functions*, which will be shown to approximate the true recourse functions (5.6). Set $\Phi^l_{J,T+1} \equiv \Phi^u_{J,T+1} \equiv 0$ and define

$$\begin{aligned} \Phi^l_{J,t}(x^{t-1}, \eta^t, \xi^t) = \max\nolimits_{x_t \in \mathbb{R}^{n_t}} \langle c_t^*(\eta^t), x_t \rangle + (E^l_{J,t}\Phi^l_{J,t+1})(x^t, \eta^t, \xi^t) \\ \text{s.t.} \quad W_t(\xi^t)x_t + T_t(\xi^t)x_{t-1} \sim_t h_t(\xi^t), \end{aligned} \tag{5.8a}$$

$$\begin{aligned} \Phi^u_{J,t}(x^{t-1}, \eta^t, \xi^t) = \max\nolimits_{x_t \in \mathbb{R}^{n_t}} \langle c_t^*(\eta^t), x_t \rangle + (E^u_{J,t}\Phi^u_{J,t+1})(x^t, \eta^t, \xi^t) \\ \text{s.t.} \quad W_t(\xi^t)x_t + T_t(\xi^t)x_{t-1} \sim_t h_t(\xi^t), \end{aligned} \tag{5.8b}$$

in a backward manner for $t = T, \ldots, 1$. Thereby, the (auxiliary) expectation functionals are constructed in the obvious way with the help of the transition probabilities $P^l_{J,t+1}$ and $P^u_{J,t+1}$:

$$(E^l_{J,t}\Phi^l_{J,t+1})(x^t, \eta^t, \xi^t) = \int \Phi^l_{J,t+1}(x^t, \eta^{t+1}, \xi^{t+1})\, dP^l_{J,t+1}(\eta_{t+1}, \xi_{t+1}|\eta^t, \xi^t),$$

$$(E^u_{J,t}\Phi^l_{J,t+1})(x^t, \eta^t, \xi^t) = \text{`} \int \Phi^u_{J,t+1}(x^t, \eta^{t+1}, \xi^{t+1})\, dP^u_{J,t+1}(\eta_{t+1}, \xi_{t+1}|\eta^t, \xi^t).$$

Note that the set Z^t defined in Section 5.3 can be interpreted as the natural domain of both $\Phi^l_{J,t}$ and $\Phi^u_{J,t}$. Moreover, unlike the true recourse function Φ_t, the auxiliary recourse functions are numerically computable. Their calculation either relies on the solution of a finite-dimensional linear program, which can, e.g., be solved by using the classical simplex algorithm (Dantzig and Thapa 1997), or on some specialized decomposition schemes (Birge and Louveaux 1997, Section 7.1).

5.5 Bounds for Stochastic Programs

In this section we will argue that—after a suitable transformation—the auxiliary recourse functions provide bounds on the true recourse functions and can be used to construct bounding sets for the optimal decisions. Furthermore, we will discuss convergence issues.

As usual, assume the regularity conditions (C1)–(C5) to hold. If, beyond that, the objective coefficients and the rhs vectors are linear functions of the random parameters, then the recourse functions are subdifferentiable and saddle shaped (Frauendorfer 1994, Section 2), while the auxiliary recourse functions can be shown to bracket the true recourse functions on their natural domains. Mathematically speaking, this translates to

$$\Phi^{l}_{J,t} \leq \Phi_t \leq \Phi^{u}_{J,t} \quad \text{on } Z^t, t = 1, \ldots, T, \; J \in \mathbb{N}. \tag{5.9}$$

Frauendorfer proved the inequalities (5.9) by backward induction with respect to t (Frauendorfer 1994, Theorem 4.1 and Lemma 4.2) using subdifferentiability and the saddle structure of the recourse functions and the universal bounding property (5.11) of the barycentric transition probabilities. If linearity of the objective and rhs vectors is abandoned, as is necessary in certain applications, the relation (5.9) fails to hold in general. Therefore, we present a stronger result in this chapter, which remains applicable if c_t^* and h_t are generic d.c. functions. In fact, we will argue that bounds on Φ_t can generally be expressed in terms of the auxiliary recourse functions shifted by specific correction terms, which have an intuitive structure and are numerically accessible.

In order to characterize these correction terms, we have to introduce some additional notation. Recall first that, by assumption (C2), the vector-valued function c_t^* is d.c. Thus, there are two convex mappings κ_t^{*+} and κ_t^{*-} on a closed convex neighborhood of Θ^t such that

$$c_t^* = \kappa_t^{*+} - \kappa_t^{*-}, \quad t = 1, \ldots, T. \tag{5.10}$$

Similarly, condition (C3) stipulates that h_t is d.c. We may thus suppose that there are two convex mappings κ_t^{+} and κ_t^{-} on a closed convex neighborhood of Ξ^t with

$$h_t = \kappa_t^{+} - \kappa_t^{-}, \quad t = 1, \ldots, T. \tag{5.11}$$

Next, let us return to the dynamic version (5.10) of the given multistage stochastic program. It should be emphasized again that this representation explicitly deals with inequality (less-or-equal, greater-or-equal) and equality constraints. Let

$$X_{\text{opt},t}(x^{t-1}, \eta^t, \xi^t) \subset \mathbb{R}^{n_t} \quad \text{and} \quad D^*_{\text{opt},t}(x^{t-1}, \eta^t, \xi^t) \subset \mathbb{R}^{r_t}$$

be the primal and dual solution sets associated with the parametric optimization problem (5.10), respectively. The dual solutions correspond to the Lagrange multipliers associated with the explicit constraints in (5.10). We will interpret $X_{\text{opt},t}$ and $D^*_{\text{opt},t}$ as multifunctions on the underlying parameter space. Under the given regularity conditions it can be proved that both $X_{\text{opt},t}$ and $D^*_{\text{opt},t}$ are non-empty-valued, bounded, and Berge upper semicontinuous on a neighborhood of the natural domain Z^t; see (Kuhn 2004, Chapter 5). Thus, there are nonnegative finite bounding vectors $X_t^+, X_t^- \in \mathbb{R}^{n_t}$ and $D_t^{*+}, D_t^{*-} \in \mathbb{R}^{r_t}$ such that

$$-X_t^- \leq x_t \leq X_t^+ \quad \text{uniformly for all} \quad x_t \in X_{\text{opt},t}(Z^t) \tag{5.12a}$$

and

$$-D_t^{*-} \leq d_t^* \leq D_t^{*+} \quad \text{uniformly for all} \quad d_t^* \in D^*_{\text{opt},t}(Z^t). \tag{5.12b}$$

Due to the assumption about the recession cone in (C3), the primal bounding vectors are usually easy to find. As for the dual bounding vectors, recall that Lagrange multipliers are nonnegative for less-or-equal constraints and nonpositive for greater-or-equal constraints; no a priori statement about the sign of Lagrange multipliers is available for equality constraints. Thus, given that $\sim_t$ only contains less-or-equal relations, we may choose $D_t^{*-} = 0$. Conversely, if $\sim_t$ is exclusively made up of greater-or-equal relations, we will set $D_t^{*+} = 0$. These basic rules persist on a componentwise level if $\sim_t$ represents a heterogeneous mixture of inequality relations. Apart from that, there are no universal a priori guidelines for how to determine the dual bounding vectors. To find them, however, a good understanding of the decision problem at hand and a physical interpretation of the involved Lagrange multipliers are usually sufficient.

By means of the d.c. components of the objective and rhs vectors on one hand and the bounding vectors for the primal and dual solutions on the other hand, it is possible to define appropriate *correction terms* α_t^{o} and α_t^{r}, respectively.

$$\alpha_t^{\text{o}}(\eta^t) = +\langle \kappa_t^{*-}(\eta^t), X_t^+ \rangle + \langle \kappa_t^{*+}(\eta^t), X_t^- \rangle \tag{5.13a}$$

$$\alpha_t^{\text{r}}(\xi^t) = -\langle D_t^{*+}, \kappa_t^+(\xi^t) \rangle - \langle D_t^{*-}, \kappa_t^-(\xi^t) \rangle \tag{5.13b}$$

These definitions reflect the intrinsic primal–dual symmetry of linear (stochastic) programs. Obviously, the correction term α_t^{o} associated with the nonconvexities in the objective function has the same general structure as the correction term α_t^{r} corresponding to the nonconvexities in the constraints. Concretely speaking, (5.13a) pairs the bounding vectors of the primal solutions with the d.c. components of the objective function coefficients, whereas (5.13b) pairs the bounding vectors of the dual solutions with the d.c. components of the rhs vector. Next, for all $t = 1, \ldots, T$ we define a combined *correction term* as

$$\alpha_t(\eta^t, \xi^t) = \alpha_t^{\mathrm{o}}(\eta^t) + \alpha_t^{\mathrm{r}}(\xi^t). \tag{5.14}$$

By construction, α_t is a continuous saddle function on a neighborhood of $\Theta^t \times \Xi^t$ being convex in η^t and concave in ξ^t. Let us now introduce three sequences of additional functions. Set $A_{T+1} \equiv A^l_{J,T+1} \equiv A^u_{J,T+1} \equiv 0$ and define for $t = T, \ldots, 1$:

$$\begin{aligned}
A_t(\eta^t, \xi^t) &= \alpha_t(\eta^t, \xi^t) + \int A_{t+1}(\eta^{t+1}, \xi^{t+1}) \;\; dP_{t+1}(\eta_{t+1}, \xi_{t+1}|\eta^t, \xi^t),\\
A^l_{J,t}(\eta^t, \xi^t) &= \alpha_t(\eta^t, \xi^t) + \int A^l_{J,t+1}(\eta^{t+1}, \xi^{t+1})\, dP^l_{J,t+1}(\eta_{t+1}, \xi_{t+1}|\eta^t, \xi^t), \quad (5.15)\\
A^u_{J,t}(\eta^t, \xi^t) &= \alpha_t(\eta^t, \xi^t) + \int A^u_{J,t+1}(\eta^{t+1}, \xi^{t+1})\, dP^u_{J,t+1}(\eta_{t+1}, \xi_{t+1}|\eta^t, \xi^t).
\end{aligned}$$

The functions (5.15) will be referred to as *conditional correction terms* below. It is important to notice that the conditional correction terms are computationally accessible. Evaluation of $A^l_{J,t}$ and $A^u_{J,t}$ requires calculation of a finite sum, while A_t can usually be evaluated by means of numerical integration techniques. Since α_t is continuous for every t, A_t, $A^l_{J,t}$, and $A^u_{J,t}$ are continuous[7] and bounded on $\Theta^t \times \Xi^t$. Moreover, by the saddle structure of the correction terms and the properties of the barycentric measures, we find $A^l_{J,t} \leq A_t \leq A^u_{J,t}$ on $\Theta^t \times \Xi^t$. Using the above definitions, we are now ready to state our main result.

Theorem 5.2 (Bounds on the Recourse Functions) *Consider a linear stochastic program satisfying the regularity conditions* (C1)–(C5) *and define the conditional correction terms as in* (5.15). *Then, we find*

$$\Phi^l_{J,t} + A^l_{J,t} - A_t \leq \Phi_t \leq \Phi^u_{J,t} + A^u_{J,t} - A_t \quad \text{on } Z^t,\, t = 1, \ldots, T,\, J \in \mathbb{N}.$$

As the refinement parameter J tends to infinity, the conditional correction terms $A^l_{J,t}$ and $A^u_{J,t}$ converge to A_t uniformly on $\Theta^t \times \Xi^t$, while the auxiliary recourse functions $\Phi^l_{J,t}$ and $\Phi^u_{J,t}$ converge to Φ_t uniformly on Z^t for $t = 1, \ldots, T$.

Proof The claim follows from Theorems 5.16 and 5.17 as well as the discussion at the end of section 5.4 in Kuhn (2004). □

This theorem tells us how to approximate the unknown original recourse functions by known quantities (i.e., quantities which can at least principally be evaluated, given sufficient computer power) and provides a nonprobabilistic error estimate as the difference of the two bounds. In particular, the theorem points out a possibility to

---

[7] Continuity of A_t follows inductively from the dominated convergence theorem.

construct numerically calculable and arbitrarily tight bounds on the optimal objective value. A decision maker, however, is interested not only in an accurate estimate of maximal expected reward, but also in the corresponding optimal policy. Thus, we will demonstrate that bounds on the recourse functions also entail *bounding sets* for the optimal decisions.

Consider again the stochastic program (5.9) subject to the regularity conditions (C1)–(C5). For notational convenience we introduce the extended real-valued functional

$$\rho_t(x^t, \eta^t, \xi^t) = \begin{cases} \langle c_t^*(\eta^t), x_t \rangle & \text{for } W_t x_t + T_t x_{t-1} \sim_t h_t(\xi^t), \\ -\infty & \text{else,} \end{cases}$$

which penalizes infeasible decisions with an infinite loss and can thus be viewed as the effective profit earned at stage t. Moreover, define

$$\begin{aligned} F_t(x^t, \eta^t, \xi^t) &= \rho_t(x^t, \eta^t, \xi^t) + (E_t \Phi_{t+1})(x^t, \eta^t, \xi^t), \\ F^l_{J,t}(x^t, \eta^t, \xi^t) &= \rho_t(x^t, \eta^t, \xi^t) + (E^l_{J,t} \Phi_{t+1})(x^t, \eta^t, \xi^t) + A^l_{J,t}(\eta^t, \xi^t) - A_t(\eta^t, \xi^t), \\ F^u_{J,t}(x^t, \eta^t, \xi^t) &= \rho_t(x^t, \eta^t, \xi^t) + (E^u_{J,t} \Phi_{t+1})(x^t, \eta^t, \xi^t) + A^u_{J,t}(\eta^t, \xi^t) - A_t(\eta^t, \xi^t). \end{aligned}$$

Given sufficient CPU speed and storage capacity, the functionals $F^l_{J,t}$ and $F^u_{J,t}$ can be numerically evaluated, whereas, in general, F_t remains computationally untractable. In the sequel, we will interpret these extended real-valued mappings as functions of x_t, while the arguments x^{t-1}, η^t, and ξ^t are interpreted as parameters. Theorem 5.2 implies that $F^l_{J,t} \le F_t \le F^u_{J,t}$ for all (feasible and infeasible) decisions x_t as well as for all parameters in Z^t. As a consequence, the optimal stage t decisions are necessarily contained in the polyhedron[8]

$$C_{J,t}(x^{t-1}, \eta^t, \xi^t) = \left\{ x_t \;\middle|\; F^u_{J,t}(x^t, \eta^t, \xi^t) \ge \max_{x'_t} F^l_{J,t}(x'_t, x^{t-1}, \eta^t, \xi^t) \right\}. \quad (5.16)$$

This set depends parametrically on the outcome and decision history, thus defining a multifunction $C_{J,t}$. Notice that the evaluation of $C_{J,t}$ at any fixed point, though possibly time consuming, is numerically feasible as it is only based on knowledge of the functionals $F^l_{J,t}$ and $F^u_{J,t}$. The most important properties of the multifunction $C_{J,t}$, all of which can be deduced from Theorem 5.2, are summarized in the following statement.

Theorem 5.3 (Bounding Sets for the Optimal Decisions) *Consider a linear multistage stochastic program subject to the regularity conditions* (C1)–(C5). *Then, the*

[8] Notice that F^l_t and F^u_t are concave polyhedral functions of x_t for all fixed parameter values, since we are dealing with *linear* stochastic programs and since the barycentric measures have *finite* supports.

multifunction $C_{J,t}$ is compact-convex-valued and we find

$$X_{\text{opt},t} \subset C_{J,t} \quad on\ Z^t,\ t = 1, \ldots, T,\ J \in \mathbb{N}.$$

As the refinement parameter J tends to infinity, $C_{J,t}$ converges to $X_{\text{opt},t}$ pointwise on Z^t in the sense of set convergence.[9]

Proof The claim is an immediate consequence of Kuhn (2004, Theorem 5.22). □

Let $X^l_{\text{opt},J,t}(x^{t-1}, \eta^t, \xi^t)$ and $X^u_{\text{opt},J,t}(x^{t-1}, \eta^t, \xi^t)$ be the solution sets of the auxiliary parametric optimization problems (5.12a) and (5.12b), respectively. As usual, we may interpret $X^l_{\text{opt},J,t}$ and $X^u_{\text{opt},J,t}$ as multifunctions valued in $\mathbb{R}^{n_t}$. Using this new terminology, we may state the following corollary to Theorem 5.3.

Corollary 5.5.1 *Let $x^\star_{J,t}$ be a selector of $X^l_{\text{opt},J,t} \cup X^u_{\text{opt},J,t}$ on Z^t and assume that $x^\star_{J,t}$ converges pointwise to $x^\star_t$ as the refinement parameter J tends to infinity. Then, $x^\star_t$ is a selector of $X_{\text{opt},t}$ on Z^t for $t = 1, \ldots, T$.*

Proof The assertion follows immediately from the inclusion

$$X^l_{\text{opt},J,t} \cup X^u_{\text{opt},J,t} \subset C_{J,t}, \quad t = 1, \ldots, T,\ J \in \mathbb{N}$$

and the fact that $\{C_{J,t}\}_{J \in \mathbb{N}}$ converges pointwise to $X_{\text{opt},t}$ on Z^t. □

Corollary 5.5.1 asserts that if J tends to infinity, all accumulation points of solutions to the auxiliary stochastic programs (5.12) solve the true recourse problem (5.10). In Frauendorfer (1992, Section 18) this qualitative convergence result is derived for the two-stage case by using the concept of epi-convergence due to Attouch and Wets (1981). The approach presented here, however, is more quantitative in nature as the bounding sets (5.16) additionally provide nonprobabilistic error estimates for the true solutions.

5.6 Application in Financial Risk Management

As an illustration of the functionality and performance of bounding methods, we now present a practical example from financial risk management. It is a simplified version of a model that was developed for a major Swiss bank (for details, see Forrest et al. 1998, Frauendorfer and Schürle 2003, 2005) to solve the following problem: A significant portion of a typical bank's balance sheet consists of liability positions with no contractual maturity such as savings deposits. Their characteristic feature is that bank clients may freely add or withdraw investments anytime at no penalty. On the other hand, the bank is allowed to adjust the rate paid on these investments at all times as a matter of policy. As a consequence, the total volume

9 For a survey of the theory of set convergence see Rockafellar and Wets (1998, Chapter 4).

of such a position may fluctuate heavily as clients react to changes in the offered deposit rate or the relative attractiveness of alternative investment opportunities. The uncertainty of the future volume complicates the task of the financial managers who have to reinvest the money on the market or internally for the funding of credits: The generation of sufficient income requires an allocation of larger amounts in long-term instruments which, on the other hand, increases the risk that a substantial portion of the savings volume will be withdrawn and the bank will run into liquidity problems (if a drop in volume cannot be compensated by maturing investments). Given the large volumes of such positions, it is obvious that the composition of the reinvested portfolio has a high impact on the bank's risk profile and, thus, a careful analysis of the problem is required.

5.6.1 Formulation as Multistage Stochastic Program

Assume that investments can be made in fixed-income securities whose maturities are given by the set $\mathcal{D} = \{1, \ldots, D\}$ (D is the longest available maturity). Let $r(\eta^t; d)$ denote the interest rate at time t for a given history of risk factors η^t. An investment of \$1 in maturity $d \in \mathcal{D}$ at time $t = 1, \ldots, T$ generates a (discounted) income of

$$\varphi_t^d(\eta^t) = \textstyle\sum_{i=1}^{\Delta} a_t^d(t+i)\big(r(\eta^t; d) - s^d\big) \tag{5.17}$$

over the lifetime of the instrument. Since the objective is the maximization of the (expected) discounted interest income during the planning horizon, payments arising after time $T+1$ are truncated by setting $\Delta := \min\{T-t+1, d\}$. The (deterministic) function $a_t^d(\tau)$ calculates the discount factor at time τ from the *initial* yield curve or is zero if an instrument with maturity d in which the model invests in t does not pay a coupon in τ. Transaction costs in terms of a bid spread are given by s^d. If a decrease in the total volume cannot be compensated by the sum of maturing tranches, the resulting gap must be refinanced in the shortest maturity. Then, a penalty p is charged in addition to the one-period market rate that represents the cost for liquidity risk, i.e., the corresponding coefficient in the objective function is $\vartheta_t(\eta^t) = a_t^1(t+1)(r(\eta^t; 1)+p)$. The volume $v_t(\xi^t)$ in the account at time t depends on the past realizations of some process $\boldsymbol{\xi}$. The specific form of the processes $\boldsymbol{\eta}$ and $\boldsymbol{\xi}$ will be introduced in the sequel.

At each point in time $t = 1, \ldots, T$, decisions on the amounts $x_{\text{inv},t}^d$, $d \in \mathcal{D}$, have to be made according to which the sum of maturing tranches, corrected by a change in the total volume of the position, is reinvested in the available instruments. Any investment $x_{\text{inv},t}^d$ increases the total volume $x_{\text{pos},t}^d$ of the positions with maturity d. An amount refinanced in the shortest maturity as compensation for large drops in volume is represented by $x_{\text{ref},t}$. Thus, the complete decision vector at stage t is defined as

$$x_t = (x_{\text{inv},t}^1, \ldots, x_{\text{inv},t}^D, x_{\text{pos},t}^1, \ldots, x_{\text{pos},t}^D, x_{\text{ref},t}) \in \mathbb{R}^{2D+1}.$$

As pointed out in Section 5.2, x_t is interpreted as the time t realization of a stochastic process $\boldsymbol{x}$ characterizing the entire scenario-dependent decision strategy, and its subprocesses $\boldsymbol{x}^1_{\text{inv}}, \ldots, \boldsymbol{x}^D_{\text{inv}}, \boldsymbol{x}^1_{\text{pos}}, \ldots, \boldsymbol{x}^D_{\text{pos}}$ and $\boldsymbol{x}_{\text{ref}}$ are defined in the obvious way. With the objective to maximize the expected income on the reinvested portfolio minus the costs for liquidity risk, our decision problem can be formally represented as the multistage stochastic program

$$\begin{aligned}
&\underset{x \in \mathcal{N}}{\text{maximize}}\; E\left[\sum_{t=1}^{T}\left(\sum_{d \in \mathcal{D}} \varphi_t^d(\boldsymbol{\eta}^t)\, \boldsymbol{x}^d_{\text{inv},t} \;-\; \vartheta_t(\boldsymbol{\eta}^t)\, \boldsymbol{x}_{\text{ref},t}\right)\right] \\
&\text{subject to:} \\
&\left.\begin{array}{lll}
\boldsymbol{x}^1_{\text{pos},t} - \boldsymbol{x}^2_{\text{pos},t-1} - \boldsymbol{x}^1_{\text{inv},t} + \boldsymbol{x}_{\text{ref},t} = 0 & & \\
\boldsymbol{x}^d_{\text{pos},t} - \boldsymbol{x}^{d+1}_{\text{pos},t-1} - \boldsymbol{x}^d_{\text{inv},t} = 0 & d = 2, \ldots, D & \\
\sum_{d \in \mathcal{D}} \boldsymbol{x}^d_{\text{pos},t} = v_t(\boldsymbol{\xi}^t) & & \\
0 \leq \boldsymbol{x}^d_{\text{inv},t} \leq \ell^d & d = 1, \ldots, D & \\
0 \leq \boldsymbol{x}_{\text{ref},t} \leq \ell^1 & &
\end{array}\right\} \begin{array}{l} P\text{-a.s.} \\ t = 1, \ldots, T. \end{array}
\end{aligned} \tag{5.18}$$

Herein, $\boldsymbol{x}^d_{\text{pos},0}$ are degenerate deterministic random variables and denote positions with maturity d held in the initial portfolio. Upper limits ℓ^d on the transaction volumes reflect liquidity restrictions in the Swiss market for certain maturities; in case of nonstandard maturities that are not traded, these limits are set to zero. We also assume here that the initial values $\boldsymbol{\eta}_1$ and $\boldsymbol{\xi}_1$ of the underlying stochastic processes, which affect the coefficients in the first stage, can be observed in the market and are thus deterministic.

5.6.2 Risk Factor Models

It now remains to specify the stochastic processes that drive the evolution of interest rates and volume. Empirical studies imply that two factors explain most of the volatility of the term structure, and that these factors can be chosen as the level $\boldsymbol{\eta}_l$ of the yield curve, e.g., in terms of the rate for an infinite maturity, and the spread $\boldsymbol{\eta}_s$ between the instantaneous short rate and the level factor (see, e.g., Frauendorfer and Schürle 2001). For the ease of exposition, we apply a simplified discrete time term structure model, in which the stochastic changes of the factors are given by

$$\begin{aligned}
\boldsymbol{\eta}_{s,t} - \boldsymbol{\eta}_{s,t-1} &= a_s(\theta_s - \boldsymbol{\eta}_{s,t-1})\,\Delta t + \hat{\boldsymbol{\varepsilon}}^{\text{o}}_{s,t} \\
\boldsymbol{\eta}_{l,t} - \boldsymbol{\eta}_{l,t-1} &= a_l\,(\theta_l - \boldsymbol{\eta}_{l,t-1})\,\Delta t + \hat{\boldsymbol{\varepsilon}}^{\text{o}}_{l,t}
\end{aligned} \qquad t = 1, \ldots, T. \tag{5.19}$$

This process specification incorporates the "mean reversion" property, i.e., there is a drift term that forces the process $\boldsymbol{\eta}_i$, $i \in \{s, l\}$, from its current value toward the long-term mean θ_i at a speed controlled by a_i. This reflects the empirical observation that interest rates fluctuate within a certain range. The disturbances $\hat{\boldsymbol{\varepsilon}}^{\mathrm{o}}_{s,t}$ and $\hat{\boldsymbol{\varepsilon}}^{\mathrm{o}}_{l,t}$ are conditionally independent of the past given $\boldsymbol{\eta}_{l,t}$. We further assume that they are conditionally normally distributed given $\boldsymbol{\eta}_{l,t}$, i.e.,

$$\hat{\boldsymbol{\varepsilon}}^{\mathrm{o}}_{s,t}, \hat{\boldsymbol{\varepsilon}}^{\mathrm{o}}_{l,t} \,|\, \boldsymbol{\eta}_{l,t} \sim \mathcal{N}(0, \Sigma^{\mathrm{o}}_t) \quad \text{where} \quad \Sigma^{\mathrm{o}}_t = \begin{pmatrix} \sigma_s^2 & \varrho\sigma_s\sigma_l\boldsymbol{\eta}^{\gamma}_{l,t} \\ \varrho\sigma_s\sigma_l\boldsymbol{\eta}^{\gamma}_{l,t} & \sigma_l^2\boldsymbol{\eta}^{2\gamma}_{l,t} \end{pmatrix} \Delta t. \tag{5.20}$$

Parameter values $\gamma > 0$ are used to reflect a possible heteroscedasticity that may be found in historical interest rate data. For $\gamma = 0$, in contrast, heteroscedasticity is lost and the disturbances $\hat{\boldsymbol{\varepsilon}}^{\mathrm{o}}_{s,t}$ and $\hat{\boldsymbol{\varepsilon}}^{\mathrm{o}}_{l,t}$ become serially independent. Note that all parameters in the above model are annualized.

A very simple approach to obtain the interest rates of the maturities relevant for investment is to model the yield curve at time t by some exponential function:

$$r(\eta_s^t, \eta_l^t; d) = (\eta_{s,t} + \beta_1 d)\, e^{-\beta_2 d} + \eta_{l,t}\,. \tag{5.21}$$

Note that the constants β_1, β_2 control the shape of the yield curve while the rates $r(\eta_s^t, \eta_l^t; d)$, $d \in \mathcal{D}$, themselves are linear in the factors. The parameters of (5.19) and (5.20) can easily be estimated when the factors are approximated by observed interest rates, e.g., the 5-year rate for $\boldsymbol{\eta}_l$ and the difference between the 1-month rate and the latter for $\boldsymbol{\eta}_s$. Then, estimates of β_1 and β_2 in the yield curve function (5.21) are derived in a second step by minimizing the differences between the rates implied by the model and those of a historical sample.

Remark In contrast to the simplified example considered here, the more general formulation of our optimization model, which is used for "real-world applications," contains also variables for short sales (negative investments). This extended model uses a term structure model which precludes arbitrage opportunities in combination with very low bid-ask spreads. If arbitrage opportunities occurred in the scenarios, the optimization model would try to exploit them by refinancing at the cheapest and investing at the highest interest rate, which would lead to unrealistic investment decisions. However, we avoid a detailed discussion of this case here for the ease of exposition.

For obvious reasons, the volume of the nonmaturing account cannot become negative since clients are not allowed to withdraw higher amounts than their previous investments. We therefore model the volume in t as an exponential function $v_t(\xi^t) = \exp(\xi_t)$ of a stochastic factor which itself follows a first-order autoregressive process:

$$\boldsymbol{\xi}_t = a_v + b\,\boldsymbol{\xi}_{t-1} + \hat{\boldsymbol{\varepsilon}}^{\mathrm{r}}_t, \qquad \hat{\boldsymbol{\varepsilon}}^{\mathrm{r}}_t \sim N(0, \sigma_\varepsilon), \qquad t = 1, \ldots, T. \tag{5.22}$$

The random variables $\hat{\boldsymbol{\varepsilon}}^{\mathrm{r}}_t$ are assumed to be serially independent but may be correlated with the noise factors $\hat{\boldsymbol{\varepsilon}}^{\mathrm{o}}_{s,t}$ and $\hat{\boldsymbol{\varepsilon}}^{\mathrm{o}}_{l,t}$ of the interest rate model. This can reflect a possible dependency between the volume and the yield curve, as is often observed for nonmaturing accounts.

In order to facilitate comparison with our theoretical results of the previous sections, we define for each $t = 1, \dots, T$ the AR(1) coefficient matrices

$$H_t^{\mathrm{o}} = \begin{pmatrix} 1 - a_s\,\Delta t & 0 \\ 0 & 1 - a_l\,\Delta t \end{pmatrix}, \qquad H_t^{\mathrm{r}} = b,$$

and introduce two random vectors

$$\boldsymbol{\varepsilon}_t^{\mathrm{o}} = \begin{pmatrix} a_s\,\theta_s\,\Delta t \\ a_l\,\theta_l\,\Delta t \end{pmatrix} + \begin{pmatrix} \hat{\boldsymbol{\varepsilon}}_{s,t}^{\mathrm{o}} \\ \hat{\boldsymbol{\varepsilon}}_{l,t}^{\mathrm{o}} \end{pmatrix}, \qquad \boldsymbol{\varepsilon}_t^{\mathrm{r}} = a_v + \hat{\boldsymbol{\varepsilon}}_t^{\mathrm{r}}.$$

Using these conventions, the risk factor processes can be recast as

$$\boldsymbol{\eta}_t = H_t^{\mathrm{o}}\boldsymbol{\eta}_{t-1} + \boldsymbol{\varepsilon}_t^{\mathrm{o}} \quad \text{and} \quad \boldsymbol{\xi}_t = H_t^{\mathrm{r}}\boldsymbol{\xi}_{t-1} + \boldsymbol{\varepsilon}_t^{\mathrm{r}}. \tag{5.23}$$

5.6.3 Check of Regularity Conditions

Let us first assume that $\gamma = 0$ in (5.20) precluding heteroscedasticity in the interest rate model. Then, the risk factors $\boldsymbol{\eta}_t$ and $\boldsymbol{\xi}_t$ are normally distributed and have infinite support. In order to satisfy condition (C1), it is thus required that the distributions of $\boldsymbol{\varepsilon}_t^{\mathrm{o}}$ and $\boldsymbol{\varepsilon}_t^{\mathrm{r}}$ are truncated outside some regular simplices which are large enough to contain most of the probability mass. For simplicity, consider the case of a one-dimensional standard normal distribution (s.n.d.). A simplex that contains a given percentage p of the mass of the s.n.d. reduces to the interval $[-\delta, \delta]$, where $\delta = \Phi(1 - \frac{1-p}{2})$ and Φ denotes the standard normal distribution function. In the two-dimensional case, the corresponding simplex is an equilateral triangle with an inner circle of radius δ which represents a δ-confidence region for the s.n.d. in $\mathbb{R}^2$. The vertices of this triangle may be chosen, e.g., as $u_0 = (-\sqrt{3}\delta, \delta)$, $u_1 = (\sqrt{3}\delta, \delta)$ and $u_2 = (0, -2\delta)$. Simplicial coverages in higher dimensions are discussed in Frauendorfer and Härtel (1995).

To determine the simplicial support of the stochastic factor $\boldsymbol{\varepsilon}_t^{\mathrm{o}}$ influencing the objective, we transform the vertices of a simplex constructed for an uncorrelated two-dimensional s.n.d. according to $\mu_t^{\mathrm{o}} + \Gamma_t\, u_i$, $i \in \{0, 1, 2\}$, where $\mu_t^{\mathrm{o}} = (a_s\theta_s\Delta t,\, a_l\theta_l\Delta t)$ and Γ_t is the Cholesky transformation of the covariance matrix Σ_t^{o} in (5.20). Analogously, the simplicial support of the risk factor $\boldsymbol{\varepsilon}_t^{\mathrm{r}}$, that controls the coefficients on the rhs, is given by the interval $[a_v - \delta\sigma_\varepsilon, a_v + \delta\sigma_\varepsilon]$. For the calculations presented below, we assume $\delta = 2$ so that any outcome within a range of at least two standard deviations around the expectation will be taken into account.

Consequently, more extremal events than $a_v \pm \delta\sigma_\varepsilon$ are ignored. This is consistent with the fact that in reality the decision maker would accept that with a certain (sufficiently small) probability a drop in volume cannot always be compensated by maturing instruments. Otherwise, he or she had to invest the complete amount into the shortest maturity and, thus, give away potential return which is not consistent with the usual practice of investing balance sheet items without contractual maturity. In this spirit, a specific choice of δ reflects the decision maker's tolerance toward liquidity risk.

Condition (C2) is satisfied since the objective function coefficients are linear in η according to (5.17) and (5.21). The rhs functions of the constraints in the optimization problem (5.18) are constant or convex in ξ and, a fortiori, d.c. Moreover, all constraint matrices are deterministic, and the feasible sets of the dynamic version of (5.18) are uniformly compact in each stage due to the bounds on the transaction volumes. This implies that condition (C3) is satisfied. It can easily be seen from (5.23) that $(\boldsymbol{\eta}, \boldsymbol{\xi})$ follows a block-diagonal autoregressive process if $\gamma = 0$, i.e., the conditional covariance matrix of $\boldsymbol{\eta}_{t+1}$ is independent of the current level of the risk factor $\boldsymbol{\eta}_t$. Regularity condition (C4) is thus fulfilled.

Condition (C5) requires that the stochastic program has relatively complete recourse. Extreme fluctuations in the $\boldsymbol{\xi}$ process might in principle lead to a situation where the volume constraint cannot hold in combination with tight transaction limits. However, this situation does not occur for reasonable parameter values which imply that the support of $\boldsymbol{\xi}$ is relatively small (partly because of the inherent "mean reversion property"; this might be different for nonstationary processes). Due to the possibility of borrowing money on a short-term basis, we can thus take condition (C5) for granted. Note also that a possible remedy to achieve relatively complete recourse might be the selection of a smaller parameter δ for the truncation of the original supports.

Since the right-hand side of the third equation in (5.18) is nonlinear in ξ_t, we must determine correction terms of the form (5.22) for each stage in order to employ Theorem 5.2. To this end, we study the dynamic version of (5.18). Considering the stage t subproblem, we must find an upper bound $D^{*+}_{\text{vol},t}$ for the dual variables associated with the constraint

$$\textstyle\sum_{d\in\mathcal{D}} x^d_{\text{pos},t} = \exp(\xi^t) \tag{5.24}$$

uniformly over all outcome and decision histories in Z^t. This dual variable can be seen as the *shadow price* of the volume in the account. A raise in the volume v_t by \$1 will increase at least one of the portfolio positions $x^1_{\text{pos},t}, \ldots, x^D_{\text{pos},t}$ which can only be achieved by increasing also some of the investments $x^d_{\text{inv},t}$, $d \in \mathcal{D}$. Therefore, the maximum gain in t per additional unit of currency is

$$D^{*+}_{\text{vol},t} = \max\{\varphi^d_t(\eta^t) \,|\, d \in \mathcal{D},\ \eta^t \in \Theta^t\}. \tag{5.25}$$

This quantity is finite since φ^d_t is linear in η^t and the domain Θ^t is bounded. Note that any increase in the volume at stage t will affect only the objective function coefficients at this stage because the additional money is invested immediately. We can thus determine the upper bounds in (5.25) from the maximum values of $\eta_{i,t}$ over Θ^t, $i \in \{s, l\}$, which are obtained by recursively using (5.19). Thereby, we assume that $\gamma = 0$ and $1 - a_i \Delta t \geq 0$:

$$\max \eta_{i,t} = a_i \theta_i \Delta t + (1 - a_i \Delta t) \max \eta_{i,t-1} + \max \hat{\varepsilon}^{\mathrm{o}}_{i,t}, \quad i \in \{s, l\}, \quad t > 1.$$

An obvious choice for the d.c. decomposition of the right-hand side in (5.24) is $\kappa_t^+(\xi^t) = \exp(\xi_t)$ and $\kappa_t^-(\xi^t) = 0$. The correction term (5.13b) then becomes

$$\alpha_t^{\mathrm{r}}(\xi^t) = -D^{*+}_{\mathrm{vol},t} \exp(\xi^t), \qquad t = 1, \ldots, T.$$

Because $\alpha_t^{\mathrm{o}}(\eta^t) = 0$ for all stages (no corrections of nonconvexities in the objective are required), this equals also the combined correction term $\alpha_t(\eta^t, \xi^t)$. Note that we do not have to estimate a lower bound $D^{*-}_{\mathrm{vol},t}$ in (5.13b) due to $\kappa_t^- = 0$. Based on this information, we can now evaluate the conditional correction terms defined in (5.15). As mentioned earlier, $A^l_{J,t}$ and $A^u_{J,t}$ are calculated by direct summation, while A_t must be evaluated either analytically or via numerical integration techniques. Notice that the recourse functions as well as the conditional correction terms of the first stage have to be evaluated only at one point, since $\boldsymbol{x}_0$, $\boldsymbol{\eta}_1$, and $\boldsymbol{\xi}_1$ are deterministic. As a consequence, we will suppress the arguments x_0, η_1, and ξ_1 below. In the simplified example presented here, we calculate A_1 approximately with respect to the unrestricted normal distribution and ignore the truncation of the state space. To this end, ξ_t is rewritten for each $t > 1$ as a linear combination of the noise terms, i.e.,

$$\xi_t = H^{\mathrm{r}}_{1,t} \xi_1 + \sum_{\tau=1}^{t} H^{\mathrm{r}}_{\tau+1,t} \varepsilon^{\mathrm{r}}_{\tau}, \quad \text{where} \quad H^{\mathrm{r}}_{\tau,t} = \begin{cases} \prod_{\tau'=\tau}^{t} H^{\mathrm{r}}_{\tau'} & \text{for } \tau \leq t, \\ 1 & \text{otherwise.} \end{cases}$$

Then, the conditional correction term A_1 becomes

$$\begin{aligned} A_1 &= -\sum_{t=1}^{T} D^{*+}_{\mathrm{vol},t} E[\exp(\xi_t)] \\ &\approx -\sum_{t=1}^{T} D^{*+}_{\mathrm{vol},t} \exp\left[H^{\mathrm{r}}_{1,t} \xi_1 + \sum_{\tau=1}^{t} \left(H^{\mathrm{r}}_{\tau+1,t} a_v + \frac{1}{2} (H^{\mathrm{r}}_{\tau+1,t} \sigma_\varepsilon)^2 \right) \right]. \end{aligned}$$

According to Theorem 5.2, the bounds on the optimal value Φ_1 of the multistage stochastic program (5.18) consist in the optimal values $\Phi^l_{J,1}$, $\Phi^u_{J,1}$ of the discretized auxiliary stochastic programs (5.12a) and (5.12b) shifted by combinations of the conditional correction terms A_1, $A^l_{J,1}$, and $A^u_{J,1}$ for the current refinement parameter $J \in \mathbb{N}_0$.

Remark We emphasized above that $\gamma > 0$ in (5.19) allows to reflect heteroscedasticity of interest rates. However, this choice is not consistent with Definition 5.3.2. It is argued in Frauendorfer and Schürle (2005) that the saddle property of the recourse functions is sometimes given also for $\gamma = 1$, since this specification still entails a linear dependency of η_t on its history of observations. In order to cope with other parameter values[10] $\gamma > 0$, one has to represent the risk factors as nonlinear combinations of some serially independent disturbances. Considering these disturbances as the fundamental data process, and packing all nonlinearities into the definition of the objective function coefficients, our bounding technique remains applicable. However, this approach involves a completely different set of correction terms and will not be further pursued in the present chapter.

5.6.4 Numerical Solution

We solved a four-stage problem with quarterly planning, investment opportunities in the maturities 3M, 6M, 1Y, 2Y, 3Y, 4Y, and 5Y, and parameters estimated from real data. The barycentric measures were refined by successively partitioning the simplicial supports $\mathcal{E}_t^{\mathrm{o}}$ and $\mathcal{E}_t^{\mathrm{r}}$ of the disturbances $\boldsymbol{\varepsilon}_t^{\mathrm{o}}$ and $\boldsymbol{\varepsilon}_t^{\mathrm{r}}$, respectively. Starting with an initial trivial partition $J = 0$ that consists of the product $\mathcal{E}_t^{\mathrm{o}} \times \mathcal{E}_t^{\mathrm{r}}$ for all stages $t = 1, \ldots, T$ we obtained (uncorrected) objective function values of $\Phi_{0,1}^l = 1011.5718$ and $\Phi_{0,1}^u = 1221.7353$ (recall that the first index represents the refinement parameter J). Taking into account the conditional correction terms $A_{0,1}^l - A_1$ and $A_{0,1}^u - A_1$, the bounds become 1011.5712 and 1221.7353, respectively. This illustrates that the magnitude of the corrections is relatively small, a result which should not be generalized to other (more complex) problems where corrections apply also to the objective, or the bounding vectors in (5.12a) and (5.12b) turn out to be very large. On the other hand, the difference between the two bounds itself is relatively large here. Furthermore, the corresponding first-stage decisions $x_{\mathrm{opt},0,1}^l$ and $x_{\mathrm{opt},0,1}^u$ do not coincide. The solution of the lower bounding problem recommends to invest the whole amount available at $t = 1$ in the longest maturity, while the suggestion of the upper bounding problem is to invest in 3M, and thus the decision maker does not obtain a unique solution.

As outlined in Section 5.4, the accuracy of the approximation can be improved by increasing the number of components in the existing partitions. This can be achieved by splitting a $\mathcal{E}_t^{\mathrm{o}}$- or a $\mathcal{E}_t^{\mathrm{r}}$-simplex (or both) at some stage $t = 1, \ldots, T$. Here we restrict the procedure to splits of the $\mathcal{E}_t^{\mathrm{o}}$-simplices only since numerical experiments have shown that this leads to the largest improvements during the first refinement steps. For our numerical example, we implemented the following refinement procedure:

(1) Fix an initial partition with N simplices and set $J = 0$.
(2) Let $\mathcal{E}_{J,t}^{\mathrm{o}(i)}$, $i = 1, \ldots, N + J$, be simplices in the current partition that cover the support of the distribution of $\boldsymbol{\varepsilon}_t^{\mathrm{o}}$, $t = 1, \ldots, T$. Solve the corresponding lower and upper bounding problems.

[10] For instance, the well-known Cox, Ingersoll, and Ross model (Cox et al. 1985) involves a "square root process" with $\gamma = 0.5$.

(3) If the difference between the objective function values after correction (or some other measure of accuracy) is sufficiently low, then terminate.
(4) Otherwise, among $\mathcal{E}^{o(1)}_{J,t}, \ldots, \mathcal{E}^{o(J+N)}_{J,t}$ determine the simplex with the longest edge and split the simplex at the midpoint of this edge. Replace it in the existing partition by the resulting (sub-) simplices.
(5) Set $J := J + 1$ and goto step (2).

It can be helpful for the determination of the splitting edge in step (4) to weight the lengths of all edges by the probabilities of the corresponding simplices. This is motivated by the fact that the partition of a subcell with low probability mass will not improve the accuracy. Note that the index J is equivalent to the number of refinements.

In this way, for $N = 1$ and with one refinement (i.e., $J = 1$) we obtain objective function values of $\Phi^l_{1,1} = 1027.9749$ and $\Phi^u_{1,1} = 1136.5517$ that become 1027.9743 and 1136.5520, respectively, after correction. The relative difference between the upper and lower bound is only half as large as for the initial calculation with $J = 0$. More importantly, the decision vectors $x^l_{\text{opt},1,1}$ and $x^u_{\text{opt},1,1}$ now coincide, i.e., the first-stage decision of the upper bounding problem switched also to the investment in the longest maturity. Thus, the decision maker obtains a unique solution at a sufficiently high accuracy, which was our original intention.

Further refinement steps (i.e., $J = 2, 3, \ldots$) improve the accuracy only marginally, while the corresponding auxiliary optimization problems, that approximate the original multistage stochastic program, become intractably large: This can immediately be attributed to the fact that, as outlined in Section 5.4, the approximation is based on the barycenters and vertices of the simplices in the partition. The size of the auxiliary problems thus may become too large as J increases. Numerical experiments have shown that the underlying value functions exhibit a relatively high degree of convexity with respect to η_t. Loosely speaking, the vertices of a (single) initial simplex $\mathcal{E}^o_t$ covering the support of the truncated normal distribution of $\boldsymbol{\varepsilon}^o_t$ have geometrically "too extreme" coordinates. Since a split of a simplex generates only one new point (the midpoint of an edge of one simplex in the given partition) but the existing vertices remain, the influence of extreme outcomes in the initial discrete approximation decreases only slowly.

To achieve tighter bounds with fewer refinement steps, we start from the consideration that the circle that covers the truncated support of a two-dimensional standard normal distribution can better be approximated by a polygon with a higher number of vertices. As an example, we consider coverages by tetragons and pentagons that are themselves partitioned into two or three simplices. Their vertices are transformed according to the expectations and covariance matrices of the actual distributions analogously to the procedure described in Section 5.6.3. The shapes of the resulting partitions are illustrated in Fig. 5.1. Then, barycentric measures are derived for each simplex individually.

Results for our example investment problem are shown in Table 5.1 for initial partitions that consist of triangles ($N = 1$), tetragons ($N = 2$), and pentagons ($N = 3$). It can be seen from the last column labeled Acc. that the unrefined problems in the latter two cases already provide an accuracy which may be achieved otherwise only

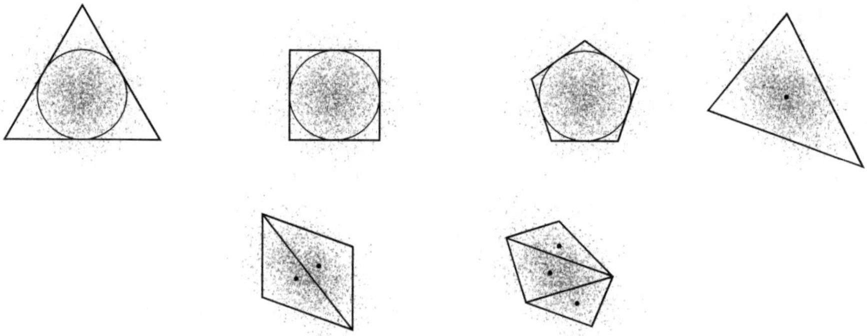

Fig. 5.1 Coverages for the uncorrelated and correlated case with barycenters

Table 5.1 Objective function values of the lower and upper auxiliary stochastic programs before and after correction for nonlinearities

N	J	$\Phi^l_{J,1}$	Corr.	LB	$\Phi^u_{J,1}$	Corr.	UB	Acc. (%)
1	0	1011.5718	–5.45	1011.5712	1221.7353	2.89	1221.7356	18.82
	1	1027.9749	–5.95	1027.9743	1136.5517	2.89	1136.5520	10.03
	2	1029.1618	–6.43	1029.1612	1128.2428	2.89	1128.2431	9.19
	3	1029.2132	–6.60	1029.2125	1127.8974	2.89	1127.8976	9.15
	4	1029.2132	–6.64	1029.2126	1127.8856	2.89	1127.8858	9.15
2	0	1034.2863	–5.41	1034.2858	1067.7401	2.17	1067.7403	3.18
	1	1037.0787	–5.89	1037.0781	1057.9300	2.17	1057.9302	1.99
	2	1037.2189	–6.19	1037.2183	1057.7964	2.17	1057.7966	1.96
	3	1037.3000	–6.31	1037.2994	1057.7764	2.17	1057.7766	1.95
	4	1037.3063	–6.38	1037.3057	1057.7740	2.17	1057.7742	1.95
3	0	1011.5534	–4.75	1011.5529	1126.2541	2.30	1126.2544	10.73
	1	1027.5144	–5.24	1027.5138	1087.5954	2.30	1087.5956	5.68
	2	1028.9230	–5.65	1028.9225	1086.2739	2.30	1086.2741	5.42
	3	1028.9583	–5.82	1028.9577	1086.1394	2.30	1086.1396	5.41
	4	1028.9589	–5.89	1028.9583	1086.1340	2.30	1086.1342	5.41

N denotes the number of simplices in the initial partition that covers the support of $\boldsymbol{\varepsilon}_t^{o}$ in each stage $t > 1$, while the index J is equivalent to the number of refinements. The column corr. contains the conditional correction terms $A^l_{J,1} - A_1$ for the lower and $A^u_{J,1} - A_1$ for the upper bounding problems multiplied by 10^4. LB and UB are the corresponding objective function values $\Phi^l_{J,1} + A^l_{J,1} - A_0$ and $\Phi^u_{J,1} + A^u_{J,1} - A_0$, respectively, after correction. The accuracy acc. of the approximation is defined as $\frac{\text{UB}-\text{LB}}{0.5(\text{LB}+\text{UB})}$.

after some refinement steps. Moreover, the first-stage decisions coincide already for $J = 0$. A suitable selection of the initial partitions is therefore of utmost importance not only to achieve a sufficient accuracy of the approximation but also to keep the overall numerical efforts moderate since each additional refinement step requires the solution of a corresponding (large-scale) auxiliary optimization problem.

5.7 Conclusions

This chapter addresses the approximation of linear multistage stochastic programs. Thereby, the original optimization problem is approximated by a lower (upper) auxiliary stochastic program that arises by substituting the true distribution of the data process by a lower (upper) discrete barycentric measure. After a suitable transformation offsetting possible nonconvexities in the random parameters, the optimal value of the lower (upper) auxiliary stochastic program provides a lower (upper) bound on the optimal value of the original stochastic program. More generally, we are able to derive arbitrarily tight bounds on the recourse functions as well as arbitrarily tight bounding sets for the optimal decisions associated with the original optimization problem.

Applicability of the presented bounding methods relies on a set of regularity conditions requiring that the feasible sets and state spaces are compact, the constraint matrices are deterministic, the objective function coefficients and rhs vectors are representable as differences of convex functions, and the random data is governed by a block-diagonal autoregressive process. Moreover, relatively complete recourse is stipulated. Focussing on block-diagonal autoregressive processes is not as restrictive as it might seem, since any absolutely continuous process is representable as an (inverse) Rosenblatt transformation of a serially independent noise process (Rosenblatt 1952). Considering this noise process as the fundamental data process, and absorbing the nonlinear Rosenblatt transformation in the definition of the objective function coefficients and rhs vectors, one usually obtains a regular reformulation of the stochastic program satisfying all conditions (C1)–(C5).

The theoretical concepts are tested on a real-life decision problem calibrated to observed data. In this example, which is slightly simplified for didactic reasons, the (conditional) correction terms hardly affect the bounds on the optimal objective value. The reason for this is that the curvature of the exponential function is small on the support of the ξ process. Of course, one could think of problems with larger correction terms. It is possible, for instance, to construct models with highly nonconvex objective function coefficients or rhs vectors such that $\Phi^u_{J,t}$ occasionally drops below $\Phi^l_{J,t}$, especially if the barycentric measures are poorly refined. In this case, the conditional correction terms will substantially impact the bounds.

The results presented in Section 5.6 illustrate that the numerical efforts required for the solution of a successively refined multistage stochastic program increase significantly as the partition size $N+J$ grows. For the problem under consideration, the size of the corresponding extensive form with refinement parameter $J = 4$ is 125, 27, or 13 times larger than for the initial state with $J = 0$, depending on the coverage setting $N = 1, 2,$ or 3, respectively. For this reason, we could solve problems with a relatively small number of stages ($T = 4$) only.

Despite the increasing problem size, the gain in accuracy soon becomes marginal as J advances. This must be attributed to the fact that we applied a rather crude refinement strategy in our example, where the simplices covering the supports of the (conditional) distributions at each stage were always split at their midpoints. A more sophisticated strategy would analyze the differences between the upper and lower bounds for each node of the scenario trees that are generated as discrete approximations of the underlying stochastic processes. Then, scenarios will only be added by a split of a simplex in the existing partition if the approximation error, which may be expressed by the difference between the bounds, is large for a certain node. On the other hand, an approximation error close to zero means that the approximation is already exact, and thus further refinements will not improve the accuracy.

For an efficient implementation of such refinement strategies, the following aspects must be considered:

(1) In which node should the scenario tree be refined (i.e., what is the threshold error above which an existing nodal partition is refined; this will have an immediate impact on the number of scenarios and the problem size, see Fig. 5.2a–c)?
(2) Does splitting $\mathcal{E}_t^{\mathrm{o}}$ or $\mathcal{E}_t^{\mathrm{r}}$ provide a higher accuracy, see Fig. 5.2d?
(3) Which edge of the simplex should be split, see Fig. 5.2e and
(4) Where should this edge be split, see Fig. 5.2f?

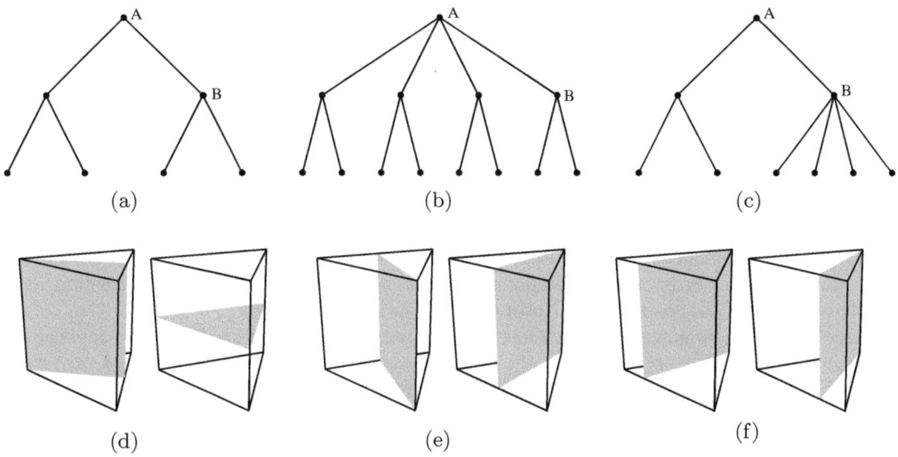

Fig. 5.2 Possible refinements of a scenario tree and simplices. (a) No refinements; (b) Refinement in A; (c) Refinement in B; (d) Split of $\mathcal{E}_t^{\mathrm{o}}$ or $\mathcal{E}_t^{\mathrm{r}}$; (e) Alternative edges; (f) Alternative points

Adding new scenarios only where profitable will keep the increase in problem size relatively moderate, which in turn allows the solution of problems with a higher number of stages (see Edirisinghe and You 1996 or Frauendorfer and Marohn 1998 for a detailed discussion of refinement techniques and their implementation). While each refinement involves the solution of successively increasing large-scale linear programs, a significant speedup can be achieved by an algorithmic technique described in Frauendorfer and Haarbrücker (2003) where the solution of the LP in refinement $J + 1$ is based on the solutions of the previous J steps. This allows us to manage numerous refinement steps (up to several hundreds) successfully with adequate time exposure.

Acknowledgments Daniel Kuhn thanks the Swiss National Science Foundation for financial support.

References

Ash, R.B.: Real Analysis and Probability. Probability and Mathematical Statistics. Academic, Berlin (1972)

Attouch, H., Wets, R.J.-B.: Approximation and convergence in nonlinear optimization. In: Mangasarian, O., Meyer, R., Robinson, S.M. (eds.) Nonlinear Programming 4, pp. 367–394. Academic, New York, NY (1981)

Beale, E.M.L.: On minimizing a convex function subject to linear inequalities. J. R. Stat. Soc. **17**B, 173–184 (1955)

Birge, J.R., Dulá, J.H.: Bounding separable recourse functions with limited distribution information. *Ann. Oper. Res.* 30, 277–298 (1991)

Birge, J.R., Louveaux, F.: Introduction to Stochastic Programming. Springer, New York, NY (1997)

Birge, J.R., Wets, R.J.-B.: Designing approximation schemes for stochastic optimization problems, in particular for stochastic programs with recourse. Math. Program. Study **27**, 54–102 (1986)

Birge, J.R., Wets, R.J.-B.: Computing bounds for stochastic programming problems by means of a generalized moment problem. Math. Oper. Res. **12**, 149–162 (1987)

Cox, J.C., Ingersoll, J.E., Ross, S.A.: A Theory of the Term Structure of Interest Rates. Econometrica **53**(2), 385–407 (1985)

Dantzig, G.B.: Linear programming under uncertainty. Manage. Sci. **1**, 197–206 (1955)

Dantzig, G.B., Thapa, M.N.: Linear Programming 1: Introduction. Springer, New York, NY (1997)

Dokov, S.P., Morton, D.P.: Higher-order upper bounds on the expectation of a convex function. The Stochastic Programming E-Print Series (SPEPS), Jan (2002)

Dokov, S.P., Morton, D.P.: Second-order lower bounds on the expectation of a convex function. Math. Oper. Res. **30**(3), 662–677 (2005) ISSN 0364-765X

Dupačová, J.: Minimax stochastic programs with nonconvex nonseparable penalty functions. In: Prékopa, A. (ed.) Progress in Operations Research, volume 1 of Colloquia Mathematica Societatis János Bolyai, pp. 303–316. North-Holland, Amsterdam (1976)

Dupačová, J.: The minimax approach to stochastic programming and an illustrative application. Stochastics **20**, 73–88 (1987)

Dupačová, J.: Applications of stochastic programming under incomplete information. Comput. Appl. Math. **56**, 113–125 (1994)

Dupačová, (as Žáčková), J.: On minimax solutions of stochastic linear programming problems. Časopis pro Pěstování Matematiky **91**, 423–429 (1966)

Edirisinghe, C.: New second-order bounds on the expectation of saddle functions with applications to stochastic linear programming. Oper. Res. **44**, 909–922 (1996)

Edirisinghe, C., You, G.-M.: Second-order scenario approximation and refinement in optimization under uncertainty. Ann. Oper. Res. **64**, 143–178 (1996)

Edirisinghe, C., Ziemba, W.T.: Bounds for two-stage stochastic programs with fixed recourse. Math. Oper. Res. **19**, 292–313 (1994a)

Edirisinghe, C., Ziemba, W.T.: Bounding the expectation of a saddle function with application to stochastic programming. Math. Oper. Res. **19**, 314–340 (1994b)

Edmundson, H.P.: Bounds on the expectation of a convex function of a random variable. Paper 982, The Rand Corporation, Santa Monica, CA (1956)

Floudas, C.A.: Deterministic Global Optimization. Kluwer, Dordrecht (2000)

Forrest, B., Frauendorfer, K., Schürle, M.: A stochastic optimization model for the investment of savings account deposits. In: Kischka, P., Lorenz, H.-W., Derigs, U., Domschke, W., Kleinschmidt, P., Mhring, R. (eds.) Operations Research Proceedings 1997, pp. 382–387. Springer, New York, NY (1998)

Frauendorfer, K.: Solving SLP recourse problems with arbitrary multivariate distributions — the dependent case. Math. Oper. Res. **13**, 377–394 (1988)

Frauendorfer, K.: Stochastic two-stage programming. Lecture Notes in Economics and Mathematical Systems, vol. 392. Springer, Berlin (1992)

Frauendorfer, K.: Multistage stochastic programming: Error analysis for the convex case. Z. Oper. Res. **39**(1), 93–122 (1994)

Frauendorfer, K.: Barycentric scenario trees in convex multistage stochastic programming. Math. Program. **75**(2), 277–294 (1996)

Frauendorfer, K., Haarbrücker, G.: Solving sequences of refined multistage stochastic linear programs. Ann. Oper. Res. **124**, 133–163 (2003)

Frauendorfer, K., Kall, P.: A solution method for SLP recourse problems with arbitrary multivariate distributions — the independent case. Probl. Contr. Inform. Theor. **17**, 177–205 (1988)

Frauendorfer, K., Marohn, C., Marti, K., Kall, P.: Refinement issues in stochastic multistage linear programm- ing. In: Stochastic Programming Methods and Technical Applications (Proceedings of the 3rd GAMM/IFIP Workshop 1996). Lecture Notes in Economics and Mathematical Systems, vol. 458, pp. 305–328. Springer, New York, NY (1998)

Frauendorfer, K., Schürle, M.: Term structure models in multistage stochastic programming: Estimation and approximation. Ann. Oper. Res. **100**, 189–209 (2001)

Frauendorfer, K., Schürle, M.: Management of non-maturing deposits by multistage stochastic programming. Eur. J. Oper. Res. **151**, 602–616 (2003)

Frauendorfer, K., Schürle, M.: Refinancing mortgages in Switzerland. In: Wallace, S.W., Ziemba, W.T. (eds.) Applications of Stochastic Programming, MPS-SIAM Series in Optimization, pp. 445–469, Philadelphia (2005)

Frauendorfer, K., Härtel, F.: On the goodness of discretizing diffusion processes for stochastic programming. Working paper, Institute of Operations Research, University of St. Gallen, St. Gallen (1995)

Gassmann, H., Ziemba, W.T.: A tight upper bound for the expectation of a convex function of a multivariate random variable. Math. Program. Study **27**, 39–53 (1986)

Hartman, P.: On functions representable as a difference of convex functions. Pac. J. Math. **9**, 707–713 (1959)

Hiriart-Urruty, J.-B., Ponstain, J.: Generalized differentiability, duality and optimization for problems dealing with differences of convex functions. In: Convexity and Duality in Optimization (Groningen, 1984). Lecture Notes in Economics and Mathematical Systems, vol. 256, pp. 37–70. Springer, New York (1985)

Horst, R., Tuy, H.: Global Optimization: Deterministic Approaches 3 ed. Springer, Berlin (1996)

Huang, C.C., Ziemba, W.T., Ben-Tal, A.: Bounds on the expectation of a convex function of a random variable: With applications to stochastic programming. Oper Res **25**, 315–325 (1977)

Jensen, J.L.: Sur les fonctions convexes et les inégalités entre les valeurs moyennes. Acta Math. **30**, 157–193 (1906)

Kall, P.: Stochastic programming with recourse: upper bounds and moment problems — a review. In: Guddat, J., Bank, B., Hollatz, H., Kall, P., Klatte, D., Kummer, B., Lommatzch, K., Tammer, K., Vlach, M., Zimmermann, K. (eds.) Advances in Mathematical Optimization, pp. 86–103. Akademie Verlag, Berlin, Germany (1988)

Kall, P.: An upper bound for SLP using first and total second moments. Ann. Oper. Res. **30**, 267–276 (1991)

Kuhn, D.: Generalized Bounds for Convex Multistage Stochastic Programs. Lecture Notes in Economics and Mathematical Systems, vol. 548. Springer, Berlin (2004)

Madansky, A.: Bounds on the expectation of a convex function of a multivariate random variable. Ann. Math. Stat. **30**, 743–746 (1959)

Madansky, A.: Inequalities for stochastic linear programming problems. Manage. Sci. **6**, 197–204 (1960)

Pardalos, P.M., Romeijn, H.E., Tuy, H.: Recent development and trends in global optimization. Comput. Appl. Math. **124**, 209–228 (2000)

Rockafellar, R.T., Wets, R.J.-B.: Nonanticipativity and L^1-martingales in stochastic optimization problems. In: Stochastic Systems: Modeling, Identification and Optimization II, Mathematical Programming Study. North-Holland, Amsterdam **6**, pp. 170–187 (1976a)
Rockafellar, R.T., Wets, R.J.-B.: Stochastic convex programming: Relatively complete recourse and induced feasibility. *SIAM J. Control Optim.* **14**(3), 574–589 (1976b)
Rockafellar, R.T., Wets, R.J.-B.: Variational Analysis. A Series of Comprehensive Studies in Mathematics vol. 317. Springer, New York, NY (1998)
Rosenblatt, M.: Remarks on a multivariate transformation. Ann. Math. Stat. **23**(3), 470–472 (1952)
R.J.-B. Wets. Induced constraints for stochastic optimization problems. In: Balakrishnan, A. (ed.) Techniques of Optimization, pp. 433–443. Academic New York, NY (1972)

Chapter 6
Stochastic Programming Approximations Using Limited Moment Information, with Application to Asset Allocation

N. Chanaka P. Edirisinghe

6.1 Introduction

In many real-world decision-making problems, given a set of possible alternative actions available at the present time, one chooses a decision in the face of uncertainty of the future states of nature. For instance, in an asset allocation problem for financial planning, a portfolio of assets must be determined under future uncertain asset returns. Upon observing the realized asset returns, the portfolio manager may revise (or rebalance) her portfolio in order to control the expected deviation from a wealth target. In this context, the portfolio manager wishes to pick an initial portfolio that would hedge well against all possible realizations of random asset returns with respect to the chosen objective criterion. Stochastic programming with recourse is an ideal modeling tool for such decision problems.

The main focus here is on problems with two decision stages. Consider the two-stage stochastic linear programming (SLP) problem with fixed recourse, e.g., Kall (1976) and Wets (1982), given by

$$\begin{aligned} Z^* := \min_x \quad & c'x + \psi(x) \\ \text{s.t.} \quad & Ax = b \\ & x \geq 0, \end{aligned} \tag{6.1}$$

where the second-stage expected recourse function is $\psi(x) = E_{P^{\text{tr}}}[\phi(x,\omega)]$ and

$$\begin{aligned} \phi(x,\omega) \equiv \phi(x,\xi,\eta) := \min_y \quad & q(\eta)'y \\ \text{s.t.} \quad & Wy = h(\xi) - T(\xi)x \\ & y \geq 0. \end{aligned} \tag{6.2}$$

N.C.P. Edirisinghe (✉)
Department of Statistics, Operations, and Management Science, College of Business Administration, University of Tennessee, Knoxville, TN 37996, USA
e-mail: chanaka@utk.edu

G. Infanger (ed.), *Stochastic Programming*, International Series in Operations Research & Management Science 150, DOI 10.1007/978-1-4419-1642-6_6,

Primes ($'$) denote the transposition of vectors. The domain of the random $(K+L)$-vector $\omega := (\xi, \eta)$ is denoted by $\Omega(\subset \Re^{K+L})$; marginal support of ξ is $\Xi(\subset \Re^K)$ and that of η is $\Theta(\subset \Re^L)$. A is a deterministic $(m_1 \times n_1)$-matrix and c is a deterministic vector in $\Re^{n_1}$. Stochasticity is present in the technology matrix $T : \Xi \to \Re^{m_2 \times n_1}$, right-hand side $h : \Xi \to \Re^{m_2}$, and objective costs $q : \Theta \to \Re^{n_2}$. $E[.] \equiv E_{P^{\mathrm{tr}}}[.]$ represents the mathematical expectation with respect to the true probability measure P^{tr} on Ω. The matrix W $(\in \Re^{m_2 \times n_2})$ is deterministic, hence the term *fixed* recourse for (6.1) and (6.2). The separation of the underlying uncertainty ω into two separate random vectors ξ and η is often motivated by the particular application; for instance, in an asset allocation problem, asset returns (ξ) and the costs (η) of funding any portfolio shortfall from a target may be separated.

In the sequel, we assume that (6.1) and (6.2) have *complete recourse*, i.e.,

$$\{z \in \Re^{m_2} \ : \ Wy = z, \ y \geq 0\} \ = \ \Re^{m_2}, \tag{6.3}$$

which also implies that the recourse problem (6.2) is feasible w.p. 1 for any $x \in X$, where

$$X := \{x \in \Re^{n_1} \ : \ Ax = b, \ x \geq 0\}, \tag{6.4}$$

and X is assumed to be nonempty. Moreover, assuming that $\pi W \leq q(\eta)$ is feasible w.p. 1, ϕ is obtained as a proper convex function in $h(\xi) - T(\xi)x$, for $x \in X$, as well as a proper concave function in $q(\eta)$. It is assumed that ϕ is finite on $\Xi \times \Theta$. Suppose q, h, and T are affine transformations of the respective random vectors, i.e.,

$$T(\xi) = T_0 + \sum_{k=1}^{K} T_k \xi_k \ , \quad h(\xi) = H\xi + h_0 \ , \quad \text{and} \quad q(\eta) = Q\eta + q_0 \ , \tag{6.5}$$

where $T_0, T_k \in \Re^{m_2 \times n_1}$, $H \in \Re^{m_2 \times K}$, and $Q \in \Re^{n_2 \times L}$ are fixed matrices while $h_0 \in \Re^{m_2}$ and $q_0 \in \Re^{n_2}$ are fixed vectors. Thus, $\phi(x, ., .)$ is a proper convex–concave saddle function for each $x \in X$. From the linear programming theory, $\phi(x, \xi, \eta)$ is polyhedral in (ξ, η) for fixed $x \in X$, thus ensuring continuity of ϕ in ω for given $x \in X$.

Properties of the program (6.1) and (6.2) are well known, such as convexity of the set X; see Kall and Wallace (1994) and Wets (1974). Since integration with respect to a probability measure preserves order, convexity of ϕ in x implies that $\psi(.)$ is convex as well, and thus (6.1) is a convex program. However, in the case of continuous distributions, direct application of convex programming methods is difficult (except possibly when W has special structure such as simple recourse; see Ziemba 1970) due to the multidimensional integration required for evaluating $\psi(x)$, of which the integrand is the optimization in (6.2). Such procedures become even more prohibitive computationally as one would need to evaluate gradients as well.

In a sampling-based approach, one iteratively draws random samples from the underlying probability distribution for computing stochastic quasi-gradients (Ermoliev 1983, Ermoliev and Gaivoronsky 1988) or for stochastic decomposition (Higle and Sen 1991), procedures that enjoy asymptotic convergence properties. Noting that the variance of the sample is key to the convergence process, various variance reduction schemes, such as importance sampling within an optimization process (Dantzig and Glynn 1990, Infanger 1992, 1994), have been employed for better algorithmic performance.

When the random parameters are described by a set of discrete outcomes (i.e., scenarios), which may possibly be very large in number, (6.1) may be written in the so-called extensive deterministic equivalent grand-LP form; however, its solution would still be difficult due to its very large size. Mathematical decomposition techniques may be applied for solution, such as the L-shaped decomposition (Van Slyke and Wets 1969) and its many variants, that employ efficient computational schemes, e.g., Birge (1988), Edirisinghe and Patterson (2007), and Gassmann (1990).

In contrast to the above approaches, this chapter focuses on a solution strategy that relies on a successive approximation procedure, based on computable bounds on the optimal objective value. See Birge and Wets (1986), Ermoliev and Wets (1988), Huang et al. (1977), and Kall and Stoyan (1988) for general details on such schemes. These methods advocate determining lower and upper bounding functions on the expected recourse function $\psi(x)$, which are then used in the outer minimization.

6.1.1 Bound-Based Approximations

The topic of bounding two-stage SLP has received significant attention in the past two decades. Two main types of bounding strategies have been employed: those based on functional approximations (Birge and Wallace 1988, Birge and Wets 1989, Morton and Wood 1999, Wallace 1987) and those based on distributional approximations of the underlying (true) probability measure (Dula 1992, Dupačová 1966, 1976, Edirisinghe and Ziemba 1992, 1994a, b, Frauendorfer 1988b, 1992, Gassmann and Ziemba 1986, Kall 1991). Generally, functional approximations are nonmoment based and they may require performing univariate integration w.r.t. the underlying distribution. On the other hand, distributional approximations are typically moment based and they may correspond to certain generalized moment problems (GMP); see Kemperman (1968). In some cases, the difference between an approximation being functional or distributional is only interpretational; e.g., Jensen's bound (Jensen 1906) or Edmundson–Madansky bound (Madansky 1959) can be interpreted either as functional approximations of the value function or as distributional approximations of the underlying probability measure.

When using bound-based approximations to solve (6.1) and (6.2), generally, a lower bounding function is first applied within an outer optimization to determine a lower approximating x-solution. An upper bounding function is then applied for this fixed lower bounding decision, x, so that the quality of the lower approximation can be estimated. In the event the latter quality is unacceptable, a further refinement of the lower approximation is desired, hopefully, leading to an improved first-stage decision. This is the essence of a sequential approximation procedure within bounding (Ermoliev and Wets 1988, Chapter 2). Toward successful computer implementation of this procedure, it is imperative that the bounding functions so used possess desirable properties such as

(i) the bounding function should be easier to evaluate (w.r.t. the probability distribution),
(ii) the bounding function should preserve such properties as convexity (in x), which may be exploited in the outer optimization,
(iii) it should be possible to develop an efficient procedure for generating a sequence of monotonically improving bounds and lower approximating decisions, and
(iv) such a sequence of decisions must converge to an optimal solution of the stochastic program.

In general, functional approximations are difficult to implement either due to property (ii), and thus the outer optimization is difficult, or due to property (iii), i.e., refining the bounds (Birge and Wallace 1986) to a user-specified level of accuracy is tedious, although property (i) may hold.

In contrast, in distributional approximations, lower and upper bounds on the expectation functional $\psi(x) := E[\phi(x,\omega)]$ of the second-stage recourse problem (6.2) are determined by approximating the underlying probability distribution P^{tr} with limited moment information. In this process, a *discrete* random vector ζ is determined to approximate P^{tr}, which implies that $E_\zeta[\phi(x,\zeta)]$ can be evaluated as a finite sum. Indeed, the computational burden, i.e., property (i), then depends on the cardinality of (i.e., the number of atoms in) the support of ζ. The latter computational burden in distributional approximations has been an important topic of research. The goal is to derive discrete distributions having a number of atoms that is manageable compared to the number of random variables, $K+L$. The elegance of this approach stems from properties (ii), (iii), and (iv) in that they can be generally used within a sequential approximation procedure to solve stochastic programs up to a desired degree of accuracy.

Computational efficiency of distributional approximations can be enhanced by the following two-fold approach: first, increase the amount of moment information utilized in determining the bounds, and second, require the domain Ω (of P^{tr}) to be of a certain polyhedral shape. In the former, the resulting bounds are expected to be tighter, as measured relative to a GMP, and in the latter, cardinality of the support of ζ can be controlled by the number of extreme points in Ω. In the sequel, the focus is precisely in these directions.

Under distributional approximations, bounds so derived are usually sharp[1] since they are computed using discrete probability measures on Ω. Nevertheless,

---

[1] If the true distribution is indeed the approximating distribution, then the bound is achieved; hence, the bound is *sharp*.

tightness of bounds must be examined with respect to an optimization criterion. The yardstick for bound tightness is a moment optimization problem with specified moment conditions.

6.1.2 Bounds Using Generalized Moment Problems

Bounds are generally examined with respect to the upper and lower approximations in

$$\min_{x\in X}\left\{c'x+\psi_{\mathrm{L}}(x)\right\}\ \le\ \min_{x\in X}\left\{c'x+\int_{\Omega}\phi(x,\omega)P^{\mathrm{tr}}(d\omega)\right\}\ \le\ \min_{x\in X}\left\{c'x+\psi_{\mathrm{U}}(x)\right\},\tag{6.6}$$

where ψ_{L} and ψ_{U} correspond to the generalized moment problems (GMP),

$$\psi(x)\ge\psi_{\mathrm{L}}(x):=\inf_{P\in\mathcal{P}}\int_{\Omega}\phi(x,\omega)P(d\omega)\tag{6.7}$$

and

$$\psi(x)\le\psi_{\mathrm{U}}(x):=\sup_{P\in\mathcal{P}}\int_{\Omega}\phi(x,\omega)P(d\omega),\tag{6.8}$$

where $\mathcal{P}$ denotes the set of all probability measures on Ω, characterized by a set of (known) moment conditions as follows:

$$\mathcal{P}:=\left\{P\text{ is a probability measure on }\Omega:\int_{\Omega}f_i(\omega)P(d\omega)=C_i,\ i=1,\ldots,N\right\}\tag{6.9}$$

for given functions $f_i:\Omega\to\Re$. Since P^{tr} is assumed to have the given moments $C_i,\ i=1,\ldots,N$, $P^{\mathrm{tr}}\in\mathcal{P}$ holds and thus, the inequalities (6.7) and (6.8) are valid. Conditions for the existence of a probability measure that solves a GMP are derived in Kemperman (1968). The set of probability measures feasible in GMP has been studied extensively. For instance, extreme points of the set of admissible probability measures of GMP are discrete measures involving no more than $N+1$ realizations, see Karr (1983). Hence, the optimizing measures in (6.7) and (6.8) are discrete measures with cardinality at most $N+1$. However, determination of these discrete measures is difficult and it may generally involve nonconvex optimization, see Birge and Wets (1986). Instead, one may attempt to solve the semi-infinite dual problem of the GMP, say for the upper bound in (6.8), given by

$$\psi_{\mathrm{U}}^*(x):=\inf_{\pi\in\Re^{N+1}}\left\{\pi_0+\sum_{i=1}^{N}C_i\pi_i\ :\ \pi_0+\sum_{i=1}^{N}f_i(\omega)\pi_i\ge\phi(x,\omega),\ \omega\in\Omega\right\}.\tag{6.10}$$

The weak duality, i.e., $\psi_{\mathrm{U}}(x)\le\psi_{\mathrm{U}}^*(x)$, holds trivially. The strong duality can be assured under mild conditions:

Proposition 6.1.1 (Kall 1988, Kemperman 1968) *If f_i, $i = 1, \ldots, N$, are continuous and Ω is compact, then (6.8) is solvable and $\psi_U(x) = \psi_U^*(x)$.*

However, when Ω is unbounded, a certain interior-type condition is required to ensure solvability. Define $\mathcal{H}$ as the convex hull of the moment conditions $f_i(\omega)$, $i = 1, \ldots, N$, for $\omega \in \Omega$, i.e.,

$$\mathcal{H} := \text{co}\{(f_1(\omega), \ldots, f_N(\omega)) \ : \ \omega \in \Omega\},$$

and let $\text{int}(\mathcal{H})$ denote its (relative) interior.

Proposition 6.1.2 (Glashoff and Gustafson 1983, Kall 1988) *If* (6.8) *is finite and the N-dimensional point $(C_1, \ldots, C_N) \in \text{int}(\mathcal{H})$, then* (6.10) *is solvable and $\psi_U(x) = \psi_U^*(x)$.*

As can be seen from the constraints of (6.10), its solvability ensures the existence of an upper bounding function for $\phi(., \omega)$ w.p. 1, for fixed x, although finding such a function is not generally easy. However, when the domain Ω is polyhedral, the saddle property of ϕ may be exploited for obtaining these upper bounds.

Consider the convex case, i.e., $\phi(x, \xi, \eta^0)$ is convex in $\xi \in \Xi$ where η is degenerate with $\Theta = \{\eta^0\}$. When the only moment condition in GMP is the first moment vector $E[\xi] = \bar{\xi} \in \Re^K$, the resulting GMP lower bound is the well-known Jensen's inequality (Jensen 1906) regardless of the shape of Ξ.

Proposition 6.1.3 *For the GMP in* (6.7) *specified with the first moment conditions $f_k(\xi, \eta^0) = \xi_k$, $k = 1, \ldots, K$, and with $\bar{\xi} \in \text{int}(\Xi)$,*

$$\psi(x) \geq \psi_L(x) = \phi(x, \bar{\xi}, \eta^0). \tag{6.11}$$

Proof Appealing to Proposition 6.1.2, $\bar{\xi} \in \text{int}(\Xi)$ is the required interior condition for gap-free duality, and thus

$$\psi_L(x) = \sup_{\pi \in \Re^{N+1}} \left\{ \pi_0 + \sum_{k=1}^{K} \bar{\xi}_k \pi_k \ : \ \pi_0 + \sum_{k=1}^{K} \xi_k \pi_k \leq \phi(x, \xi, \eta^0), \ \xi \in \Xi \right\}.$$

Feasibility of the above supremum problem is assured since there always exists a hyperplane that is dominated by the convex function $\phi(x, \xi, \eta^0)$ for all $\xi \in \Xi$. Let $\hat{\pi}$ be any feasible solution. Then, $\hat{\pi}_0 + \sum_{k=1}^{K} \xi_k \hat{\pi}_k \leq \phi(x, \xi, \eta^0)$ holds for $\xi \in \Xi$, and in particular for $\bar{\xi} (\in \Xi)$, $\hat{\pi}_0 + \sum_{k=1}^{K} \bar{\xi}_k \hat{\pi}_k \leq \phi(x, \bar{\xi}, \eta^0)$, implying that $\psi_L(x) \leq \phi(x, \bar{\xi}, \eta^0)$. However, $\psi_L(x) = \phi(x, \bar{\xi}, \eta^0)$ can be attained by choosing $\hat{\pi}$ corresponding to the supporting hyperplane of the convex function $\phi(x, \xi, \eta^0)$ at $\xi = \bar{\xi}$. □

As for upper bounds under first moments for convex ϕ, the resulting GMP upper bound is as follows:

Proposition 6.1.4 (Kall 1988) *Suppose Ξ is compact polyhedral with extreme points u^j, $j = 1, \ldots, J$. For the GMP in (6.8) specified with $f_k(\xi, \eta^0) = \xi_k$, $k = 1, \ldots, K$, i.e., the first moment conditions $\bar{\xi}_k = E[\xi_k]$,*

$$\psi_{\mathrm{U}}(x) = \psi^*_{\mathrm{U}}(x) = \max_{\mu \geq 0} \left\{ \sum_{j=1}^{J} \phi(x, u^j, \eta^0)\mu_j \ : \ \sum_{j=1}^{J} u^j \mu_j = \bar{\xi}, \ \sum_{j=1}^{J} \mu_j = 1 \right\}. \tag{6.12}$$

The above first-moment (convex) upper bound, also developed in Dupačová (1966), Gassmann and Ziemba (1986), when specialized to univariate ξ with $\Xi = [\hat{a}_0, \hat{a}_1]$ leads to the well-known Edmundson–Madansky inequality (Madansky 1959):

$$\psi(x) \leq \left(\frac{\hat{a}_1 - \bar{\xi}}{\hat{a}_1 - \hat{a}_0} \right) \phi(x, \hat{a}_0, \eta^0) + \left(\frac{\bar{\xi} - \hat{a}_1}{\hat{a}_1 - \hat{a}_0} \right) \phi(x, \hat{a}_1, \eta^0). \tag{6.13}$$

When Ω is not polyhedral, Ω may be enclosed within suitably chosen polyhedra, although this may weaken the upper bound in (6.10) since additional constraints involving ω are then added. Increasing the number of moment conditions, N, tightens the bound since $\psi^*_{\mathrm{U}}(x)$ is nonincreasing in N. Both these ideas are exploited in the sequel. However, rather than solving the semi-infinite dual directly as above, we take an alternate route.

The case of unbounded Ω is investigated and bounds are developed in Birge and Wets (1987) and Edirisinghe and Ziemba (1994b). For the remainder of the chapter, however, it is assumed that Ω is compact (and polyhedral). This is not a serious drawback in view of applications because when stochastic programs are developed under discrete scenarios, this condition can always be satisfied. In Section 6.2, first-order moment bounds are developed and compared. Second (and higher)-order bounds are discussed in Section 6.3. A sequential approximation solution scheme is described under a certain simplicial decomposition procedure in Section 6.4. An application in asset allocation is in Section 6.5, where the performance of the solution procedure and pertinent sensitivity analyses are discussed. Concluding remarks are in Section 6.6.

6.2 Bounding the Expected Recourse Function

The joint domain Ω is assumed to be a compact polyhedron with extreme points $w^i \equiv (u^i, v^i)$, $i = 1, \ldots, I$. Thus, the extreme points of $\mathrm{co}\{u^i \ : \ i = 1, \ldots, I\}$ are the vertices of the marginal domain Ξ, herein denoted by the vertex set $\{u^i \ : \ i \in V_\xi\}$. Extreme points of Θ are given by the vertex set $\{v^i \ : \ i \in V_\eta\}$, where $\Theta = \mathrm{co}\{v^i \ : \ i = 1, \ldots, I\}$. Note that $V_\xi, V_\eta \subseteq \{1, \ldots, I\}$. The strategy here is to use convex representations of the underlying polyhedral domains in developing bounds and then to investigate their relationship to GMPs. In so doing, one of two basic approaches is used: either use a convex hull representation of the marginal domains Ξ and Θ separately or use a single convex hull representation of the joint domain Ω. In the former case, the resulting bounds are shown to use only first-order moment information of P^{tr}, while in the latter case, the resulting bounds require certain second-order moments of P^{tr}. As we shall see later, the amount of moment information required for computing bounds is very closely related to the shape (and hence the number of vertices) of the above polyhedral domains. This is an important factor because the computational effort required in evaluating the bounds is directly linked to the number of vertices in the polyhedral domain.

6.2.1 First-Order Moment Bounds

Moments of P^{tr} are called first-order type if each random variable is raised to a power of no more than 1 in the moment computation. Define

$$M(\nu) := \int_{\Omega} \left[\prod_{r=1}^{K+L} (\omega_r)^{\nu_r} \right] P^{\mathrm{tr}}(d\omega), \tag{6.14}$$

where the components of the $(K+L)$-vector ν are integral and nonnegative. It follows then that $M(\nu)$ is a first-order moment if $\nu_r \in \{0,1\}$ for $r = 1, \ldots, K+L$. For example, when $\nu = \boldsymbol{e}_i$, the ith elementary vector, (6.14) determines the ith first moment $\bar{\omega}_i$. Also, when one sets $\nu_k \nu_{K+l} = 1$ if $1 \le k \le K$, $1 \le l \le L$, and 0 otherwise, (6.14) determines the cross moment between ξ_k and η_l, i.e., $m_{kl} := E_{P^{\mathrm{tr}}}[\omega_k \omega_l]$. All first-order moments are assumed to be finite.

The following approach determines bounds under first-order moments. Consider each realization of the random vector as a convex combination of the extreme points of the respective polyhedral domain. Thus, by the convexity of Ξ, for given $\xi \in \Xi$, there exists a (measurable) multiplier function $\lambda(\xi)$ that satisfy

$$\sum_{i \in V_\xi} u^i \lambda_i(\xi) = \xi, \ \sum_{i \in V_\xi} \lambda_i(\xi) = 1, \ \lambda_i(\xi) \ge 0, \ i \in V_\xi. \tag{6.15}$$

Similarly, by the convexity of Θ, for given $\eta \in \Theta$, there exists a (measurable) multiplier function $\mu(\eta)$ satisfying

$$\sum_{j \in V_\eta} v^j \mu_j(\eta) = \eta, \ \sum_{j \in V_\eta} \mu_j(\eta) = 1, \ \mu_j(\eta) \ge 0, \ j \in V_\eta. \tag{6.16}$$

Then, by the saddle property of $\phi(x, ., .)$,

$$\sum_{j \in V_\eta} \mu_j(\eta) \phi(x, \xi, v^j) \le \phi(\xi, \eta) \le \sum_{i \in V_\xi} \lambda_i(\xi) \phi(x, u^i, \eta) \quad \text{for all } (\xi, \eta) \in \Xi \times \Theta. \tag{6.17}$$

Upon integration of (6.17), with respect to P^{tr}, the following bounding inequalities on $\psi(x)$ are obtained:

$$\sum_{j\in V_\eta} E_{P^{\mathrm{tr}}}[\mu_j(\eta)\phi(x,\xi,v^j)] =: \mathcal{L}_1(x) \;\le\; \psi(x)$$
$$\le\; \mathcal{U}_1(x) := \sum_{i\in V_\xi} E_{P^{\mathrm{tr}}}[\lambda_i(\xi)\phi(x,u^i,\eta)]. \tag{6.18}$$

However, $\mathcal{L}_1(x)$ or $\mathcal{U}_1(x)$ is not generally easy to evaluate. The following procedure leads to valid bounds. The lower bound is illustrated below; an analogous procedure may be followed for the upper bound:

$$\begin{aligned}
\mathcal{L}_1(x) &= E_{P^{\mathrm{tr}}} \sum_{j\in V_\eta} \mu_j(\eta) \min_{y^j\ge 0} \left\{ q(v^j)'y^j \;:\; Wy^j = h(\xi) - T(\xi)x \right\} \\
&= E_{P^{\mathrm{tr}}} \sum_{j\in V_\eta} \min_{y^j\ge 0} \left\{ q(v^j)'y^j \;:\; Wy^j = \mu_j(\eta)[h(\xi) - T(\xi)x] \right\} \\
&\ge \sum_{j\in V_\eta} \min_{y^j\ge 0} \left\{ q(v^j)'y^j \;:\; Wy^j = E_{P^{\mathrm{tr}}}\left[\mu_j(\eta)[h(\xi) - T(\xi)x]\right] \right\},
\end{aligned} \tag{6.19}$$

where the last inequality follows from applying Jensen's inequality in (6.11) on the inner minimization. Noting the linear affine structure in (6.5), along with (6.15) and (6.16),

$$\begin{aligned}
&E_{P^{\mathrm{tr}}}\left\{\mu_j(\eta)[h(\xi) - T(\xi)x]\right\} \\
&= (h_0 - T_0x)\, E[\mu_j(\eta)] + H\, E[\xi\mu_j(\eta)] - \sum_{k=1}^{K} T_k E[\xi_k\mu_j(\eta)] \\
&= (h_0 - T_0x)\, E\left[\sum_{i\in V_\xi} \lambda_i(\xi)\mu_j(\eta)\right] + H\, E\left[\sum_{i\in V_\xi} u^i\lambda_i(\xi)\mu_j(\eta)\right] \\
&\quad - \sum_{k=1}^{K} T_k E\left[\sum_{i\in V_\xi} u^i_k\lambda_i(\xi)\mu_j(\eta)\right] \\
&= (h_0 - T_0x)\sum_{i\in V_\xi}\rho_{ij} + H\sum_{i\in V_\xi} u^i\rho_{ij} - \sum_{k=1}^{K} T_k \sum_{i\in V_\xi} u^i_k\rho_{ij}\,,
\end{aligned} \tag{6.20}$$

where we have defined $\rho_{ij} := E_{P^{\mathrm{tr}}}[\lambda_i(\xi)\mu_j(\eta)]$ and λ and μ are determined according to (6.15) and (6.16), respectively. Hence, ρ_{ij} is determined on the joint vertex set $V_\xi \times V_\eta$. Let $\hat{\xi}^j$ be determined as a solution to

$$\sum_{i \in V_\xi} u^i \rho_{ij} = \hat{\xi}^j \sum_{i \in V_\xi} \rho_{ij}, \quad j \in V_\eta. \tag{6.21}$$

Observe that if $\sum_{i \in V_\xi} \rho_{ij} \neq 0$, then $\hat{\xi}^j$ is computed by taking convex combinations of the extreme points u^i of Ξ and thus $\hat{\xi}^j \in \Xi$ follows; otherwise, one may set (arbitrarily) $\hat{\xi}^j$ to be any point in Ξ. Consequently,

$$E_{P^{\mathrm{tr}}}\left[\mu_j(\eta)[h(\xi) - T(\xi)x]\right] = \left(\sum_{i \in V_\xi} \rho_{ij}\right)\left[h(\hat{\xi}^j) - T(\hat{\xi}^j)x\right] \tag{6.22}$$

holds and thus

$$\begin{aligned} \mathcal{L}_1(x) &\geq \sum_{j \in V_\eta} \left(\sum_{i \in V_\xi} \rho_{ij}\right) \inf_{y^j \geq 0} \left\{q(v^j)'y^j \ : \ Wy^j = h(\hat{\xi}^j) - T(\hat{\xi}^j)x\right\} \\ &= \sum_{j \in V_\eta} \sum_{i \in V_\xi} \rho_{ij} \phi(x, \hat{\xi}^j, v^j), \end{aligned} \tag{6.23}$$

where $\rho_{ij} = E[\lambda_i(\xi)\mu_j(\eta)]$ must be determined from (6.15) and (6.16).

Proposition 6.2.1 *For $\lambda(\xi)$ and $\mu(\eta)$ feasible in* (6.15) *and* (6.16)*, respectively, $\rho_{ij} := E[\lambda_i(\xi)\mu_j(\eta)]$ must satisfy the following set of constraints, denoted by the set $\mathcal{C}_{1f}$,*

$$\begin{aligned} \mathcal{C}_{1f} := \Bigg\{\rho \in V_\xi \times V_\eta : &\sum_{i \in V_\xi} \sum_{j \in V_\eta} (u^i, v^j)\rho_{ij} = (\bar{\xi}, \bar{\eta}), \\ &\sum_{i \in V_\xi} \sum_{j \in V_\eta} \rho_{ij} = 1, \quad \rho_{ij} \geq 0\Bigg\}. \end{aligned} \tag{6.24}$$

Proof Multiplying the constraints in (6.15) by η_l and $\mu_j(\eta)$, and also those in (6.16) by ξ_k and $\lambda_i(\xi)$, the following expressions are implied by (6.15) and (6.16):

$$\begin{aligned} &\sum_{i \in V_\xi} u^i \lambda_i(\xi)\mu_j(\eta) = \xi\mu_j(\eta), \ \sum_{i \in V_\xi} \lambda_i(\xi)\mu_j(\eta) = \mu_j(\eta), \ \lambda_i(\xi)\mu_j(\eta) \geq 0, \\ &\sum_{j \in V_\eta} v^j \lambda_i(\xi)\mu_j(\eta) = \eta\lambda_i(\xi), \ \sum_{i \in V_\xi} \eta\lambda_i(\xi) = \eta, \ \sum_{j \in V_\eta} \xi\mu_j(\eta) = \xi. \end{aligned}$$

First, taking a row aggregation of the above constraints (over the indices i and j), and then taking the expectation w.r.t. P^{tr}, it follows that $\rho \in \mathcal{C}_{1f}$. □

Since every feasible λ and μ in (6.15) and (6.16) imply that the corresponding ρ is in $\mathcal{C}_{1f}$, the latter feasible set is a relaxation. Hence, for some arbitrary $\rho \in \mathcal{C}_{1f}$, the expression in (6.23) is not guaranteed to give a lower bound on $\mathcal{L}_1(x)$; however, the infimum of (6.23) over $\mathcal{C}_{1f}$ does indeed provide a valid lower bound, as given below:

Proposition 6.2.2 *(First-Moment Lower Bound)*

$$\begin{aligned}\psi(x) \geq \mathcal{L}_1(x) \geq \mathcal{L}_{1f}(x) := \inf_{\rho} & \sum_{i \in V_\xi} \sum_{j \in V_\eta} \rho_{ij} \phi(x, \hat{\xi}^j, v^j) \\ \text{s.t.} & \sum_{i \in V_\xi} u^i \rho_{ij} - \hat{\xi}^j \sum_{i \in V_\xi} \rho_{ij} = 0, \quad j \in V_\eta \\ & \rho \in \mathcal{C}_{1f}.\end{aligned}$$

Using an analogous procedure, the following first-moment upper bound (UBFM) is derived:

Proposition 6.2.3 *(First-Moment Upper Bound)*

$$\begin{aligned}\psi(x) \leq \mathcal{U}_1(x) \leq \mathcal{U}_{1f}(x) := \sup_{\rho} & \sum_{i \in V_\xi} \sum_{j \in V_\eta} \rho_{ij} \phi(x, u^i, \hat{\eta}^i) \\ \text{s.t.} & \sum_{j \in V_\eta} v^j \rho_{ij} - \hat{\eta}^i \sum_{j \in V_\eta} \rho_{ij} = 0, \quad i \in V_\xi \\ & \rho \in \mathcal{C}_{1f}.\end{aligned}$$

The above upper bound is illustrated in Fig. 6.1 for the case of univariate ξ and η. Consider the cross sections of $\phi(., \xi, \eta)$ that pass through the first moment point $\bar{\omega}$,

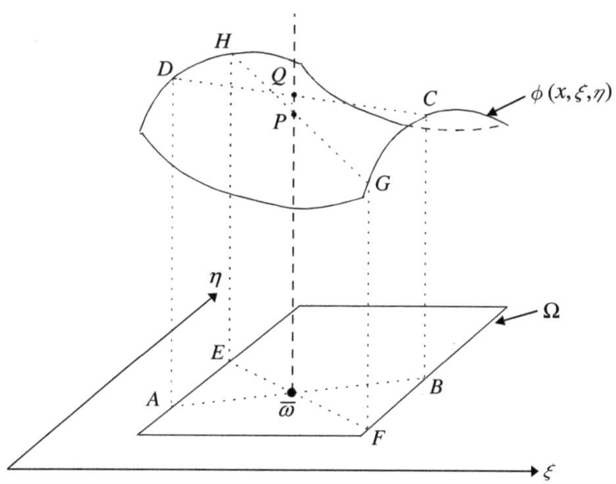

Fig. 6.1 First-moment upper bound

such as ABCD or EFGH; one with the *largest height* at $\bar{\omega}$ is the UBFM; hence, the required optimization for bound computation. The points A, B, E, and F correspond to $\hat{\eta}^i$ in Proposition 6.2.3.

Two important observations follow. First, when η is degenerate, i.e., the domain Θ is a singleton, ϕ is a convex function of the random variables (ξ), and then the upper bound UBFM is precisely the bound in (6.12). Moreover, in this case, the lower bound LBFM coincides with the Jensen's bound in (6.11). Second, the above first-moment upper and lower bounds are *sharp* as they correspond to probability measures ρ on $\Xi \times \Theta$ that satisfy the first moment conditions. Furthermore, these bounds are *tight* because they can be proven to be solutions of GMPs with first moment conditions.

Proposition 6.2.4 (Edirisinghe and Ziemba 1994a) *Under the assumption that* $\Omega = \Xi \times \Theta$,

$$\mathcal{L}_{1f}(x) = \inf_{P \in \mathcal{P}} \left\{ \int_{\Omega} \phi(x,\omega) P(d\omega) \ : \int_{\Omega} \omega P(d\omega) = \bar{\omega}, \ P \text{ is a prob. measure on } \Omega \right\} \tag{6.25}$$

and

$$\mathcal{U}_{1f}(x) = \sup_{P \in \mathcal{P}} \left\{ \int_{\Omega} \phi(x,\omega) P(d\omega) : \int_{\Omega} \omega P(d\omega) = \bar{\omega}, \ P \text{ is a prob. measure on } \Omega \right\}. \tag{6.26}$$

6.2.2 Tightening First-Moment Bounds

From the foregoing derivation it is evident that one method of developing improved lower and upper bounds on $\psi(x)$ is to tighten the relaxation due to $\rho \in \mathcal{C}_{1f}$. This can be done by specifying more moment conditions that are (naturally) implied by the systems in (6.15) and (6.16). For instance, multiplying the first equality in (6.15) by η_l and multiplying the first equality in (6.16) by $\lambda_i(\xi)$,

$$\sum_{i \in V_\xi} u_k^i \eta_l \lambda_i(\xi) = \xi_k \eta_l \ \text{ and } \sum_{j \in V_\eta} v_l^j \mu_j(\eta) \lambda_i(\xi) = \eta_l \lambda_i(\xi),$$

which thus implies that

$$\begin{aligned} &\sum_{i \in V_\xi} u_k^i \left[\sum_{j \in V_\eta} v_l^j \mu_j(\eta) \lambda_i(\xi) \right] = \xi_k \eta_l \\ \Rightarrow &\sum_{i \in V_\xi} \sum_{j \in V_\eta} u_k^i v_l^j \rho_{ij} = E_{P^{\mathrm{tr}}}[\xi_k \eta_l] =: m_{kl}, \ k = 1, \ldots, K, \ l = 1, \ldots, L, \end{aligned} \tag{6.27}$$

where m_{kl} are first-order *cross* moments. The inclusion of (6.27) into the constraints defining C_{1f} in Proposition 6.2.1 yields a restricted feasible set on ρ, denoted by $\mathcal{C}_{1c}$, and thus, $\mathcal{C}_{1c} \subseteq \mathcal{C}_{1f}$. Replacing C_{1f} in the optimization problems in Propositions 6.2.2 and 6.2.3 with the more restrictive set C_{1c}, therefore, yields tighter lower and upper bounds on $\psi(x)$ under first and cross moments, denoted herein by $\mathcal{L}_{1c}(x)$ and $\mathcal{U}_{1c}(x)$, respectively. The bounds $\mathcal{L}_{1c}(x)$ and $\mathcal{U}_{1c}(x)$ are sharp and they can also be shown to provide optimal solutions to the corresponding GMPs with first and cross moment conditions (Edirisinghe and Ziemba 1994a).

How much further can the above first-order bounds be improved by including additional moment conditions that are implied by (6.15) and (6.16)? It is possible to saturate the relaxation on ρ resulting from (6.15) and (6.16), if an appropriate amount of moment conditions is specified and the required conditions depend on the geometric shapes of Ξ and Θ. In such a case, the mathematical programs defining the bounds, see Propositions 6.2.2 and 6.2.3, would admit unique (*singleton*) solutions. That is, the lower and upper bounds are then computed trivially without any optimization on ρ. Such a practice has enormous computational implications since the bounding measures are then computed independent of the first-stage decisions x.

Such a unique bounding solution results quite straightforwardly under first and cross moments provided Ξ and Θ are simplices—Ξ is a K-dimensional polyhedron with $K+1$ vertices and Θ is an L-dimensional polyhedron with $L+1$ vertices. To see this, for instance, for the lower bound, denoting

$$\sum_{i \in V_\xi} u^i \rho_{ij} =: t^j \in \Re^K \quad \text{and} \quad \sum_{j \in V_\eta} \rho_{ij} =: p_j \ ,$$

it follows that the constraint system in $\mathcal{C}_{1c}$ implies

$$\left.\begin{array}{ll} \sum_{j \in V_\eta} v_l^j t_k^j = m_{kl} \ (l = 1, \ldots, L), & \sum_{j \in V_\eta} t_k^j = \bar{\xi}_k \ , \\ \sum_{j \in V_\eta} v_l^j p_j = \bar{\eta}_l \ (l = 1, \ldots, L), & \sum_{j \in V_\eta} p_j = 1, \quad \text{and} \quad p_j \geq 0. \end{array}\right\} \tag{6.28}$$

Given that $\Theta = \text{co}(v^j \ : \ j \in V_\eta)$ is a simplex, the system in (6.28), for each $k = 1, \ldots, K$, has a unique solution t_k^{*j} and p_j^*, which thus leads to the unique solution $\hat{\xi}_k^{*j} = \frac{t_k^{*j}}{p_j^*}$, and hence the lower bound,

$$\mathcal{L}_{1c}^{\text{sim}}(x) := \sum_{j \in V_\eta} p_j^* \phi(x, \hat{\xi}^{*j}, v^j). \tag{6.29}$$

Similarly, the upper bound $\mathcal{U}_{1c}(x)$ under simplicial domains simplifies as

$$\mathcal{U}_{1c}^{\text{sim}}(x) := \sum_{i \in V_\xi} p_i^* \phi(x, u^i, \hat{\eta}^{*i}), \tag{6.30}$$

where $\hat{\eta}_l^{*i} = \frac{t_l^{*i}}{p_i^*}$ and (t_l^{*i}, p_i^*) are the unique solution in

$$\left.\begin{array}{c} \sum_{i \in V_\xi} u_k^i t_l^i = m_{kl} \ (k = 1, \ldots, K), \quad \sum_{i \in V_\xi} t_l^i = \bar{\eta}_l \ , \\ \sum_{i \in V_\xi} u_k^i p_i = \bar{\xi}_k \ (k = 1, \ldots, K), \quad \sum_{i \in V_\xi} p_i = 1, \quad \text{and} \ p_i \geq 0. \end{array}\right\} \tag{6.31}$$

The bounds in (6.29) and (6.30) are the barycentric approximations developed in Frauendorfer (1992). When ξ and η are (bounded) univariate, hence Ξ and Θ are one-dimensional simplices, these first and cross moment bounds are as follows:

Proposition 6.2.5 (Edirisinghe and Ziemba 1994a) *If ξ and η are both univariate (with interval domains $\Xi := [\hat{a}_0, \hat{a}_1]$, $\Theta := [\tilde{a}_0, \tilde{a}_1]$ such that $\hat{a}_0 < \hat{a}_1$ and $\tilde{a}_0 < \tilde{a}_1$), and denoting the cross moment $m_{11} := E[\xi\eta]$,*

$$\mathcal{L}_{1c}(x) = \left(\frac{\tilde{a}_1 - \bar{\eta}}{\tilde{a}_1 - \tilde{a}_0}\right) \phi\left(x, \frac{\tilde{a}_1 \bar{\xi} - m_{11}}{\tilde{a}_1 - \bar{\eta}}, \tilde{a}_0\right) + \left(\frac{\bar{\eta} - \tilde{a}_0}{\tilde{a}_1 - \tilde{a}_0}\right) \phi\left(x, \frac{m_{11} - \tilde{a}_0 \bar{\xi}}{\bar{\eta} - \tilde{a}_0}, \tilde{a}_1\right) \tag{6.32}$$

and

$$\mathcal{U}_{1c}(x) = \left(\frac{\hat{a}_1 - \bar{\xi}}{\hat{a}_1 - \hat{a}_0}\right) \phi\left(x, \hat{a}_0, \frac{\hat{a}_1 \bar{\eta} - m_{11}}{\hat{a}_1 - \bar{\xi}}\right) + \left(\frac{\bar{\xi} - \hat{a}_0}{\hat{a}_1 - \hat{a}_0}\right) \phi\left(x, \hat{a}_1, \frac{m_{11} - \hat{a}_0 \bar{\eta}}{\bar{\xi} - \hat{a}_0}\right). \tag{6.33}$$

A unique bounding solution exists also in the following case. Suppose Ξ and Θ are multidimensional interval supports (hence, hyper-rectangles). Then, every feasible solution in (6.15) and (6.16) can be shown to imply a feasible solution $\rho_{ij} := E[\lambda_i(\xi)\mu_j(\eta)] \geq 0$ under the following set of (first-order) joint moment conditions, denoted by the set C_{1j}:

$$\sum_{i \in V_\xi} \sum_{j \in V_\eta} (u_k^i)^\nu \left(\prod_{l \in \Lambda} v_l^j\right) \rho_{ij} = E\left[(\xi_k)^\nu \left(\prod_{l \in \Lambda} \eta_l\right)\right],$$

$$\nu = 0, 1, \ \Lambda \in B_L, \ k = 1, \ldots, K, \tag{6.34}$$

$$\sum_{i \in V_\xi} \sum_{j \in V_\eta} \left(\prod_{k \in \Lambda} u_k^i\right) (v_l^j)^\nu \rho_{ij} = E\left[\left(\prod_{k \in \Lambda} \xi_k\right) (\eta_l)^\nu\right],$$

$$\nu = 0, 1, \ \Lambda \in B_K, \ l = 1, \ldots, L \ , \tag{6.35}$$

where B_K is the power set of $\{1, \ldots, K\}$ and B_L is the power set of $\{1, \ldots, L\}$. Under (6.34) and (6.35), the corresponding $\hat{\xi}^i$ and $\hat{\eta}^j$ are uniquely determined, independent of x, see Kall (1987a) for a proof of a similar result. Note that the above moments qualify as "first-order" conditions, see (6.14), because each random variable within the respective expectation operator is raised to the power of at most 1. Detailed expressions for these first-order (joint) moment bounds are derived in Frauendorfer (1988a). Quite interestingly, when this first-order (joint moment) lower bound is specialized to a convex function (i.e., η is degenerate), the result is Jensen's first-moment lower bound in (6.11). In contrast, the upper bound in this convex case leads to a generalization of the upper bound in (6.12) under rectangular domains and using all first-order joint moments of ξ; see Frauendorfer (1988b) for details. In the event the components ξ_k are all statistically independent of one another, the latter bound indeed is the multivariate extension of the Edmundson–Madansky upper bound in (6.13), see Kall and Stoyan (1988) for details.

It is worthwhile noting that the first and cross moment (polyhedral) lower bound $\mathcal{L}_{1c}(x)$ requires a computational effort that depends on the number of vertices in Θ, which can possibly be large. However, $\mathcal{L}_{1c}(x)$ can be used to develop a slightly weaker lower bound (Edirisinghe and Ziemba 1994a), denoted herein by $\mathcal{L}^*_{1c}(x)$, that requires only an effort that is linear in L even under general polyhedral domains. In particular, $\mathcal{L}^*_{1c}(x)$ can be applied under a rectangular domain for Θ, but as is evident below, the bounding model would still only be linear in L, as opposed to the exponential size in the preceding rectangular domain-based bounds under joint moments.

Proposition 6.2.6 (Edirisinghe and Ziemba 1996) *If Θ is a multidimensional rectangle of the form $X_{l=1}^{L}[\tilde{a}_{l0}, \tilde{a}_{l1}]$,*

$$\mathcal{L}_{1c}(x) \geq \mathcal{L}^*_{1c}(x) := \min_{y^0} \left\{ q_0' y^0 + \sum_{l=1}^{L} \varphi_l(x, y^0) : W y^0 = h(\bar{\xi}) - T(\bar{\xi})x, \; y^0 \geq 0 \right\} \tag{6.36}$$

and for $d^l := \bar{\eta}_l(h_0 - T_0 x) + \sum_{k=1}^{K} m_{kl}(h_k - T_k x)$ and $q^{(l)}$ being the lth column of Q, $\varphi_l(x, y^0)$ is defined by

$$\varphi_l(x, y^0) := \min_{y^l} \left\{ q^{(l)'} y^l \; : \; W y^l = d^l, \; \tilde{a}_{l0} y^0 \leq y^l \leq \tilde{a}_{l1} y^0 \right\}. \tag{6.37}$$

The lower bound $\mathcal{L}^*_{1c}(x)$ uses first and cross moments on a rectangular domain for Θ; however, the computational effort it requires within the stochastic program does not grow exponentially in L; it grows only linearly in L. This bound was iteratively refined under rectangular partitioning of the domain toward solving stochastic programs in Edirisinghe and Ziemba (1996). $\mathcal{L}^*_{1c}(x)$ cannot be claimed as tight since it does not correspond to solution of the GMP, nor can it be declared sharp since it does not correspond to an approximating probability measure.

6.3 Second-Order Moment Bounds

It must be noted that under simplicial domains for Ξ and Θ and using cross moments, any correlations in the ξ vector or the correlations in the η vector are ignored—only $E[\xi_k \eta_l]$ are used as higher moments. In contrast, the rectangular bounds with first-order joint moments in (6.34) and (6.35) allow the use of the latter correlations. However, the application of these rectangular domain bounds becomes computationally intractable because the number of vertices is exponential—2^K and 2^L—and thus an exponential amount of recourse function evaluations is necessary. For instance, for the lower bound, ϕ needs to be evaluated at $(\hat{\xi}^j, v^j)$ for each $j = 1, \ldots, 2^L$.

An important question, therefore, is whether it is possible to compute bounds using moment information of (possible) correlations in ξ (and η as well), while not requiring an exponential number of function evaluations. The answer is positive when a certain second-order moment information is utilized, as developed in Edirisinghe (1996). Second-order moments are defined by $M(\nu)$ in (6.14) when $\nu_r \in \{0, 1, 2\}$. In order to utilize such moment conditions, one has to proceed slightly differently from that in the previous section.

Consider each realization of the random (K+L)-vector $\omega \in \Omega$ as a convex combination of the extreme points $w^i \equiv (u^i, v^i), i \in V_\omega$, of the convex polyhedral domain Ω. Due to the convexity of Ω, there must exist (measurable) multipliers $\lambda(\omega)$, for given $\omega \in \Omega$, such that

$$\sum_{i \in V_\omega} w^i \lambda_i(\omega) = \omega, \quad \sum_{i \in V_\omega} \lambda_i(\omega) = 1, \quad \lambda_i(\omega) \geq 0, \qquad i \in V_\omega. \tag{6.38}$$

For a fixed value of η, denote the conditional domain of ω by Ω_η, i.e., $\Omega_\eta := \{\omega \in \Omega \; : \; \eta \text{ fixed}\}$. Similarly, let $\Omega_\xi := \{\omega \in \Omega \; : \; \xi \text{ fixed}\}$. Then by the saddle property of $\phi(x, ., .)$,

$$\phi(x, \xi, \eta) \geq \sum_{i \in V_\omega} \lambda_i(\omega) \phi(x, \xi, v^i), \qquad \forall \omega \in \Omega_\eta, \text{ for fixed } \eta \tag{6.39}$$

$$\text{and} \quad \phi(x, \xi, \eta) \leq \sum_{i \in V_\omega} \lambda_i(\omega) \phi(x, u^i, \eta), \qquad \forall \omega \in \Omega_\xi, \text{ for fixed } \xi. \tag{6.40}$$

Since the recourse function ϕ is finite on $\Xi \times \Theta$, upon integration with respect to P^{tr},

$$\begin{aligned} \sum_{i \in V_\omega} E_{P^{\text{tr}}}[\lambda_i(\omega) \phi(x, \xi, v^i)] =: \mathcal{L}_2(x) \; &\leq \; \psi(x) \\ \leq \mathcal{U}_2(x) := \sum_{i \in V_\omega} E_{P^{\text{tr}}}[\lambda_i(\omega) \phi(x, u^i, \eta)]&. \end{aligned} \tag{6.41}$$

We follow the basic approach of the preceding section to obtain computable bounds from $\mathcal{L}_2(x)$ and $\mathcal{U}_2(x)$; the lower bound is illustrated here. Following the idea in (6.19),

$$\mathcal{L}_2(x) \geq \sum_{i \in V_\omega} \inf_{y^i \geq 0} \left\{ q(v^i)' y^i \ : \ W y^i = E_{P^{\text{tr}}} [\lambda_i(\omega)[h(\xi) - T(\xi)x]] \right\}, \tag{6.42}$$

and similar to (6.20), and defining $\rho_i := E[\lambda_i(\omega)]$ and $\theta^i_k := E[\xi_k \lambda_i(\omega)]$,

$$\begin{aligned} & E_{P^{\text{tr}}} [\lambda_i(\omega)[h(\xi) - T(\xi)x]] \\ &= (h_0 - T_0 x)\, E[\lambda_i(\omega)] + H\, E[\xi \lambda_i(\omega)] - \sum_{k=1}^{K} T_k E[\xi_k \lambda_i(\omega)] \\ &= (h_0 - T_0 x)\rho_i + H\theta^i - \sum_{k=1}^{K} T_k \theta^i_k \\ &= \rho_i \left[h(\hat{\xi}^i) - T(\hat{\xi}^i)x \right], \end{aligned} \tag{6.43}$$

where $\hat{\xi}^i_k := \theta^i_k / \rho_i$ if $\rho_i \neq 0$; otherwise, $\hat{\xi}^i_k$ is any arbitrary point in Ξ. This leads to the lower bound

$$\mathcal{L}_2(x) \geq \sum_{i \in V_\omega} \rho_i \phi(x, \hat{\xi}^i, v^i), \tag{6.44}$$

provided that $\lambda(\omega)$ is chosen feasible in (6.38). For any such feasible $\lambda(\omega)$, the inequalities in (6.38) imply that

$$\sum_{i \in V_\omega} w^i E[\lambda_i(\omega)] = E[\omega],\ \sum_{i \in V_\omega} E[\lambda_i(\omega)] = 1,\ E[\lambda_i(\omega)] \geq 0,\ i \in V_\omega$$

$$\text{and} \quad \sum_{i \in V_\omega} w^i E[\xi_k \lambda_i(\omega)] = E[\xi_k \omega],\ \sum_{i \in V_\omega} E[\xi_k \lambda_i(\omega)] = E[\xi_k]$$

and thus a relaxation of the system in (6.38) is the following set of constraints on (ρ, θ):

$$\begin{aligned} \mathcal{C}_{2c} := \Bigg\{ (\rho, \theta) : & \sum_{i \in V_\omega} w^i \rho_i = \bar{\omega},\ \sum_{i \in V_\omega} \rho_i = 1,\ \rho_i \geq 0,\ i \in V_\omega, \\ & \sum_{i \in V_\omega} w^i \theta^i_k = E[\xi_k \omega],\ \sum_{i \in V_\omega} \theta^i_k = \bar{\xi}_k \Bigg\}. \end{aligned} \tag{6.45}$$

Observe that the rhs of (6.44) is a valid lower bound on $\mathcal{L}_2(x)$ so long as $\lambda(\omega)$ is chosen feasible in (6.38). However, the rhs of (6.44) may not be a valid lower bound for any arbitrary (ρ, θ) feasible in the (relaxed) set $\mathcal{C}_{2c}$. Nevertheless, the infimum of the rhs of (6.44) over the set $\mathcal{C}_{2c}$ in (6.45) is indeed a valid and computable lower bound, herein referred to as the mean–covariance lower bound (LBMC):

Proposition 6.3.1 *(Mean–Covariance Lower Bound, LBMC)*

$$\psi(x) \geq \mathcal{L}_2(x) \geq \mathcal{L}_{2c}(x) := \inf_{\rho} \sum_{i \in V_\omega} \rho_i \phi(x, \hat{\xi}^i, v^i)$$
$$\text{s.t. } \theta_k^i - \hat{\xi}_k^j \rho_i = 0, \; k = 1, \ldots, K, \quad i \in V_\omega$$
$$(\rho, \theta) \in \mathcal{C}_{2c}.$$

Observe that this lower bound uses, in addition to the first moments, the complete variance–covariance information of the random vector ξ, as well as the cross moments between the two vectors ξ and η. An analogous procedure can be followed to derive a *mean–covariance upper bound* that uses the complete variance–covariance information of η, as well as the cross moments between ξ and η.

6.3.1 Simplicial Mean–Covariance Approximation

An important issue is whether it is possible to obtain a *singleton* solution in $\mathcal{C}_{2c}$, denoted by (ρ^*, θ^*), so that the optimization in Proposition 6.3.1 is fictitious. It turns out that when the polyhedral (joint) Ω is a $(K + L)$-dimensional simplex, there is a unique solution satisfying the constraints in (6.45).

To compute this simplicial lower bound, let the nonsingular $(K+L+1) \times (K+L+1)$ vertex matrix of the simplex Ω be denoted by M, whose ith column is given by

$$(w_1^i, \ldots, w_{K+L}^i, 1)'.$$

The inverse matrix of M is denoted by M^{-1} whose ith row is M_i^{-1}. Then, the unique solution of the system (6.45) is given by

$$\rho_i^* = M_i^{-1} \begin{pmatrix} \bar{\omega} \\ 1 \end{pmatrix} \geq 0, \text{ and } \theta_k^{i*} = M_i^{-1} \begin{pmatrix} E[\xi_k \omega] \\ \bar{\xi}_k \end{pmatrix}, \quad \text{for } k = 1, \ldots, K. \tag{6.46}$$

Analogously, for the simplicial upper bound, define

$$\beta_l^{i*} := M_i^{-1} \begin{pmatrix} E[\eta_l \omega] \\ \bar{\eta}_l \end{pmatrix}, \quad \text{for } l = 1, \ldots, L. \tag{6.47}$$

Proposition 6.3.2 (Edirisinghe 1996) *(Simplicial Mean–Covariance Approximation)*

$$\sum_{i=1}^{K+L+1} \rho_i^* \phi(x, \hat{\xi}^{*i}, v^i) =: \mathcal{L}_{2c}^{sim}(x) \le \psi(x)$$
$$\le \mathcal{U}_{2c}^{sim}(x) := \sum_{i=1}^{K+L+1} \rho_i^* \phi(x, u^i, \hat{\eta}^{*i}), \tag{6.48}$$

where $\hat{\xi}_k^{*i} = \theta_k^{i*}/\rho_i^*$ *and* $\hat{\eta}_l^{*i} = \beta_l^{i*}/\rho_i^*$*, for* $k = 1, \ldots, K$ *and* $l = 1, \ldots, L$*, respectively.*

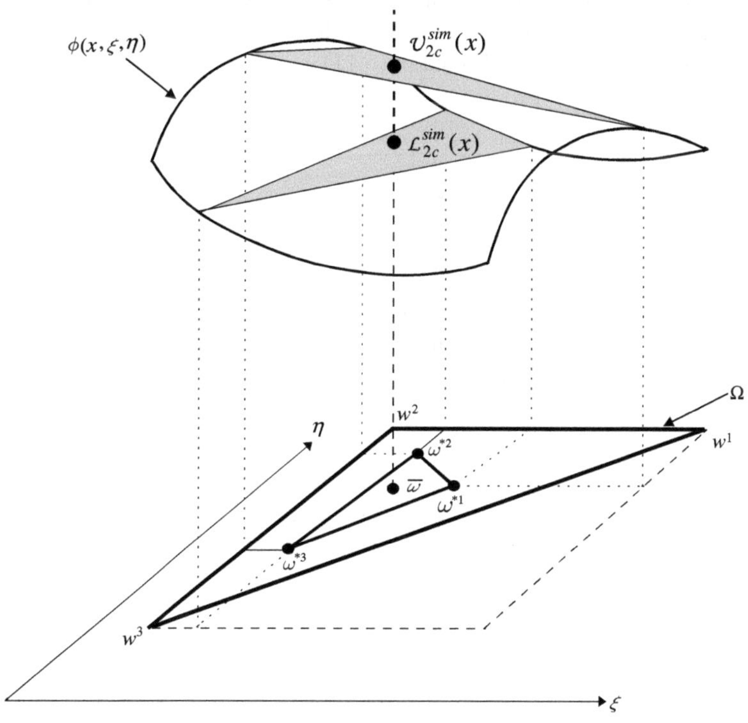

Fig. 6.2 Mean–covariance lower/upper bounds

It can be shown that $\omega^{*i} := (\hat{\xi}^{*i}, \hat{\eta}^{*i}) \in \Omega$, see Edirisinghe (1996). These approximations for univariate ξ and η are illustrated in Fig. 6.2.

Consider the specialization of the bounds in (6.48) when η is degenerate (say $\eta \equiv \eta^0$), that is, the recourse problem has no uncertainty in the objective coefficients and thus ϕ is only convex in the random parameters. The resulting upper bound is then

$$\sum_{i=1}^{K+1} \rho_i^* \phi(x, u^i, \eta^0) , \tag{6.49}$$

which is the upper bound in (6.12), but specialized to simplicial domains, and it is determined completely using only the first moment information $\bar{\xi}$. That is, for the convex case, covariance information cannot be captured in the upper bound. The situation is quite different for the lower bound in the convex case and it is given by

$$\sum_{i=1}^{K+1} \rho_i^* \phi(x, \hat{\xi}^{*i}, \eta^0). \tag{6.50}$$

Thus, the convex lower bound does utilize both the mean and the variance–covariance information of the random vector ξ. Figure 6.3 depicts the lower bound (denoted EB) in two dimensions, along with the upper bound in (6.49), denoted as GZ; JB denotes Jensen's lower bound in (6.11).

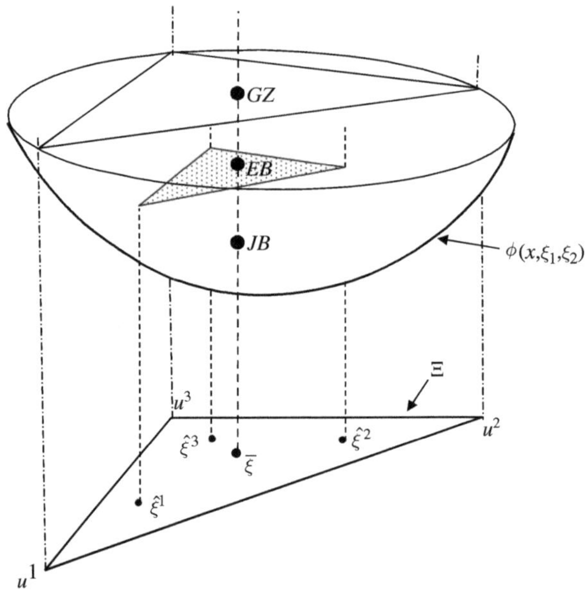

Fig. 6.3 Mean–covariance lower bound for convex case

For the convex recourse function $\phi(x, \xi, \eta^0)$ under univariate ξ (where $\hat{a}_0 \leq \xi \leq \hat{a}_1$), the resulting upper bound is the first moment-based Edmundson–Madansky inequality in (6.13). However, the lower bound does not simplify to the first moment-based Jensen's inequality in (6.11); instead, a mean–variance-based lower bound results (Edirisinghe 1996):

$$\psi(x) \geq \left(\frac{\hat{a}_1 - \bar{\xi}}{\hat{a}_1 - \hat{a}_0} \right) \phi(x, \tilde{\xi}^0, \eta^0) + \left(\frac{\bar{\xi} - \hat{a}_0}{\hat{a}_1 - \hat{a}_0} \right) \phi(x, \tilde{\xi}^1, \eta^0), \tag{6.51}$$

where

$$\tilde{\xi}^0 := \bar{\xi} - \frac{\sigma^2}{\hat{a}_1 - \bar{\xi}} \quad \text{and} \quad \tilde{\xi}^1 := \bar{\xi} + \frac{\sigma^2}{\bar{\xi} - \hat{a}_0} \tag{6.52}$$

and σ^2 is the variance of ξ. It must be stated that the mean–covariance lower (or upper) bounds in (6.48) do not provide solutions to their respective GMPs nor can they be shown to provide approximating probability measures for which all of the specified moment conditions are satisfied. For instance, considering the lower bound $\mathcal{L}_{2c}^{\text{sim}}(x)$ in (6.48) and denoting the corresponding lower approximating probability distribution by

$$P_{2c}^{\text{sim}} := \left\{ (\rho_i^*, \hat{\xi}^{*i}) \ : \ i \in V_\omega \right\},$$

it can be shown that while $E_{P_{2c}^{\text{sim}}}[\omega] = E_{P^{\text{tr}}}[\omega]$ holds, the second moments may not match generally, i.e., $E_{P_{2c}^{\text{sim}}}[\xi_k \omega_l] \neq E_{P^{\text{tr}}}[\xi_k \omega_l]$. Thus, the second-order lower bounds are not *sharp*.

6.3.2 Tightening Second-Order Bounds

The above mean–covariance bounds can be further tightened by incorporating additional second-order joint moment information, in addition to mean and variance–covariance information. For the case when Ω is rectangular, lower and upper bounds using second-order joint moments are developed in Edirisinghe (1996) for the expectation of $\phi(x, \xi, \eta)$. These bounds are computed without requiring an optimization to determine the approximating measure. While the bounds are tighter, their computations require 2^{K+L} recourse function evaluations and thus their application to solve stochastic programs becomes computationally tedious. In comparison, the preceding simplicial bounds require only $(K + L + 1)$ recourse function evaluations.

An alternative approach to tighten the univariate (convex) lower bound in (6.51), but still using only mean and variance information, is developed in Dokov and Morton (2005). It is based on the following inequality:

Proposition 6.3.3 (Dokov and Morton 2005) *For the convex recourse function $\varphi(x, \xi) \equiv \phi(x, \xi, \eta^0)$ under univariate ξ (where $\hat{a}_0 \leq \xi \leq \hat{a}_1$) and for mean $\bar{\xi}$ and variance σ^2,*

$$\psi(x) \geq \ell(t_1, t_2) := \min \left\{ \ell_1(t_1), \hat{\ell}_1(t_1), \ell_2(t_2), \hat{\ell}_2(t_2) \right\}, \quad \forall\, t_1 \in [\tilde{\xi}^1, \hat{a}^1],\ t_2 \in [\hat{a}^0, \tilde{\xi}^0], \tag{6.53}$$

where $\tilde{\xi}^0$ and $\tilde{\xi}^1$ are given by (6.52) and

$$\ell_1(t_1) := \left(\frac{\tilde{\xi}^1 - \bar{\xi}}{\tilde{\xi}^1 - \zeta^0(t_1)}\right) \varphi\left(x, \zeta^0(t_1)\right) + \left(\frac{\bar{\xi} - \zeta^0(t_1)}{\tilde{\xi}^1 - \zeta^0(t_1)}\right) \varphi\left(x, \tilde{\xi}^1\right),$$

$$\hat{\ell}_1(t_1) := \left(\frac{\tilde{\xi}^1 - \bar{\xi}}{\tilde{\xi}^1 - \tilde{\xi}^0}\right) \varphi\left(x, \tilde{\xi}^0\right) + \left(\frac{\bar{\xi} - \tilde{\xi}^0}{\tilde{\xi}^1 - \tilde{\xi}^0}\right) \left[\left(\frac{t_1 - \tilde{\xi}^1}{t_1 - \bar{\xi}}\right) \varphi\left(x, \bar{\xi}\right) + \left(\frac{\tilde{\xi}^1 - \bar{\xi}}{t_1 - \bar{\xi}}\right) \varphi\left(x, t_1\right)\right],$$

$$\ell_2(t_2) := \left(\frac{\zeta^1(t_2) - \bar{\xi}}{\zeta^1(t_2) - \tilde{\xi}^0}\right) \varphi\left(x, \tilde{\xi}^0\right) + \left(\frac{\bar{\xi} - \tilde{\xi}^0}{\zeta^1(t_2) - \tilde{\xi}^0}\right) \varphi\left(x, \zeta^1(t_2)\right),$$

$$\hat{\ell}_2(t_2) := \left(\frac{\tilde{\xi}^1 - \bar{\xi}}{\tilde{\xi}^1 - \tilde{\xi}^0}\right) \left[\left(\frac{\bar{\xi} - \tilde{\xi}^0}{\bar{\xi} - t_2}\right) \varphi\left(x, t_2\right) + \left(\frac{\tilde{\xi}^0 - t_2}{\bar{\xi} - t_2}\right) \varphi\left(x, \bar{\xi}\right)\right] + \left(\frac{\bar{\xi} - \tilde{\xi}^0}{\tilde{\xi}^1 - \tilde{\xi}^0}\right) \varphi\left(x, \tilde{\xi}^1\right),$$

where

$$\zeta^0(t) := \bar{\xi} - \frac{\sigma^2}{t - \bar{\xi}} \quad \text{and} \quad \zeta^1(t) := \bar{\xi} + \frac{\sigma^2}{\bar{\xi} - t}. \tag{6.54}$$

Given the valid inequality in (6.53), the strongest bound in this class is determined by solving the optimization problem (see Dokov and Morton 2005),

$$\max_{t_1 \in [\tilde{\xi}^1, \hat{a}^1], t_2 \in [\hat{a}^0, \tilde{\xi}^0]} \ell(t_1, t_2). \tag{6.55}$$

It is straightforward to see by setting $t_1 = \tilde{\xi}^1$ and $t_2 = \tilde{\xi}^0$ that $\ell(\tilde{\xi}^1, \tilde{\xi}^0)$ is the lower bound in (6.51) for the univariate convex case. Thus, the lower bound in (6.55) is generally tighter than that in (6.51). However, computation of this tighter bound in (6.55) is more expensive as additional recourse function evaluations are involved. Furthermore, similar to (6.51), the lower approximating distribution in this case does not match the given variance σ^2, although the first moment is matched, and thus one cannot guarantee that the bound will be sharp. It remains unknown at this point how this improved mean–variance lower bound in one dimension may be generalized to multivariate ξ with dependent components.

6.3.3 Higher Order Moment Upper Bounds in Convex Case

So far, the computations require an upper (or lower) bounding distribution (that approximates P^{tr}) and the computational burden is dependent on the number of vertices in the domain. Simplicial domains are preferred to rectangular domains since the latter involve an exponential number of function evaluations. However, in the convex case, simplicial domains do not allow the use of more than the first moment in an upper bound computation and thus the upper bound can be weak. In contrast, the simplicial lower bound in the convex case allows the use of variance–covariance information. Can higher moments be used to compute an upper bound under dependent ξ components in the convex case?

One approach is to specify higher order moment conditions in a GMP and determine solutions or upper bounds on it. Information on second moment is introduced in Dupačová (1966) to obtain improvements over first-moment bounds. An upper bound for the expectation of a *simplicial* function of a random vector using first moments and the sum total of all variances is developed in Dula (1992). The latter moment conditions are also used in Kall (1991) to demonstrate how the related moment problems can be solved to obtain the upper bound by solving nonsmooth optimization problems along with simple linear constraint systems. Nevertheless, how such second moment upper bounds can be used within a solution procedure for stochastic programs, requiring bound refinement based on their error analysis, remains to be developed.

An alternative approach is taken in Dokov and Morton (2002) where a class of higher order moment upper bounds is developed for the case of convex $\varphi(x, \xi)$, using simplicial (or rectangular) domains, under dependent ξ components. Moreover, over simplicial domains, this procedure allows the use of more than just the first moments in computing upper bounds on the expectation of $\varphi(x, \xi)$. Given the simplex $\Xi = \text{co}(u^i \; : \; i = 1, \ldots, K + 1)$, the essential idea is to use (6.15) to determine the unique convex multipliers $\lambda_i(\xi)$ and then to express each $\xi \in \Xi$ as a certain nth-order representation (for $n = 1, 2, \ldots$) by

$$\xi = \sum_{r \in \mathcal{R}} \binom{n}{r_1, \ldots, r_{K+1}} \left(\prod_{i=1}^{K+1} \lambda_i(\xi) \right) \left[\frac{\sum_{i=1}^{K+1} r_i u^i}{n} \right], \tag{6.56}$$

where the index set $\mathcal{R}$ is given by

$$\mathcal{R} := \left\{ r \in \Re^{K+1} \; : \; \sum_{k=1}^{K+1} r_k = n, \; r_k \geq 0, \; r_k \text{ integer} \right\}. \tag{6.57}$$

Then, due to convexity of $\varphi(x, \xi)$ in ξ,

$$\varphi(x, \xi) \leq \sum_{r \in \mathcal{R}} \binom{n}{r_1, \ldots, r_{K+1}} \left(\prod_{i=1}^{K+1} \lambda_i(\xi) \right) \varphi \left(x, \left[\frac{\sum_{i=1}^{K+1} r_i u^i}{n} \right] \right), \tag{6.58}$$

which is known as the nth-order Bernstein polynomial for a function defined on a simplex, see, e.g., Lorentz (1986). Then, upon taking expectations on both sides of (6.58) w.r.t. P^{tr},

$$\psi(x) \le \sum_{r\in\mathcal{R}} \binom{n}{r_1,\ldots,r_{K+1}} E_{P^{\rm tr}}\left[\prod_{i=1}^{K+1} \lambda_i(\xi)\right] \varphi\left(x, \left[\frac{\sum_{i=1}^{K+1} r_i u^i}{n}\right]\right) \quad (6.59)$$

$$=: EM_n^{\rm sim}. \quad (6.60)$$

Since $\lambda_i(\xi)$ is linear in the components of the ξ vector, computation of $EM_n^{\rm sim}$ requires higher order moments of ξ depending upon the value of n. When $n = 1$, note that only the first moments $\bar{\xi}$ are required for its computation, and for $n = 2$, $EM_2^{\rm sim}$ uses variance–covariance information in ξ. It is also shown in Dokov and Morton (2002) that

$$\psi(x) \le EM_n^{\rm sim} \le EM_{n-1}^{\rm sim}, \quad (6.61)$$

thus allowing for the use of higher order moment information for computing improved upper bounds. However, the number of function evaluations required for the bound computation is $\binom{K+n}{n}$. Setting $n = 2$, the resulting mean–covariance upper bound thus requires $\frac{1}{2}(K+1)(K+2)$ number of recourse function evaluations, rather than the $(K + 1)$ number of evaluations necessary for the mean–covariance lower bound computation in (6.50).

6.4 Simplicial Sequential Solution (SSS)

An iterative procedure for solving the stochastic program in (6.1) using the mean–covariance bounds $\mathcal{L}_{2c}^{\rm sim}(x)$ and $\mathcal{U}_{2c}^{\rm sim}(x)$ in (6.48) is discussed here. The basic idea is to apply bounds within a sequential approximation scheme (Frauendorfer and Kall 1988, Kall and Wallace 1994). First, at iteration $\nu = 0$, bounds are applied on the complete simplicial domain Ω to determine $\mathcal{L}_{2c}^{\rm sim}(x)$ and $\mathcal{U}_{2c}^{\rm sim}(x)$. Thereafter, upon partitioning the domain Ω in a sub-simplicial decomposition $\mathcal{S}^\nu$, at each iteration $\nu = 1, 2, \ldots$, an improving sequence of approximations $\mathcal{L}_{2c,\nu}^{\rm sim}(x)$ and $\mathcal{U}_{2c,\nu}^{\rm sim}(x)$ are computed based on *conditional* mean–covariance information within each cell of the partition. The first-order (mean and cross moment) simplicial bounds in (6.29) and (6.30) can also be applied with two separate simplicial partitioning schemes on Ξ and Θ; see Frauendorfer (1992).

In the sequel, $P^{\rm tr}$ is treated as a discrete distribution where a set of (discrete) scenarios are prespecified, denoted by ω^s (with probability p_s), $s \in S$, where $\sum_{s\in S} p_s = 1$. The number of scenarios, $|S|$, can be extremely large, which is possibly determined by sampling from a continuous multivariate distribution. The essential simplicity due to this setup is that all required moments can be computed as finite sums; moreover, $co\{\omega^s : s \in S\}$ is a bounded (convex) polyhedral domain.

6.4.1 Partitioning and Convergence

Let $\mathcal{S}^\nu := \{\Omega_r \ : \ r = 1, \ldots, R_\nu\}$ represent, at some iteration ν, a partition of the joint simplicial domain Ω into subsimplices Ω_r such that

$$\bigcup_{r=1}^{R_\nu} \Omega_r = \Omega, \quad \Omega_r \bigcap \Omega_{r'} = \emptyset, \quad \forall r \neq r', \ r, r' \in \{1, \ldots, R_\nu\}. \tag{6.62}$$

Let the corresponding index sets of scenarios be $S_1, \ldots, S_{R_\nu}$ where $S_r := \{s \ : \ \omega^s \in \Omega_r, \ s \in S\}$. The dependence of S_r on the iteration count ν is suppressed for convenience. The corresponding *cell probabilities* are evaluated as $P_r := \sum_{s \in S_r} p_s$ where $\sum_{r=1}^{R_\nu} P_r = 1$. Under the cell-conditional first and second moments computed by $\frac{1}{P_r} \sum_{s \in S_r} p_s \omega_k^s [\omega_l^s]^\theta, \quad \theta \in \{0, 1\}, \ k, l = 1, \ldots, K + L$, the corresponding lower and upper approximations for cell Ω_r, due to $\mathcal{L}_{2c}^{\text{sim}}(x)$ and $\mathcal{U}_{2c}^{\text{sim}}(x)$ in (6.48), are

$$\phi_{\text{L}}^{\nu,r}(x) := \sum_{i=1}^{K+L+1} \rho_i^*(r) \phi(x, \hat{\xi}^{*i}(r), v^i(r)) \ \text{ and}$$

$$\phi_{\text{U}}^{\nu,r}(x) := \sum_{i=1}^{K+L+1} \rho_i^*(r) \phi(x, u^i(r), \hat{\eta}^{*i}(r)).$$

Thus, for iteration ν,

$$\sum_{r=1}^{R_\nu} P_r \phi_{\text{L}}^{\nu,r}(x) =: \psi_{\text{L}}^\nu(x) \ \leq \ \psi(x) \ \leq \ \psi_{\text{U}}^\nu(x) := \sum_{r=1}^{R_\nu} P_r \phi_{\text{U}}^{\nu,r}(x). \tag{6.63}$$

Corresponding to the above sequence of partitioning, lower and upper approximating objective value sequences $\{Z_{\text{L}}^\nu\}$ and $\{Z_{\text{U}}^\nu\}$ can be obtained by solving the linear programs

$$Z_L^\nu := \min_{x \in X} \left\{ c'x \ + \ \psi_{\text{L}}^\nu(x) \right\} \quad \text{and} \quad Z_{\text{U}}^\nu := \min_{x \in X} \left\{ c'x \ + \ \psi_{\text{U}}^\nu(x) \right\}, \tag{6.64}$$

along with their optimal first-stage solutions x_{L}^ν and x_{U}^ν, respectively. It can be shown that the sequence $\{Z_{\text{L}}^\nu\}$ is monotonically increasing while $\{Z_{\text{U}}^\nu\}$ is monotonically decreasing. Furthermore, as long as $\max_{r=1,\ldots,R_\nu} \mathcal{D}_r \to 0$ when $\nu \to \infty$, where $\mathcal{D}_r$ denotes the diameter of the subsimplex Ω_r, the approximating scenario distributions $\rho^*(r)|_{r=1}^{R_\nu}$ can be shown to converge *weakly* to the original distribution $p_s|_{s=1}^{S}$. This yields $\lim_{\nu \to \infty} Z_{\text{L}}^\nu = \lim_{\nu \to \infty} Z_{\text{U}}^\nu = Z^*$. Also, see e.g., Birge and Wets (1986), Kall (1987b), and Robinson and Wets (1987).

An important question is whether the approximate (first-stage) solutions x_{L}^ν and x_{U}^ν so generated would converge to true optimal solutions of (6.1). This question is answered by the theory of *epiconvergence*, which can loosely be described as the convergence of the epigraphical sets of the approximating functions (convex in x, in our case) to the epigraph of the original recourse function $\psi(.)$. With weak convergence of the approximating measures along with the joint domain Ω being a (compact) simplex (see Birge and Wets 1986, Section 2.11), one has the required epiconvergence of the approximating functions. Under epiconvergence (see Wets 1980, Theorem 9), the set of approximate solutions belongs to the set of optimal solutions, i.e.,

$$\limsup[\operatorname{argmin} g^{\nu}] \subset \operatorname{argmin} g,$$

where $g^{\nu}(x) := c'x + \psi_{\mathrm{L}}^{\nu}(x)$ (respectively, $g^{\nu}(x) := c'x + \psi_{\mathrm{U}}^{\nu}(x)$) and $g(x) := c'x + \psi(x)$. Hence, $\{x_{\mathrm{L}}^{\nu}\} \to x^*$ and $\{x_{\mathrm{U}}^{\nu}\} \to x^{**}$ hold as desired, where x^* and x^{**} are optimal solutions of the stochastic program (6.1).

There are at least two related issues: stability and speed of convergence. While the above approximations may be sharpened to a user-specified accuracy through a partitioning procedure as above, that the solution sets so generated behave in a continuous manner with respect to small perturbations in the approximating measures can be ensured by appealing to the stability results in Kall (1987b) and Robinson and Wets (1987). The second important issue is how fast can the convergence (or a given accuracy for the error bound) be achieved. Obviously, this depends on many factors such as the initial simplex that contains the given set of scenarios, the manner in which simplicial partitioning is performed, as well as how efficiently one can solve the second-order approximations.

Solving the upper and lower approximations in (6.64) for separate first-stage solution vectors x_{L}^{ν} and x_{U}^{ν} does not help much in identifying cells (i.e., clusters of scenarios) of the current partition for further refinement. Observing that the lower approximating solution x_{L}^{ν} is feasible in the upper approximating problem, the standard strategy is to compute a weaker upper bound by

$$\bar{Z}_{\mathrm{U}}^{\nu} := c'x_{\mathrm{L}}^{\nu} + \psi_{\mathrm{U}}^{\nu}(x_{\mathrm{L}}^{\nu}). \tag{6.65}$$

Consequently, the accuracy of the second-order approximation for estimating the objective value Z^* of (6.1), at some iteration ν, can be evaluated on the basis of the current lower bounding solution x_{L}^{ν} by (provided $Z_{\mathrm{L}}^{\nu} > 0$)

$$\left|\frac{\bar{Z}_{\mathrm{U}}^{\nu} - Z_{\mathrm{L}}^{\nu}}{Z_{\mathrm{L}}^{\nu}}\right| < \varepsilon, \tag{6.66}$$

i.e., if the *relative gap* of the upper and lower bounds for the current solution x_{L}^{ν} is within a certain tolerance ε, then no further refinement of the current approximation is needed.

6.4.2 Partitioning Strategies and Cell Redefining

When the termination rule in (6.66) is not satisfied at some iteration ν, the current approximation is refined by choosing a cell for further partitioning by the maximum (probabilistically) weighted difference (WDIFF) criterion, i.e.,

$$r^* := \arg\max\{P_r\left[\phi_U^{\nu,r}(x_L^\nu) - \phi_L^{\nu,r}(x_L^\nu)\right] \;:\; r = 1, \ldots, R_\nu\}. \qquad (6.67)$$

Having chosen a cell Ω_{r*} for further subdivision, the way in which partitioning is actually carried out can have a major impact on solution efficiency. This was demonstrated for rectangular partitioning in Edirisinghe and Ziemba (1996) and for simplicial domains by Edirisinghe and You (1996), Frauendorfer (1994), and Kuhn (2004). Consider a *bipartitioning* procedure where the cell Ω_{r*} is split into only two sub-simplices by choosing a single *partition plane* that divides an edge of the simplex and passes through the remaining $(K + L - 1)$ vertices. In so doing, one needs to choose

1. the partitioned edge and
2. the point at which the chosen edge is partitioned.

The standard practice in edge selection is to use the notion of *degree of nonlinearity* (DON) of the recourse function along edges of the domain and to partition along directions which are perpendicular to such edges; see Frauendorfer and Kall (1988). In our case, it is motivated by the fact that the second-order moment approximations are exact if the recourse function is linear in its domain. Since partitioning is carried out here in the joint domain Ω, nonlinearity of ϕ along joint directions (ξ, η) in Ω can be exploited in sharpening the approximations.

6.4.2.1 Edge Selection

In order to apply the DON measure for edge selection, let us drop the iteration count ν and the chosen cell index r^* for expositional simplicity. Suppose the vertices $w^i \equiv (u^i, v^i)$ and $w^j \equiv (u^j, v^j)$ represent a candidate edge to be divided and that $u^i \neq u^j$ and $v^i \neq v^j$. The degree of nonlinearity associated with the edge (i, j) is measured as follows.

Consider the displacement from w^i to w^j in two possible two-step movements: $w^i \to \hat{w} := (u^j, v^i) \to w^j$ and $w^i \to \tilde{w} := (u^i, v^j) \to w^j$. Then, for each possible two-step displacement, the recourse function is approximated by two-piece linear functions, determined by primal and dual solutions of the recourse problem at w^i and w^j. Then the functional displacement due to this two-piece linear approximation is computed at both $\hat{w}$ and $\tilde{w}$, of which the minimum is taken as the nonlinearity measure; see Fig. 6.4 for an illustration for univariate ξ and η.

That is, denoting the optimal primal and dual solutions corresponding to $\phi(x_L, w^i)$ and $\phi(x_L, w^j)$ by (y^i, π^i) and (y^j, π^j), respectively, we have

$$\phi(x_L, \hat{w}) \;\geq\; (\pi^i)'[h(u^j) - T(u^j)x_L] \quad \text{and} \quad \phi(x_L, \hat{w}) \;\leq\; q(v^i)'y^j. \qquad (6.68)$$

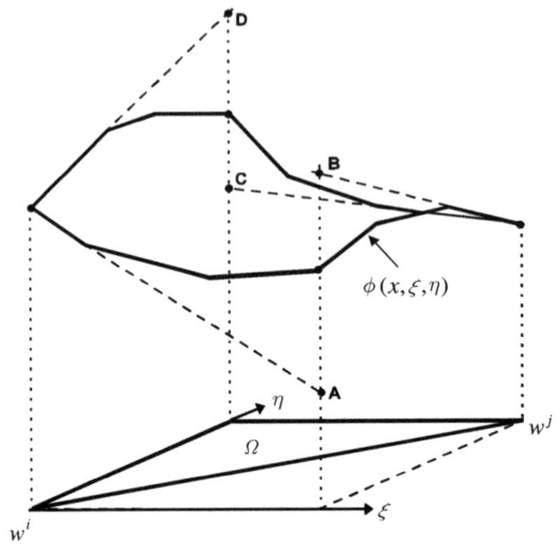

Fig. 6.4 Nonlinearity measure $\Delta(i, j) = \min\{AB, CD\}$

Furthermore,

$$\phi(x_L, \tilde{w}) \geq (\pi^j)'[h(u^i) - T(u^i)x_L] \quad \text{and} \quad \phi(x_L, \tilde{w}) \leq q(v^j)'y^i. \tag{6.69}$$

Then, the DON measure for the edge (i, j) is defined by

$$\Delta(i, j) := \min\Big\{q(v^i)'y^j - (\pi^i)'[h(u^j) - T(u^j)x_L], \\ q(v^j)'y^i - (\pi^j)'[h(u^i) - T(u^i)x_L]\Big\}. \tag{6.70}$$

The intuition behind the measure Δ is that if the recourse function is almost linear in the rectangle determined by the points w^i, $\hat{w}$, w^j, and $\tilde{w}$, then Δ is expected to be smaller. See Edirisinghe and You (1996) for details where directional derivatives are also used in measuring DON. If either $u^i = u^j$ or $v^i = v^j$ occurs, then the nonlinearity measure Δ reduces to that proposed in Frauendorfer and Kall (1988).

Once an edge (i, j) is selected according to the $\max_{(i,j)} \Delta(i, j)$ rule, the current cell (simplex Ω) can be partitioned by determining a partitioning plane that passes through the cell-conditional mean and the vertices w^k, $k \notin \{i, j\}$, $k = 1, \ldots, K + L + 1$; see Fig. 6.5. The partitioning point on the edge (i, j) is given by

$$w^{ij} := \frac{\rho_i^*}{\rho_i^* + \rho_j^*} w^i + \frac{\rho_j^*}{\rho_i^* + \rho_j^*} w^j. \tag{6.71}$$

Conditional mean partitioning divides a cell based upon probability mass and this idea is consistent with that done under rectangular domains, see Edirisinghe and Ziemba (1996) and Frauendorfer and Kall (1988).

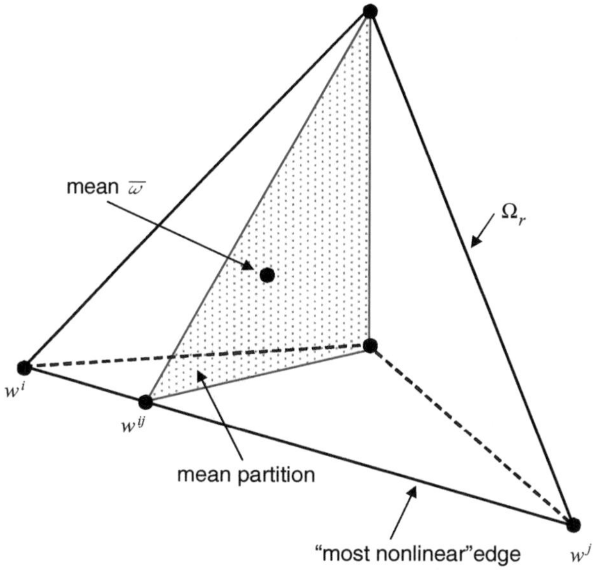

Fig. 6.5 Conditional mean partitioning of a simplicial cell

6.4.2.2 Cell Redefining

To apply the simplicial bounds, an initial simplex Ω covering co$\{\omega^s,\ s \in S\}$ must be determined, see Edirisinghe and You (1996) for details. Such an initial simplex generally consists of significant volume of *zero measure*, which typically contributes to weakening the approximations. Moreover, at some iteration ν, when Ω_r is bipartitioned, the resulting two simplices may also contain *removable* volumes of zero measure. The *cell-redefining* (C-R) procedure is aimed at obtaining every simplicial cell as compactly as possible.

A C-R procedure is introduced and implemented in Edirisinghe and Ziemba (1996) in order to further tighten first-order bounds when partitioning in rectangular domains. A C-R procedure under simplicial partitioning is developed in Edirisinghe and You (1996), where the essential idea is to consider each vertex in Ω for relocation to form a new simplex that is more compact. Let the nonsingular $(K+L+1)\times(K+L+1)$ vertex matrix of the simplex Ω be M where the ith column of M is $(w_1^i, \ldots, w_{K+L}^i, 1)'$. The inverse matrix of M is denoted by M^{-1}, whose ith row is M_i^{-1}. Then, a vertex w^j of Ω is relocated to the $(K+L)$-dimensional (new) point $w^* \in \Omega$ such that

$$\Omega' := \text{conv}\{_r w^1, \ldots, w^{j-1}, w^*, w^{j+1}, \ldots, w^{K+L+1}\} \subset \Omega, \tag{6.72}$$

where w^* is a solution of the linear program,

$$\min_{w^* \in \Re^{K+L}} \quad M_j^{-1} \begin{pmatrix} w^* \\ 1 \end{pmatrix}$$

$$\text{s.t.} \quad M^{-1} \begin{pmatrix} w^* \\ 1 \end{pmatrix} \geq 0$$

$$\left(\lambda_i^s M_j^{-1} - \lambda_j^s M_i^{-1}\right) \begin{pmatrix} w^* \\ 1 \end{pmatrix} \geq 0, \ \forall i \neq j, \ i = 1, \ldots, \text{K+L+1}; \ s \in S, \tag{6.73}$$

where the convex multipliers λ^s are defined for each scenario ω^s, $s \in S$, in simplex Ω by

$$\lambda^s := M^{-1} \begin{pmatrix} \omega^s \\ 1 \end{pmatrix}. \tag{6.74}$$

The LP in (6.73) has $(K + L)$ variables and $(|S| + 1)(K + L) + 1$ number of constraints. Since stochastic programs are typically formulated with a large sample of outcomes, the C-R LP is large in size. However, this LP can be solved trivially (in linear time) as discussed in Edirisinghe (2005). Figure 6.6 illustrates the C-R procedure graphically.

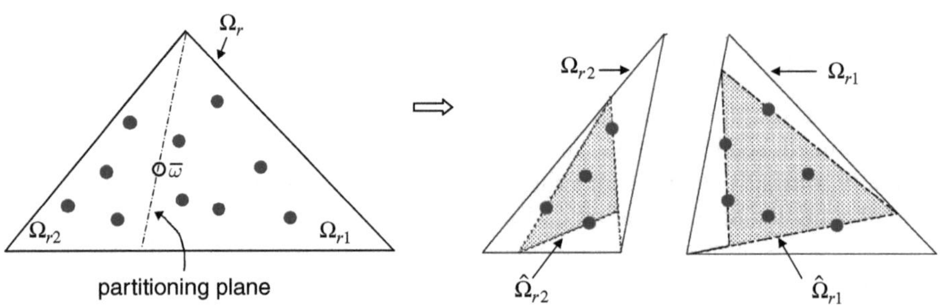

Fig. 6.6 Cell-redefining procedure for a scenario-based simplex in two dimensions

6.5 Asset Allocation Application

The mean–covariance approximation-based solution scheme outlined in the preceding sections will be applied to a certain financial asset allocation model, developed for a large bank (Edirisinghe et al. 1996). The basic problem is to allocate a financial budget among competing six asset classes, identified as US equities large capitalization stocks (UsEqLC), US equities small capitalization stocks (UsEqSC), non-US equities large capitalization stocks (NUsEqLC), emerging markets (EmerMkt), US bonds (UsBond), and cash/cash equivalents (UsCash), indexed as $k = 1, 2, 3, 4, 5,$ and $6(= K)$, respectively. The annual asset returns ξ_k, $k = 1, \ldots, 6$ are uncertain and modeled by random variables.

The bank has set an annual total wealth (W) target of T percent due to its obligations to its customers, and therefore, any shortfall from this target leads to an "embarrassment" (and thus, risk). Cost of such risk is quantified by a piecewise linear (convex) function where a negative wealth surprise from the terminal wealth target leads to penalty costs according to one of three levels determined by the embarrassment type: normal type (NT), higher type (HT), and severe type (ST). NT applies up to β_1 percent of T, HT applies from β_1 percent to β_2 percent of T, and ST applies for under β_2 percent of wealth target, where $0 < \beta_2 < \beta_1 < 1$. Penalty cost in each region is linear with cost rates q_1, q_2, and q_3, respectively, see Fig. 6.7. The initial positions in each asset class are denoted by x_k^0 and assigned 20% for all assets except for EmerMkt where it is 0%.

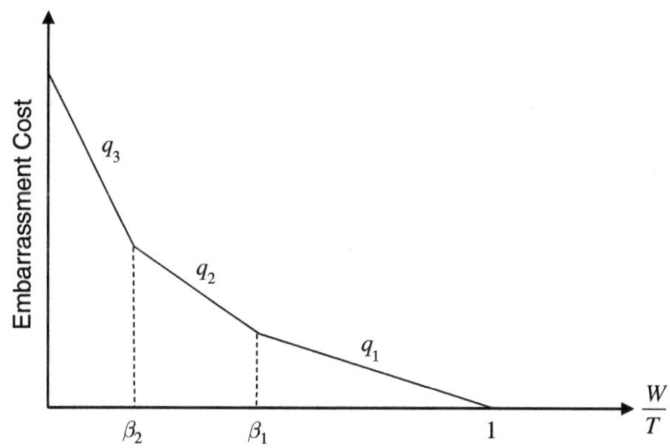

Fig. 6.7 Embarrassment penalty by wealth target shortfall

The objective is to determine optimal asset allocations for the upcoming year such that the expected total wealth, less the expected cost of embarrassment (for not meeting the specified wealth target), is maximized. Revision of asset holdings from their initial positions incurs (variable) transaction costs (TC) at the rate of a_k percent for purchases and b_k percent for sales (for asset class k). Asset holdings for the year are denoted by the decision variables x_k, asset purchases are denoted by x_{K+k}, and asset sales are denoted by x_{2K+k} for $k = 1, \ldots, K = 6$. The problem can be formulated as the following minimization two-stage stochastic program:

$$Z^* := \quad \min \quad -\sum_{k=1}^{K}(1+\bar{\xi}_k)x_k + \lambda\,\psi(x) \qquad \text{(AAM)}$$

$$\begin{aligned}
\text{s.t.}&\\
\text{asset revision:}\quad & x_k - x_{K+k} + x_{2K+k} = x_k^0,\ k = 1, \ldots, K\\
\text{budget constraint:}\quad & \sum_{k=1}^{K}(1+a_k)x_{K+k} - \sum_{k=1}^{K}(1-b_k)x_{2K+k} = 0\\
\text{asset class bounds:}\quad & \underline{x}_k \le x_k \le \bar{x}_k,\ k = 1, \ldots, K\\
\text{nonnegativity:}\quad & x_{K+k}, x_{2K+k} \ge 0,\ k = 1, \ldots, K\ ,
\end{aligned}$$

where λ is a risk aversion parameter that trades off the expected total wealth with penalty costs of falling short of the wealth target. $\psi(x)$ is the expectation of the cost of embarrassment $\phi(x, \xi)$ under the random asset return vector ξ, given by

$$
\begin{aligned}
\phi(x,\xi) := \quad & \min \quad q_1y_1 + q_2y_2 + q_3y_3 \\
& \text{s.t.} \\
& \text{Target shortfall:} \quad y_1 + y_2 + y_3 \geq T - \sum_{k=1}^{K}(1+\xi_k)x_k \\
& \text{embarrassment NT: } 0 \leq y_1 \leq (1-\beta_1)T \\
& \text{embarrassment HT: } 0 \leq y_2 \leq (\beta_1-\beta_2)T \\
& \text{embarrassment ST: } 0 \leq y_3.
\end{aligned}
$$

The case considered here is when the penalty rate q applicable to a specific embarrassment level being fixed (nonrandom) and thus the second-stage recourse function ϕ is a convex function of the random vector ξ. Penalty rates are set at $q_1 = 1$, $q_2 = 3$, and $q_3 = 5$, along with $\beta_1 = 0.9$ and $\beta_2 = 0.8$. Annual asset returns vector ξ is modeled by a multivariate lognormal distribution with the estimated means, standard deviations, and correlations given in Table 6.1. The stochastic program AAM is specified with discrete scenario samples drawn from this distribution. For the results reported here, $T = 104\%$ is set.

Table 6.1 Asset return parameters

Asset	Mean	Std. Dev	Correlations					
k	$\bar{\xi}_k$	σ_k	α_{1k}	α_{2k}	α_{3k}	α_{4k}	α_{5k}	α_{6k}
1	0.110	0.169	1					
2	0.109	0.248	0.801	1				
3	0.108	0.208	0.499	0.302	1			
4	0.106	0.249	0.300	0.307	0.296	1		
5	0.070	0.069	0.396	0.296	0.186	−0.015	1	
6	0.057	0.010	−0.006	−0.006	−0.016	−0.006	0.303	1

The mean–covariance bounds in (6.49) and (6.50) for the convex case are applied within the SSS iterative solution procedure to solve the AAM model for various sample sizes that approximate the lognormal distribution. Indeed, the accuracy of model solution depends on how closely the sample approximates the six-variate lognormal distribution. For a given sample, when the solution accuracy reaches $\varepsilon = \frac{1}{4}\,\%$, see (6.66), the SSS iteration is terminated with the lower bounding x-solution termed the optimal asset holdings. The progress of the SSS procedure for a large sample of 500,000 scenarios with no transaction costs and $\lambda = 0.6$ is depicted in Fig. 6.8. Observe that the bound improvement early in the iterative process is quite dramatic, but it tapers off after the bound accuracy has significantly improved. Furthermore, the computational effort is near linear with the partitioning iteration. All CPU times are on an IBM RS/6000 workstation. The trade-off between the

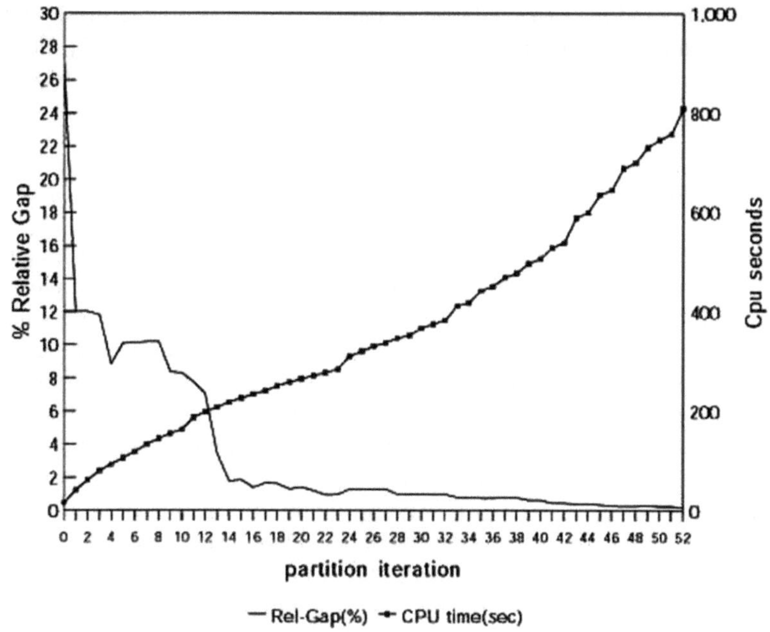

Fig. 6.8 SSS performance on AAM with 500,000 scenarios (0% TC)

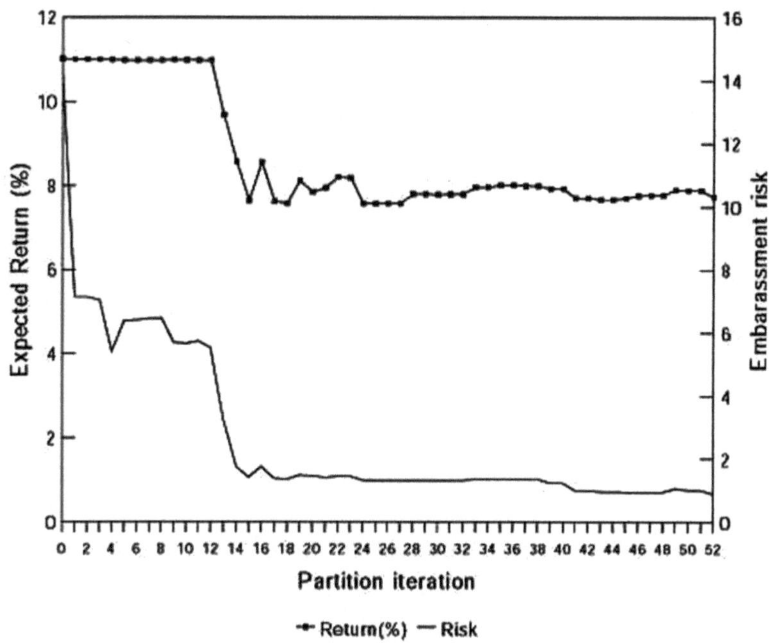

Fig. 6.9 Risk–return trade-off in AAM with 500,000 scenarios (0% TC)

expected return and the expected cost of target shortfall is plotted in Fig. 6.9. The optimal allocations in this case are `UsEqLC: 13.04%, UsEqSC: 0%, NUsEqLC: 7.93%, EmerMkt: 7.64%, UsBond: 39.83%, and UsCash: 31.56%`.

6.5.1 Scenario Clustering and Sensitivity Analysis

When the SSS procedure is applied, scenarios are partitioned into some cell structure, $\mathcal{S}^{\nu}$ at every iteration ν, thus forming clusters of scenarios. These clusters are then refined in subsequent partitioning iterations until the prescribed bound tolerance is satisfied. When the SSS procedure terminates, the location of these scenario clusters can provide useful information on sensitivity of the optimal solution on possible new scenarios. Suppose the average (mean) of the scenarios in cluster r, having $|S_r|$ scenarios, is $\bar{\xi}(r)$. Also, define the maximum distance among the scenarios in cell (simplex) Ξ_r by $D(r) := \max\{||\xi^{s_1} - \xi^{s_2}|| \ : \ s_1, s_2 \in S_r\}$, the cluster diameter. The pair, $\bar{\xi}(r)$ and $D(r)$, would provide information on not only where cells are located, but also how dispersed (or dense) a given scenario cluster is.

For instance, for an AAM model run with 100,000 scenarios, no transaction costs, and $\lambda = 0.6$, the optimal allocations are determined as $x^* \equiv$ `UsEqLC: 13.88%, UsEqSC: 0%, NUsEqLC: 6.79%, EmerMkt: 7.64%, UsBond: 24.28%, and UsCash: 47.41%` by the SSS procedure, which results in 28 simplicial cells. Table 6.2 lists how the 100,000 scenarios are clustered at termination into the 28 cells; there are many cells with fairly large number of scenarios. For example, the cell #3 having a cluster of 15,327 scenarios has not been further split because at the current optimal solution, the lower and upper bounds in this cell lead to WDIFF in (6.67) below the specified tolerance. Thus, the mean–covariance information of cell #3 alone is sufficient to verify the optimality of x^* with respect to this scenario cluster. This observation leads to a quick test to check the sensitivity of x^* if a new scenario, say ξ^{new}, is incorporated—determine first if ξ^{new} can be located in one of the 28 clusters. Suppose $\xi^{\text{new}} \in \Xi_r$. Then, lower and upper bounds need to be re-evaluated (with the modified moment information) only for cell Ξ_r to verify if the WDIFF criterion is still satisfied for the currently optimal allocation x^*—if not, cell Ξ_r should be further partitioned to improve the solution x^* in the presence of the new scenario ξ^{new}.

6.5.2 Efficient Frontiers and Transaction Costs

The above idea can be used in parametric analysis of the risk tolerance parameter λ. Given the preceding optimal allocation x^* for $\lambda = 0.6$, when λ is increased to (say) 0.7, it is only necessary to re-evaluate the bounds, fixing x^*, to see in which cell(s) WDIFF condition is possibly violated. Then, such cells may be further partitioned in parametric style for finding a new optimal solution. This practice can substantially reduce the total solution time, compared to starting "from scratch" every time.

Table 6.2 Scenario clustering in SSS procedure with 100,000 scenarios

Cell #	# of scen.	Scenario average $\bar{\xi}_k(r)$ UsqLC	UsEqSC	NUsEqLC	EmerMkt	UsBond	UsCash	Diameter $D(r)$
1	217	0.606	0.775	0.741	0.649	0.143	0.058	1.277
2	4	−0.234	−0.21	−0.508	0.102	0.033	0.062	0.27
3	15327	0.288	0.313	0.38	0.273	0.101	0.057	1.513
4	2590	0.159	0.139	0.367	0.211	0.081	0.057	0.978
5	11021	0.119	0.114	0.219	0.231	0.069	0.057	1.226
6	6373	0.051	0.012	0.241	0.117	0.063	0.057	1.014
7	2150	0.062	0.034	0.147	0.083	0.077	0.058	0.976
8	122	0.734	1.045	0.471	0.572	0.161	0.059	1.038
9	32	0.673	0.956	0.356	0.567	0.154	0.056	0.684
10	4	−0.116	−0.22	−0.066	0.061	−0.188	0.047	0.277
11	184	0.644	0.909	0.257	0.441	0.15	0.057	1.018
12	8120	0.379	0.473	0.141	0.233	0.112	0.057	1.459
13	3037	0.265	0.338	−0.032	0.122	0.098	0.057	1.18
14	1885	0.212	0.253	−0.008	0.1	0.104	0.058	1.049
15	3844	0.17	0.203	−0.042	0.07	0.094	0.058	0.948
16	3742	0.11	0.137	−0.089	0.118	0.064	0.057	1.044
17	6167	−0.014	−0.127	0.095	0.096	0.09	0.059	1.101
18	2324	−0.019	−0.061	0.037	−0.037	0.066	0.057	0.77
19	2266	−0.027	−0.041	0.017	−0.06	0.048	0.057	0.659
20	3889	0.046	0.074	0.096	0.088	0.046	0.056	0.861
21	5475	−0.079	−0.128	−0.027	−0.107	0.047	0.057	0.643
22	3675	0.059	0.063	−0.134	−0.027	0.072	0.057	0.82
23	4548	−0.034	−0.044	−0.211	−0.08	0.025	0.056	0.893
24	3950	−0.133	−0.187	−0.091	−0.151	0.017	0.056	0.669
25	1573	0.017	0.089	0.041	−0.025	0.008	0.054	0.792
26	4110	−0.019	−0.018	0.041	0.15	0.007	0.055	0.823
27	1584	−0.191	−0.251	−0.15	−0.182	−0.041	0.055	0.573
28	1787	−0.055	−0.135	0.013	−0.072	0.076	0.058	0.59

In particular, such a parametric approach can be applied when *efficient frontiers* between risk and return are desired. Figure 6.10 illustrates the savings in CPU time under the risk tolerance parametric analysis for the 100,000 scenario problem under zero transaction costs. Here, starting from $\lambda = 0.2$, the cumulative (total) CPU time to get the new optimal solution is plotted as λ is increased.

Complete efficient frontiers under various levels of transaction costs (see Fig. 6.11) are generated under parametric analysis—$a_k = b_k = 0\%, \frac{1}{4}\%, \frac{1}{2}\%, \frac{3}{4}\%,$ and 1%. Each point in the figure corresponds to an AAM model run with the same sample of 100,000 scenarios and a solution accuracy of 0.25% in SSS. In the same figure, several optimal asset allocations are indicated to show the impact on asset allocation due to TC. At moderate risk levels, the loss of return is quite limited as TC increases, although optimal asset allocations do vary considerably under different TC levels. In contrast, at very high or low risk levels, TC does lead to a significant erosion in return for the same level of risk.

How difficult is it for the SSS procedure to solve the AAM stochastic program as TC increases? As portfolio revision becomes more expensive (under increased

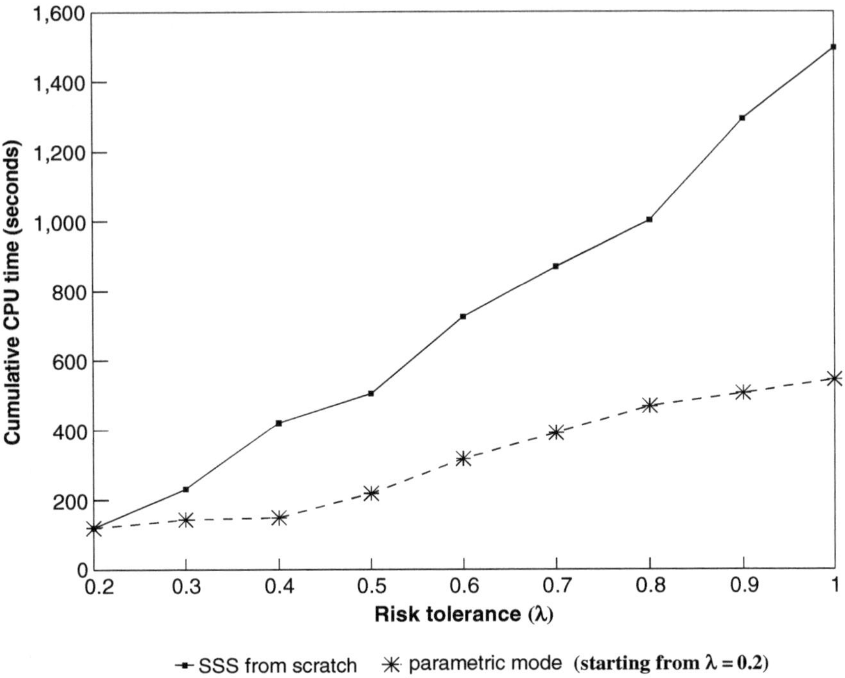

Fig. 6.10 CPU time comparison with risk tolerance parametric analysis (0% TC)

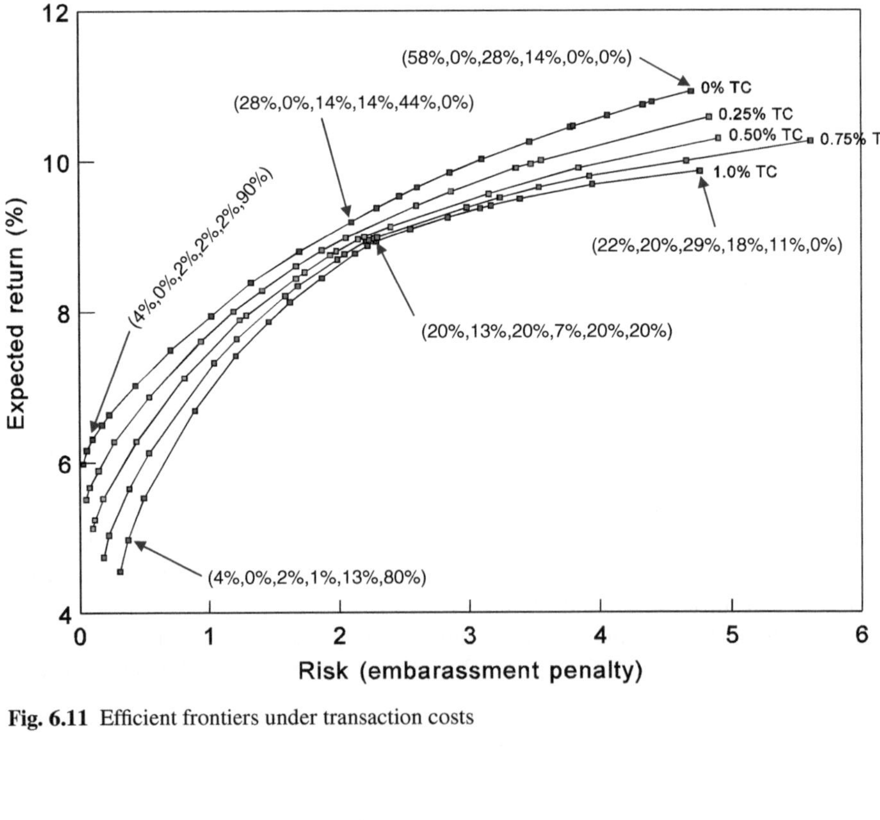

Fig. 6.11 Efficient frontiers under transaction costs

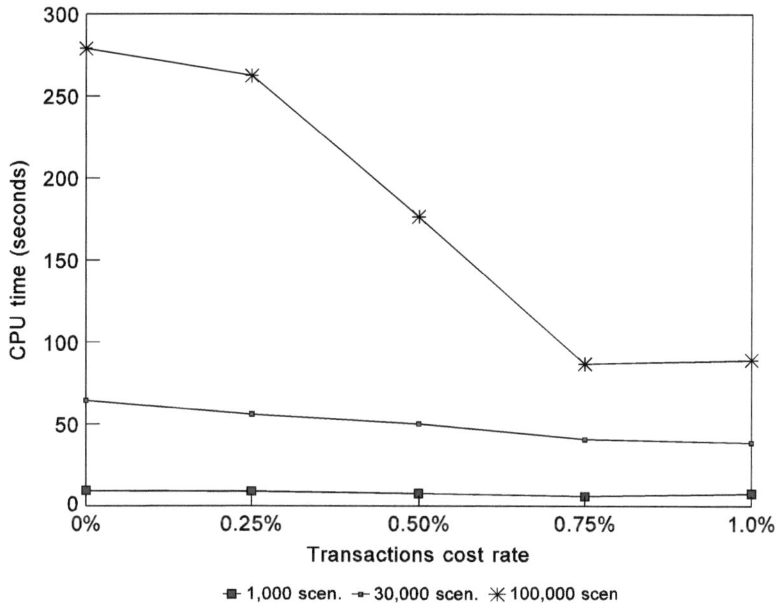

Fig. 6.12 CPU time sensitivity with transaction costs

TC), an optimal solution is attained with less clustering effort, i.e., the SSS scheme creates fewer partitions in the scenario domain. This is evident in Fig. 6.12, where the reported CPU times are averaged over four runs for each sample size (using $\lambda = 0.6$). Upon increasing TC, the computational burden diminishes monotonically, more prominently so for large sample sizes.

Using TC=0%, the sensitivity of the CPU time on increasing sample size is depicted in Fig. 6.13. In the same figure, solution time using (Bender's) L-shaped decomposition (LSD) algorithm (Van Slyke and Wets 1969) is also plotted. LSD also sets the termination criterion at 0.25%, the value set in the SSS scheme. Each CPU time is the result of averaging over three sample runs. Clearly, the SSS procedure is quite insensitive to increases in sample size, in comparison to the LSD algorithm which indeed is severely affected by the sample size.

Often, one needs to address the sample bias in the optimal asset allocation. Indeed, optimal allocations are sensitive to the sample size, and sample-based optimal solutions asymptotically converge to the true optimal asset allocations as sample sizes increase—convergence rate is typically of the order $O(\frac{1}{\sqrt{|S|}})$. Figure 6.14 illustrates the stability of the optimal allocation for various TC levels. For each asset class k, for a (fixed) sample size $n := |S|$, standard deviation, $SD_k(n)$, is computed by solving AAM for different samples of size n. Then, the metric concerning allocation sensitivity on sample size is defined by $SD^{\max}(n) := \max\{SD_k(n) : k = 1, \ldots, K = 6\}$. $SD^{\max}(n)$ is plotted against n at various transaction cost

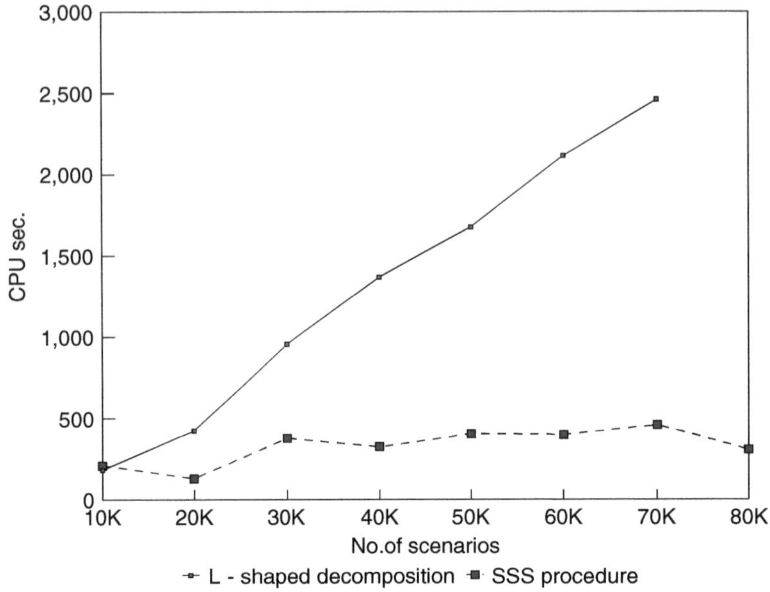

Fig. 6.13 CPU time comparison with Bender's decomposition

levels in Figure 6.14. It can be concluded that at least for low transaction cost levels, it is necessary to have a sufficiently large sample size to gain stability in the asset allocations. In this sense, the SSS procedure becomes an invaluable solution tool.

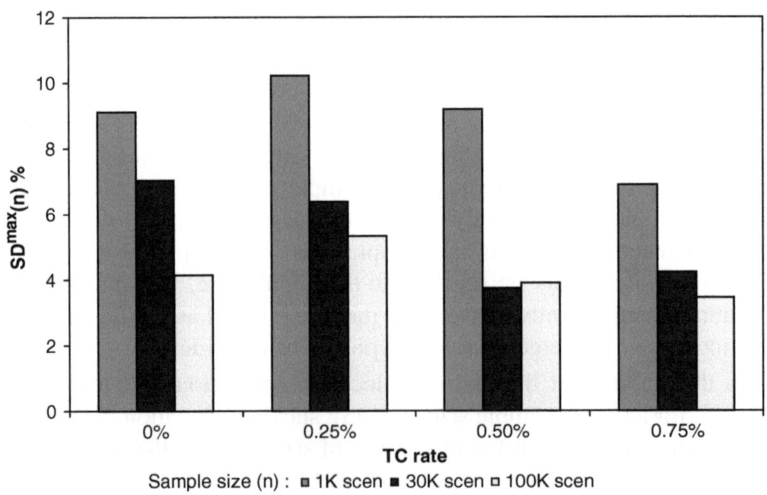

Fig. 6.14 Optimal allocation sensitivity with sample size

6.6 Concluding Remarks

This chapter focused on using limited moment information of the underlying random vector to obtain bounds on the expectation of the recourse function in stochastic programming. These bounds are then categorized as first-order or second (and higher)-order moment approximations. In particular, the mean–covariance bound under simplicial domains was exploited within a sequential refinement scheme to develop an efficient solution procedure.

Bounds based on functional (as opposed to distributional) approximations are developed in the literature for several special cases, notably for, e.g., separable, restricted, or network recourse, see Birge and Dula (1991), Morton and Wood (1999), and Wallace (1987). Such approximations are not easy to embed within a sequential approximation scheme toward solving the underlying stochastic program, except possibly when specialized structures exist. These bounds are generally more appealing under a large number of random variables. Nevertheless, with distributional approximations such as the simplicial mean–covariance bounds, stochastic programs can be solved with hundreds of random variables. In contrast, evaluating moment-based bounds that use rectangular domains would require astronomical amount of computations, let alone their refinement.

It must also be noted that the success of the SSS scheme in the reported financial asset allocation application is partly due to the two-stage nature of the problem. In practice, such asset allocations must be made dynamically over time in multiple future time periods, thus leading to multiperiod stochastic programming models. These involve random vectors (of successive time periods) that are nested to form scenario trees. In such cases, at least conceptually, the simplicial bounds can be applied within each period and nested from period to period to form approximating scenario trees. Needless to say that as the period-based distributions are refined and nested together in a sequential approximation procedure, solving even the bounding multistage program may become an onerous task in itself. This is especially the case when there are many decision stages in the stochastic program. The approach in Edirisinghe and Ziemba (1992) and the barycentric scenario tree development in Frauendorfer (1996) and Frauendorfer (1994) follow such a logic. For a theoretical exposition of the multistage case, see Kuhn (2004) and the Chapter 5 in this book by Frauendorfer, Kuhn, and Schürle. With T stages of random observations, each having N random variables, the resulting barycentric scenario tree will have $(N+1)^T$ number of scenarios before any refinement of the bounds is contemplated. An alternative approach is proposed in Edirisinghe (1999) where bounds are used in an aggregation scheme that does not suffer from the above dimensionality dilemma. Also, see Casey and Sen (2005) for an algorithm that generates an iterative sequence of scenario trees with asymptotic convergence properties.

References

Birge, J.R.: An l-shaped method computer code for multistage linear programs. In: Ermoliev, Y., Wets, R.J-B. (eds.) Numerical Techniques for Stochastic Optimization. Springer, New York, NY (1988)

Birge, J.R., Dula, J.: Bounding separable recourse functions with limited moment information. Ann. Oper. Res. **30**, 277–298 (1991)

Birge, J.R., Wallace, S.W.: Refining bounds for stochastic linear programs with linearly transformed independent random variables. Oper. Res. Lett. **5**, 73–77 (1986)

Birge, J.R., Wallace, S.W.: A separable piecewise linear upper bound for stochastic linear programs. SIAM J. Control Optim. **26**, 725–739 (1988)

Birge, J.R., Wets, R.J.-B.: Designing approximation schemes for stochastic optimization problems, in particular for stochastic programs with recourse. Math. Program. Study. **27**, 54–102 (1986)

Birge, J.R., Wets, R.J.-B.: Computing bounds for stochastic programming problems by means of a generalized moment problem. Math. Oper. Res. **12**, 149–162 (1987)

Birge, J.R., Wets, R.J.-B.: Sublinear upper bounds for stochastic programs with recourse. Math. Program. **43**, 131–149 (1989)

Casey, M.S., Sen, S.: The scenario generation algorithm for multistage stochastic linear programming. Math. Oper. Res. **30**, 615–631 (2005)

Dantzig, G.B., Glynn, P.W.: Parallel processors for planning under uncertainty. Ann. Oper. Res. **22**, 1–21 (1990)

Dokov, S.P., Morton, D.P.: Higher-order upper bounds on the expectation of a convex function, University of Texas at Austin, Austin, TX (2002)

Dokov, S.P., Morton, D.P.: Second-order lower bounds on the expectation of a convex function. Math. Oper. Res. **30**, 662–677 (2005)

Dula, J.H.: An upper bound on the expectation of simplicial functions of multivariate random variables. Math. Program. **55**, 69–80 (1992)

Dupačová, J.: On minimax solutions of stochastic linear programming problems. Časopis Pro Pětováni Matematiky. **91**, 423–430 (1966) (as Žáčková)

Dupačová, J.: Minimax stochastic programs with nonconvex nonseparable penalty functions. In: Prékopa, A. (ed.) Progress in Operations Research, pp. 303–316. North-Holland, Amsterdam (1976)

Edirisinghe, C., Carino, D., Fan Y.-A., Murray, S., Stacey, C.: Implementation of stochastic programming approximations for asset allocation two-stage models, 1996. Invited talk at INFORMS national meeting, Atlanta, GA, (1996)

Edirisinghe, N.C.P.: New second-order bounds on the expectation of saddle functions with applications to stochastic linear programming. Oper. Res. **44**, 909–922 (1996)

Edirisinghe, N.C.P.: Bound-based approximations in multistage stochastic programming: Nonanticipativity aggregation. Ann. Oper. Res. **85**, 103–127 (1999)

Edirisinghe, N.C.P.: A Linear-Time Algorithm for Enclosing Multivariate Data in a Compact Simplex. University of Tennessee, Knoxville, TN (2005)

Edirisinghe, N.C.P., Patterson, E. I.: Multiperiod stochastic portfolio optimization: Block-separable decomposition. *Ann. Oper. Res.* **152**, 367–394 (2007)

Edirisinghe, N.C.P., You, G-M.: Second-order scenario approximation and refinement in optimization under uncertainty. *Ann. Oper. Res.* **19**, 314–340 (1996)

Edirisinghe, N.C.P., Ziemba, W.T.: Tight bounds for stochastic convex programs. Oper. Res. **40**, 660–677 (1992)

Edirisinghe, N.C.P., Ziemba, W.T.: Bounds for two-stage stochastic programs with fixed recourse. Math. Oper. Res. **19**, 292–313 (1994a)

Edirisinghe, N.C.P., Ziemba, W.T.: Bounding the expectation of a saddle function with application to stochastic programming. Math. Oper. Res. **19**, 314–340 (1994b)

Edirisinghe, N.C.P., Ziemba, W.T.: Implementing bounds-based approximations in convex-concave two-stage stochastic programming. Math. Program. **19**, 314–340 (1996)

Ermoliev, Y.: Stochastic quasigradient methods and their application to systems optimization. Stochastics **9**, 1–36 (1983)

Ermoliev, Y., Gaivoronsky, A.: Stochastic quasigradient methods and their implementation. In: Ermoliev, Y., Wets, R.J-B., (eds.) Numerical Techniques for Stochastic Optimization. Springer, New York, NY (1988)

Ermoliev, Y., Wets, R.J-B.: Numerical Techniques for Stochastic Optimization. Springer, New York, NY (1988)
Frauendorfer, K.: Slp problems: Objective and right-hand side being stochastically dependent—part ii, University of Zurich (1988a)
Frauendorfer, K.: Slp recourse problems with arbitrary multivariate distributions—the dependent case. Math. Oper. Res. **13**, 377–394 (1988b)
Frauendorfer, K.: Stochastic Two-Stage Programming. Springer, Berlin (1992)
Frauendorfer, K.: Multistage stochastic programming: Error analysis for the convex case. Z. Oper. Res. **39**, 93–122 (1994)
Frauendorfer, K.: Barycentric scenario trees in convex multistage stochastic programming. Math. Program. **75**, 277–293 (1996)
Frauendorfer, K., Kall, P.: A solution method for slp problems with arbitrary multivariate distributions—the independent case. Prob. Cont. Inf. Theor. **17**, 177–205 (1988)
Gassmann, H.I.: Mslip: A computer code for the multistage stochastic linear programming problem. Math. Program. **47**, 407–423 (1990)
Gassmann, H.I., Ziemba, W.T.: A tight upper bound for the expectation of a convex function of a multivariate random variable. Math. Program. Study. **27**, 39–53 (1986)
Glashoff, K., Gustafson, S-A.: Linear Optimization and Approximation. Springer, New York, NY (1983)
Higle, J.L., Sen, S.: Stochastic decomposition: An algorithm for two-stage stochastic linear programs with recourse. Math. Oper. Res. **16**, 650–669 (1991)
Huang, C.C., Ziemba, W.T., Ben-Tal, A.: Bounds on the expectation of a convex function of a random variable: With applications to stochastic programming. Oper. Res. **25**, 315–325 (1977)
Infanger, G.: Monte carlo (importance) sampling within a benders decomposition algorithm for stochastic linear programs. Ann. Oper. Res. **39**, 69–95 (1992)
Infanger, G.: Planning Under Uncertainty, Solving Large-scale Stochastic Programs. The Scientific Press Series, Danvers, MA (1994)
Jensen, J.L.: Sur les fonctions convexes et les inégalités entre les valeurs moyennes. *Acta. Math.* **30**, 173–177 (1906)
Kall, P.: Stochastic Linear Programming. Springer, Berlin (1976)
Kall, P.: Stochastic programs with recourse: An upper bound and the related moment problem. Z. Oper. Res. **31**, A119–A141 (1987a)
Kall, P.: On approximations and stability in stochastic programming. In: Guddat, J., Jongen, H.Th., Kummer, B., Nozicka, F., (eds.) Parametric Optimization and Related Topics, pp. 387–407 Akademie, Berlin (1987b)
Kall, P.: Stochastic programming with recourse : Upper bounds and moment problems—a review. In: Guddat, J., Bank, B., Hollatz, H., Kall, P., Klatte, D., Kummer, B., Lommatzsch, K., Vlach, M., Zimmermann, K. (eds.) Advances in Mathematical Optimization and Related Topics, pp. 86–103. Akademie, Berlin (1988)
Kall, P.: An upper bound for slp using first and total second moments. Ann. Oper. Res. **30**, 267–276 (1991)
Kall, P., Stoyan, D.: Solving stochastic programming problems with recourse including error bounds. Math. Oper. schung Stat. Ser. Optim. **13**, 431–447 (1988)
Kall, P., Wallace, S.W.: Stochastic Programming. Wiley, New York, NY (1994)
Karr, A.F.: Extreme points of certain sets of probability measures, with applications. Math. Oper. Res. **8**, 74–85 (1983)
Kemperman, J.: The general moment problem. a geometric approach. Ann. Math. Stat. **39**, 93–112 (1968)
Kuhn, D.: Generalized Bounds for Convex Multistage Stochastic Programs. Lecture Notes in Economics and Mathematical Systems, Vol. 548. Springer, Berlin (2004)
Lorentz, G.G.: Bernstein Polynomials, 2nd edn. Chelsea, New York, NY (1986)
Madansky, A.: Bounds on the expectation of a convex function of a multivariate random variable. Ann. Math. Stat. **30**, 743–746 (1959)

Morton, D.P., Wood, R.K.: Restricted-recourse bounds for stochastic linear programming. Oper. Res. **47**, 943–956 (1999)

Robinson, S.M., Wets, R.J.-B.: Stability in two-stage stochastic programming. SIAM J. Control Optim. **25**, 1409–1416 (1987)

Van Slyke, R., Wets, R.J.-B.: L-shaped linear programs with application to optimal control and stochastic optimization. SIAM J. App. Math. **17**, 638–663 (1969)

Wallace, S. W.: A piecewise linear upper bound on the network recourse function. Math. Program. **38**, 133–146 (1987)

Wets, R.J-B.: Stochastic programs with fixed recourse: The equivalent deterministic problem. SIAM Rev. **16**, 309–339 (1974)

Wets, R.J-B.: Convergence of convex functions, variational inequalities and convex optimization problems. In: Cottle, R.W., Giannesi, F., Lious, J-L., (eds.) Variational Inequalities and Complimentary Problems, pp. 375–403. Wiley, New York, NY (1980)

Wets, R.J-B.: Stochastic programming: Solution techniques and approximation schemes. In: Bachem, A., Groetschel, M., Korte, B., (eds.) Mathematical Programming: State-of-the-Art, pp. 566–603 Springer, Berlin (1982)

Ziemba, W.T.: Computational algorithms for convex stochastic programs with simple recourse. Oper. Res. **18**, 414–431 (1970)

Chapter 7
Stability and Scenario Trees for Multistage Stochastic Programs

Holger Heitsch and Werner Römisch

7.1 Introduction

Multistage stochastic programs are often used to model practical decision processes over time and under uncertainty, e.g., in finance, production, energy, and logistics. We refer to the pioneering work of Dantzig (1955, 1963) and to the recent books by Ruszczyński and Shapiro (2003) and Wallace and Ziemba (2005), and the monograph by Kall and Mayer (2005) for the state of the art of the theory and solution methods for multistage models and for a variety of applications.

The inputs of multistage stochastic programs are multivariate stochastic processes $\{\xi_t\}_{t=1}^T$ defined on some probability space $(\Omega, \mathcal{F}, \mathbb{P})$ and with ξ_t taking values in some $\mathbb{R}^d$. The decision x_t at t belonging to $\mathbb{R}^{m_t}$ is assumed to be *nonanticipative*, i.e., to depend only on $(\xi_1, \ldots, \xi_t)$. This property is equivalent to the measurability of x_t with respect to the σ-field $\mathcal{F}_t(\xi) \subseteq \mathcal{F}$ which is generated by $(\xi_1, \ldots, \xi_t)$. Clearly, we have $\mathcal{F}_t(\xi) \subseteq \mathcal{F}_{t+1}(\xi)$ for $t = 1, \ldots, T-1$. Since at time $t = 1$ the input is known, we assume that $\mathcal{F}_1 = \{\emptyset, \Omega\}$.

The multistage stochastic program is assumed to be of the form

$$\min \left\{ \mathbb{E}\left[\sum_{t=1}^{T} \langle b_t(\xi_t), x_t \rangle \right] \;\middle|\; \begin{array}{l} x_t \in X_t, t = 1, \ldots, T, A_{1,0}x_1 = h_1(\xi_1), \\ x_t \text{ is } \mathcal{F}_t(\xi)\text{-measurable}, \ t = 1, \ldots, T, \\ A_{t,0}x_t + A_{t,1}(\xi_t)x_{t-1} = h_t(\xi_t), t = 2, \ldots, T \end{array} \right\}, \tag{7.1}$$

where the sets $X_t \subseteq \mathbb{R}^{m_t}$ are polyhedral cones, the cost coefficients $b_t(\xi_t)$ and right-hand sides $h_t(\xi_t)$ belong to $\mathbb{R}^{m_t}$ and $\mathbb{R}^{n_t}$, respectively, the fixed recourse matrices $A_{t,0}$ and the technology matrices $A_{t,1}(\xi_t)$ are (n_t, m_t)- and (n_t, m_{t-1})-matrices, respectively. The costs $b_t(\cdot)$, technology matrices $A_{t,1}(\cdot)$, and right-hand sides $h_t(\cdot)$ are assumed to depend affinely linear on ξ_t.

While the first and third groups of constraints in (7.1) have to be satisfied pointwise with probability 1, the second group, the measurability or *information*

H. Heitsch (✉)
Institute of Mathematics, Humboldt-University Berlin, Berlin, Germany
e-mail: heitsch@math.hu-berlin.de

G. Infanger (ed.), *Stochastic Programming*, International Series in Operations Research & Management Science 150, DOI 10.1007/978-1-4419-1642-6_7,

constraints, are functional and non-pointwise at least if $T > 2$ and $\mathcal{F}_2 \subsetneqq \mathcal{F}_t \subseteq \mathcal{F}$ for some $2 < t \leq T$. The presence of such qualitatively different constraints constitutes the origin of both the theoretical and computational challenges of multistage models. Recent results (see Shapiro 2006, 2008) indicate that multistage stochastic programs have higher computational complexity than two-stage models.

The main computational approach to multistage stochastic programs consists in approximating the stochastic process $\xi = \{\xi_t\}_{t=1}^T$ by a process having finitely many scenarios exhibiting tree structure and starting at a fixed element ξ_1 of $\mathbb{R}^d$. This leads to linear programming models that are very large scale in most cases and can be solved by linear programming techniques, in particular by decomposition methods that exploit specific structures of the model. We refer to Ruszczyński and Shapiro (2003, Chapter 3) for a recent survey.

Presently, there exist several approaches to generate scenario trees for multistage stochastic programs (see Dupačová et al. 2000 for a survey). They are based on several different principles. We mention here (i) bound-based constructions by Casey and Sen (2005), Edirisinghe (1999), Frauendorfer (1996), and Kuhn (2005); (ii) Monte Carlo-based schemes by Chiralaksanakul and Morton (2005) and Shapiro (2003, 2008) or quasi-Monte Carlo-based methods by Pennanen (2005, 2009); (iii) (EVPI-based) sampling within decomposition schemes by Corvera Poiré (2005), Dempster (2004), Higle et al. (to appear), and Infanger (1994); (iv) the target/moment-matching principle by Høyland and Wallace (2001) and Høyland et al. (2003); and (v) probability metric-based approximations by Gröwe-Kuska et al. (2003), Heitsch and Römisch (2009), Hochreiter (2005), Hochreiter and Pflug (2007) and Pflug (2001).

We add a few more detailed comments on some of the recent work. The approach of (i) relies on constructing discrete probability measures that correspond to lower and upper bounds (under certain assumptions on the model and the stochastic input) and on refinement strategies. The recent paper/monograph by Casey and Sen (2005) and Kuhn (2005) belonging to (i) also offers convergence arguments (restricted to linear models containing only stochasticity in right-hand sides in Casey and Sen (2005) and to convex models whose stochasticity is assumed to follow some linear block-diagonal autoregressive process with compact supports in Kuhn 2005). The Monte Carlo-based methods in (ii) utilize conditional sampling schemes and lead to a large number of (pseudo) random number generator calls for conditional distributions. Consistency results are shown in Shapiro (2003), and the complexity is discussed in Shapiro (2006). The quasi-Monte Carlo-based methods in Pennanen (2005, 2009) are developed for convex models and for stochastic processes driven by time series models with uniform innovations. While the general theory on epi-convergent discretizations in Pennanen (2005) also applies to conditional sampling procedures, a general procedure for generating scenario trees of such time series-driven stochastic processes is developed in Pennanen (2009) by approximating each of the (independent) uniform random variables using quasi-Monte Carlo methods (see Niederreiter 1992). The motivation of using quasi-Monte Carlo schemes originates from their remarkable convergence properties and good performance for the computation of high-dimensional integrals while "generating random samples is difficult" (Niederreiter 1992, p. 7). The approach of (v) is based on probability distances that are relevant for the stability of multistage models. While the papers by Gröwe-Kuska et al. (2003), Hochreiter and Pflug (2007), and Pflug (2001) employ Fortet–Mourier or Wasserstein distances, our recent work (Heitsch and Römisch 2009) is based on the rigorous stability result for linear multistage stochastic programs in Heitsch et al. (2006). Most of the methods for generating scenario trees require to prescribe (at least partially) the tree structure. Finally, we also mention the importance of evaluating the quality of scenario trees and of a postoptimality analysis (Dupačová et al. 2000, Kaut and Wallace 2007).

In the present chapter we extend the theoretical results obtained in Heitsch et al. (2006) by proving an existence result for solutions of (7.1) (Theorem 7.1), a Lipschitz stability result for ε-approximate solution sets, and a (qualitative) stability result for solutions of multistage models. In addition, we review the forward technique of Heitsch and Römisch (2009) for generating scenario trees. Its idea is to start with an initial finite scenario set with given probabilities which represents a "good" approximation of the underlying stochastic input process ξ. Such a finite set of scenarios may be obtained by sampling or resampling techniques based on parametric or nonparametric stochastic models of ξ or by optimal quantization techniques (Luschgy 2000). Starting from the initial scenario set, a tree is constructed recursively by scenario reduction (Dupačová et al. 2003; Heitsch and Römisch 2003) and bundling (Algorithm 7.1). We review an error estimate for Algorithm 7.1 in terms of the L_r-distance (Theorem 7.6) and a convergence result (Theorem 7.8). Algorithm 7.1 represents a stability-based heuristic for generating scenario trees. It has been implemented and tested on real-life data in several practical applications. Numerical experience was reported in Heitsch and Römisch (2009) on generating inflow demand scenario trees based on real-life data provided by the French electricity company EdF. Algorithm 7.1 or a modified version was used in Schmöller (2005) to generate scenario trees in power engineering models and in Möller et al. (2008) on generating passenger demand scenario trees in airline revenue management.

Section 7.2 presents extensions of the stability result of Heitsch et al. (2006), which provide the basis of our tree constructions. Section 7.3 reviews some results of Heitsch and Römisch (2009), in particular, the forward tree construction and error estimates in terms of the L_r-distances and a convergence result. In Section 7.4 we discuss some numerical experience on generating load-price scenario trees for an electricity portfolio optimization model based on real-life data of a municipal German power company.

7.2 Stability of Multistage Models

We assume that the stochastic input process $\xi = \{\xi_t\}_{t=1}^T$ belongs to the linear space $\times_{t=1}^T L_r(\Omega, \mathcal{F}, \mathbb{P}; \mathbb{R}^d)$ for some $r \in [1, +\infty]$. The model (7.1) is regarded as optimization problem in the space $\times_{t=1}^T L_{r'}(\Omega, \mathcal{F}, \mathbb{P}; \mathbb{R}^{m_t})$ for some $r' \in [1, \infty]$, where both linear spaces are Banach spaces when endowed with the norms

$$\|\xi\|_r := \left(\sum_{t=1}^{T} \mathbb{E}[|\xi_t|^r]\right)^{\frac{1}{r}} \quad \text{for } r \in [1,\infty) \text{ and } \quad \|\xi\|_\infty := \max_{t=1,\dots,T} \operatorname{ess\,sup} |\xi_t|,$$

$$\|x\|_{r'} := \left(\sum_{t=1}^{T} \mathbb{E}[|x_t|^{r'}]\right)^{\frac{1}{r'}} \quad \text{for } r' \in [1,\infty) \text{ and } \quad \|x\|_\infty := \max_{t=1,\dots,T} \operatorname{ess\,sup} |x_t|,$$

respectively. Here, $|\cdot|$ denotes some norm on the relevant Euclidean spaces and r' is defined by

$$r' := \begin{cases} \frac{r}{r-1} & \text{if costs are random,} \\ r & \text{if only right-hand sides are random,} \\ r = 2 & \text{if only costs and right-hand sides are random,} \\ \infty & \text{if all technology matrices are random and } r = T. \end{cases} \tag{7.2}$$

The definition of r' is justified by the proof of Heitsch et al. (2006, Theorem 2.1), which we record as Theorem 7.2. Since r' depends on r and our assumptions will depend on both r and r', we will add some comments on the choice of r and its interplay with the structure of the underlying stochastic programming model. To have the stochastic program well defined, the existence of certain moments of ξ has to be required. This fact is well known for the two-stage situation (see, e.g., Chapter 2 in Ruszczyński and Shapiro 2003). If either right-hand sides or costs in a multistage model (7.1) are random, it is sufficient to require $r \geq 1$. The flexibility in case that the stochastic process ξ has moments of order $r > 1$ may be used to choose r' as small as possible in order to weaken the condition (A3) (see below) on the feasible set. If the linear stochastic program is fully random (i.e., costs, right-hand sides, and technology matrices are random), one needs $r \geq T$ to have the model well defined and no flexibility on r' remains.

Let us introduce some notation. Let F denote the objective function defined on $L_r(\Omega, \mathcal{F}, \mathbb{P}; \mathbb{R}^s) \times L_{r'}(\Omega, \mathcal{F}, \mathbb{P}; \mathbb{R}^m) \to \overline{\mathbb{R}}$ by

$$F(\xi, x) := \begin{cases} \mathbb{E}\Big[\sum_{t=1}^{T} \langle b_t(\xi_t), x_t \rangle\Big], & x \in \mathcal{X}(\xi), \\ +\infty, & \text{otherwise,} \end{cases}$$

where

$$\mathcal{X}(\xi) := \{x \in L_{r'}(\Omega, \mathcal{F}, \mathbb{P}; \mathbb{R}^m) : x_1 \in \mathcal{X}_1(\xi_1), x_t \in \mathcal{X}_t(x_{t-1}; \xi_t),\ t = 2, \dots, T\}$$

is the set of feasible elements of (7.1) and

$$\mathcal{X}_1(\xi_1) := \{x_1 \in X_1 : A_{1,0} x_1 = h_1(\xi_1)\},$$
$$\mathcal{X}_t(x_{t-1}; \xi_t) := \{x_t \in \mathbb{R}^{m_t} : x_t \in X_t,\ A_{t,0} x_t + A_{t,1}(\xi_t) x_{t-1} = h_t(\xi_t)\}$$

the tth feasibility set for every $t = 2, \ldots, T$. Denoting by

$$\mathcal{N}_{r'}(\xi) := \times_{t=1}^{T} L_{r'}(\Omega, \mathcal{F}_t(\xi), \mathbb{P}; \mathbb{R}^{m_t})$$

the nonanticipativity subspace of ξ allows to rewrite the stochastic program (7.1) in the form

$$\min\{F(\xi, x) : x \in \mathcal{N}_{r'}(\xi)\}. \tag{7.3}$$

Let $v(\xi)$ denote the optimal value of (7.3) and, for any $\alpha \geq 0$, let

$$S_\alpha(\xi) := \{x \in \mathcal{N}_{r'}(\xi) : F(\xi, x) \leq v(\xi) + \alpha\}$$

denote the *α-approximate solution set* of the stochastic program (7.3). Since, for $\alpha = 0$, the set $S_\alpha(\xi)$ coincides with the set solutions to (7.3), we will also use the notation

$$S(\xi) := S_0(\xi).$$

The following conditions are imposed on (7.3):
(A1) $\xi \in L_r(\Omega, \mathcal{F}, \mathbb{P}; \mathbb{R}^s)$, i.e., $\int_\Omega |\xi(\omega)|^r d\mathbb{P}(\omega) < \infty$.
(A2) There exists a $\delta > 0$ such that for any $\tilde{\xi} \in L_r(\Omega, \mathcal{F}, \mathbb{P}; \mathbb{R}^s)$ with $\|\tilde{\xi} - \xi\|_r \leq \delta$, any $t = 2, \ldots, T$ and any $x_1 \in \mathcal{X}_1(\tilde{\xi}_1)$, $x_\tau \in \mathcal{X}_\tau(x_{\tau-1}; \tilde{\xi}_\tau)$, $\tau = 2, \ldots, t-1$, there exists an $\mathcal{F}_t(\tilde{\xi})$-measurable $x_t \in \mathcal{X}_t(x_{t-1}; \tilde{\xi}_t)$ (*relatively complete recourse locally around* ξ).
(A3) The optimal values $v(\tilde{\xi})$ of (7.3) with input $\tilde{\xi}$ are finite for all $\tilde{\xi}$ in a neighborhood of ξ and the objective function F is *level-bounded locally uniformly at* ξ, i.e., for some $\alpha > 0$ there exists a $\delta > 0$ and a bounded subset B of $L_{r'}(\Omega, \mathcal{F}, \mathbb{P}; \mathbb{R}^m)$ such that $S_\alpha(\tilde{\xi})$ is contained in B for all $\tilde{\xi} \in L_r(\Omega, \mathcal{F}, \mathbb{P}; \mathbb{R}^s)$ with $\|\tilde{\xi} - \xi\|_r \leq \delta$.

For any $\tilde{\xi} \in L_r(\Omega, \mathcal{F}, \mathbb{P}; \mathbb{R}^s)$ with $\|\tilde{\xi} - \xi\|_r \leq \delta$, condition (A2) implies the existence of some feasible $\tilde{x}$ in $\mathcal{X}(\tilde{\xi})$ and (7.2) implies the finiteness of the objective $F(\tilde{\xi}, \cdot)$ at any feasible $\tilde{x}$. A sufficient condition for (A2) to hold is the *complete recourse condition* on every recourse matrix $A_{t,0}$, i.e., $A_{t,0}X_t = \mathbb{R}^{n_t}$, $t = 1, \ldots, T$. The locally uniform level-boundedness of the objective function F is quite standard in perturbation results for optimization problems (see, e.g., Rockafellar and Wets 1998 Theorem 1.17). The finiteness condition on the optimal value $v(\xi)$ is not implied by the level-boundedness of F for all relevant pairs (r, r'). In general, the conditions (A2) and (A3) get weaker for increasing r and decreasing r', respectively.

To state our first result on the existence of solutions to (7.3) in full generality, we need two additional conditions:
(A4) There exists a feasible element z in $\times_{t=1}^{T} L_{\hat{r}}(\Omega, \mathcal{F}, \mathbb{P}; \mathbb{R}^{n_t})$, $\frac{1}{r} + \frac{1}{\hat{r}} = 1$, of the dual stochastic program to (7.3), i.e., it holds that

$$A_{t,0}^\top z_t + A_{t+1,1}^\top(\xi_{t+1}) z_{t+1} - b_t(\xi_t) \in X_t^*,\ t = 1, \ldots, T-1,\ A_{T,0}^\top z_T - b_T(\xi_T) \in X_T^*, \tag{7.4}$$

where X_t^* denotes the polar to the polyhedral cone X_t, $t = 1, \ldots, T$.
(A5) If $r' = 1$ we require that, for each $c \geq 0$, there exists $g \in L_1(\Omega, \mathcal{F}, \mathbb{P})$ such that

$$\sum_{t=1}^{T} \langle b_t(\xi_t(\omega)), x_t \rangle \geq c|x| - g(\omega)$$

for all $x \in \mathbb{R}^m$ such that $x_t \in X_t$, $t = 1, \ldots, T$, $A_{1,0}x_1 = h_1(\xi_1)$, $A_{t,0}x_t + A_{t,1}(\xi_t(\omega))x_{t-1} = h_t(\xi_t(\omega))$, $t = 2, \ldots, T$, and for $\mathbb{P}$-almost all $\omega \in \Omega$.

To use Weierstrass' result on the existence of minimizers, we need a topology $\mathcal{T}$ on $L_{r'}(\Omega, \mathcal{F}, \mathbb{P}; \mathbb{R}^m)$ such that some approximate solution set $S_\alpha(\xi)$ is compact with respect to $\mathcal{T}$. Since, in general, the norm topology is too strong for infinite-dimensional optimization models in L_p-spaces, we resort to the weak topologies $\sigma(L_p, L_q)$ on the spaces $L_p(\Omega, \mathcal{F}, \mathbb{P}; \mathbb{R}^m)$, where $p \in [1, \infty]$ and $\frac{1}{p} + \frac{1}{q} = 1$. They are Hausdorff topological spaces and generated by a basis consisting of the sets

$$\mathcal{O} = \left\{ x \in L_p(\Omega, \mathcal{F}, \mathbb{P}; \mathbb{R}^m) : \left| \mathbb{E}\left[\sum_{t=1}^{T} \langle x_t - x_t^0, y_t^i \rangle \right] \right| < \varepsilon,\ i = 1, \ldots, n \right\}$$

for all $x^0 \in L_p(\Omega, \mathcal{F}, \mathbb{P}; \mathbb{R}^m)$, $n \in \mathbb{N}$, $\varepsilon > 0$, and $y^i \in L_q(\Omega, \mathcal{F}, \mathbb{P}; \mathbb{R}^m)$, $i = 1, \ldots, n$. For $p \in [1, \infty)$, the weak topology $\sigma(L_p, L_q)$ is of the form $\sigma(E, E^*)$ with some Banach space E and its topological dual E^*. For $p = \infty$, the weak topology $\sigma(L_\infty, L_1)$ on the Banach space $L_\infty(\Omega, \mathcal{F}, \mathbb{P}; \mathbb{R}^m)$ is sometimes called weak* topology since it is of the form $\sigma(E^*, E)$. If Ω is finite, the weak topologies coincide with the norm topology. If the space $L_p(\Omega, \mathcal{F}, \mathbb{P}; \mathbb{R}^m)$ is infinite dimensional, its weak topology $\sigma(L_p, L_q)$ is even not metrizable. For $p \in [1, \infty)$, subsets of $L_p(\Omega, \mathcal{F}, \mathbb{P}; \mathbb{R}^m)$ are (relatively) weakly compact iff they are (relatively) weakly sequentially compact due to the Eberlein–Šmulian theorem. For $p = \infty$ the latter property is lost in general. However, if a subset B of $L_p(\Omega, \mathcal{F}, \mathbb{P}; \mathbb{R}^m)$ is compact with respect to the weak topology $\sigma(L_p, L_q)$, its restriction to B is metrizable if $L_q(\Omega, \mathcal{F}, \mathbb{P}; \mathbb{R}^m)$ is separable. We note that the Banach space $L_p(\Omega, \mathcal{F}, \mathbb{P}; \mathbb{R}^m)$ with $p \in [1, \infty)$ is separable if there exists a countable set $\mathcal{G}$ of subsets of Ω such that $\mathcal{F}$ is the smallest σ-field containing $\mathcal{G}$ (Zaanen 1953). A σ-field $\mathcal{F}$ contains such a countable generator if it is generated by a $\mathbb{R}^m$-valued random vector. For these and related results we refer to Fabian et al. (2001, Sections 3 and 4).

Now, we are ready to state our existence result for solutions of (7.3).

Theorem 7.1 *Let (A1)–(A5) be satisfied for some pair (r, r') satisfying (7.2). Then the solution set $S(\xi)$ of (7.3) is nonempty, convex, and compact with respect to the weak topology $\sigma(L_{r'}, L_q)$ ($\frac{1}{r'} + \frac{1}{q} = 1$). Here, the conditions (A4) and (A5) are only needed for $r' \in \{1, \infty\}$.*

Proof We define the integrand $f : \Omega \times \mathbb{R}^m \to \overline{\mathbb{R}}$

$$f(\omega,x) := \begin{cases} \sum_{t=1}^{T} \langle b_t(\xi_t(\omega)), x_t\rangle, & x_1 \in \mathcal{X}_1(\xi_1), x_t \in \mathcal{X}_t(x_{t-1},\xi_t(\omega)), t = 2,\ldots,T, \\ +\infty, & \text{otherwise.} \end{cases}$$

Then f is a proper normal convex integrand (cf. Rockafellar 1976, Rockafellar and Wets 1998, Chapter 14).
Let $(\omega, x) \in \Omega \times \mathbb{R}^m$ be such that $x_1 \in \mathcal{X}_1(\xi_1), x_t \in \mathcal{X}_t(x_{t-1}, \xi_t(\omega)), t = 2,\ldots,T$. Then we conclude from (A4) the existence of $z \in \times_{t=1}^{T} L_{\hat{r}}(\Omega, \mathcal{F}, \mathbb{P}; \mathbb{R}^{n_t})$ such that (7.4) is satisfied. Hence, for each $t = 1,\ldots,T$, there exists $x_t^*(\omega) \in X_t^*$ such that

$$b_t(\xi_t(\omega)) = A_{t,0}^{\top} z_t(\omega) + A_{t+1,1}^{\top}(\xi_{t+1}(\omega)) z_{t+1}(\omega) - x_t^*(\omega) \quad (t = 1,\ldots,T-1)$$
$$b_T(\xi_T(\omega)) = A_{T,0}^{\top} z_T(\omega) - x_T^*(\omega)\,.$$

Inserting the latter representation of $b_t(\xi_t(\omega))$ into the integrand f (defining $F(\xi, x) = \mathbb{E}[f(\omega, x)]$) leads to

$$\begin{aligned} f(\omega,x) &= \sum_{t=1}^{T-1} \langle A_{t,0}^{\top} z_t(\omega) + A_{t+1,1}^{\top}(\xi_{t+1}(\omega)) z_{t+1}(\omega) - x_t^*(\omega), x_t\rangle \\ &\quad + \langle A_{T,0}^{\top} z_T(\omega) - x_T^*(\omega), x_T\rangle \\ &\geq \sum_{t=1}^{T-1} \langle A_{t,0}^{\top} z_t(\omega) + A_{t+1,1}^{\top}(\xi_{t+1}(\omega)) z_{t+1}(\omega), x_t\rangle + \langle A_{T,0}^{\top} z_T(\omega), x_T\rangle \\ &= \sum_{t=1}^{T} \langle z_t(\omega), A_{t,0} x_t\rangle + \sum_{t=1}^{T-1} \langle z_{t+1}(\omega), A_{t+1,1}(\xi_{t+1}(\omega)) x_t\rangle \\ &= \sum_{t=1}^{T} \langle z_t(\omega), h_t(\xi_t(\omega))\rangle. \end{aligned}$$

Hence, we have

$$f(\omega,x) \geq g(\omega)\,, \quad \text{where} \quad g := \sum_{t=1}^{T} \langle z_t, h_t(\xi_t)\rangle \in L_1(\Omega, \mathcal{F}, \mathbb{P}).$$

This implies for the conjugate normal convex integrand $f^* : \Omega \times \mathbb{R}^m \to \overline{\mathbb{R}}$ given by

$$f^*(\omega, y) := \sup_{x \in \mathbb{R}^m} \{\langle y, x\rangle - f(\omega, x)\}$$

that the estimate $f^*(\omega, 0) \leq -g(\omega)$ holds. Hence, the assumption of Rockafellar (1976, Corollary 3D) is satisfied and we conclude that the integral functional

$F(\xi,\cdot) = \mathbb{E}[f(\omega,\cdot)]$ is lower semicontinuous on $L_{r'}(\Omega,\mathcal{F},\mathbb{P};\mathbb{R}^m)$ with respect to the weak topology $\sigma(L_{r'},L_q)$.

The nonanticipativity subspace $\mathcal{N}_{r'}(\xi)$ is closed with respect to the weak topology $\sigma(L_{r'},L_q)$ for all $r' \in [1,\infty]$. For $r' \in [1,\infty)$ this fact is a consequence of the norm closedness and convexity of $\mathcal{N}_{r'}(\xi)$. For $r' = \infty$, let $(x_\alpha)_{\alpha\in I}$ be a net in $\mathcal{N}_\infty(\xi)$ with some partially ordered set $(I,\le)$ that converges to some $x^* \in L_\infty(\Omega,\mathcal{F},\mathbb{P};\mathbb{R}^m)$. Any neighborhood $U(x^*)$ of x^* with respect to the weak topology $\sigma(L_\infty,L_1)$ is of the form

$$U(x^*) = \left\{ x \in L_\infty(\Omega,\mathcal{F},\mathbb{P};\mathbb{R}^m) : \left|\mathbb{E}\left[\sum_{t=1}^T \langle x_t - x_t^*, y_t^i\rangle\right]\right| < \varepsilon_i,\ i=1,\dots,n \right\},$$

where $n\in\mathbb{N}$, $y^i \in L_1(\Omega,\mathcal{F},\mathbb{P};\mathbb{R}^m)$, $\varepsilon_i > 0$, $i=1,\dots,n$. Since the net $(x_\alpha)_{\alpha\in I}$ converges to x^*, there exists $\alpha_0 \in I$ such that $x_\alpha \in U(x^*)$ whenever $\alpha_0 \le \alpha$. If the elements y^i belong to $\times_{t=1}^T L_1(\Omega,\mathcal{F}_t,\mathbb{P};\mathbb{R}^{m_t})$ for each $i=1,\dots,n$, we obtain

$$\begin{aligned}\left|\mathbb{E}\left[\sum_{t=1}^T \langle x_{\alpha,t} - x_t^*, y_t^i\rangle\right]\right| &= \left|\mathbb{E}\left[\sum_{t=1}^T \mathbb{E}[\langle x_{\alpha,t} - x_t^*, y_t^i\rangle|\mathcal{F}_t]\right]\right| \\ &= \left|\mathbb{E}\left[\sum_{t=1}^T \langle x_{\alpha,t} - \mathbb{E}[x_t^*|\mathcal{F}_t], y_t^i\rangle\right]\right| < \varepsilon_i\end{aligned}$$

due to the fact that $\mathbb{E}[x_{\alpha,t}|\mathcal{F}_t] = x_{\alpha,t}$ for each $t=1,\dots,T$ and $\alpha\in I$. Hence, we have in this case,

$$U(x^*) = U(\mathbb{E}[x_1^*|\mathcal{F}_1],\dots,\mathbb{E}[x_T^*|\mathcal{F}_T]).$$

Since the net $(x_\alpha)_{\alpha\in I}$ converges to x^* and the weak topology is Hausdorff, we conclude $x_t^* = \mathbb{E}[x_t^*|\mathcal{F}_t]$, $t=1,\dots,T$, and, thus, $x^*\in\mathcal{N}_\infty(\xi)$.

It remains to show that, for some $\alpha > 0$, the α-approximate solution set $S_\alpha(\xi)$ is compact with respect to the weak topology $\sigma(L_{r'},L_q)$. For $r'\in(1,\infty)$ the Banach space $L_{r'}(\Omega,\mathcal{F},\mathbb{P};\mathbb{R}^m)$ is reflexive. Furthermore, any α-approximate solution set $S_\alpha(\xi)$ is closed and convex. For some $\alpha > 0$ the level set is also bounded due to (A3) and, hence, compact with respect to $\sigma(L_{r'},L_q)$. For $r' = 1$ the compactness of any α-level set with respect to $\sigma(L_1,L_\infty)$ follows from Rockafellar (1976, Theorem 3K) due to condition (A5). For $r'=\infty$, some α-level set is bounded due to (A3) and, hence, relatively compact with respect to $\sigma(L_\infty,L_1)$ due to Alaoglu's theorem (Fabian et al. 2001, Theorem 3.21). Since the objective function $F(\xi,\cdot)$ is lower semicontinuous and $\mathcal{N}_\infty(\xi)$ weakly closed with respect to $\sigma(L_\infty,L_1)$, the α-level set is even compact with respect to $\sigma(L_\infty,L_1)$.

Altogether, $S(\xi)$ is nonempty due to Weierstrass' theorem and compact with respect to $\sigma(L_{r'},L_q)$. The convexity of $S(\xi)$ is an immediate consequence of the convexity of the objective $F(\xi,\cdot)$ of the stochastic program (7.3).

Finally, we note that assumptions (A4) and (A5) are not needed for proving that $S(\xi)$ is nonempty and compact with respect to the topology $\sigma(L_{r'}, L_q)$ in case $r' \in (1, \infty)$. This fact is an immediate consequence of minimizing a linear continuous functional on a closed, convex, bounded subset of a reflexive Banach space. □

To state our next result we introduce the functional $D_{\mathrm{f}}(\xi, \tilde{\xi})$ depending on the filtrations of ξ and of its perturbation $\tilde{\xi}$, respectively. It is defined by

$$D_{\mathrm{f}}(\xi, \tilde{\xi}) := \sup_{\varepsilon \in (0,\alpha]} \inf_{\substack{x \in S_\varepsilon(\xi) \\ \tilde{x} \in S_\varepsilon(\tilde{\xi})}} \sum_{t=2}^{T-1} \max\{\|x_t - \mathbb{E}[x_t|\mathcal{F}_t(\tilde{\xi})]\|_{r'}, \|\tilde{x}_t - \mathbb{E}[\tilde{x}_t|\mathcal{F}_t(\xi)]\|_{r'}\}. \tag{7.5}$$

In the following, we call the functional D_{f} *filtration distance*, although it fails to satisfy the triangle inequality in general. If solutions of (7.3) for the inputs ξ and $\tilde{\xi}$ exist (see Theorem 7.1), the filtration distance is of the simplified form

$$D_{\mathrm{f}}(\xi, \tilde{\xi}) = \inf_{\substack{x \in S(\xi) \\ \tilde{x} \in S(\tilde{\xi})}} \sum_{t=2}^{T-1} \max\{\|x_t - \mathbb{E}[x_t|\mathcal{F}_t(\tilde{\xi})]\|_{r'}, \|\tilde{x}_t - \mathbb{E}[\tilde{x}_t|\mathcal{F}_t(\xi)]\|_{r'}\}.$$

We note that the conditional expectations $\mathbb{E}[x_t|\mathcal{F}_t(\tilde{\xi})]$ and $\mathbb{E}[\tilde{x}_t|\mathcal{F}_t(\xi)]$ may be written equivalently in the form $\mathbb{E}[x_t|\tilde{\xi}_1, \ldots, \tilde{\xi}_t]$ and $\mathbb{E}[\tilde{x}_t|\xi_1, \ldots, \xi_t]$, respectively.

The following stability result for optimal values of program (7.3) is essentially (Heitsch et al. 2006, Theorem 2.1).

Theorem 7.2 *Let (A1), (A2), and (A3) be satisfied and the sets $\mathcal{X}_1(\tilde{\xi}_1)$ be nonempty and uniformly bounded in $\mathbb{R}^{m_1}$ if $|\tilde{\xi}_1 - \xi_1| \leq \delta$ (where $\delta > 0$ is taken from (A3)). Then there exist positive constants L and δ such that the estimate*

$$|v(\xi) - v(\tilde{\xi})| \leq L(\|\xi - \tilde{\xi}\|_r + D_{\mathrm{f}}(\xi, \tilde{\xi})) \tag{7.6}$$

holds for all random elements $\tilde{\xi} \in L_r(\Omega, \mathcal{F}, \mathbb{P}; \mathbb{R}^s)$ with $\|\tilde{\xi} - \xi\|_r \leq \delta$.

The proof of Heitsch et al. (2006, Theorem 2.1) extends easily to constraints for x_1 that depend on ξ_1 (via the right-hand side of the equality constraint $A_{1,0}x_1 = h(\xi_1)$). We note that the constant L depends on $\|\xi\|_r$ in all cases.

To prove a stability result for (approximate) solutions of (7.3), we need a stronger version of the filtration distance D_{f}, namely

$$D_{\mathrm{f}}^*(\xi, \tilde{\xi}) = \sup_{\|x\|_{r'} \leq 1} \sum_{t=2}^{T} \|\mathbb{E}[x_t|\mathcal{F}_t(\xi)] - \mathbb{E}[x_t|\mathcal{F}_t(\tilde{\xi})]\|_{r'}. \tag{7.7}$$

Notice that the sum is extended by the additional summand for $t = T$ and that the former infimum is replaced by a supremum with respect to a sufficiently large bounded set (the unit ball in $L_{r'}$). Clearly, the conditions (A1)–(A3) imply the estimate

$$D_{\mathrm{f}}(\xi,\tilde{\xi}) \leq \sup_{x\in B} \sum_{t=2}^{T-1} \|\mathbb{E}[x_t|\mathcal{F}_t(\xi)] - \mathbb{E}[x_t|\mathcal{F}_t(\tilde{\xi})]\|_{r'} \leq C\, D_{\mathrm{f}}^*(\xi,\tilde{\xi}) \tag{7.8}$$

for all ξ and $\tilde{\xi}$ in $L_r(\Omega,\mathcal{F},\mathbb{P};\mathbb{R}^s)$ with $\|\tilde{\xi}-\xi\|_r \leq \delta$, where $\delta > 0$ and B are the constant and $L_{r'}$-bounded set appearing in (A2) and (A3), respectively, and the constant $C > 0$ is chosen such $\|x\|_{r'} \leq C$ for all $x \in B$.

Sometimes, the unit ball in $L_{r'}$ in the definition of D_{f}^* is too large. It may be replaced by the smaller set $\mathcal{B}_\infty := \{x : \Omega \to \mathbb{R}^m : x \text{ is measurable}, |x(\omega)| \leq 1 \text{ for all } \omega \in \Omega\}$ if the following stronger condition (A3)$'$ is satisfied.

(A3)$'$ The optimal values $v(\tilde{\xi})$ of (7.3) with input $\tilde{\xi}$ are finite for all $\tilde{\xi}$ in a neighborhood of ξ and for some $\alpha > 0$ there exist constants $\delta > 0$ and $C > 0$ such that $|\tilde{x}(\omega)| \leq C$ for $\mathbb{P}$-almost every $\omega \in \Omega$ and all $\tilde{x} \in S_\alpha(\tilde{\xi})$ with $\tilde{\xi} \in L_r(\Omega,\mathcal{F},\mathbb{P};\mathbb{R}^s)$ and $\|\tilde{\xi}-\xi\|_r \leq \delta$.

If (A3)$'$ is satisfied, we define

$$D_{\mathrm{f}}^*(\xi,\tilde{\xi}) := \sup_{x\in\mathcal{B}_\infty} \sum_{t=2}^{T} \|\mathbb{E}[x_t|\mathcal{F}_t(\xi)] - \mathbb{E}[x_t|\mathcal{F}_t(\tilde{\xi})]\|_{r'} \tag{7.9}$$

and have $D_{\mathrm{f}}(\xi,\tilde{\xi}) \leq C\, D_{\mathrm{f}}^*(\xi,\tilde{\xi})$. We note that D_{f}^* always satisfies the triangle inequality.

In the next result we derive a (local) Lipschitz property of the feasible set-valued mapping $\mathcal{X}(\cdot)$ from $L_r(\Omega,\mathcal{F},\mathbb{P};\mathbb{R}^s)$ into $L_{r'}(\Omega,\mathcal{F},\mathbb{P};\mathbb{R}^m)$ in terms of a "truncated" Pompeiu–Hausdorff-type distance

$$\hat{dl}_\rho(B,\tilde{B}) = \inf\left\{\eta \geq 0 : B\cap\rho\mathbb{B} \subset \tilde{B}+\eta\mathbb{B},\ \ \tilde{B}\cap\rho\mathbb{B} \subset B+\eta\mathbb{B}\right\}$$

of closed subsets B and $\tilde{B}$ of the space $L_{r'}(\Omega,\mathcal{F},\mathbb{P};\mathbb{R}^m)$ with $\mathbb{B}$ denoting its unit ball. The Pompeiu–Hausdorff distance may be defined by

$$dl_\infty(B,\tilde{B}) = \lim_{\rho\to\infty} \hat{dl}_\rho(B,\tilde{B})$$

(see Rockafellar and Wets 1998, Corollary 4.38).

Proposition 7.3 *Let (A1), (A2), and (A3) be satisfied with $r' \in [1,\infty)$ and the sets $\mathcal{X}_1(\tilde{\xi}_1)$ be nonempty and uniformly bounded in $\mathbb{R}^{m_1}$ if $|\tilde{\xi}_1-\xi_1| \leq \delta$ (with $\delta > 0$ from (A3)). Then there exist positive constants L and δ such that the estimate*

$$\hat{dl}_\rho(\mathcal{X}(\xi),\mathcal{X}(\tilde{\xi})) \leq L(\|\xi-\tilde{\xi}\|_r + \rho D_{\mathrm{f}}^*(\xi,\tilde{\xi}))$$

holds for any $\rho > 0$ and any $\tilde{\xi} \in L_r(\Omega,\mathcal{F},\mathbb{P};\mathbb{R}^s)$ with $\|\xi-\tilde{\xi}\|_r \leq \delta$. If (A3)$'$ is satisfied instead of (A3), the estimate is valid with $\hat{dl}_\rho$ denoting the "truncated" Pompeiu–Hausdorff distance in $L_\infty(\Omega,\mathcal{F},\mathbb{P};\mathbb{R}^m)$ and D_{f}^* defined by (7.9).

Proof Let $\rho > 0$, $\delta > 0$ be selected as in (A2) and (A3), $x \in \mathcal{X}(\xi) \cap \rho\mathbb{B}$ and $\tilde{\xi} \in L_r(\Omega, \mathcal{F}, \mathbb{P}; \mathbb{R}^s)$ be such that $\|\tilde{\xi} - \xi\|_r < \delta$. With the same arguments as in the proof of Heitsch et al. (2006, Theorem 2.1), there exists $\tilde{x} \in \mathcal{X}(\tilde{\xi})$ such that the estimate

$$|\mathbb{E}[x_t|\mathcal{F}_t(\tilde{\xi})]-\tilde{x}_t| \leq \hat{L}_t \left(\sum_{\tau=1}^{t} \mathbb{E}[|\xi_\tau - \tilde{\xi}_\tau| \, |\mathcal{F}_\tau(\tilde{\xi})] + \sum_{\tau=2}^{t-1} \mathbb{E}[|x_\tau - \mathbb{E}[x_\tau|\mathcal{F}_\tau(\tilde{\xi})]| \, |\mathcal{F}_{\tau+1}(\tilde{\xi})] \right) \tag{7.10}$$

holds $\mathbb{P}$-almost surely with some positive constant $\hat{L}_t$ for $t = 1, \ldots, T$. Note that $r' < \infty$ means that only costs and/or right-hand sides in (7.3) are random and that the first sum on the right-hand side of (7.10) disappears if only costs are random. From the definition of r' we know that $r \neq r'$ may occur only in the latter case.

Hence, together with the estimate

$$|x_t - \tilde{x}_t| \leq |x_t - \mathbb{E}[x_t|\mathcal{F}_t(\tilde{\xi})]| + |\mathbb{E}[x_t|\mathcal{F}_t(\tilde{\xi})] - \tilde{x}_t|$$

$\mathbb{P}$-almost surely and for all $t = 1, \ldots, T$, (7.10) implies for all pairs (r, r') with $r' \in [1, \infty)$ that

$$\mathbb{E}[|x_t - \tilde{x}_t|^{r'}] \leq L_t \left(\sum_{\tau=1}^{t} \mathbb{E}[|\xi_\tau - \tilde{\xi}_\tau|^{r}] + \sum_{\tau=2}^{t} \mathbb{E}[|x_\tau - \mathbb{E}[x_\tau|\mathcal{F}_\tau(\tilde{\xi})]|^{r'}] \right)$$

holds with certain constants L_t, $t = 1, \ldots, T$. We conclude

$$\|x - \tilde{x}\|_{r'} \leq L(\|\xi - \tilde{\xi}\|_r + \rho D_{\mathrm{f}}^*(\xi, \tilde{\xi})),$$

with some constant $L > 0$. The second estimate follows by interchanging the role of the pairs (x, ξ) and $(\tilde{x}, \tilde{\xi})$. If (A3)$'$ is satisfied instead of (A3), the changes are obvious. □

Now, we are ready to establish a Lipschitz property of approximate solution sets.

Theorem 7.4 *Let (A1), (A2), and (A3) be satisfied with $r' \in [1, \infty)$ and the sets $\mathcal{X}_1(\tilde{\xi}_1)$ be nonempty and uniformly bounded in $\mathbb{R}^{m_1}$ if $|\tilde{\xi}_1 - \xi_1| \leq \delta$. Assume that the solution sets $S(\xi)$ and $S(\tilde{\xi})$ are nonempty for some $\tilde{\xi} \in L_r(\Omega, \mathcal{F}, \mathbb{P}; \mathbb{R}^s)$ with $\|\xi - \tilde{\xi}\|_r \leq \delta$ (with $\delta > 0$ from (A3)). Then there exist $\bar{L} > 0$ and $\bar{\varepsilon} > 0$ such that*

$$dl_\infty(S_\varepsilon(\xi), S_\varepsilon(\tilde{\xi})) \leq \frac{\bar{L}}{\varepsilon}(\|\xi - \tilde{\xi}\|_r + D_{\mathrm{f}}^*(\xi, \tilde{\xi})) \tag{7.11}$$

holds for any $\varepsilon \in (0, \bar{\varepsilon})$.

Proof Let $\rho_0 \geq 1$ be chosen such that the $L_{r'}$-bounded set B in (A3) is contained in $\rho_0\mathbb{B}$ (with $\mathbb{B}$ denoting the unit ball in $L_{r'}$) and $\min\{v(\xi), v(\tilde{\xi})\} \geq -\rho_0$. Let $\rho > \rho_0$,

$\bar{\varepsilon} = \min\{\alpha, \rho - \rho_0\}$, and $0 < \varepsilon < \bar{\varepsilon}$. Let $\tilde{\xi} \in L_r(\Omega, \mathcal{F}, \mathbb{P}; \mathbb{R}^s)$ with $\|\xi - \tilde{\xi}\|_r \leq \delta$. Then the assumptions of Theorem 7.69 in Rockafellar and Wets (1998) are satisfied for the functions $F(\xi, \cdot)$ and $F(\tilde{\xi}, \cdot)$. We note that most of the results in Rockafellar and Wets (1998) are stated in finite-dimensional spaces. However, the proof of Rockafellar and Wets (1998, Theorem 7.69) carries over to linear normed spaces (see also Attouch and Wets 1993 Theorem 4.3). We obtain from the proof the inclusion

$$S_\varepsilon(\xi) = S_\varepsilon(\xi) \cap \rho\mathbb{B} \subseteq S_\varepsilon(\tilde{\xi}) + \frac{2\eta}{\varepsilon + 2\eta} 2\rho\mathbb{B} \subseteq S_\varepsilon(\tilde{\xi}) + \frac{4\rho}{\varepsilon}\eta\mathbb{B}, \tag{7.12}$$

for all $\eta > \hat{dl}^+_{\rho+\varepsilon}(F(\xi, \cdot), F(\tilde{\xi}, \cdot))$, where the auxiliary epi-distance $\hat{dl}^+_\rho(F(\xi, \cdot), F(\tilde{\xi}, \cdot))$ is defined as the infimum of all $\eta \geq 0$ such that for all $x, \tilde{x} \in \rho\mathbb{B}$,

$$\min_{\tilde{y} \in \mathbb{B}(x,\eta)} F(\tilde{\xi}, \tilde{y}) \leq \max\{F(\xi, x), -\rho\} + \eta \tag{7.13}$$

$$\min_{y \in \mathbb{B}(\tilde{x},\eta)} F(\xi, y) \leq \max\{F(\tilde{\xi}, \tilde{x}), -\rho\} + \eta. \tag{7.14}$$

The estimate (7.12) implies

$$S_\varepsilon(\xi) \subseteq S_\varepsilon(\tilde{\xi}) + \frac{4\rho}{\varepsilon} \hat{dl}^+_{\rho+\varepsilon}(F(\xi, \cdot), F(\tilde{\xi}, \cdot))\mathbb{B}.$$

Since the same argument works with ξ and $\tilde{\xi}$ interchanged, we obtain

$$dl_\infty(S_\varepsilon(\xi), S_\varepsilon(\tilde{\xi})) \leq \frac{4\rho}{\varepsilon} \hat{dl}^+_{\rho+\varepsilon}(F(\xi, \cdot), F(\tilde{\xi}, \cdot))$$

and it remains to estimate $\hat{dl}^+_{\rho+\varepsilon}(F(\xi, \cdot), F(\tilde{\xi}, \cdot))$. Let $\eta > \hat{dl}^+_{\rho+\varepsilon}(F(\xi, \cdot), F(\tilde{\xi}, \cdot))$ and $x \in \mathcal{X}(\xi)$. Proposition 7.3 implies the existence of $\tilde{x} \in \mathcal{X}(\tilde{\xi})$ such that

$$\|x - \tilde{x}\|_{r'} \leq L(\|\xi - \tilde{\xi}\|_r + \|x\|_{r'} D^*_{\mathrm{f}}(\xi, \tilde{\xi})).$$

In order to check condition (7.13), we have to distinguish three cases, namely that randomness appears in costs and right-hand sides, only in costs, and only in right-hand sides. Next we consider the first case, i.e., $r = r' = 2$, and obtain as in the proof of Heitsch et al. (2006, Theorem 2.1) the estimate

$$
\begin{aligned}
F(\tilde{\xi},\tilde{x}) &\leq F(\xi,x) + |F(\tilde{\xi},\tilde{x}) - F(\tilde{\xi},x)| + |F(\tilde{\xi},x) - F(\xi,x)| \\
&\leq F(\xi,x) + \left|\mathbb{E}\left[\sum_{t=1}^{T}\langle b_t(\tilde{\xi}_t), \tilde{x}_t - \mathbb{E}[x_t|\mathcal{F}_t(\tilde{\xi})]\rangle\right]\right| \\
&\qquad + \left|\mathbb{E}\left[\sum_{t=1}^{T}\langle b_t(\tilde{\xi}_t) - b_t(\xi_t), x_t\rangle\right]\right| \\
&\leq F(\xi,x) + \hat{K}\left(\left(\sum_{t=1}^{T}(1+\mathbb{E}[|\tilde{\xi}_t|^2])\right)^{\frac{1}{2}}\left(\sum_{t=1}^{T}\mathbb{E}[|\tilde{x}_t - \mathbb{E}[x_t|\mathcal{F}_t(\tilde{\xi})]|^2]\right)^{\frac{1}{2}}\right. \\
&\qquad \left. + \|\tilde{\xi}-\xi\|_2\|x\|_2\right) \\
&\leq F(\xi,x) + \hat{L}\left(\rho\|\tilde{\xi}-\xi\|_2 + \sum_{t=2}^{T-1}\|x_t - \mathbb{E}[x_t|\mathcal{F}_t(\tilde{\xi})]\|_2\right) \\
&\leq F(\xi,x) + L\rho(\|\tilde{\xi}-\xi\|_2 + D_{\mathrm{f}}^*(\xi,\tilde{\xi}))
\end{aligned}
$$

with certain constants $\hat{K}$, $\hat{L}$, and L (depending on $\|\xi\|_2$), where the Cauchy–Schwarz inequality, (A3), and the estimate (7.10) are used. Hence, condition (7.13) is satisfied if

$$\eta = L\rho(\|\tilde{\xi}-\xi\|_2 + D_{\mathrm{f}}^*(\xi,\tilde{\xi}))$$

holds with certain constant $L > 0$. The same estimate holds in the remaining two cases and when checking condition (7.14) (possibly with different constants). Taking the maximal constant $L > 0$ we conclude

$$\hat{d\!l}^{+}_{\rho+\varepsilon}(F(\xi,\cdot), F(\tilde{\xi},\cdot)) \leq L\rho(\|\tilde{\xi}-\xi\|_r + D_{\mathrm{f}}^*(\xi,\tilde{\xi}))$$

and, hence,

$$d\!l_{\infty}(S_{\varepsilon}(\xi), S_{\varepsilon}(\tilde{\xi})) \leq \frac{4L\rho^2}{\varepsilon}(\|\tilde{\xi}-\xi\|_r + D_{\mathrm{f}}^*(\xi,\tilde{\xi})).$$

Setting $\bar{L} = 4L\rho^2$ completes the proof. □

For solution sets the situation is less comfortable. Stability of solutions can only be derived with respect to the weak topology $\sigma(L_{r'}, L_r)$.

Theorem 7.5 *Let (A1), (A2), and (A3) be satisfied with $r' \in (1,\infty)$ and the sets $\mathcal{X}_1(\tilde{\xi}_1)$ be nonempty and uniformly bounded in $\mathbb{R}^{m_1}$ if $|\tilde{\xi}_1 - \xi_1| \leq \delta$ (with $\delta > 0$ from (A3)). If $(\xi^{(n)})$ is a sequence in $L_r(\Omega,\mathcal{F},\mathbb{P};\mathbb{R}^s)$ converging to ξ in L_r and with respect to D_{f}^* and if $(x^{(n)})$ is a sequence of solutions of the approximate* *problems, i.e.,* $x^{(n)} \in S(\xi^{(n)})$*, then there exists a subsequence* $(x^{(n_k)})$ *of* $(x^{(n)})$ *that converges with respect to the weak topology* $\sigma(L_{r'}, L_r)$ *to some element of* $S(\xi)$*. If* $S(\xi)$ *is a singleton, the sequence* $(x^{(n)})$ *converges with respect to the weak topology* $\sigma(L_{r'}, L_r)$ *to the unique solution of (7.3).*

Proof Let $(\xi^{(n)})$ and $(x^{(n)})$ be selected as above. Since there exists $n_0 \in \mathbb{N}$ such that $\|\xi^{(n)} - \xi\|_r \le \delta$ and $x^{(n)} \in S_\alpha(\xi^{(n)})$ for any $n \ge n_0$, where $\alpha > 0$ and $\delta > 0$ are chosen as in (A3), the sequence $(x^{(n)})$ is contained in a bounded set of the reflexive Banach space $L_{r'}(\Omega, \mathcal{F}, \mathbb{P}; \mathbb{R}^m)$. Hence, there exists a subsequence $(x^{(n_k)})$ of $(x^{(n)})$ that converges with respect to the weak topology $\sigma(L_{r'}, L_r)$ to some element x^* in $L_{r'}(\Omega, \mathcal{F}, \mathbb{P}; \mathbb{R}^m)$. Theorem 7.2 implies

$$v(\xi^{(n_k)}) = F(\xi^{(n_k)}, x^{(n_k)}) = \mathbb{E}\left[\sum_{t=1}^{T} \langle b_t(\xi_t^{(n_k)}), x_t^{(n_k)} \rangle\right] \to v(\xi).$$

Due to the norm convergence of $(\xi^{(n_k)})$ and the weak convergence of $(x^{(n_k)})$, we also obtain

$$\mathbb{E}\left[\sum_{t=1}^{T} \langle b_t(\xi_t^{(n_k)}), x_t^{(n_k)} \rangle\right] \to \mathbb{E}\left[\sum_{t=1}^{T} \langle b_t(\xi_t), x_t^* \rangle\right].$$

Hence, it remains to show that x^* is feasible for (7.3), i.e., $x^* \in \mathcal{X}(\xi)$ and $x^* \in \mathcal{N}_{r'}(\xi)$.

In the present situation, the set $\mathcal{X}(\xi)$ is of the form

$$\mathcal{X}(\xi) = \{x \in L_{r'}(\Omega, \mathcal{F}, \mathbb{P}; \mathbb{R}^m) : x \in X, Ax = h(\xi)\}, \tag{7.15}$$

where $X := \times_{t=1}^{T} X_t$, $h(\xi) := (h_1(\xi_1), \ldots, h_T(\xi_T))\}$ and

$$A := \begin{pmatrix} A_{1,0} & 0 & 0 & \cdots & 0 & 0 & 0 \\ A_{2,1} & A_{2,0} & 0 & \cdots & 0 & 0 & 0 \\ \vdots & \vdots & \vdots & \vdots & \vdots & \vdots & \vdots \\ 0 & 0 & 0 & \cdots & 0 & A_{T,1} & A_{T,0} \end{pmatrix}.$$

The graph of $\mathcal{X}$, i.e., graph $\mathcal{X} = \{(x, \xi) \in L_{r'}(\Omega, \mathcal{F}, \mathbb{P}; \mathbb{R}^m) \times L_r(\Omega, \mathcal{F}, \mathbb{P}; \mathbb{R}^s) | x \in \mathcal{X}(\xi)\}$ is closed and convex. Since $(\xi^{(n_k)})$ norm converges in $L_r(\Omega, \mathcal{F}, \mathbb{P}; \mathbb{R}^s)$ to ξ and $(x^{(n_k)})$ weakly converges to x^*, the sequence $((x^{(n_k)}, \xi^{(n_k)}))$ of pairs in graph $\mathcal{X}$ converges weakly to (x^*, ξ). Due to the closedness and convexity of graph $\mathcal{X}$, Mazur's theorem (Fabian et al. 2001 Theorem 3.19) implies that graph $\mathcal{X}$ is weakly closed and, thus, $(x^*, \xi) \in$ graph $\mathcal{X}$ or $x^* \in \mathcal{X}(\xi)$.

Finally, we have to show that x^* belongs to $\mathcal{N}_{r'}(\xi)$. For any $y \in L_r(\Omega, \mathcal{F}, \mathbb{P}; \mathbb{R}^m)$ we obtain the estimate

$$
\begin{aligned}
\left|\mathbb{E}\left[\sum_{t=1}^{T}\langle y_t, x_t^* - \mathbb{E}[x_t^*|\mathcal{F}_t(\xi)]\rangle\right]\right| \le & \left|\sum_{t=1}^{T}\mathbb{E}[\langle y_t, x_t^* - x_t^{(n_k)}\rangle]\right| \\
& + \left|\sum_{t=1}^{T}\mathbb{E}[\langle y_t, x_t^{(n_k)} - \mathbb{E}[x_t^{(n_k)}|\mathcal{F}_t(\xi)]\rangle]\right| \\
& + \left|\sum_{t=1}^{T}\mathbb{E}[\langle y_t, \mathbb{E}[x_t^{(n_k)} - x_t^*|\mathcal{F}_t(\xi)]\rangle]\right| \\
\le & 2\left|\sum_{t=1}^{T}\mathbb{E}[\langle y_t, x_t^* - x_t^{(n_k)}\rangle]\right| \\
& + \max_{t=1,\ldots,T}\|y_t\|_r\sum_{t=2}^{T}\|x_t^{(n_k)} - \mathbb{E}[x_t^{(n_k)}|\mathcal{F}_t(\xi)]\|_{r'}.
\end{aligned}
$$

The first term on the right-hand side converges to 0 for k tending to ∞ as the sequence $(x^{(n_k)})$ converges weakly to x^*. The second term converges to 0 due to the estimate (7.8) since $(D_{\mathrm{f}}^*(\xi, \xi^{(n_k)}))$ also converges to 0. We conclude that

$$
\mathbb{E}\left[\sum_{t=1}^{T}\langle y_t, x_t^* - \mathbb{E}[x_t^*|\mathcal{F}_t(\xi)]\rangle\right] = 0
$$

holds for any $y \in L_r(\Omega, \mathcal{F}, \mathbb{P}; \mathbb{R}^m)$ and, hence, that $x_t^* = \mathbb{E}[x_t^*|\mathcal{F}_t(\xi)]$ for each $t = 1, \ldots, T$. This means $x^* \in \mathcal{N}_{r'}(\xi)$. □

Remark 7.1 Theorem 7.5 remains true if the filtration distance D_{f}^* is replaced by the weaker distance

$$
\hat{D}_{\mathrm{f}}(\xi, \tilde{\xi}) = \sup_{\tilde{x}\in S(\tilde{\xi})}\sum_{t=2}^{T}\|\tilde{x}_t - \mathbb{E}[\tilde{x}_t|\mathcal{F}_t(\xi)]\|_{r'}.
$$

Furthermore, if the solutions $x^{(n)} \in S(\xi^{(n)})$ are adapted to the filtration $\mathcal{F}_t(\xi)$, $t = 1, \ldots, T$, of the original process ξ (as in Heitsch and Römisch 2009 Proposition 5.5), the convergence of $(\xi^{(n)})$ to ξ in L_r is sufficient for the weak convergence of some subsequence of $(x^{(n)})$ to some element of $S(\xi)$ (in the sense of $\sigma(L_{r'}, L_r)$).

Remark 7.2 The stability analysis of (linear) two-stage stochastic programs (see, e.g., Rachev and Römisch 2002 Section 3.1, Römisch and Wets 2007) mostly studied the continuity behavior of *first-stage* (approximate) solution sets. Hence, for the specific case $T = 2$, our stability results in Theorems 7.4 and 7.5 extend earlier work because they concern first- *and* second-stage solutions. The new important assumption is (A3), i.e., the level-boundedness of the objective (locally uniformly at ξ) with respect to both first- and second-stage variables.

Remark 7.3 In many applications of stochastic programming it is of interest to develop risk-averse models (e.g., in electricity risk management and in finance). For example, this can be achieved if the expectation in the objective of (7.1) is replaced by a (convex) risk functional (measure). Typically, risk functionals are inherently nonlinear. If, however, a multiperiod *polyhedral* risk functional (Eichhorn and Römisch 2005) replaces the expectation in (7.1), the resulting risk-averse stochastic program may be reformulated as a linear multistage stochastic program of the form (7.1) by introducing new state variables and (linear) constraints (see Eichhorn and Römisch 2005 Section 4). Moreover, it is shown in Eichhorn (2008) that the stability behavior of the reformulation does not change (when compared with the original problem with expectation objective) if the multiperiod polyhedral (convex) risk functional has bounded L_1-level sets. The latter property is shared by the conditional or average value-at-risk and several of its multiperiod extensions (Eichhorn 2008 Section 4).

7.3 Generating Scenario Trees

Let ξ be the original stochastic process on a probability space $(\Omega, \mathcal{F}, \mathbb{P})$ with parameter set $\{1, \ldots, T\}$ and state space $\mathbb{R}^d$. We aim at generating a scenario tree ξ_{tr} such that the distances

$$\|\xi - \xi_{\mathrm{tr}}\|_r \quad \text{and} \quad D_{\mathrm{f}}^*(\xi, \xi_{\mathrm{tr}}) \tag{7.16}$$

are small and, hence, the optimal values $v(\xi)$ and $v(\xi_{\mathrm{tr}})$ and the approximate solution sets $S_\varepsilon(\xi)$ and $S_\varepsilon(\xi_{\mathrm{tr}})$ are close to each other according to Theorems 7.2 and 7.4, respectively.

The idea is to start with a good initial approximation $\hat{\xi}$ of ξ having a finite number of scenarios $\xi^i = (\xi_1^i, \ldots, \xi_T^i) \in \mathbb{R}^{Td}$ with probabilities $p_i > 0$, $i = 1, \ldots, N$, and common root, i.e., $\xi_1^1 = \cdots = \xi_1^N =: \xi_1^*$. These scenarios might be obtained by quantization techniques (Luschgy 2000) or by sampling or resampling techniques based on parametric or nonparametric stochastic models of ξ.

In the following we assume that

$$\|\xi - \hat{\xi}\|_r + D_{\mathrm{f}}^*(\xi, \hat{\xi}) \leq \varepsilon \tag{7.17}$$

holds for some given (initial) tolerance $\varepsilon > 0$. For example, condition (7.17) may be satisfied for D_{f}^* given by (7.9) and for any tolerance $\varepsilon > 0$ if $\hat{\xi}$ is obtained by sampling from a finite set with sufficiently large sample size (see Heitsch and Römisch 2009 Example 5.3).

Next we describe an algorithmic procedure that starts from $\hat{\xi}$ and ends up with a scenario tree process ξ_{tr} having the same root node ξ_1^*, less nodes than $\hat{\xi}$, and allowing for constructive estimates of

$$\|\hat{\xi} - \xi_{\mathrm{tr}}\|_r \,.$$

The idea of the algorithm consists in forming clusters of scenarios based on scenario reduction on the time horizon $\{1, \ldots, t\}$ recursively for increasing time t.

To this end, the L_r-seminorm $\|\cdot\|_{r,t}$ on $L_r(\Omega, \mathcal{F}, \mathbb{P}; \mathbb{R}^s)$ (with $s = Td$) given by

$$\|\xi\|_{r,t} := \left(\mathbb{E}[|\xi|_t^r]\right)^{\frac{1}{r}} \tag{7.18}$$

is used at step t, where $|\cdot|_t$ is a seminorm on $\mathbb{R}^s$ which, for each $\xi = (\xi_1, \ldots, \xi_T) \in \mathbb{R}^s$, is given by $|\xi|_t := |(\xi_1, \ldots, \xi_t, 0, \ldots, 0)|$.

The following procedure determines recursively stochastic processes $\hat{\xi}^t$ having scenarios $\hat{\xi}^{t,i}$ endowed with probabilities p_i, $i \in I := \{1, \ldots, N\}$, and, in addition, partitions $\mathcal{C}_t = \{C_t^1, \ldots, C_t^{K_t}\}$ of the index set I, i.e.,

$$C_t^k \cap C_t^{k'} = \emptyset \quad (k \neq k') \quad \text{and} \quad \bigcup_{k=1}^{K_t} C_t^k = I. \tag{7.19}$$

The index sets $C_t^k \in \mathcal{C}_t$, $k = 1, \ldots, K_t$, characterize clusters of scenarios. The initialization of the procedure consists in setting $\hat{\xi}^1 := \hat{\xi}$, i.e., $\hat{\xi}^{1,i} = \xi^i$, $i \in I$, and $\mathcal{C}_1 = \{I\}$. At step t (with $t > 1$) we consider each cluster C_{t-1}^k, i.e., each scenario subset $\{\hat{\xi}^{t-1,i}\}_{i \in C_{t-1}^k}$, separately and delete scenarios $\{\hat{\xi}^{t-1,j}\}_{j \in J_t^k}$ by the forward selection algorithm of Heitsch and Römisch (2003) such that

$$\left(\sum_{k=1}^{K_{t-1}} \sum_{j \in J_t^k} p_j \min_{i \in I_t^k} |\hat{\xi}^{t-1,i} - \hat{\xi}^{t-1,j}|_t^r\right)^{\frac{1}{r}}$$

is bounded from above by some prescribed tolerance. Here, the index set I_t^k of remaining scenarios is given by

$$I_t^k = C_{t-1}^k \setminus J_t^k.$$

As in the general scenario reduction procedure in Heitsch and Römisch (2003), the index set J_t^k is subdivided into index sets $J_{t,i}^k$, $i \in I_t^k$, such that

$$J_t^k = \bigcup_{i \in I_t^k} J_{t,i}^k, \quad J_{t,i}^k := \{j \in J_t^k : i = i_t^k(j)\}, \quad \text{and} \quad i_t^k(j) \in \arg\min_{i \in I_t^k} |\hat{\xi}^{t-1,i} - \hat{\xi}^{t-1,j}|_t^r.$$

Next we define a mapping $\alpha_t : I \to I$ such that

$$\alpha_t(j) = \begin{cases} i_t^k(j), & j \in J_t^k,\ k = 1, \ldots, K_{t-1} \\ j, & \text{otherwise.} \end{cases} \tag{7.20}$$

Then the scenarios of the stochastic process $\hat{\xi}^t = \{\hat{\xi}^t_\tau\}_{\tau=1}^T$ are defined by

$$\hat{\xi}^{t,i}_\tau = \begin{cases} \xi_\tau^{\alpha_\tau(i)}, & \tau \le t \\ \xi_\tau^i, & \text{otherwise}, \end{cases} \tag{7.21}$$

with probabilities p_i for each $i \in I$. The processes $\hat{\xi}^t$ are illustrated in Fig. 7.1, where $\hat{\xi}^t$ corresponds to the tth picture for $t = 1, \ldots, T$. The partition $\mathcal{C}_t$ at t is defined by

$$\mathcal{C}_t = \{\alpha_t^{-1}(i) : i \in I_t^k,\ k = 1, \ldots, K_{t-1}\}, \tag{7.22}$$

i.e., each element of the index sets I_t^k defines a new cluster and the new partition $\mathcal{C}_t$ is a refinement of the former partition $\mathcal{C}_{t-1}$.

The scenarios and their probabilities of the final scenario tree $\xi_{\text{tr}} := \hat{\xi}^T$ are given by the structure of the final partition $\mathcal{C}_T$, i.e., they have the form

$$\xi_{\text{tr}}^k = (\xi_1^*, \xi_2^{\alpha_2(i)}, \ldots, \xi_t^{\alpha_t(i)}, \ldots, \xi_T^{\alpha_T(i)}) \quad \text{and} \quad \pi_T^k = \sum_{j \in C_T^k} p_j \quad \text{if } i \in C_T^k \tag{7.23}$$

for each $k = 1, \ldots, K_T$. The index set I_t of realizations of ξ_t^{tr} is given by

$$I_t := \bigcup_{k=1}^{K_{t-1}} I_t^k.$$

For each $t \in \{1, \ldots, T\}$ and each $i \in I$ there exists an unique index $k_t(i) \in \{1, \ldots, K_t\}$ such that $i \in C_t^{k_t(i)}$. Moreover, we have $C_t^{k_t(i)} = \{i\} \cup J_{t,i}^{k_{t-1}(i)}$ for each $i \in I_t$. The probability of the ith realization of ξ_t^{tr} is $\pi_t^i = \sum_{j \in C_t^{k_t(i)}} p_j$. The branching degree of scenario $i \in I_{t-1}$ coincides with the cardinality of $I_t^{k_t(i)}$.

The next result quantifies the relative error of the tth construction step and is proved in Heitsch and Römisch (2009 Theorem 3.4).

Theorem 7.6 *Let the stochastic process $\hat{\xi}$ with fixed initial node ξ_1^*, scenarios ξ^i, and probabilities p_i, $i = 1, \ldots, N$, be given. Let ξ_{tr} be the stochastic process with scenarios $\xi_{\text{tr}}^k = (\xi_1^*, \xi_2^{\alpha_2(i)}, \ldots, \xi_t^{\alpha_t(i)}, \ldots, \xi_T^{\alpha_T(i)})$ and probabilities π_T^k if $i \in C_T^k$, $k = 1, \ldots, K_T$. Then we have*

$$\|\hat{\xi} - \xi_{\text{tr}}\|_r \le \sum_{t=2}^T \left(\sum_{k=1}^{K_{t-1}} \sum_{j \in J_t^k} p_j \min_{i \in I_t^k} |\xi_t^i - \xi_t^j|^r \right)^{\frac{1}{r}}. \tag{7.24}$$

Next, we provide a flexible algorithm that allows to generate a variety of scenario trees satisfying a given approximation tolerance with respect to the L_r-distance.

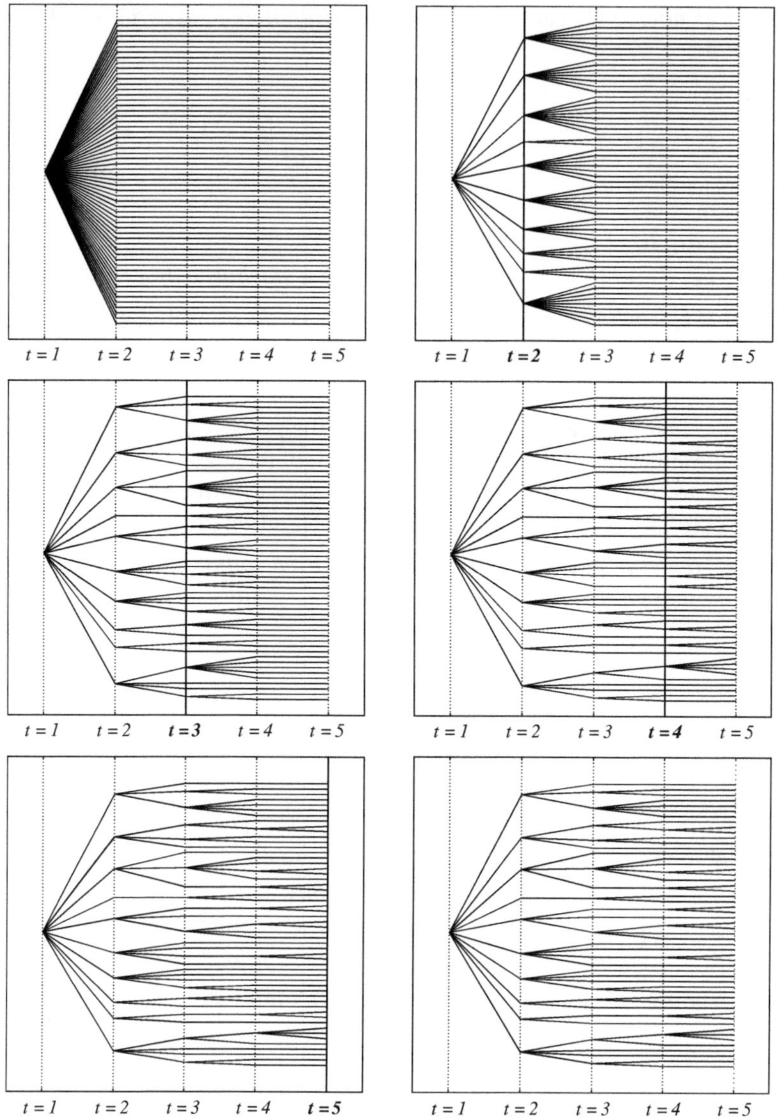

Fig. 7.1 Illustration of the tree construction for an example with $T = 5$ time periods

Algorithm 7.1 (Forward tree construction) *Let N scenarios ξ^i with probabilities p_i, $i = 1, \ldots, N$, fixed root $\xi_1^* \in \mathbb{R}^d$, and probability distribution P, $r \geq 1$, and tolerances ε_r, ε_t, $t = 2, \ldots, T$, be given such that $\sum_{t=2}^{T} \varepsilon_t \leq \varepsilon_r$.*

Step 1: *Set $\hat{\xi}^1 := \hat{\xi}$ and $\mathcal{C}_1 = \{\{1, \ldots, N\}\}$.*

Step *t*: *Let $\mathcal{C}_{t-1} = \{C_{t-1}^1, \ldots, C_{t-1}^{K_{t-1}}\}$. Determine disjoint index sets I_t^k and J_t^k such that $I_t^k \cup J_t^k = C_{t-1}^k$, the mapping $\alpha_t(\cdot)$ according to (7.20) and a stochastic*

process $\hat{\xi}^t$ having N scenarios $\hat{\xi}^{t,i}$ with probabilities p_i according to (7.21) and such that

$$\|\hat{\xi}^t - \hat{\xi}^{t-1}\|_{r,t}^r = \sum_{k=1}^{K_{t-1}} \sum_{j \in J_t^k} p_j \min_{i \in I_t^k} |\xi_t^i - \xi_t^j|^r \le \varepsilon_t^r.$$

Set $C_t = \{\alpha_t^{-1}(i) : i \in I_t^k,\ k = 1 \ldots, K_{t-1}\}$.

Step *T*+1: *Let $C_T = \{C_T^1, \ldots, C_T^{K_T}\}$. Construct a stochastic process ξ_{tr} having K_T scenarios ξ_{tr}^k such that $\xi_{\text{tr},t}^k := \xi_t^{\alpha_t(i)}$, $t = 1, \ldots, T$, if $i \in C_T^k$ with probabilities π_T^k according to (7.23), $k = 1, \ldots, K_T$.*

While the first picture in Fig. 7.1 illustrates the process $\hat{\xi}$, the tth picture corresponds to the situation after Step t, $t = 2, 3, 4, 5$, of the algorithm. The final picture corresponds to Step 6 and illustrates the final scenario tree ξ_{tr}. The proof of the following corollary is also given in Heitsch and Römisch (2009).

Corollary 7.7 *Let a stochastic process $\hat{\xi}$ with fixed initial node ξ_1^*, scenarios ξ^i, and probabilities p_i, $i = 1, \ldots, N$, be given. If ξ_{tr} is constructed by Algorithm 7.1, we have*

$$\|\hat{\xi} - \xi_{\text{tr}}\|_r \le \sum_{t=2}^{T} \varepsilon_t \le \varepsilon_r.$$

The next results state that the distance $|v(\xi) - v(\xi_{\text{tr}})|$ of optimal values gets small if the initial tolerance ε in (7.17) as well as ε_r is small.

Theorem 7.8 *Let (A1), (A2), and (A3) be satisfied with $r' \in [1, \infty)$ and the sets $\mathcal{X}_1(\tilde{\xi}_1)$ be nonempty and uniformly bounded in $\mathbb{R}^{m_1}$ if $|\tilde{\xi}_1 - \xi_1| \le \delta$. Let $L > 0$, $\delta > 0$, and $C > 0$ be the constants appearing in Theorem 7.2 and (7.8). If $(\varepsilon_r^{(n)})$ is a sequence tending to 0 such that the corresponding tolerances $\varepsilon_t^{(n)}$ in Algorithm 7.1 are nonincreasing for all $t = 2, \ldots, T$, the corresponding sequence $(\xi_{\text{tr}}^{(n)})$ has the property*

$$\limsup_{n \to \infty} |v(\xi) - v(\xi_{\text{tr}}^{(n)})| \le L \max\{1, C\}\varepsilon, \tag{7.25}$$

where $\varepsilon > 0$ is the initial tolerance in (7.17).

Proof It is shown in Heitsch and Römisch (2009 Proposition 5.2) that the estimate

$$|v(\xi) - v(\xi_{\text{tr}}^{(n)})| \le L(\varepsilon_r^{(n)} + \|\xi - \hat{\xi}\|_r + C\, D_{\text{f}}^*(\xi, \hat{\xi}) + C\, D_{\text{f}}^*(\hat{\xi}, \xi_{\text{tr}}^{(n)})) \tag{7.26}$$

is valid and that $D_{\text{f}}^*(\hat{\xi}, \xi_{\text{tr}}^{(n)})$ tends to 0 as $n \to \infty$. We conclude that the estimate (7.26) implies (7.25). □

7.4 Numerical Experience

We consider a mean-risk optimization model for electricity portfolios of a German municipal electricity company which consist of own (thermal) electricity production, the spot market contracts, supply contracts, and electricity futures. Stochasticity enters the model via the electricity demand, heat demand, spot prices, and future prices (cf. Eichhorn et al. 2005). Our approach of generating input scenarios in the form of a scenario tree consists in developing a statistical model for all stochastic components and in using Algorithm 7.1 started with a finite number of scenarios which are simulated from the statistical model.

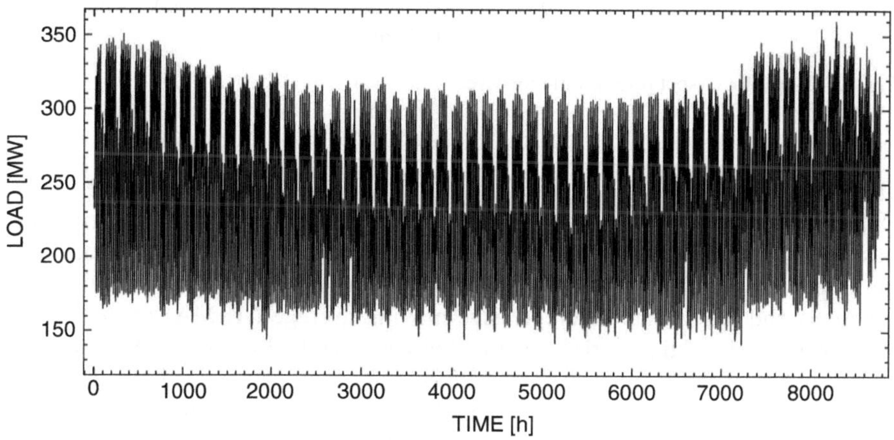

Fig. 7.2 Time plot of load profile for 1 year

7.4.1 Adapting a Statistical Model

For the stochastic input data of the optimization model (namely electricity demand, heat demand, and electricity spot prices), we had access to historical data (from a yearly period of hourly observations, cf. Figure 7.3). Due to climatic influences the demands are characterized by typical yearly cycles with high (low) demand during winter (summer) time. Furthermore, the demands contain weekly cycles due to varying consumption behavior of private and industrial customers on working days and weekends. The intraday profiles reflect a characteristic consumption behavior of the customers with seasonal differences. Outliers can be observed on public holidays, on days between holidays, and on days with extreme climatic conditions. Spot prices are affected by climatic conditions, economic activities, local power producers, customer behavior, etc. An all-embracing modeling is hardly possible. However, spot prices are also characterized by typical yearly cycles with high (lower) prices during winter (summer) time, and they show weekly and daily cycles, too. Hence, the (price

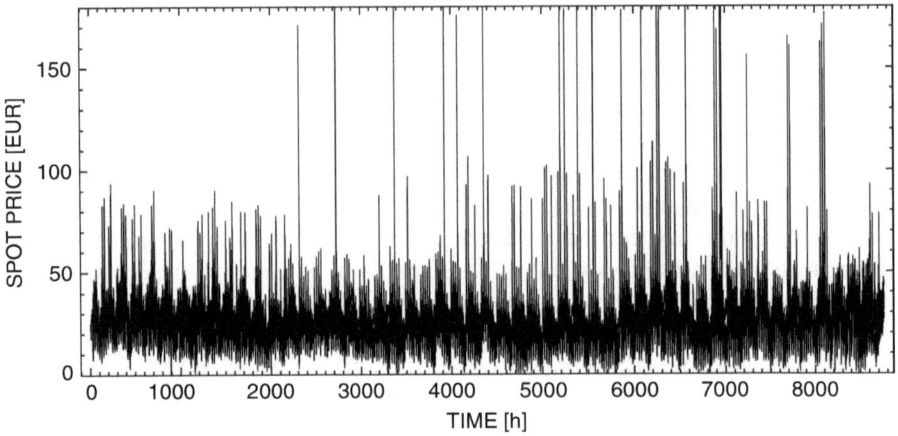

Fig. 7.3 Time plot of spot price profile for 1 year

and demand) data were decomposed into intraday profiles and daily average values. While the intraday profiles are modeled by a distribution-free resampling procedure based on standard clustering algorithms, a three-dimensional time series model was developed for the daily average values. The latter consists of deterministic trend functions and a trivariate autoregressive moving average (ARMA) model for the (stationary) residual time series (see Eichhorn et al. 2005, for details). Then an arbitrary number of three-dimensional scenarios can easily be obtained by simulating white noise processes for the ARMA model and by adding on afterward the trend functions, the matched intraday profiles from the clusters, and extreme price outliers modeled by a discrete jump diffusion process with time-varying jump parameters. Future price scenarios are directly derived from those for the spot prices.

7.4.2 Construction of Input Scenario Trees

The three-dimensional (electricity demand, heat demand, spot price) scenarios form the initial scenario set and serve as inputs for the forward tree construction (Algorithm 7.1). In our test series we started with a total number of 100 sample scenarios for a 1-year time horizon with hourly discretization. Table 7.1 displays the dimension of the simulated input scenarios. Due to the fact that electricity future

Table 7.1 Dimension of simulated input scenarios

Components	Horizon	Scenarios	Time steps	Nodes
3 (trivariate)	1 year	100	8,760	875,901

products can only be traded monthly, branching was allowed only at the end of each month. Scenario trees were generated by Algorithm 7.1 for $r = r' = 2$ and different relative reduction levels $\varepsilon_{\mathrm{rel}}$. The relative levels are given by

$$\varepsilon_{\mathrm{rel}} := \frac{\varepsilon}{\varepsilon_{\max}} \quad \text{and} \quad \varepsilon_{\mathrm{rel},t} := \frac{\varepsilon_t}{\varepsilon_{\max}},$$

where $\varepsilon_{\max}$ is given as the maximum of the best possible L_r-distance of $\hat{\xi}$ and of one of its scenarios endowed with unit mass. The individual tolerances ε_t at branching points were chosen such that

$$\varepsilon_t^r = \frac{\varepsilon^r}{T}\left[1 + \overline{q}\left(\frac{1}{2} - \frac{t}{T}\right)\right], \quad t = 2, \ldots, T, \quad r = 2, \tag{7.27}$$

where $\overline{q} \in [0, 1]$ is a parameter that affects the branching structure of the constructed trees. For the test runs we used $\overline{q} = 0.2$ which results in a slightly decreasing sequence ε_t. All test runs were performed on a PC with a 3 GHz Intel Pentium CPU and 1 GByte main memory.

Table 7.2 displays the results of our test runs with different relative reduction levels. As expected, for very small reduction levels, the reduction affects only a few scenarios. Furthermore, the number of nodes decreases considerably if the reduction level is increased. The computing times of less than 30 s already include approximately 20 s for computing distances of all scenario pairs that are needed in all calculations. Figure 7.4 illustrates the scenario trees obtained for reduction levels of 40% and 55%.

Table 7.2 Numerical results of Algorithm 7.1 for yearly demand–price scenario trees

	Scenarios		Nodes			
$\varepsilon_{\mathrm{rel}}$	Initial	Tree	Initial	Tree	Stages	Time (s)
0.20	100	100	875, 901	775, 992	4	24.53
0.25	100	100	875, 901	752, 136	5	24.54
0.30	100	100	875, 901	719, 472	7	24.55
0.35	100	97	875, 901	676, 416	8	24.61
0.40	100	98	875, 901	645, 672	10	24.64
0.45	100	96	875, 901	598, 704	10	24.75
0.50	100	95	875, 901	565, 800	9	24.74
0.55	100	88	875, 901	452, 184	10	24.75
0.60	100	87	875, 901	337, 728	11	25.89

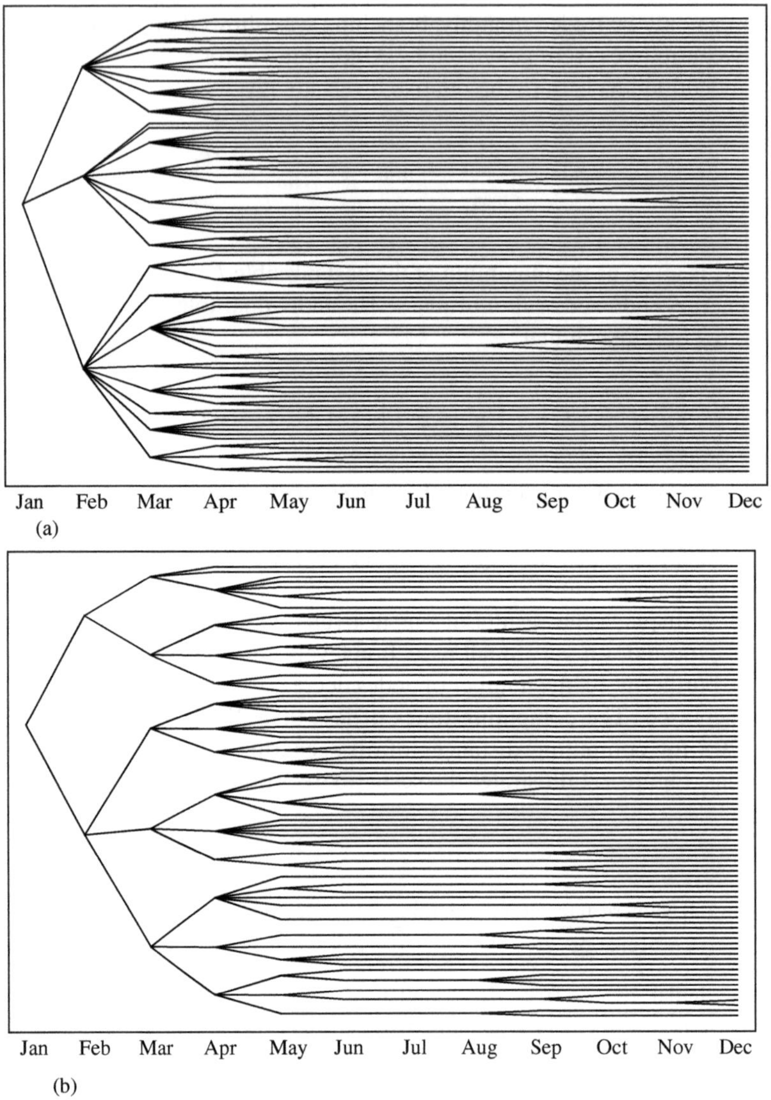

Fig. 7.4 Yearly demand–price scenario trees obtained by Algorithm 7.1. (a) Forward constructed scenario tree with reduction level $\varepsilon_{\mathrm{rel}} = 0.4$. (b) Forward constructed scenario tree with reduction level $\varepsilon_{\mathrm{rel}} = 0.55$

Acknowledgments This work was supported by the DFG Research Center MATHEON "Mathematics for key technologies" in Berlin, the BMBF under the grant 03SF0312E, and a grant of EDF – Electricité de France.

References

Attouch, H., Wets, R.J.-B.: Quantitative stability of variational systems: III. ε-approximate solutions. Math. Program. **61**(1–3), 197–214 (1993)

Casey, M., Sen, S.: The scenario generation algorithm for multistage stochastic linear programming. Math. Oper. Res. **30**(3), 615–631 (2005)

Chiralaksanakul, A., Morton, D.P.: Assessing policy quality in multi-stage stochastic programming. 12–2004, Stochastic Programming E-Print Series, ¡www.speps.org¿ (2005)

Corvera Poiré, X.: The scenario generation algorithm for multistage stochastic linear programming. PhD thesis, Department of Mathematics, University of Essex (2005)

Dantzig, G.B,: Linear programming under uncertainty. Manage. Sci. **1**(3–4), 197–206 (1955)

Dantzig, G.B,: Linear Programming and Extensions. Princeton University Press, Princeton, NJ (1963)

Dempster, M. A.H.: Sequential importance sampling algorithms for dynamic stochastic programming. Pomi 312, Zap. Nauchn. Semin (2004)

Dupačová, J., Consigli, G., Wallace, S.W.: Scenarios for multistage stochastic programs. Ann. Oper. Res. **100**(1–4), 25–53 (2000)

Dupačová, J., Gröwe-Kuska, N., Römisch, W.: Scenario reduction in stochastic programming: An approach using probability metrics. Math. Program, **95**(3), 493–511 (2003)

Edirisinghe, N. C.P,: Bound-based approximations in multistage stochastic programming: Nonanticipativity aggregation. Ann. Oper. Res. **85**(0), 103–127 (1999)

Eichhorn, A., Römisch, W.: Polyhedral risk measures in stochastic programming. SIAM J. Optim. **16**(1), 69–95 (2005)

Eichhorn, A., Römisch, W., Wegner, I.: Mean-risk optimization of electricity portfolios using multiperiod polyhedral risk measures. IEEE St. Petersburg Power Tech. 1–7 (2005)

Eichhorn, W., Römisch, A.: Stability of multistage stochastic programs incorporating polyhedral risk measures. Optimization **57**(2), 295–318 (2008)

Fabian, M., Habala, P., Hájek, P., Montesinos Santalucia, V., Pelant, J., Zizler, V.: Functional Analysis and Infinite-Dimensional Geometry. CMS Books in Mathematics. Springer, New York, NY (2001)

Frauendorfer, K.: Barycentric scenario trees in convex multistage stochastic programming. Math. Program. **75**(2), 277–293 (1996)

Gröwe-Kuska, N., Heitsch, H., Römisch, W.: Scenario reduction and scenario tree construction for power management problems. In: Borghetti, A., Nucci, C.A., Paolone, M., (eds.) IEEE Bologna Power Tech Proceedings. IEEE (2003)

Heitsch, H., Römisch, W.: Scenario reduction algorithms in stochastic programming. Comput. Optim. Appl. **24**(2–3), 187–206 (2003)

Heitsch, H., Römisch, W.: Scenario tree modelling for multistage stochastic programs. Math. Program. **118**(2), 371–406 (2009)

Heitsch, H., Römisch, W., Strugarek, C.: Stability of multistage stochastic programs. SIAM J. Optim. **17**(2), 511–525 (2006)

Higle, J.L., Rayco, B., Sen, S.: Stochastic scenario decomposition for multistage stochastic programs. IMA J. Manage. Math. **21**, 39–66 (2010)

Hochreiter, R.: Computational Optimal Management Decisions – The case of Stochastic Programming for Financial Management. PhD thesis, University of Vienna, Vienna, Austria (2005)

Hochreiter, R., Pflug, G.: Financial scenario generation for stochastic multi-stage decision processes as facility location problem. Ann. Oper. Res. **152**(1), 257–272 (2007)

Høyland, K., Wallace, S.W.: Generating scenario trees for multi-stage decision problems. Manag. Sci. **47**(2), 295–307 (2001)

Høyland, K., Kaut, M., Wallace, S.W.: A heuristic for moment-matching scenario generation. Comput. Optim. Appl. **24**(2–3), 169–185 (2003)

Infanger, G.: Planning under Uncertainty – Solving Large-Scale Stochastic Linear Programs. Boyd & Fraser Danvers, Massachusetts (1994)

Kall, P., Mayer, J.: Stochastic Linear Programming. Springer, New York, NY (2005)

Kaut, M., Wallace, S.W.: Evaluation of scenario-generation methods for stochastic programming. Pacific J. Optim. **3**(2), 257–271 (2007)

Kuhn, D.: Generalized Bounds for Convex Multistage Stochastic Programs. Lecture Notes in Economics and Mathematical Systems, vol. 548. Springer, Berlin (2005)

Luschgy, H.: Foundations of Quantization for Probability Distributions. Lecture Notes in Mathematics, vol. 1730. Springer, Berlin (2000)

Möller, A., Römisch, W., Weber, K.: Airline network revenue management by multistage stochastic programming. Comput. Manage. Sci. **5**(4), 355–377 (2008)

Niederreiter, H.: Random Number Generation and Quasi-Monte Carlo Methods. CMBS-NSF Regional Conference Series in Applied Mathematics, vol. 63. SIAM, Philadelphia (1992)

Pennanen, T.: Epi-convergent discretizations of multistage stochastic programs. Math. Oper. Res. **30**(1), 245–256 (2005)

Pennanen, T.: Epi-convergent discretizations of multistage stochastic programs via integration quadratures. Math. Program. **116**(1–2), 461–479 (2009)

Pflug, G.Ch.: Scenario tree generation for multiperiod financial optimization by optimal discretization. Math. Program. **89**(2), 251–271 (2001)

Rachev, S.T., Römisch, W.: Quantitative stability in stochastic programming: The method of probability metrics. Math. Oper. Res. **27**(3), 792–818 (2002)

Rockafellar, R.T.: Integral functionals, normal integrands and measurable selections. In: Gossez, J.P., et al., editor, Nonlinear Operators and the Calculus of Variations. Lecture Notes in Mathematics, vol. 543, pp. 157–207. Springer, Berlin (1976)

Rockafellar, R.T., Wets, R.J.-B.: Variational Analysis. Springer, Berlin (1998)

Römisch, W., Wets, R.J.-B.: Stability of ε-approximate solutions to convex stochastic programs. SIAM J. Optim. **18**(3), 961–979 (2007)

Ruszczyński, A., Shapiro, A., (eds.).: Stochastic Programming. Handbooks in Operations Research and Management Science, vol. 10. Elsevier, Amsterdam (2003)

Schmöller, H.: Modellierung von Unsicherheiten bei der mittelfristigen Stromerzeugungs- und Handelsplanung. Aachener Beiträge zur Energieversorgung, Aachen (2005) Band 103

Shapiro, A.: Inference of statistical bounds for multistage stochastic programming problems. Math. Methods. Oper. Res. **58**(1), 57–68 (2003)

Shapiro, A.: On complexity of multistage stochastic programs. Oper. Res. Lett. **34**(1), 1–8 (2006)

Shapiro, A.: Stochastic programming approach to optimization under uncertainty. Math. Program. **112**(1), 183–220 (2008)

Wallace, S.W., Ziemba, W.T., (eds.): Applications of Stochastic Programming. Series in Optimization. MPS-SIAM Philadelphia (2005)

Zaanen, A.C.: Linear Analysis. North-Holland, Amsterdam (1953)

Chapter 8
Risk Aversion in Two-Stage Stochastic Integer Programming

Rüdiger Schultz

8.1 Introduction

Stochastic programs are derived from random optimization problems usually involving information constraints. In this chapter we start out from the following model:

$$\min\{c^\top x+q^\top y+q'^\top y' \;:\; Tx+Wy+W'y' = z(\omega),\; x \in X,\; y \in \mathbb{Z}_+^{\bar{m}},\; y' \in \mathbb{R}_+^{m'}\}, \tag{8.1}$$

together with the information constraint that x must be selected without anticipation of $z(\omega)$. This leads to a two-stage scheme of alternating decision and observation: the decision on x is followed by observing $z(\omega)$ and then (y, y') is taken, thus depending on x and $z(\omega)$. Accordingly, x and (y, y') are called first- and second-stage decisions, respectively.

Assume that the ingredients of (8.1) have conformable dimensions, W, W' are rational matrices, and $X \subseteq \mathbb{R}^m$ is a nonempty polyhedron, possibly involving integer requirements to components of x.

The mentioned two-stage dynamics becomes explicit by the following reformulation of (8.1):

$$\begin{aligned}\min_x \Big\{c^\top x + \min_{y,y'}\{q^\top y + q'^\top y' \;:\; & Wy + W'y' \\ &= z(\omega) - Tx,\; y \in \mathbb{Z}_+^{\bar{m}},\; y' \in \mathbb{R}_+^{m'}\} \;:\; x \in X\Big\} \\ &= \min_x\{c^\top x + \Phi(z(\omega) - Tx) \;:\; x \in X\},\end{aligned}$$

where

$$\Phi(t) := \min\{q^\top y + q'^\top y' \;:\; Wy + W'y' = t,\; y \in \mathbb{Z}_+^{\bar{m}},\; y' \in \mathbb{R}_+^{m'}\}. \tag{8.2}$$

R. Schultz (✉)
Department of Mathematics, University of Duisburg-Essen, Campus Duisburg, Duisburg, Germany
e-mail: schultz@math.uni-duisburg.de

G. Infanger (ed.), *Stochastic Programming*, International Series in Operations Research & Management Science 150, DOI 10.1007/978-1-4419-1642-6_8,

The fundamental problem of finding a best nonanticipative decision x in (8.1) then turns into finding a best member in the indexed family of random variables

$$\Big(c^\top x + \Phi(z(\omega) - Tx)\Big)_{x \in X}. \tag{8.3}$$

The random optimization problem (8.1) thus induces a family of random variables and finding an "optimal" x means finding an "optimal" random variable. This immediately raises the point of how to rank random variables. In the literature there are different approaches to accomplish this. One possibility is to rank according to statistical parameters, with the expectation being the most popular one. Expectation models have been and still are the method of choice in much of the existing stochastic programming literature, cf. Birge and Louveaux (1997), Kall and Wallace (1994), Prékopa (1995), and Ruszczyński and Shapiro (2003). The expectation model resulting from (8.1) reads

$$\min\{Q_{I\!E}(x) \ : \ x \in X\}, \tag{8.4}$$

where

$$Q_{I\!E}(x) \ := \ I\!E\big[c^\top x + \Phi(z - Tx)\big] \ = \ \int_{I\!R^s} (c^\top x + \Phi(z - Tx))\, \mu(dz).$$

Here, μ denotes the Borel probability measure on $I\!R^s$ reflecting the distribution of z. Moreover the following assumptions are made to have the above integral well defined:

(A1) (complete recourse) $W(Z\!\!Z_+^{\bar{m}}) + W'(I\!R_+^{m'}) = I\!R^s$,
(A2) (sufficiently expensive recourse) $\{u \in I\!R^s \ : \ W^\top u \leq q, \ W'^\top u \leq q'\} \neq \emptyset$,
(A3) (finite first moment) $\int_{I\!R^s} \|z\| \mu(dz) < +\infty$.

Indeed, (A1) and (A2) imply that the mixed-integer program in (8.2) is solvable for any $t \in I\!R^s$. Properties of the value function Φ—see Bank and Mandel (1988), Blair and Jeroslow (1977), and Proposition 8.1 below—guarantee that the arising integrand is measurable and grows affinely linearly. Hence (A3) ensures that the integral is finite.

Risk aversion is addressed by modifying the expectation in (8.4) with the help of statistical parameters reflecting risk. For the moment, let $\mathcal{R}$ denote such an abstract parameter (or risk measure) of which specifications will be given later on. The expectation model is extended into a mean–risk model by ranking the relevant random variables according to a weighted sum of $I\!E$ and $\mathcal{R}$:

$$\min\{Q_{MR}(x) \ : \ x \in X\}, \tag{8.5}$$

where

$$\begin{aligned} Q_{MR}(x) &:= (I\!E + \rho \cdot \mathcal{R})\big[c^\top x + \Phi(z - Tx)\big] \\ &= I\!E\big[c^\top x + \Phi(z - Tx)\big] \; + \; \rho \cdot \mathcal{R}\big[c^\top x + \Phi(z - Tx)\big] \\ &= Q_{I\!E}(x) + \rho \cdot Q_{\mathcal{R}}(x) \end{aligned}$$

with some fixed $\rho > 0$. In the present chapter we will survey structural and algorithmic results for optimization problems of the type (8.5).

Before continuing the discussion of (8.5) let us come back to principal possibilities of ranking random variables. Besides ranking with the help of statistical parameters, partial orders of random variables, more specifically notions of stochastic dominance, have gained substantial attention in decision making under uncertainty; see, e.g., Levy (1992), Müller and Stoyan (2002), and Ogryczak and Ruszczyński (1999, 2002). When preferring small outcomes to big ones, as we do in our minimization setting, a (real-valued) random variable X is said to dominate a random variable Y to second degree ($\mathsf{X} \preceq_2 \mathsf{Y}$) if $I\!Eh(\mathsf{X}) \leq I\!Eh(\mathsf{Y})$ for all nondecreasing convex functions h for which both expectations exist. It is said to dominate to first degree ($\mathsf{X} \preceq_1 \mathsf{Y}$) if $I\!Eh(\mathsf{X}) \leq I\!Eh(\mathsf{Y})$ for all nondecreasing functions h for which both expectations exist.

Now those risk measures $\mathcal{R}$ are of particular interest for which a possible dominance relation among random variables is inherited by the mean–risk model (8.5). This is captured by the notion of consistency of mean–risk models with stochastic dominance (Müller and Stoyan 2002, Ogryczak and Ruszczyński 1999, 2002). In our setting, i.e., for random variables

$$f(x, z) \; := \; c^\top x + \Phi(z - Tx), \qquad x \in X,$$

and with preference of small outcomes, this results in calling the mean–risk model (8.5) consistent with stochastic dominance (to degree $i = 1, 2$) if $f(x_1, z) \preceq_i f(x_2, z)$ implies $Q_{MR}(x_1) \leq Q_{MR}(x_2)$.

Since risk can be specified by a variety of statistical parameters one may ask for guidelines according to which $\mathcal{R}$ should be specified in (8.5). On the one hand, these guidelines may be strictly user driven, for instance, if a certain application requires risk to be measured with a predesigned parameter $\mathcal{R}$. On the other hand, guidelines may result from the wish to arrive at models (8.5) that are sound in the following more theoretical respects:

- Stochastically sound, meaning that (8.5) is consistent with stochastic dominance.
- Structurally sound, meaning that $Q_{MR}(.)$ obeys desirable analytical properties.
- Algorithmically sound, meaning that (8.5) is accessible to computations.

To see that these requirements in general are nontrivial let us consider the variance

$$V(\mathsf{X}) := I\!E(\mathsf{X} - I\!E\mathsf{X})^2.$$

It is well known—see, e.g., Example 8.2.1. on page 274 in Müller and Stoyan (2002)—that the variance does not lead to consistent mean–risk models: consider the random variables $\mathsf{X}(\omega)$, $\mathsf{Y}(\omega)$ where $\mathsf{Y}(\omega) := 2\mathsf{X}(\omega)$ and $\mathsf{X}(\omega)$ is the uniform distribution on the two-point set $\{0, a\}$ with $a < 0$. For any nondecreasing function h it holds that

$$\mathbb{E}h(\mathsf{Y}) \;=\; \frac{1}{2}h(0) + \frac{1}{2}h(2a) \;\leq\; \frac{1}{2}h(0) + \frac{1}{2}h(a) \;=\; \mathbb{E}h(\mathsf{X}),$$

and hence $\mathsf{Y} \preceq_1 \mathsf{X}$. On the other hand, it holds for $a < -\frac{2}{3\rho}$ that

$$\mathbb{E}\mathsf{Y} + \rho \cdot V(\mathsf{Y}) \;>\; \mathbb{E}\mathsf{X} + \rho \cdot V(\mathsf{X}).$$

Hence, for any $\rho > 0$, there exist random variables X, Y violating the consistency requirement.

In classical portfolio selection models (Markowitz 1952), the variance is taken over expressions, e.g., portfolio returns, that are linear in the decision variables. This leads to favorable properties of the objective, convexity, for instance, in the resulting portfolio optimization models. In our setting, however, the dependence of $c^\top x + \Phi(z - Tx)$ on the decision x is essentially governed by the mixed-integer value function Φ, which is a discontinuous object (see Proposition 8.1 below). This may lead to substantial ill posedness when using the variance in (8.5), as is shown by the following example from page 120 in Schultz and Tiedemann (2003):

In (8.5) we specify $\mathcal{R}$ as the variance, $m = s = 1, c = 1, T = -1, \rho = 4, X = \{x \in \mathbb{R} : x \geq 0\}$, $\Phi(t) = \min\{y : y \geq t, y \in \mathbb{Z}\}$, and $z(\omega)$ attaining the values 0 and $\frac{1}{2}$ each with probability $\frac{1}{2}$. One computes that $Q_{\mathbb{E}}(x) = x + \frac{1}{2}\lceil x \rceil + \frac{1}{2}\lceil x + \frac{1}{2} \rceil$ and $Q_V(x) = \frac{1}{4}(\lceil x \rceil - \lceil x + \frac{1}{2} \rceil)^2$.

Then (8.5) has the infimum 1, and any sequence $(x_n)_{n \in \mathbb{N}}$ with $x_n \downarrow 0, x_n \neq 0$ is a minimizing sequence. However, the infimum is not attained since the objective value for $x = 0$ is $\frac{3}{2}$. The reason for this deficiency is that $Q_V(.)$ lacks a desirable analytical property, namely lower semicontinuity. Therefore, infima over compact sets, although finite, need no longer be attained in general.

8.2 Structural Results

8.2.1 Prerequisites: Value Functions and Expectation Models

The mixed-integer value function defined in (8.2) has crucial impact on the structure of two-stage mixed-integer linear stochastic programs (with or without risk aversion). Its main properties are summarized in the following statement, see Bank and Mandel (1988) and Blair and Jeroslow (1977) for original results.

Proposition 8.1 *Under (A1), (A2) the following is valid:*

(i) Φ *is real valued and lower semicontinuous on* $I\!R^s$, *i.e.,* $\liminf_{t_n \to t} \Phi(t_n) \geq \Phi(t)$ *for all* $t \in I\!R^s$*;*
(ii) there exists a countable partition $I\!R^s = \cup_{i=1}^{\infty} \mathcal{T}_i$ *such that the restrictions of* Φ *to* $\mathcal{T}_i$ *are piecewise linear and Lipschitz continuous with a uniform constant not depending on* i*;*
(iii) each of the sets $\mathcal{T}_i$ *has a representation* $\mathcal{T}_i = \{t_i + \mathcal{K}\} \setminus \cup_{j=1}^{N} \{t_{ij} + \mathcal{K}\}$ *where* $\mathcal{K}$ *denotes the polyhedral cone* $W'(I\!R_+^{m'})$ *and* t_i, t_{ij} *are suitable points from* $I\!R^s$*; moreover,* N *does not depend on* i*;*
(iv) there exist positive constants β, γ *such that* $|\Phi(t_1) - \Phi(t_2)| \leq \beta \|t_1 - t_2\| + \gamma$ *whenever* $t_1, t_2 \in I\!R^s$.

At first sight, these properties may look rather technical. Their main value, however, is that they provide insights into the *global* behavior of Φ. This is important, since specifications of $\mathcal{R}$, in one way or another, incorporate the global behavior of Φ, for instance, by referring to integrals over Φ or to probabilities of level sets of Φ. The study of value functions is one of the central topics in parametric optimization. For value functions of nonlinear programs, mainly local and hardly any global results are available up to now; see e.g., Bank et al. (1982) and Bonnans and Shapiro (2000). This somehow explains why structural investigations in two-stage stochastic programming so far have not gone very much beyond the mixed-integer linear setting.

From a stochastic programming perspective the following implications of Proposition 8.1 will be particularly important: Part (i) yields the measurability of Φ. Moreover, it already provides some intuition about more and about less promising specifications of $\mathcal{R}$ to arrive (at least) at a lower semicontinuous $Q_{M\mathcal{R}}$. If $\mathcal{R}$ constitutes operations inheriting lower semicontinuity (possibly under mild additional assumptions), such as taking the maximum or an integral, then prospects are good. If $\mathcal{R}$ involves operations destroying lower semicontinuity, such as taking the square, then $Q_{M\mathcal{R}}$ may lose this property, as seen above with the variance. Parts (ii) and (iii) imply that level sets of Φ are related to countable unions of polyhedra and that the set of discontinuity points of Φ is contained in a set of Lebesgue measure zero, namely a countable union of hyperplanes. Part (iv) is important when majorizing or minorizing Φ. It says that affinely linear functions serve these purposes. Let us also mention that (A2) implies $\Phi(0) = 0$.

The structure of the expectation model (8.4), explored in Schultz (1995), is reflected by the following statement.

Proposition 8.2

(i) Assume (A1)–(A3). Then $Q_{I\!E}(.)$ *is real valued and lower semicontinuous on* $I\!R^m$.
(ii) Assume (A1)–(A3) and that $\mu(E(x)) = 0$ *where* $E(x) = \{z \in I\!R^s : \Phi$ *is discontinuous at* $z - Tx\}$. *Then* $Q_{I\!E}(.)$ *is continuous at* x.

8.2.2 Risk Measures

In the context of mixed-integer linear stochastic programming those specifications of $\mathcal{R}$ are particularly attractive that can be represented by mixed-integer linear expressions. Later on, in the algorithmic part of this chapter, it will be shown that this allows for numerical treatment using mixed-integer linear programming techniques. We distinguish two classes of risk measures: risk measures defined via quantiles and deviation measures, which are given by expectations of deviations of the relevant random variable from its mean or from some prescribed target.

From the class of quantile measures we will discuss the following examples:

- *Excess probability,*
  which reflects the probability of exceeding a prescribed target level $\eta \in I\!R^s$:

$$Q_{I\!P^\eta}(x) \quad := \quad \mu\big(\{z \in I\!R^s \ : \ f(x,z) > \eta\}\big). \tag{8.6}$$

- *Conditional value-at-risk,*
  which, for a given probability level $\alpha \in]0,1[$, reflects the expectation of the $(1-\alpha)\cdot 100\%$ worst outcomes. There are different ways to formalize this quantity; see, e.g., Ogryczak and Ruszczyński (2002), Pflug (2000), and Rockafellar and Uryasev (2002). Of the equivalent representations we will employ the following variational formulation:

$$Q_{\alpha CVaR}(x) \quad := \quad \min_{\eta \in I\!R} g(\eta, x), \tag{8.7}$$

  where

$$g(\eta, x) \quad := \quad \eta + \frac{1}{1-\alpha} \int_{I\!R^s} \max\{f(x,z) - \eta, 0\}\, \mu(dz). \tag{8.8}$$

The following deviation measures will be considered.

- *Expected excess,*
  which reflects the expected value of the excess over a given target $\eta \in I\!R^s$:

$$Q_{\mathcal{D}^\eta}(x) \quad := \quad \int_{I\!R^s} \max\{f(x,z) - \eta, 0\}\, \mu(dz). \tag{8.9}$$

- *Semideviation,*
  which is similar in spirit to the expected excess, but with the prefixed target replaced by the mean:

$$Q_{\mathcal{D}^+}(x) \quad := \quad \int_{I\!R^s} \max\{f(x,z) - Q_{I\!E}(x), 0\}\, \mu(dz). \tag{8.10}$$

These examples do not aim at completeness, but rather are selected to discuss some paradigmatic techniques for structural analysis and algorithmic treatment of mean–risk stochastic integer programs. All examples fulfil consistency requirements with stochastic dominance. With $\mathcal{R} = Q_{I\!P^\eta}$, the mean–risk model (8.5) is consistent with stochastic dominance to degree 1, Schultz and Tiedemann (2003). With $\mathcal{R} = Q_{\alpha CVaR}$ it is consistent with stochastic dominance to degree 2, Ogryczak and Ruszczyński (2002). The same holds true with $\mathcal{R} = Q_{\mathcal{D}^\eta}$, Müller and Stoyan (2002), Ogryczak and Ruszczyński (2002), and, if $\rho \in [0, 1]$, with $\mathcal{R} = Q_{\mathcal{D}^+}$, Ogryczak and Ruszczyński (1999).

When embedding the semideviation and the expected excess into mean–risk models (8.5) the following equivalent representations are useful:

$$\begin{aligned} & Q_{I\!E}(x) + \rho Q_{\mathcal{D}^+}(x) \\ & \quad = I\!E f(x,z) + \rho I\!E \max\left\{f(x,z) - I\!E f(x,z), 0\right\} \\ & \quad = (1-\rho) I\!E f(x,z) \; + \; \rho I\!E \max\left\{f(x,z), I\!E f(x,z)\right\}, \qquad (8.11) \\ & Q_{I\!E}(x) + \rho Q_{\mathcal{D}^\eta}(x) \\ & \quad = I\!E f(x,z) + \rho I\!E \max\{f(x,z) - \eta, 0\} \\ & \quad = I\!E f(x,z) \; + \; \rho I\!E \max\{f(x,z), \eta\} \; - \; \rho\eta. \end{aligned}$$

For $\rho \in [0, 1]$ and $\rho \geq 0$ these objective functions thus practically are nonnegative linear combinations of $Q_{I\!E}$ and the functionals

$$Q_{\max}(x) \;:=\; \int_{I\!R^s} \max\left\{f(x,z), Q_{I\!E}(x)\right\} \mu(dz), \qquad (8.12)$$

and

$$Q_{\max,\eta}(x) \;:=\; \int_{I\!R^s} \max\{f(x,z), \eta\} \, \mu(dz), \qquad (8.13)$$

respectively. Properties of the mean–risk objectives in (8.5) hence readily follow from respective properties of these functionals and Proposition 8.2.

8.2.3 Structure of Objective Functions

The objective functions introduced in the previous subsection are given via (parametric) probabilities and integrals. The objective $Q_{\alpha CVaR}$ has a special role in that it is the pointwise minimum over a family of parametric integrals. Therefore, $Q_{\alpha CVaR}$ will be discussed separately.

A first observation yields that, by the lower semicontinuity of Φ (Proposition 8.1(i)), the relevant sets and integrands in (8.6), (8.9), (8.10), (8.12), (8.13) all are measurable. The following is known about (semi-) continuity of the corresponding functionals.

Proposition 8.3 *Assume* (A1), (A2). *Then* $Q_{\mathbb{P}^\eta}$ *is real valued and lower semicontinuous on* $\mathbb{R}^m$. *Let furthermore* (A3) *be valid. Then* $Q_{\max}$, $Q_{\max,\eta}$ *are real valued and lower semicontinuous on* $\mathbb{R}^m$.

Sketch of Proof Let $x \in \mathbb{R}^m$ and $x_n \to x$. Denote $M_>(x) := \{z \in \mathbb{R}^s : f(x,z) > \eta\}$, $M_=(x) := \{z \in \mathbb{R}^s : f(x,z) = \eta\}$ and recall the definition of $E(x)$ from Proposition 8.2. The lower semicontinuity of Φ then implies

$$M_>(x) \subseteq \liminf_{x_n \to x} M_>(x_n) \subseteq \limsup_{x_n \to x} M_>(x_n) \subseteq M_>(x) \cup M_=(x) \cup E(x), \tag{8.14}$$

where the limes inferior and limes superior are the sets of points belonging to all but a finite number of sets $M_>(x_n)$, $n \in \mathbb{N}$ and to infinitely many of these sets, respectively. The lower semicontinuity of $Q_{\mathbb{P}^\eta}$ then follows from the first inclusion and the semicontinuity of the probability measure on sequences of sets.

Finiteness of the integral defining $Q_{\max}$ follows from (A3) and the estimate

$$\begin{aligned} |\max\{f(x,z), Q_{\mathbb{E}}(x)\}| &\le |f(x,z)| + |Q_{\mathbb{E}}(x)| \\ &\le |\Phi(z - Tx) - \Phi(0)| + |c^\top x| + |Q_{\mathbb{E}}(x,\mu)| \\ &\le \beta\|z\| + \beta\|Tx\| + \gamma + |c^\top x| + |Q_{\mathbb{E}}(x)|, \end{aligned} \tag{8.15}$$

where Proposition 8.1(iv) was used in the last inequality. The lower semicontinuity of $Q_{\max}$ follows from the lower semicontinuity of $f(.,z)$ together with an interchange of the lim inf and the integral. The latter is justified by Fatou's Lemma for which an integrable minorant of the integrand is needed. The following estimate, again essentially relying on Proposition 8.1(iv), provides such a minorant

$$\max\{f(x_n,z), Q_{\mathbb{E}}(x_n)\} \ge f(x_n,z) \ge -(\beta\|z\| + \beta\|Tx_n\| + \gamma + |c^\top x_n|) \ge -\beta\|z\| + \kappa,$$

with some suitable bound $\kappa \in \mathbb{R}$. The proof for $Q_{\max,\eta}$ is similar. □

Proposition 8.4 *Assume* (A1), (A2) *and let* $\mu(M_=(x) \cup E(x)) = 0$. *Then* $Q_{\mathbb{P}^\eta}$ *is continuous at* x. *If* (A1)–(A3) *are valid and* $\mu(E(x)) = 0$, *then* $Q_{\max}$, $Q_{\max,\eta}$ *are continuous at* x.

Sketch of Proof For $Q_{\mathbb{P}^\eta}$ the assertion follows from (8.14) and the continuity of the probability measure on sequences of sets. For $Q_{\max}$, $Q_{\max,\eta}$ the assertion is a consequence of Lebesgue's Dominated Convergence Theorem. The integrable majorant needed for this theorem is obtained analogously to estimate (8.15) above. □

Remark 8.1 As discussed in Section 8.2.1, the set $E(x)$ is always contained in a countable union of hyperplanes and thus a set of Lebesgue measure zero. The set $M_=(x)$, however, may have nonzero Lebesgue measure: Let $c = -1, T = 0, \eta = 0$ and $\Phi(t) = \min\{y : y \ge t, y \in \mathbb{Z}_+\} = \lceil t \rceil$. Then $M_=(x) = \{z \in \mathbb{R} : \lceil z \rceil = x\}$ which is either empty or the half-open interval $]x - 1, x]$ for $x \in \mathbb{Z}$. A sufficient condition for $M_=(x)$ to have Lebesgue measure zero is that all vertices of $\{u \in I\!R^s : W'^\top u \leq q'\}$ are distinct from zero; see Tiedemann (2005) for details and further sufficient conditions. In case μ has a density, Proposition 8.4 hence always allows to conclude that $Q_{\max}, Q_{\max,\eta}$ are continuous on $I\!R^m$. For $Q_{I\!P^\eta}$ additional assumptions such as the mentioned one are needed.

According to (8.7) the relevant objective function for the conditional value-at-risk is defined as the optimal value of a parametric optimization problem. For its analysis the following lemma from parametric optimization, cf. Bank et al. (1982) and Schultz and Tiedemann (2006), is useful.

Lemma 8.5 *Consider an abstract optimization problem*

$$P(\lambda) \qquad \inf\{\mathcal{F}(\theta,\lambda) : \theta \in \Theta\}, \qquad \lambda \in \Lambda,$$

where Λ and Θ are nonempty sets each endowed with some notion of convergence and $\mathcal{F} : \Theta \times \Lambda \to I\!R$. Let $\varphi(\lambda)$ and $\psi(\lambda)$ denote the infimum and optimal set, respectively, of $P(\lambda)$. Then the following holds:

(i) φ is lower semicontinuous at $\lambda_o \in \Lambda$ if $\mathcal{F}$ is lower semicontinuous on $\Theta \times \{\lambda_o\}$ and if for each sequence $\lambda_n \to \lambda_o$ there exists a compact subset K of Θ such that $\psi(\lambda_n) \cap K \neq \emptyset$ holds for all $n \in I\!N$.
(ii) φ is upper semicontinuous at λ_o if $\mathcal{F}(\theta,\cdot)$ is upper semicontinuous at λ_o for all $\theta \in \Theta$.

Subsequently, Θ always will be a subset of some Euclidean space with the usual convergence. For the parameter set Λ different settings will occur, either in Euclidean spaces with usual convergence or in spaces of probability measures with weak convergence.

Proposition 8.6 *Assume (A1)–(A3). Then $Q_{\alpha CVaR}$ is real valued and lower semicontinuous on $I\!R^m$. If, in addition, $\mu(E(x)) = 0$ then $Q_{\alpha CVaR}$ is continuous at x. If μ has a density, then $Q_{\alpha CVaR}$ is continuous on $I\!R^m$.*

Sketch of Proof From basic statements about the conditional value-at-risk (Ogryczak and Ruszczyński 2002, Pflug 2000, Rockafellar and Uryasev 2002), it follows that $Q_{\alpha CVaR}(x) \in I\!R$ provided $|f(x,z)|$ is integrable. The latter, however, is a specific consequence of (A3) and (8.15). Moreover, it is known from the mentioned references that

$$\eta_\alpha(x) := \min\{\eta : \mu(\{z \in I\!R^s : f(x,z) \leq \eta\} \geq \alpha\},$$

also called the α-value-at-risk, is always a minimizer in (8.7). With this information the compactness assumption in Lemma 8.5(i) can be verified as follows:

Let $x_n \to x$, and $B \subseteq I\!R^m$ be compact and containing all x_n. By Proposition 8.1(iv), there exists a constant $\kappa > 0$ such that $|f(x',z)| \leq \beta\|z\| + \kappa$ for all $x' \in B$ and all $z \in I\!R^s$. Then the following estimates are valid for all $n \in I\!N$:

$$\beta \cdot \min\{\eta : \mu(\{z \in I\!R^s : \|z\| \geq -\eta\}) \geq \alpha\} - \kappa \quad \leq \quad \eta_\alpha(x_n) \tag{8.16}$$

and

$$\eta_\alpha(x_n) \quad \leq \quad \beta \cdot \min\{\eta : \mu(\{z \in I\!R^s : \|z\| \leq \eta\}) \geq \alpha\} + \kappa. \tag{8.17}$$

As a probability measure, μ is tight, meaning that for any $\varepsilon \in]0, 1[$ there exists a compact set $C \subset I\!R^s$ such that $\mu(C) > 1 - \varepsilon$. Hence, there exists a constant $R_1 > 0$ such that $\mu(\{z \in I\!R^s : \|z\| \leq R_1\}) > \alpha$. This implies

$$\min\{\eta : \mu(\{z \in I\!R^s : \|z\| \leq \eta\}) \geq \alpha\} \quad \leq \quad R_1.$$

Again by tightness of μ, there exists a constant $R_2 > 0$ such that $\mu(\{z \in I\!R^s : \|z\| \leq R_2\}) > 1 - \alpha$. Hence

$$\begin{aligned} \min\{\eta : \mu(\{z \in I\!R^s : \|z\| \geq -\eta\}) \geq \alpha\} \quad &= \quad -\max\{\eta : \mu(\{z \in I\!R^s : \|z\| < \eta\}) \\ &\leq 1 - \alpha\} \geq -2R_2. \end{aligned} \tag{8.18}$$

Together with (8.16) and (8.17) this provides that all $\eta_\alpha(x_n), n \in I\!N$, belong to the compact interval $[-2\beta R_2 - \kappa, \beta R_1 + \kappa]$.

Now Lemma 8.5 is applied: Let $x \in I\!R^m$. By (A1)–(A3) the statement regarding $Q_{\max,\eta}$ in Proposition 8.3 is valid. A straightforward argument now yields that the function g from (8.8) is lower semicontinuous on $I\!R \times \{x\}$. Together with Lemma 8.5(i) this implies that $Q_{\alpha CVaR}$ is lower semicontinuous at x.

If, in addition, $\mu(E(x)) = 0$ then, by Proposition 8.4, $g(\eta, .)$ is continuous at x for all $\eta \in I\!R$, and Lemma 8.5(ii) yields continuity of $Q_{\alpha CVaR}$ at x. If μ has a density then $\mu(E(x)) = 0$ for all $x \in I\!R^m$; see Remark 8.1, and $Q_{\alpha CVaR}$ is continuous on $I\!R^m$. □

8.2.4 Joint Continuity and Stability

Here the objective Q_{MR} of (8.5) is considered as a joint function in the decision variable x and the integrating probability measure μ. This is motivated by stability analysis under perturbations of μ. Joint continuity of Q_{MR} in (x, μ) then is the essential prerequisite for deriving continuity properties of optimal values and optimal solutions. In a more abstract parametric optimization setting, Lemma 8.5 already gives a taste for this.

The reasons for studying the dependence of stochastic programs on the integrating probability measures are twofold. First, in the course of modeling the probability measure is often selected on the basis of approximation, estimation, or subjective (expert) knowledge. Second, the multivariate integrals over implicit integrands arising in the objectives of stochastic programs are difficult to access numerically if μ is multivariate and continuous. Approximation of μ by discrete measures is the method of choice to tackle these numerical difficulties. In both situations, the question arises whether "small" perturbations in μ cause only "small" perturbations in the optimal value and optimal solution set of the stochastic program—a topic dealt with in stability analysis. For recent surveys on stability analysis of stochastic programs see Römisch (2003) and Schultz (2000).

When studying Q_{MR} as a function of μ, some notion of convergence on the space $\mathcal{P}(I\!R^s)$ of Borel probability measures on $I\!R^s$ is needed. Weak convergence of probability measures Billingsley (1968) serves well in this respect, for detailed arguments see again Römisch (2003), and Schultz (2000). A sequence $\{\mu_n\}_{n\in I\!N}$ of probability measures in $\mathcal{P}(I\!R^s)$ is said to converge weakly to $\mu \in \mathcal{P}(I\!R^s)$ ($\mu_n \xrightarrow{w} \mu$), if for any bounded continuous function $h : I\!R^s \to I\!R$ it holds that $\int_{I\!R^s} h(z)\mu_n(dz) \to \int_{I\!R^s} h(z)\mu(dz)$ as $n \to \infty$.

For the expectation model (8.4) the following is known about joint continuity of $Q_{I\!E}$, cf. Schultz (1995).

Proposition 8.7 *Assume (A1), (A2).*
Let $\mu \in \Delta_{p,K}(I\!R^s) := \{\nu \in \mathcal{P}(I\!R^s) : \int_{I\!R^s} \|z\|^p \, \nu(dz) \leq K\}$ *for some* $p > 1$ *and* $K > 0$ *and* $\mu(E(x)) = 0$. *Then* $Q_{I\!E} : I\!R^m \times \Delta_{p,K}(I\!R^s) \longrightarrow I\!R$ *is continuous at* (x, μ).

We start our discussion of joint continuity of risk functionals with the excess probability defined in (8.6).

Proposition 8.8 *Assume (A1), (A2) and let* $\mu(M_=(x) \cup E(x)) = 0$ *where* $M_=(x)$ *is as in the proof sketch of Proposition 8.3 and* $E(x)$ *is as in Proposition 8.2. Then* $Q_{I\!P^\eta} : I\!R^m \times \mathcal{P}(I\!R^s) \longrightarrow I\!R$ *is continuous at* (x, μ).

Sketch of Proof Let $x_n \to x$ and $\mu_n \xrightarrow{w} \mu$. By $\chi_n, \chi : I\!R^s \longrightarrow \{0, 1\}$ denote the indicator functions of the sets $M_>(x_n)$, $M_>(x)$, $n \in I\!N$, defined in the proof sketch of Proposition 8.3. Moreover, consider the exceptional set

$$\mathcal{E} := \{z \in I\!R^s : \exists z_n \to z \text{ such that } \chi_n(z_n) \not\to \chi(z)\}.$$

Then it holds that $\mathcal{E} \subseteq M_=(x) \cup E(x)$ and hence $\mu(\mathcal{E}) = 0$. A theorem on weak convergence of image measures attributed to Rubin in Billingsley (1968), page 34, now yields that the weak convergence $\mu_n \xrightarrow{w} \mu$ implies the weak convergence $\mu_n \circ \chi_n^{-1} \xrightarrow{w} \mu \circ \chi^{-1}$. Since $\mu_n \circ \chi_n^{-1}, \mu \circ \chi^{-1}, n \in I\!N$, are probability measures on $\{0, 1\}$, their weak convergence particularly implies that $\mu_n \circ \chi_n^{-1}(\{1\}) \to \mu \circ \chi^{-1}(\{1\})$. In other words, $\mu_n(M(x_n)) \to \mu(M(x))$ or $Q_{I\!P^\eta}(x_n, \mu_n) \to Q_{I\!P^\eta}(x, \mu)$. □

The joint continuity of the semideviation and the expected excess again is obtained via the functionals $Q_{\max}$, $Q_{\max,\eta}$ defined in (8.12) and (8.13).

Proposition 8.9 *Assume (A1), (A2). Let* $\mu \in \Delta_{p,K}(I\!R^s)$ *for some* $p > 1$ *and* $K > 0$ *and* $\mu(E(x)) = 0$. *Then* $Q_{\max}, Q_{\max,\eta} : I\!R^m \times \Delta_{p,K}(I\!R^s) \longrightarrow I\!R$ *are continuous at* (x, μ).

Sketch of Proof Let $x_n \to x$ and $\mu_n \xrightarrow{w} \mu$. Again the result of Rubin quoted in the proof sketch of Proposition 8.8 is applied. To do this for $Q_{\max}$ define $h_n(z) := \max\{f(x_n, z), Q_{I\!E}(x_n, \mu_n)\}$ and $h(z) := \max\{f(x, z), Q_{I\!E}(x, \mu)\}$. The exceptional set $\mathcal{E}$ now reads $\mathcal{E} := \{z \in I\!R^s \,:\, \exists z_n \to z \text{ such that } h_n(z_n) \not\to h(z)\}$. Then $\mathcal{E} \subseteq E(x)$. Hence $\mu(\mathcal{E}) = 0$, and Rubin's theorem yields $\mu_n \circ h_n^{-1} \xrightarrow{w} \mu \circ h^{-1}$. According to Theorem 5.4 in Billingsley (1968), see also Schultz (2003) for a detailed presentation, the desired relation $\lim_{n\to\infty} \int_{I\!R^s} h_n(z)\, \mu_n(dz) = \int_{I\!R^s} h(z)\, \mu(dz)$ now would follow if the uniform integrability

$$\lim_{a\to\infty} \sup_n \int_{|h_n(z)|\geq a} |h_n(z)|\, \mu_n(dz) = 0 \tag{8.19}$$

was to hold. To show (8.19) notice that

$$\int_{I\!R^s} |h_n(z)|^p\, \mu_n(dz) \geq a^{p-1} \int_{|h_n(z)|\geq a} |h_n(z)|\, \mu_n(dz) \tag{8.20}$$

and

$$\begin{aligned} |h_n(z)|^p &= |\max\{c^\top x_n + \Phi(z - T x_n), Q_{I\!E}(x_n, \mu_n)\}|^p \\ &\leq \max\left\{|c^\top x_n| + |\Phi(z - T x_n) - \Phi(0)|, |Q_{I\!E}(x_n, \mu_n)|\right\}^p \\ &\leq \left(\beta\|z\| + \beta\|T x_n\| + \gamma + |c^\top x_n| + |Q_{I\!E}(x_n, \mu_n)|\right)^p. \end{aligned}$$

By the continuity of $Q_{I\!E}$ at (x, μ), see Proposition 8.7, hence there exists a constant $\kappa_1 > 0$ such that $|h_n(z)|^p \leq (\beta\|z\|^p + \kappa_1)^p$. In view of $\mu_n \in \Delta_{p,K}(I\!R^s)$ and (8.20) then there exists a constant $\kappa_2 > 0$ such that

$$\int_{|h_n(z)|\geq a} |h_n(z)|\, \mu_n(dz) \leq \frac{\kappa_2}{a^{p-1}}.$$

Since $p > 1$, this verifies (8.19). □

Proposition 8.10 *Assume (A1), (A2). Let $\mu \in \Delta_{p,K}(I\!R^s)$ for some $p > 1$ and $K > 0$, and $\mu(E(x)) = 0$. Then $Q_{\alpha CVaR} : I\!R^m \times \Delta_{p,K}(I\!R^s) \longrightarrow I\!R$ is continuous at (x, μ).*

Sketch of Proof As with Proposition 8.6 the proof employs the variational representation (8.7), the validity of the result for $Q_{min,\eta}$ and hence for $Q_{\mathcal{D}^\eta}$, and Lemma 8.5. What remains to show is that the compactness assumption in Lemma 8.5(i) holds for sequences $\lambda_n := (x_n, \mu_n)$ with $x_n \to x$ and $\mu_n \xrightarrow{w} \mu$. The argument is the same as in the proof sketch of Proposition 8.6 with the extension that Prohorov's theorem—see Theorem 6.2 in Billingsley (1968)—ensures that the weakly converging sequence $\{\mu_n\}_{n\in I\!N}$ is uniformly tight, i.e., for any $\varepsilon \in]0, 1[$ there exists a compact set $C \subset I\!R^s$ such that $\mu_n(C) > 1 - \varepsilon$ for all $n \in I\!N$. □

As announced at the beginning of this section, there is a direct way from the above propositions to stability results. To illustrate this consider the generic problem

$$P(\mu) \qquad \min\{Q_{MR}(x,\mu) \ : \ x \in X\},$$

where the risk term is specified by either (8.6), (8.7), (8.9), or (8.10).

The presence of integer variables in the second stage leads to nonconvexity already in the expectation model (8.4); for example, see Louveaux and Schultz (2003). Therefore, the following localized concepts of optimal values and optimal solution sets are appropriate:

$$\varphi_V(\mu) := \inf\{Q_{MR}(x,\mu) \ : \ x \in X \cap cl\ V\},$$
$$\Psi_V(\mu) := \{x \in X \cap cl\ V \ : \ Q_{MR}(x,\mu) = \varphi_V(\mu)\},$$

where $V \subset I\!R^m$. Given $\mu \in \mathcal{P}(I\!R^s)$, a nonempty set $C \subset I\!R^m$ is called a complete local minimizing set (CLM set) of $P(\mu)$ with respect to V if V is open and $C = \Psi_V(\mu) \subset V$. Roughly speaking, a set of local minimizers has the CLM property if it contains all "nearby" local minimizers. Without this property pathologies under perturbations of μ may occur. Isolated local minimizers and the set of global minimizers are examples for CLM sets, while strict local minimizers not necessarily obey the CLM property; see Robinson (1987) for details. A typical stability result that can be derived via Propositions 8.7–8.10 now looks as follows:

Proposition 8.11 *According to the specification of $\mathcal{R}$ in $P(\mu)$ let the assumptions in the relevant Proposition 8.7–8.10 be satisfied for all $x \in X$. Suppose further that there exists a subset $C \subset I\!R^m$ which is a CLM set for $P(\mu)$ with respect to some bounded open set $V \subset I\!R^m$. Then it holds*

(i) the function $\varphi_V : \Delta_{p,K}(I\!R^s) \longrightarrow I\!R$ is continuous at μ, where $\Delta_{p,K}(I\!R^s)$ is equipped with weak convergence of probability measures,
(ii) the multifunction $\Psi_V : \Delta_{p,K}(I\!R^s) \longrightarrow 2^{I\!R^m}$ is Berge upper semicontinuous at μ, i.e., for any open set $\mathcal{O}$ in $I\!R^m$ with $\mathcal{O} \supseteq \Psi_V(\mu)$ there exists a neighborhood $\mathcal{N}$ of μ in $\Delta_{p,K}(I\!R^s)$, again equipped with the topology of weak convergence of probability measures, such that $\Psi_V(\nu) \subseteq \mathcal{O}$ for all $\nu \in \mathcal{N}$,
(iii) there exists a neighborhood $\mathcal{N}'$ of μ in $\Delta_{p,K}(I\!R^s)$ such that for all $\nu \in \mathcal{N}'$ the set $\Psi_V(\nu)$ is a CLM set for $P(\nu)$ with respect to V.

8.3 Algorithms

8.3.1 Block Structures

It is well known that the expectation model (8.4) can be represented as a mixed-integer linear program if the underlying probability measure μ is discrete. This remains valid for the mean–risk extensions of (8.4) introduced in Section 8.2.2. In the present section it is assumed that μ is discrete with realizations z_j and probabilities π_j, $j = 1, \ldots, J$. Hence, (A3) is always satisfied. (A1) and (A2), in principle, need no longer be imposed since the subsequent algorithms detect infeasibility or unboundedness of second-stage subproblems.

Proposition 8.12 *For the different specifications of $\mathcal{R}$ the following representations are valid:*

(i) Let $\mathcal{R} := \mathcal{Q}_{IP^\eta}$ and $\rho \geq 0$. If X is bounded then there exists a constant $M > 0$ such that (8.5) is equivalent to the following mixed-integer linear program

$$
\begin{aligned}
\min\{c^\top x + \sum_{j=1}^{J} \pi_j (q^\top y_j + q'^\top y_j') + \rho \cdot \sum_{j=1}^{J} \pi_j \theta_j \ : \\
Tx + Wy_j + W'y_j' &= z_j, \\
c^\top x + q^\top y_j + q'^\top y_j' - \eta &\leq M \cdot \theta_j, \\
x \in X,\ y_j \in \mathbb{Z}_+^{\bar{m}},\ y_j' \in \mathbb{R}_+^{m'},\ \theta_j \in \{0,1\},\ j = 1, \ldots, J\}.
\end{aligned} \tag{8.21}
$$

(ii) Let $\mathcal{R} := \mathcal{Q}_{\alpha CVaR}$ and $\rho \geq 0$. Then (8.5) is equivalent to the following mixed-integer linear program:

$$
\begin{aligned}
\min\{c^\top x + \sum_{j=1}^{J} \pi_j (q^\top y_j + q'^\top y_j') + \rho \cdot (\eta + \frac{1}{1-\alpha} \sum_{j=1}^{J} \pi_j v_j) \ : \\
Tx + Wy_j + W'y_j' &= z_j, \\
c^\top x + q^\top y_j + q'^\top y_j' - \eta &\leq v_j, \\
x \in X,\ \eta \in \mathbb{R},\ y_j \in \mathbb{Z}_+^{\bar{m}},\ y_j' \in \mathbb{R}_+^{m'},\ v_j \in \mathbb{R}_+,\ j = 1, \ldots, J\}.
\end{aligned} \tag{8.22}
$$

(iii) Let $\mathcal{R} := \mathcal{Q}_{D^\eta}$ and $\rho \geq 0$. Then (8.5) is equivalent to the following mixed-integer linear program

$$
\begin{aligned}
\min\{c^\top x + \sum_{j=1}^{J} \pi_j (q^\top y_j + q'^\top y_j') + \rho \cdot \sum_{j=1}^{J} \pi_j v_j \ : \\
Tx + Wy_j + W'y_j' &= z_j, \\
c^\top x + q^\top y_j + q'^\top y_j' - \eta &\leq v_j, \\
x \in X,\ y_j \in \mathbb{Z}_+^{\bar{m}},\ y_j' \in \mathbb{R}_+^{m'},\ v_j \in \mathbb{R}_+,\ j = 1, \ldots, J\}.
\end{aligned} \tag{8.23}
$$

(iv) *Let* $\mathcal{R} := \mathcal{Q}_{\mathcal{D}^+}$ *and* $\rho \in [0,1]$. *Then (8.5) is equivalent to the following mixed-integer linear program*

$$\begin{aligned}\min\{(1-\rho)c^\top x + (1-\rho)\sum_{j=1}^{J}\pi_j(q^\top y_j + q'^\top y'_j) + \rho \cdot \sum_{j=1}^{J}\pi_j v_j \ :& \\ Tx + Wy_j + W'y'_j \ &= \ z_j, \\ c^\top x + q^\top y_j + q'^\top y'_j \ &\le \ v_j, \\ c^\top x + \sum_{i=1}^{J}\pi_i(q^\top y_i + q'^\top y'_i) \ &\le \ v_j, \\ x \in X, \ y_j \in \mathbb{Z}_+^{\bar{m}}, \ y'_j \in \mathbb{R}_+^{m'}, \ v_j \in \mathbb{R}, \ j = 1,\dots,J\}.& \end{aligned} \tag{8.24}$$

Sketch of Proof In (i) the constant M can be selected as $\sup\{c^\top x + \Phi(z_j - Tx) \ : \ x \in X, \ j \in \{1,\dots,J\}\}$ which is finite by the boundedness of X and Proposition 8.1(iv). The equivalences in (ii) and (iii) are direct consequences of the definitions in (8.7) and (8.9). The representation in (iv) follows from (8.11). □

In principle, any mixed-integer linear programming (MILP) solver could be used for tackling (8.21)–(8.24). Problem size, however, grows with the number J of scenarios and in many real-life situations becomes too big for these solvers. This motivates the study of problem decomposition.

From the expectation model it is known that the MILP equivalent has a staircase block structure: Second-stage variables for different $j \in \{1,\dots,J\}$ never occur in the same constraint, but are linked through first-stage variables only. An alternative way of expressing this property is by relaxation of nonanticipativity. Nonanticipativity is captured implicitly in (8.21)–(8.24) by the fact that x does not depend on j. An explicit statement of nonanticipativity is obtained by introducing copies x_j, $j = 1,\dots,J$, of x and claiming that

$$x_1 \ = \ x_2 \ = \ \cdots \ = x_J$$

or equivalently

$$\sum_{j=1}^{J} H_j x_j \ = \ 0 \tag{8.25}$$

with suitable $l \times m$ matrices H_j. The mentioned property then says that in a model with explicit representation of nonanticipativity its relaxation leads to a model whose constraints are separable in the scenarios. Of (8.21)–(8.24), the models (8.21)–(8.23) obviously share this property of the expectation model. For (8.24) this is not the case due to presence of the constraints

$$c^\top x + \sum_{i=1}^{J} \pi_i (q^\top y_i + q'^\top y_i') \leq v_j, \quad j = 1, \ldots, J. \tag{8.26}$$

As a consequence, algorithmic techniques for the expectation model now can be transferred to (8.21)–(8.23) fairly directly, while additional effort is needed for (8.24).

8.3.2 Decomposition Methods

The mean–risk extensions (8.5) for the different specifications of $\mathcal{R}$ all are nonconvex nonlinear optimization problems in general. Even when alleviating numerical integration problems by using discrete probability distributions the problem remains to find a global minimizer in a nonconvex optimization problem. According to the results in Section 8.2, analytical properties of Q_{MR} are particularly poor when imposing a discrete μ. For lack of smoothness (even lack of continuity), hence, local (sub-) gradient based descent approaches to minimizing Q_{MR} do not seem very promising.

The expanded problem formulations of the previous section provide an alternative. Although problem dimension was increased considerably, now there is the possibility of resorting to the well-developed algorithmic methodology of mixed-integer linear programming.

Despite its poor analytical properties the compound model (8.5) serves well as a conceptual starting point. The idea is to solve (8.5) by a branch-and-bound procedure in the spirit of global optimization. To this end, the set X is partitioned with increasing granularity. To maintain the (mixed-integer) *linear* description linear inequalities are used for the partitioning. On the current elements of the partition upper and lower bounds for the optimal objective function value are sought. This is embedded into a coordination procedure to guide the partitioning and to prune elements due to infeasibility, optimality, or inferiority.

In this context, upper bounding and coordination will be done quite conventionally. Upper bounds are generated via objective values of feasible points whose quality may or may not be enhanced by search heuristics. Coordination is based on established rules or their analogues. The specific nature of (8.5) and of the specifications (8.21)–(8.24) becomes essential in the way lower bounding is achieved.

Generally speaking, lower bounds result from relaxations. Consider (8.21)–(8.23) for which it was observed that relaxation of nonanticipativity decomposes the constraints according to the scenarios. Moreover, the objective functions of these models are sums over the scenarios and, in view of (8.25), this persists when forming a Lagrangian function. This motivates Lagrangian relaxation of nonanticipativity which is now demonstrated at the conditional-value-at-risk model (8.22).

In (8.22) the specific feature arises that in addition to the decision variable x the auxiliary variable η is independent on j and, hence, can be understood as a first-stage variable that has to meet nonanticipativity. This leads to modifying (8.25) as follows:

$$\sum_{j=1}^{J} H'_j x_j = 0, \quad \sum_{j=1}^{J} H''_j \eta_j = 0$$

with suitable $l' \times m$ matrices H'_j and $l'' \times 1$ vectors H''_j, $j = 1, \ldots, J$. The following Lagrangian function results

$$L(x, y, y', v, \eta, \lambda) := \sum_{j=1}^{J} L_j(x_j, y_j, y'_j, v_j, \eta_j, \lambda),$$

where $\lambda = (\lambda', \lambda'')$ and

$$L_j(x_j, y_j, y'_j, v_j, \eta_j, \lambda', \lambda'') := \pi_j \left(c^\top x_j + q^\top y_j + q'^\top y'_j + \rho \eta_j + \rho \frac{1}{1-\alpha} v_j \right) + \lambda'^\top H'_j x_j + \lambda''^\top H''_j \eta_j.$$

The Lagrangian dual reads

$$\max \left\{ D(\lambda) \; : \; \lambda \in \mathbb{R}^{l'+l''} \right\}, \tag{8.27}$$

where

$$\begin{aligned} D(\lambda) = \min \Bigg\{ & \sum_{j=1}^{J} L_j(x_j, y_j, y'_j, v_j, \eta_j, \lambda) \; : \\ & T x_j + W y_j + W' y'_j = z_j, \\ & c^\top x_j + q^\top y_j + q'^\top y'_j - \eta_j \le v_j, \\ & x_j \in X, \; \eta_j \in \mathbb{R}, \; y_j \in \mathbb{Z}_+^{\bar{m}}, \; y'_j \in \mathbb{R}_+^{m'}, \; v_j \in \mathbb{R}_+, \; j = 1, \ldots, J \Bigg\}. \end{aligned}$$

This optimization problem is separable with respect to the individual scenarios, i.e.,

$$D(\lambda) = \sum_{j=1}^{J} D_j(\lambda)$$

with

$$\begin{aligned}D_j(\lambda) = \min\{L_j(x_j, y_j, y'_j, v_j, \eta_j, \lambda) \;:\; & \\ Tx_j + Wy_j + W'y'_j \;&=\; z_j, \\ c^\top x_j + q^\top y_j + q'^\top y'_j - \eta_j \;&\leq\; v_j, \\ x_j \in X,\ \eta_j \in I\!R,\ y_j \in \mathbb{Z}_+^{\bar{m}},\ y'_j \in I\!R_+^{m'},\ v_j \in I\!R_+\}, &\quad j = 1, \ldots, J.\end{aligned} \tag{8.28}$$

The Lagrangian dual (8.27) is a nonsmooth concave maximization (or convex minimization) problem with piecewise linear objective for whose solution advanced bundle methods—see, e.g., Helmberg and Kiwiel (2002) and Kiwiel (1990)—can be applied. In this way, solving the dual, or in other words, finding a desired lower bound reduces to function value and subgradient computations for $-D(\lambda)$ (when adopting a convex minimization setting in (8.27)). A subgradient of $-D$ at λ is given by

$$-\left(\sum_{j=1}^{J} H'_j x_j^\lambda\,,\ \sum_{j=1}^{J} H''_j \eta_j^\lambda\right),$$

where x_j^λ and η_j^λ are the corresponding components in an optimal solution vector to the optimization problem defining $D_j(\lambda)$. As a consequence, the desired lower bound can be computed by solving the single-scenario mixed-integer linear programs in (8.28) instead of working with the full-size model (8.22). This decomposition is often instrumental since (8.28) may be tractable for MILP solvers while (8.22) is not.

Obviously, for any $\lambda \in I\!R^{l'+l''}$ the value $D(\lambda)$ provides a lower bound to the optimal value of (8.22). The dual optimization aims at finding a best such bound. Integer programming theory says that the optimal value of (8.27) is never worse the bound that were obtained by solving the LP relaxation of (8.22).

Moreover, it is well known that in integer programming applications of Lagrangian relaxation there is a duality gap in general, i.e., the optimal value of (8.27) is strictly below the optimal value of (8.22). The reason is that optimal solution points to (8.28) for optimal λ in (8.27) need not fulfil nonanticipativity. Results of the dual optimization typically provide the basis for Lagrangian heuristics aiming at the identification of "promising" feasible points for the original problem.

With Lagrangian relaxation of nonanticipativity these heuristics benefit from the simplicity of the relaxed constraints. Starting from an optimal or nearly optimal λ the components $x_j^\lambda, \eta_j^\lambda$, $j = 1, \ldots, J$, of an optimal solution to (8.28) are seen as J proposals for a nonanticipative first-stage solution (x, η). A "promising" point then is selected, for instance, by deciding for the most frequent one or by averaging and rounding to integers if necessary.

The specific nature of η as an auxiliary variable and argument in an optimization problem, see (8.7), allows to improve this heuristic. Instead of selecting right away a candidate for (x, η), a candidate $\bar{x}$ for x is fixed first, and then η is computed as the best possible value, namely as optimal solution to

$$\min\{g(\eta, \bar{x}) \ : \ \eta \in I\!R\},$$

which is equivalent to

$$\min\left\{\eta + \frac{1}{1-\alpha}\sum_{j=1}^{J}\pi_j v_j \ : \ c^\top \bar{x} + \Phi(z_j - T\bar{x}) - \eta \le v_j, \right.$$
$$\left. \eta \in I\!R, \ v_j \in I\!R_+, \ j = 1, \ldots, J\right\}.$$

The input quantities $\Phi(z_j - T\bar{x})$, $j = 1, \ldots, J$ are readily computed via (8.2). In case $\Phi(z_j - T\bar{x}) = +\infty$ (infeasibility of a second-stage subproblem), $\bar{x}$ is discarded. This concludes the consideration of (8.22) as an illustration for how to achieve decomposition in the lower bounding of (8.21)–(8.23).

Lagrangian relaxation of nonanticipativity, of course, is possible for (8.24) too. However, the counterpart to (8.28) then is no longer separable in the scenarios due to the presence of the constraints (8.26). As an alternative one may think about lower bounds to (8.24) that decompose into scenario specific subproblems after relaxation of nonanticipativity. An immediate such bound is the expectation problem (8.4), although neglecting risk effects completely. The following lemma provides another bound with the desired separability that incorporates risk and strengthens the expectation bound.

Lemma 8.13 *Assume (A1)–(A3), fix $x \in X$, and let $\eta \le Q_{I\!E}(x)$. Then the following is valid for all $\rho \in [0, 1]$*

$$Q_{I\!E}(x) \ \le \ (1-\rho)Q_{I\!E}(x) + \rho Q_{\mathcal{D}^\eta}(x) + \rho\eta \ \le \ Q_{I\!E}(x) + \rho Q_{\mathcal{D}^+}(x).$$

Sketch of Proof: The first inequality follows from

$$\begin{aligned} Q_{\mathcal{D}^\eta}(x) + \eta &= I\!E \max\{f(x,z) - \eta, 0\} + \eta \\ &= I\!E \max\{f(x,z), \eta\} \ \ge \ I\!E f(x,z) \ = \ Q_{I\!E}(x). \end{aligned} \qquad (8.29)$$

The second inequality is obtained from

$$\begin{aligned} Q_{I\!E}(x) + \rho Q_{\mathcal{D}^+}(x) \ &= \ (1-\rho)Q_{I\!E}(x) + \rho I\!E \max\{f(x,z), Q_{I\!E}(x)\} \\ &\ge \ (1-\rho)Q_{I\!E}(x) + \rho I\!E \max\{f(x,z), \eta\}, \end{aligned} \qquad (8.30)$$

where the last estimate is a consequence of $\eta \le Q_{I\!E}(x)$. □

A possible specification of η fulfilling $\eta \le Q_{I\!E}(x)$ is given by the wait-and-see solution of the expectation model (8.4), Birge and Louveaux (1997), Kall and

Wallace (1994), Prékopa (1995), and Ruszczyński and Shapiro (2003). This is the expected value of

$$\Phi_{WS}(z) := \min \Big\{ c^\top x + q^\top y + q'^\top y' \; : \; Tx + Wy + W'y' \\ = z, \; x \in X, \; y \in \mathbb{Z}_+^{\bar{m}}, \; y' \in I\!R_+^{m'} \Big\}.$$

Clearly, $\Phi_{WS}(z) \leq f(x, z)$ for all $x \in X$ and all $z \in I\!R^s$, such that $\eta := I\!E\Phi_{WS}(z) \leq Q_{I\!E}(x)$ for all $x \in X$. If $\{z \in I\!R^s \; : \; I\!E\Phi_{WS}(z) > f(x,z)\}$ has positive μ-measure, then the bound in Lemma 8.13 strengthens the straightforward expectation bound; see (8.29).

Lemma 8.13 says that the semideviation model (8.24) can be tackled by expected-excess methodology if the excess level is properly selected. The branch-and-bound algorithm, whose framework was sketched above, now proceeds as follows: Lower bounds are obtained via Lagrangian relaxation of nonanticipativity applied to

$$\min\{(1-\rho)Q_{I\!E}(x) + \rho Q_{\mathcal{D}^\eta}(x) + \rho\eta \; : \; x \in X\} \tag{8.31}$$

(with X, of course, replaced by partition elements at later stages of the algorithm). Since $Q_{\mathcal{D}^\eta}$ leads to decoupled single-scenario models after relaxation of nonanticipativity—see (8.23)—the counterpart to (8.28) again enjoys decomposition in the scenarios. The results of the dual optimization again serve as inputs for the generation of upper bounds. Nonanticipative proposals, however, are not inserted into the objective of (8.31) but into the original objective $Q_{I\!E}(x) + \rho Q_{\mathcal{D}^+}(x)$.

Bibliographical Notes

The study of risk-related objects and risk aversion in the context of stochastic programming seemingly dates back to Bereanu (1964) and Charnes and Cooper (1963)—see also Bereanu (1981)—where probability objectives occur in linear two-stage models with continuous variables. In Raik (1971, 1972), see also the monographs Kibzun and Kan (1996) and Prékopa (1995), some early results on continuity of probability functionals can be found. In Mulvey et al. (1995) and Takriti and Ahmed (2004) the authors propose robustified versions of linear two-stage stochastic programs that aim at controlling variability of second-stage costs. The conceptual difference to the approach taken in the present paper is that second-stage optimality of optimal solutions to the robustified models in Mulvey et al. (1995) and Takriti and Ahmed (2004) cannot be ensured in general.

Stochastic dominance and its interrelations with stochastic programming are discussed in Dentcheva and Ruszczyński (2003, 2004), Ogryczak and Ruszczyński (1999, 2002). In Ogryczak and Ruszczyński (1999, 2002) consistency of mean–risk models with stochastic dominance is studied in detail. In Dentcheva and Ruszczyński (2003, 2004) an innovative class of optimization models on spaces of random variables is investigated whose feasible sets are determined by dominance relations. For a recent textbook on mathematical basics of comparison methods for risks see Müller and Stoyan (2002), a related survey is Levy (1992).

Risk aversion in linear two-stage stochastic programs with continuous variables is addressed in Ahmed (2006) with accent on convexity properties and decomposition, in Kristoffersen (2005) for structural and algorithmic properties of deviation measures, and in Riis and Schultz (2003) for the excess probability. The structural and algorithmic results for mixed-integer linear models surveyed in Sections 8.2 and 8.3 were taken from Märkert and Schultz (2005), Schultz and Tiedemann (2003, 2006) and Tiedemann (2005). In Märkert and Schultz (2005) and Tiedemann (2005) stochastic integer programs with risk functionals based on the semideviation and the value-at-risk are studied.

Mean–risk models (8.5) can be understood as scalarizations of bicriteria optimization problems which raises the issue of tracking efficient points by variation of the weight parameter ρ. The presence of integer variables induces nonconvexity such that, in general, there exist efficient points not representable as optimal solutions to scalarizations. Those who are, usually are referred to as supported efficient points. Numerical experiments to compute the supported parts of efficient frontiers with the help of the algorithms discussed in Section 8.3 are reported in Schultz and Tiedemann (2003, 2006) and Tiedemann (2005). A finance application of tracing efficient points in a stochastic bicriteria optimization model is presented in Ruszczyński and Vanderbei (2003).

With risk aversion in multistage stochastic programs an essential conceptual difference occurs. Instead of comparing random numbers as in (8.3) induced by a single stochastic event, (multicomponent) random vectors induced by a sequence in time of random events must be compared. This has launched research on proper formalization of multiperiod risk, see, for instance, Artzner et al. (2007), Eichhorn and Römisch (2005), Pflug (2006), and Pflug and Ruszczyński (2004). In Eichhorn and Römisch (2005) polyhedral risk measures are introduced that are defined as optimal values of multistage linear stochastic programs and, therefore, enable the use of stochastic programming methodology for structural and algorithmic investigations.

The scenario decomposition methods for mean–risk models in Section 8.3 are extensions of the method in Carøe and Schultz (1999) for expectation models. In Alonso-Ayuso et al. (2003a,b) the authors start out from relaxation of nonanticipativity and integrality and continue with an LP-based branch-and-bound scheme where nonanticipativity is regained via constraint branching. This method is particularly suited for multistage models where it removes bottlenecks of scenario decomposition (intractable dimension of the dual and non-straightforward Lagrangian heuristics). On the other hand, the coordination effort is increased such that efficient overall handling of this method still is a field of active research. An analysis of duality gaps for different Lagrangian relaxation approaches in stochastic integer programming is presented in Dentcheva and Römisch (2004). First algorithmic approaches to multistage mean–risk models with polyhedral risk measures are discussed in Eichhorn and Römisch (2005).

References

Ahmed, S.: Convexity and decomposition of mean-risk stochastic programs. Math. Program. **106**(3), 433–446 (2006)

Alonso-Ayuso, A., Escudero, L. F., Garín, A., Ortuño, M. T., Pérez, G.: An approach for strategic supply chain planning under uncertainty based on stochastic 0-1 programming. J. Global Optim. **26**(1), 97–124 (2003a)

Alonso-Ayuso, A., Escudero, L.F., Ortuño, M.T.: BFC: A branch-and-fix coordination algorithmic framework for solving some types of stochastic pure and mixed 0-1 programs. Eur. J. Oper. Res. **151**(3), 503–519 (2003b)

Artzner, P., Delbaen, F., Eber, J.-M., Heath, D., Ku, H.: Coherent multiperiod risk adjusted values and bellman's principle. Ann. Oper. Res. **152**(1), 5–22 (2007)

Bank, B., Mandel, R.: Parametric Integer Optimization. Akademie-Verlag, Berlin (1988)

Bank, B., Guddat, J., Klatte, D., Kummer, B., Tammer, K.: Non-linear Parametric Optimization. Akademie-Verlag, Berlin (1982)

Bereanu, B.: Programme de risque minimal en programmation linéaire stochastique. Comptes Rendus de l' Académie des Sciences Paris **259**(5), 981–983 (1964)

Bereanu, B.: Minimum risk criterion in stochastic optimization. Econ. Comput. Econ. Cybern. Stud. Res. **2**, 31–39 (1981)

Billingsley, P.: Convergence of Probability Measures. Wiley, New York, NY (1968)

Birge, J.R., Louveaux, F.: Introduction to Stochastic Programming. Springer, New York, NY (1997)

Blair, C.E., Jeroslow, R.G.: The value function of a mixed integer program: I. Discr. Math. **19**, 121–138 (1977)

Bonnans, J.F., Shapiro, A.: Perturbation Analysis of Optimization Problems. Springer, New York, NY (2000)

Carøe, C.C., Schultz, R.: Dual decomposition in stochastic integer programming. Oper. Res. Lett. **24**(1–2), 37–45 (1999)

Charnes, A., Cooper, W.W.: Deterministic equivalents for optimizing and satisficing under chance constraints. Oper. Res. **11**(1), 18–38 (1963)

Dentcheva, D., Römisch, W.: Duality gaps in nonconvex stochastic optimization. Math. Program. **101**(3), 515–535 (2004)

Dentcheva, D., Ruszczyński, A.: Stochastic optimization with dominance constraints. SIAM J. Optim. **14**(2), 548–566 (2003)

Dentcheva, D., Ruszczyński, A.: Optimality and duality theory for stochastic optimization with nonlinear dominance constraints. Math. Program. **99**(2), 329–350 (2004)

Eichhorn, A., Römisch, W.: Polyhedral risk measures in stochastic programming. SIAM J. Optim. **16**(1), 69–95 (2005)

Helmberg, C., Kiwiel, K.C.: A spectral bundle method with bounds. Math. Program. **93**(7), 173–194 (2002)

Kall, P., Wallace, S.W.: Stochastic Programming. Wiley, Chichester (1994)

Kibzun, A.I., Kan, Y.S.: Stochastic Programming Problems with Probability and Quantile Functions. Wiley, Chichester, (1996)

Kiwiel, K.C.: Proximity control in bundle methods for convex nondifferentiable optimization. Math. Program. **46**(1–2), 105–122 (1990)

Kristoffersen, T.: Deviation measures in linear two-stage stochastic programming. Math. Methods Oper. Res. **62**(2), 255–274 (2005)

Levy, H.: Stochastic dominance and expected utility: survey and analysis. Manage. Sci. **38**(4), 555–593 (1992)

Louveaux, F.V., Schultz, R.: Stochastic integer programming. In: Ruszczyński, A., Shapiro, A., (eds.) Stochastic Programming, *Handbooks in Operations Research and Management Science*, pp. 213–266. Elsevier, Amsterdam (2003)

Märkert, A., Schultz, R.: On deviation measures in stochastic integer programming. Oper. Res. Lett. **33**(5), 441–449 (2005)

Markowitz, H.M.: Portfolio selection. J. Finance **7**(1), 77–91 (1952)
Müller, A., Stoyan, D.: Comparison Methods for Stochastic Models and Risks. Wiley, Chichester (2002)
Mulvey, J.M., Vanderbei, R.J., Zenios, S.A.: Robust optimization of large-scale systems. Oper. Res. **43**(2), 264–281 (1995)
Ogryczak, W., Ruszczyński, A.: From stochastic dominance to mean-risk models: Semideviations as risk measures. Eur. J. Oper. Res. **116**(1), 33–50 (1999)
Ogryczak, W., Ruszczyński, A.: Dual stochastic dominance and related mean-risk models. SIAM J. Optim. **13**(1), 60–78 (2002)
Pflug, G.C.: Some remarks on the value-at-risk and the conditional value-at-risk. In: Uryasev, S., (ed.) Probabilistic Constrained Optimization: Methodology and Applications, pp. 272–281. Kluwer, Dordrecht (2000)
Pflug, G.C.: A value-of-information approach to measuring risk in multiperiod economic activity. J. Bank. Finance. **30**(2), 695–715 (2006)
Pflug, G.C., Ruszczyński, A.: Risk measures for income streams. In: Szegö, G., (ed.) Risk Measures for the 21st. Century, pp. 249–269. Wiley, Chichester (2004)
Prékopa, A.: Stochastic Programming. Kluwer, Dordrecht (1995)
Raik, E.: Qualitative research into the stochastic nonlinear programming problems. Eesti NSV Teaduste Akademia Toimetised / Füüsika, Matemaatica (News of the Estonian Academy of Sciences / Physics, Mathematics), **20**, 8–14 (1971) In Russian
Raik, E.: On the stochastic programming problem with the probability and quantile functionals. Eesti NSV Teaduste Akademia Toimetised / Füüsika, Matemaatica (News of the Estonian Academy of Sciences / Physics, Mathematics), **21**, 142–148 (1972) In Russian
Riis, M., Schultz, R.: Applying the minimum risk criterion in stochastic recourse programs. Comput. Optim. Appl. **24**(2–3), 267–287 (2003)
Robinson, S.M.: Local epi-continuity and local optimization. Math. Program. **37**(2), 208–222 (1987)
Rockafellar, R.T., Uryasev, S.: Conditional value-at-risk for general loss distributions. J. Bank. Finance. **26**(7), 1443–1471 (2002)
Römisch, W.: Stability of stochastic programming problems. In: Ruszczyński, A.A., Shapiro, A., (eds.) Stochastic Programming, *Handbooks in Operations Research and Management Science*, pp. 483–554. Elsevier, Amsterdam (2003)
Ruszczyński, A., Shapiro, A.: Stochastic Programming. *Handbooks in Operations Research and Management Science*. Elsevier, Amsterdam (2003)
Ruszczyński, A., Vanderbei, R.J.: Frontiers of stochastically nondominated portfolios. Econometrica **71**(4), 1287–1297 (2003)
Schultz, R.: On structure and stability in stochastic programs with random technology matrix and complete integer recourse. Math. Program. **70**(1–3), 73–89 (1995)
Schultz, R.: Some aspects of stability in stochastic programming. Ann. Oper. Res. **100**(1–4), 55–84 (2000)
Schultz, R.: Mixed-integer value functions in stochastic programming. In: Jünger, M., Reinelt, G., Rinaldi, G., (eds.) Combinatorial Optimization – Eureka, You Shrink!, Papers Dedicated to Jack Edmonds, LNCS 2570, pp. 171–184. Springer, Berlin (2003)
Schultz, R., Tiedemann, S.: Risk aversion via excess probabilities in stochastic programs with mixed-integer recourse. SIAM J. Optim. **14**(1), 115–138 (2003)
Schultz, R., Tiedemann, S.: Conditional value-at-risk in stochastic programs with mixed-integer recourse. Math. Program. **105**(2–3), 365–386 (2006)
Takriti, S., Ahmed, S.: On robust optimization of two-stage systems. Math. Program. **99**(1), 109–126 (2004)
Tiedemann, S.: Risk measures with preselected tolerance levels in two-stage stochastic mixed-integer programming. PhD thesis, University of Duisburg-Essen, Cuvillier Verlag, Göttingen (2005)

Chapter 9
Portfolio Optimization with Risk Control by Stochastic Dominance Constraints

Darinka Dentcheva and Andrzej Ruszczyński

9.1 Introduction

Recently, much attention is devoted to the development of risk models, risk-averse optimization, and their application to portfolio problems. In this work, we focus on risk shaping based on stochastic dominance relations. We analyze and compare this approach to utility-based approaches and to mean–risk models.

A popular classical approach to model risk aversion in decision problems is the utility optimization approach. Von Neumann and Morgenstern (1947) in their book developed the *expected utility theory*: for every rational decision maker there exists a utility function $u(\cdot)$ such that the decision maker prefers outcome R over outcome Y if and only if $\mathbb{E}[u(R)] > \mathbb{E}[u(Y)]$. This approach can also be implemented very efficiently; however, it is almost impossible to elicit the utility function of a decision maker explicitly. More difficulties arise when a group of decision makers with different utility functions who have to reach a consensus. Recently, the *dual utility theory* (or *rank-dependent expected utility theory*) has attracted much attention in economics. This approach was first presented in Quiggin (1982) and later developed and analyzed in Yaari (1987). From a different system of axioms, than those of von Neumann and Morgenstern, one derives that every decision maker has a certain *rank dependent utility function*, also called *distortion*, $w : [0, 1] \to \mathbb{R}$. Then a nonnegative outcome R is preferred over a nonnegative outcome Y, if and only if

$$-\int_0^1 w(p)\, dF_{(-1)}(R; p) \geq -\int_0^1 w(p)\, dF_{(-1)}(Y; p), \tag{9.1}$$

where $F_{(-1)}(R; \cdot)$ is the inverse distribution function of R. For a comprehensive treatment of the rank-dependent utility theory, we refer to Quiggin (1993) and for

D. Dentcheva (✉)
Department of Mathematical Sciences, Stevens Institute of Technology,
Hoboken, NJ 07030, USA
e-mail: darinka.dentcheva@stevens.edu

G. Infanger (ed.), *Stochastic Programming*, International Series in Operations Research & Management Science 150, DOI 10.1007/978-1-4419-1642-6_9,

its application in actuarial mathematics, see Wang et al. (1997) and Wang and Yong (1998).

The theory of stochastic orders (or stochastic dominance) has a universal character, with respect to families of utility functions. A random variable R dominates the random variable Y if $\mathbb{E}\big[u(R)\big] \geq \mathbb{E}\big[u(Y)\big]$ for all functions $u(\cdot)$ from a certain set of functions, called the generator of the order. Stochastic dominance of the first order (the usual stochastic order) is defined by generator consisting of all nondecreasing functions. If R dominates Y in the first order, then no rational decision maker will prefer a portfolio with return rate Y over a portfolio with return rate R. Stochastic dominance of the second order is defined by the generator comprising all nondecreasing concave functions. If R dominates Y in the second order, then no risk-averse decision maker will prefer a portfolio with return rate Y over a portfolio with return rate R. The concept of stochastic dominance originated from the theory of majorization (Hardy et al. 1934 and Marshall and Olkin 1979) for the discrete case, was later extended to general distributions (Quirk and Saposnik 1962, Hadar and Russell 1969, Hanoch and Levy 1969, and Rothschild and Stiglitz 1970). It is very popular and widely used in economics and finance (Fishburn 1964, Mosler and Scarsini 1991, and Whitmore and Findlay 1978).

Mean–risk models represent another classical approach, which is pioneered by Markowitz Markowitz (1952, 1959, 1987). This approach compares two characteristics of the possible portfolios: the expected return rate (*the mean*) and *the risk*, which are given by some scalar measure of the uncertainty of the portfolio return rate. The mean–risk approach recommends the selection of Pareto-efficient portfolios with respect to these two criteria. In a mean–risk portfolio model we combine these criteria by specifying some parameter as a trade-off between them. As a parametric optimization problem the mean–risk model can be solved numerically very efficiently, which makes this approach very attractive (Konno and Yamazaki 1991, Ruszczyński and Vanderbei 2003).

In this work we formulate a model for risk-averse portfolio optimization which includes stochastic orders as constraints. We optimize the portfolio performance or another characteristic of the portfolio return rate under the additional constraint that the portfolio return rate stochastically dominates a benchmark return rate. The model is based on our earlier publications Dentcheva and Ruszczyński (2003, 2004a,b). This approach has a fundamental advantage over mean–risk models and utility function models. All data for our model are readily available. In mean–risk models the choice of the risk measure has an arbitrary character and it is difficult to argue for one measure against another. Similarly, optimization of expected utility requires the form of the utility function to be specified.

We formulate conditions of optimality for this optimization problem which are formulated in terms of utility function and measures of risk. Our analysis derives a utility function, which is implied by the benchmark used and by the problem under consideration. We provide two problem formulations in which the stochastic dominance has a primal or inverse form: a Lorenz curve. The primal form has a dual problem related to the expected utility theory and the inverse form has a dual problem related to the rank-dependent utility theory (distortions). In this way our model provides also a link between this two competing economic approaches. Furthermore, we show that mean–risk optimization models can be viewed as Lagrangian relaxation of a dominance constrained optimization model.

9.2 Portfolio Problem

We want to invest our capital in n assets, with random return rates $R_1, R_2, \ldots, R_n$. We assume that $\mathbb{E}\big[|R_j|\big] < \infty$ for all $j = 1, \ldots, n$. Our objective is to shape the distribution of the total return rate on the investment. Denoting by $x_1, x_2, \ldots, x_n$ the fractions of the initial capital invested in assets $1, 2, \ldots, n$ we derive the formula for the total return rate:

$$R(x) = R_1 x_1 + R_2 x_2 + \cdots + R_n x_n. \tag{9.2}$$

The set of possible asset allocations is defined as follows:

$$X = \{x \in \mathbb{R}^n : x_1 + x_2 + \cdots + x_n = 1,\ x_j \geq 0,\ j = 1, 2, \ldots, n\}.$$

In some applications one may introduce the possibility of *short positions*, i.e., allow some x_j's to become negative. Other restrictions may limit the exposure to particular assets or their groups, by imposing upper bounds on the x_j's or on their partial sums. One can also limit the absolute differences between the x_j's and some reference investments $\bar{x}_j$, which may represent the existing portfolio, etc. Our analysis does not depend on the detailed way this set is defined; we only use the fact that it is a convex polyhedron. All modifications discussed above define some convex polyhedral feasible sets, and are, therefore, covered by our approach.

The main challenge in formulating a meaningful portfolio optimization problem is the definition of the preference structure among feasible portfolios. If we use only the mean return rate $\mathbb{E}\big[R(x)\big]$, then the resulting optimization problem has a trivial and very risky solution: invest everything in assets that have the maximum expected return rate. To account for risk, several approaches are possible.

In the first approach we associate with portfolio x some risk quantifier $\upsilon[R(x)]$ representing the uncertainty of the return rate $R(x)$. In the classical Markowitz model it is the variance of the return rate, $\mathbb{V}ar\big[R(x)\big]$, but many other functionals are possible here as well.

The mean–risk portfolio optimization problem is formulated as follows:

$$\operatorname*{maximize}_{x \in X} \mathbb{E}[R(x)] - \lambda \upsilon[R(x)]. \tag{9.3}$$

Here, λ is a nonnegative parameter representing our desirable exchange rate of mean for risk. If $\lambda = 0$, the risk has no value and the problem reduces to the problem of maximizing the mean. If $\lambda > 0$ we look for a compromise between the mean and the risk. More generally, we can consider the combined objective in (9.3) as a certain *measure of risk*,

$$\varrho[R(x)] = -\mathbb{E}[R(x)] + \lambda \upsilon[R(x)].$$

The general question of constructing mean–risk models which are in harmony with the stochastic dominance relations has been the subject of the analysis of the papers Ogryczak and Ruszczyński (1999, 2001, 2002). Popular examples of uncertainty measures that enjoy this property are central semideviations of order $p \geq 1$,

$$\upsilon[R(x)] = \mathbb{E}\big[\{(\mathbb{E}[R(x)] - R(x))^+\}^p\big]^{\frac{1}{p}}$$

and *weighted mean deviations from quantiles*:

$$\upsilon[R(x)] = \min_{\eta \in \mathbb{R}} \mathbb{E}\Big[\max\Big(\frac{1-\beta}{\beta}(\eta - R(x)), (R(x) - \eta)\Big)\Big], \quad \beta \in (0, 1).$$

The second approach is to select a certain *utility function* $u : \mathbb{R} \to \mathbb{R}$ and to solve the following optimization problem

$$\underset{x \in X}{\text{maximize}} \ \mathbb{E}\big[u(R(x))\big]. \tag{9.4}$$

It is usually required that the function $u(\cdot)$ is concave and nondecreasing, thus representing preferences of a risk-averse decision maker (Fishburn 1964, 1970).

Recently, a *dual (rank-dependent) utility* model attracts much attention. It is based on distorting the cumulative probability distribution of the random variable $R(x)$ rather than applying a nonlinear function $u(\cdot)$ to the realizations of $R(x)$. The corresponding portfolio optimization problem has the form

$$\underset{x \in X}{\text{maximize}} \int_0^1 F_{(-1)}(R(x), p)\, dw(p). \tag{9.5}$$

Here $F_{(-1)}(R(x), p)$ is the p-quantile of the random variable $R(x)$, and $w(\cdot)$ is the *rank-dependent utility function*, which distorts the probability distribution. We discuss this in section 9.3.2.

The challenge in both utility approaches is to select the appropriate utility function or rank-dependent utility function that represents our preferences and whose application leads to non-trivial and meaningful solutions of (9.4) or (9.5).

We propose an alternative approach, by introducing a comparison to a benchmark return rate into our optimization problem. The comparison is based on the stochastic dominance relation. More specifically, we consider only portfolios whose return rates stochastically dominate a certain benchmark return rate.

9.3 Stochastic Dominance Relations

9.3.1 Direct Forms

Stochastic dominance relates distribution function of random variables. For a real-random variable V, its right-continuous cumulative distribution function is given by the following formula:

$$F_1(V;\eta) = \mathbb{P}\{V \le \eta\} \quad \text{for } \eta \in \mathbb{R}.$$

A random return V is said (Lehmann 1955, Quirk and Saposnik 1962) to *stochastically dominate* another random return S in the first order, denoted $V \succeq_{(1)} S$, if

$$F_1(V;\eta) \le F_1(S;\eta) \quad \text{for all } \eta \in \mathbb{R}.$$

The second-order distribution function F_2 is given by areas below the distribution function F,

$$F_2(V;\eta) = \int_{-\infty}^{\eta} F_1(V;\xi)\,d\xi \quad \text{for } \eta \in \mathbb{R}$$

and defines the weak relation of the *second-order stochastic dominance* (SSD). That is, random return V stochastically dominates S in the second order, denoted $V \succeq_{(2)} S$, if

$$F_2(V;\eta) \le F_2(S;\eta) \quad \text{for all } \eta \in \mathbb{R}.$$

(see Hadar and Russell (1969) and Rothschild and Stiglitz (1970)).

We can express the function $F_2(V;\cdot)$ as the expected shortfall (Ogryczak and Ruszczyński 1999): for each target value η we have

$$F_2(V;\eta) = \mathbb{E}\big[(\eta - V)_+\big], \tag{9.6}$$

where $(\eta - V)_+ = \max(\eta - V, 0)$. The function $F_{(2)}(V;\cdot)$ is continuous, convex, nonnegative, and nondecreasing. It is well defined for all random variables V with finite expected value. Due to this representation, the second-order stochastic dominance relation $V \succeq_{(2)} S$ can be equivalently characterized by the following infinite system of inequalities:

$$\mathbb{E}\big[(\eta - V)_+\big] \le \mathbb{E}\big[(\eta - S)_+\big] \quad \text{for all } \eta \in \mathbb{R}. \tag{9.7}$$

An equivalent characterization of the relation using expected utility of von Neumann and Morgenstern (see, e.g., Müller and Stoyan 2002) is available:

(i) For any two random variables V, S the relation $V \succeq_{(1)} S$ holds if and only if

$$\mathbb{E}\big[u(V)\big] \geq \mathbb{E}\big[u(S)\big] \tag{9.8}$$

for all nondecreasing functions $u(\cdot)$ defined on $\mathbb{R}$.

(ii) For any two random variables V, S with finite expectations, the relation $V \succeq_{(2)} S$ holds if and only if (9.8) is satisfied for all nondecreasing concave functions $u(\cdot)$.

In the context of portfolio optimization, we compare random return rates by stochastic dominance relations. Thus, we say that portfolio x *dominates* portfolio y *in the first order*, if

$$F_1(R(x);\eta) \leq F_1(R(y);\eta) \quad \text{for all } \eta \in \mathbb{R}.$$

This is illustrated in Fig. 9.1.

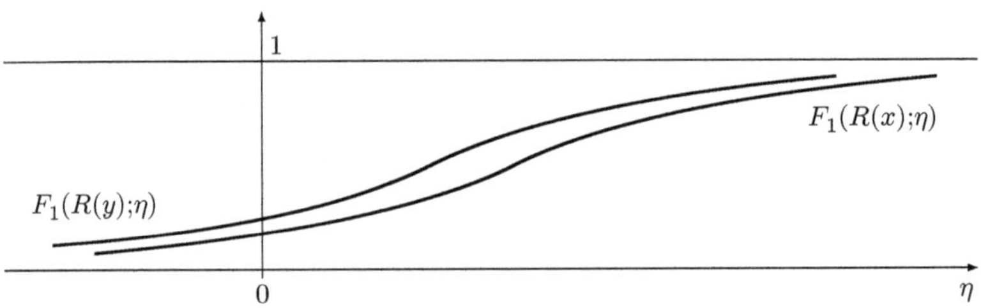

Fig. 9.1 First-degree stochastic dominance: $R(x) \succeq_{(1)} R(y)$

Similarly, we say that x *dominates* y *in the second order* ($R(x) \succeq_{(2)} R(y)$), if

$$F_2(R(x);\eta) \leq F_2(R(y);\eta) \quad \text{for all } \eta \in \mathbb{R}.$$

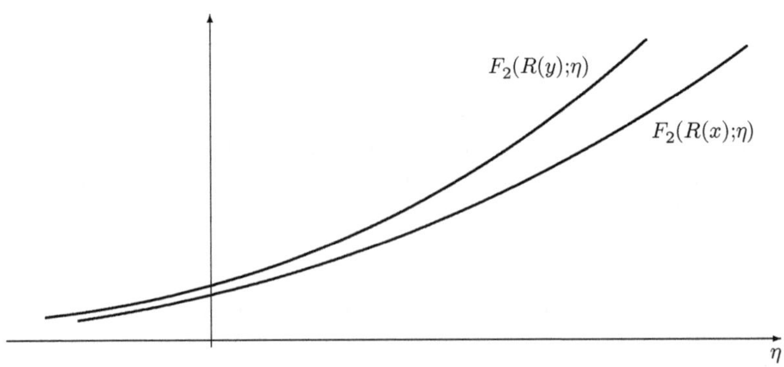

Fig. 9.2 Second-order dominance: $R(x) \succeq_{(2)} R(y)$

Recall that the individual return rates R_j have finite expected values and thus the function $F_2(R(x);\cdot)$ is well defined. The second order relation is illustrated in Fig. 9.2.

9.3.2 Inverse Forms

The inverse stochastic dominance relation compares the Lorenz curves of two random variables and it is frequently referred to as *Lorenz dominance*. For a real random variable V (e.g., a random return rate) we define the left-continuous inverse of the cumulative distribution function $F_1(V;\cdot)$ as follows:

$$F_{(-1)}(V;p) = \inf\{\eta : F_1(V;\eta) \geq p\} \quad \text{for} \quad 0 < p < 1.$$

Given $p \in (0,1)$, the number $q = q(V;p)$ is a p-quantile of the random variable V if

$$\mathbb{P}\{V < q\} \leq p \leq \mathbb{P}\{V \leq q\}.$$

For $p \in (0,1)$ the set of p-quantiles is a closed interval and $F_{(-1)}(V;p)$ represents its left end. We take the convention that $F_{(-1)}(X;1) = +\infty$, if the 1-quantile does not exist.

First-order stochastic dominance can be characterized equivalently as follows:

$$V \succeq_{(1)} S \quad \Leftrightarrow \quad F_{(-1)}(V;p) \geq F_{(-1)}(S;p) \quad \text{for all} \quad 0 < p < 1. \tag{9.9}$$

The first-order dominance constraint can be interpreted as a continuum of probabilistic (chance) constraints, studied in stochastic optimization (see, Prékopa (2003) and Shapiro et al. (2009)).

We recall that the *absolute Lorenz function* $F_{(-2)}(V;\cdot) : [0,1] \to \mathbb{R}$, introduced in Lorenz (1905) is defined as the cumulative quantile function:

$$F_{(-2)}(V;p) = \int_0^p F_{(-1)}(V;\alpha)\,d\alpha \quad \text{for} \quad 0 < p \leq 1. \tag{9.10}$$

We set $F_{(-2)}(V;0) = 0$.

Notice that

$$F_{(-2)}(V;1) = \int_0^1 F_{(-1)}(V;\alpha)\,d\alpha = \mathbb{E}[V].$$

Therefore, $F_{(-2)}(V;\cdot)$ is well defined for all random variables that have finite expectations. The Lorenz function is convex by construction.

In economics, relative Lorenz functions are commonly used $p \mapsto F_{(-2)}(V;p)/\mathbb{E}[V]$, where the expectation acts as a norming factor. Income inequalities and comparison of payments are the most frequent application (see Arnold 1980, Gastwirth 1971, Muliere and Scarsini 1989). The relative Lorenz function is convex and nondecreasing. The absolute Lorenz function, though, is not monotone, when negative outcomes are possible.

It is well known (see, e.g., Ogryczak and Ruszczyński 2002) that we may fully characterize the second-order dominance relation by using the Lorenz function:

$$V \succeq_{(2)} S \quad \Leftrightarrow \quad F_{(-2)}(V; p) \geq F_{(-2)}(S; p) \quad \text{for all} \quad 0 \leq p \leq 1. \tag{9.11}$$

Following Dentcheva and Ruszczyński (2006a), we provide an equivalent characterization of the stochastic dominance relation using rank-dependent utility functions (distortions), which mirrors the expected utility characterization.

(i) For any two random variables V, S the relation $V \succeq_{(1)} S$ holds if and only if

$$\int_0^1 F_{(-1)}(V; p)\, dw(p) \geq \int_0^1 F_{(-1)}(S; p)\, dw(p). \tag{9.12}$$

for all nondecreasing functions $w(\cdot)$ defined on $[0, 1]$.

(ii) For any two random variables V, S with finite expectations, the relation $V \succeq_{(2)} S$ holds if and only if (9.12) is satisfied for all nondecreasing concave functions $w(\cdot)$.

9.3.3 Relations to Value at Risk and Conditional Value at Risk

The representation of the stochastic dominance relations by quantile functions have the following implications for the portfolio optimization problem. Portfolio x *dominates* portfolio y *in the first order*, if

$$F_{(-1)}(R(x); p) \geq F_{(-1)}(R(y); p) \quad \text{for all } p \in (0, 1). \tag{9.13}$$

This is illustrated in Fig. 9.3.

Similarly, we say that x *dominates* y *in the second order* ($R(x) \succeq_{(2)} R(y)$), if

$$F_{(-2)}(R(x); p) \geq F_{(-2)}(R(y); p) \quad \text{for all } p \in [0, 1]. \tag{9.14}$$

Recall that the individual return rates R_j have finite expected values and thus the function $F_{(-2)}(R(x); \cdot)$ is well defined. The second-order relation is illustrated in Fig. 9.4.

The infinite systems of inequalities (9.13) and (9.13) have fundamental relations to the concepts of value at risk (VaR) and conditional value at risk (CVaR), which are fundamental characteristics of portfolio returns. The VaR constraint in the portfolio

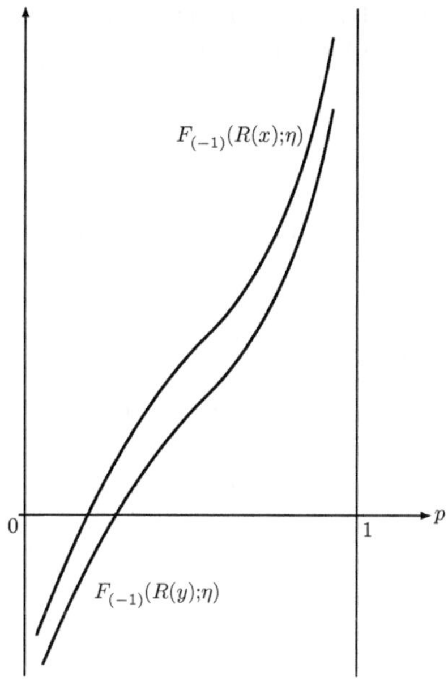

Fig. 9.3 First-degree stochastic dominance: $R(x) \succeq_{(1)} R(y)$ in the inverse form

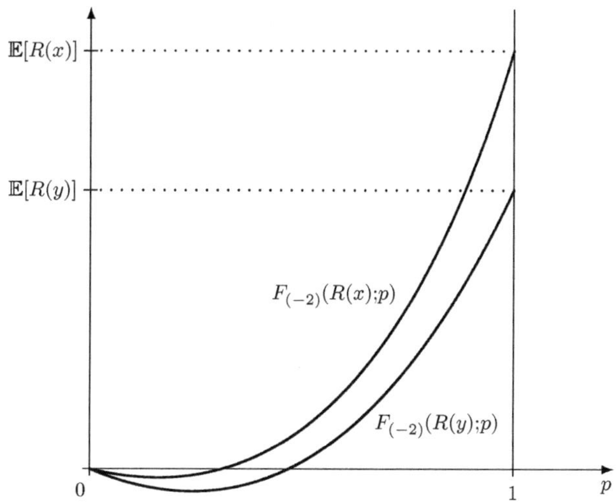

Fig. 9.4 Second-order dominance $R(x) \succeq_{(2)} R(y)$ in the inverse form

context is formulated as follows. We define the loss rate $L(x) = -R(x)$. We specify the maximum fraction ω_p of the initial capital allowed for risk exposure at risk level $p \in (0, 1)$, and we require that

$$\mathbb{P}\big[L(x) \le \omega_p\big] \ge 1 - p.$$

Denoting by $\mathrm{VaR}_p(L(x))$ the left $(1-p)$-quantile of the random variable $L(x)$, we can equivalently formulate the VaR constraint as

$$\mathrm{VaR}_p(L(x)) \le \omega_p.$$

We notice that the first-order stochastic dominance relation (9.13) between two portfolios is equivalent to the continuum of VaR constraints. Portfolio x dominates portfolio y in the first order, if

$$\mathrm{VaR}_p(L(x)) \le \mathrm{VaR}_p(L(y)) \ \text{ for all } \ p \in (0,1).$$

The CVaR at level p for continuous distributions has the form

$$\mathrm{CVaR}_p(L(x)) = \mathbb{E}\big[L(x)|L(x) \ge \mathrm{VaR}_p(L(x))\big].$$

For general distributions of the return rate, it is given by the following formula:

$$\mathrm{CVaR}_p(L(x)) = \frac{1}{p}\int_0^p \mathrm{VaR}_\alpha(L(x))d\alpha.$$

Notice that

$$\mathrm{CVaR}_p(L(x)) = -\frac{1}{p}F_{(-2)}(R(x), p). \tag{9.15}$$

Another description uses extremal properties of quantiles and equivalently represents CVaR_p as follows (see Rockafellar and Uryasev 2000):

$$\mathrm{CVaR}_p(L(x)) = \inf_\eta \left\{\frac{1}{p}\mathbb{E}\big[(\eta - R(x))_+\big] - \eta\right\}. \tag{9.16}$$

A Conditional value-at-risk constraint on the portfolio x can be formulated as follows:

$$\mathrm{CVaR}_p(L(x)) \le \omega_p. \tag{9.17}$$

Using (9.15) and (9.14) we conclude that the second-order stochastic dominance relation for two portfolios x and y is equivalent to the continuum of CVaR constraints:

$$\mathrm{CVaR}_p(L(x)) \le \mathrm{CVaR}_p(L(y)) \text{ for all } p \in (0,1]. \tag{9.18}$$

Assume that we compare the performance of a portfolio x with a random benchmark Y (e.g., an index return rate or another portfolio return rate) requiring $R(x) \succeq_{(2)} Y$. Then the fraction ω_p of the initial capital allowed for risk exposure at level p is given by the benchmark Y:

$$\omega_p = \mathrm{CVaR}_p(-Y), \quad p \in (0, 1].$$

Assume that Y has a discrete distribution with realizations y_i, $i = 1, \ldots, m$. Then relation (9.7) is equivalent to

$$\mathbb{E}\big[(y_i - R(x))_+\big] \le \mathbb{E}\big[(y_i - Y)_+\big], \quad i = 1, \ldots, m. \tag{9.19}$$

This result *does not* imply that the continuum of CVaR constraints (9.18) can be replaced by finitely many constraints of form

$$\mathrm{CVaR}_{p_i}(-R(x)) \le \mathrm{CVaR}_{p_i}(-Y) \quad i = 1, \ldots, m,$$

with some fixed probabilities p_i, $i = 1, \ldots, m$. The reason is that we do not know at which probability levels the CVaR constraints have to be imposed.

9.4 Portfolio Problem with Stochastic Dominance Constraints

9.4.1 Direct Formulation

We assume that a benchmark random return rate Y is acceptable for the decision maker. It may have the form $Y = R(\bar{z})$, for some benchmark portfolio $\bar{z}$, which may be an index or our current portfolio. Our goal is to find a portfolio whose return rate $R(x)$ is preferable over Y. Therefore, we introduce the following optimization problem:

$$\text{maximize} \quad f(x) \tag{9.20}$$

$$\text{subject to} \quad R(x) \succeq_{(2)} Y, \tag{9.21}$$

$$x \in X. \tag{9.22}$$

The objective function can take various forms: $f(x) = \mathbb{E}\big[R(x)\big]$ or $f(x) = -\varrho\big[R(x)\big]$ with $\varrho[\cdot]$ being a certain measure of risk. In this chapter, for the sake of simplicity of the presentation of our results, we focus on

$$f(x) = \mathbb{E}\big[R(x)\big].$$

However, the analysis remains valid for more general functions. Even if we maximize the expected value of the return rate, our model still leads to nontrivial solutions due to the presence of the dominance constraint (9.21).

An advantage of our problem formulation is that all data in it are readily available. Moreover, the set defined by (9.21) is convex (see Dentcheva and Ruszczyński 2003, 2004a,c).

In the special case of discrete distributions, problem (9.20)–(9.22) has an equivalent linear programming formulation. Assume that Y has realizations y_i attained with probabilities π_i, $i = 1, \ldots, m$. We also assume that the return rates of the assets have a discrete joint distribution with realizations r_{jt}, $t = 1, \dots, T$, $j = 1, \dots, n$, attained with probabilities p_t, $t = 1, 2, \dots, T$. Introducing variables s_{it} representing shortfall of $R(x)$ below y_i in realization t, $i = 1, \dots, m$, $t = 1, \dots, T$, we can formulate a linear programming equivalent of the dominance-constrained portfolio problem. The equivalent has the following form:

$$\text{maximize} \sum_{t=1}^{T} p_t \sum_{j=1}^{n} x_j r_{jt} \tag{9.23}$$

$$\text{subject to} \sum_{j=1}^{n} x_j r_{jt} + s_{it} \geq y_i, \quad i = 1, \dots, m, \quad t = 1, \dots, T, \tag{9.24}$$

$$\sum_{t=1}^{T} p_t s_{it} \leq \sum_{k=1}^{m} \pi_k (y_i - y_k)_+, \quad i = 1, \dots, m, \tag{9.25}$$

$$s_{it} \geq 0, \quad i = 1, \dots, m, \quad t = 1, \dots, T. \tag{9.26}$$

$$x \in X. \tag{9.27}$$

This problem may be solved by general-purpose linear programming solvers. However, the size of the problem increases dramatically with the number of assets n, their return realizations T, and benchmark realizations m. For this reason, we have developed efficient special methods presented in Dentcheva and Ruszczyński (2006b) and Rudolf and Ruszczyński (2008).

9.4.2 Inverse Formulation

Assume that the benchmark return rate has a discrete distribution and the return rates of the assets have a joint discrete distribution, with all notation as in the previous section. Relations (9.14) imply that problem (9.20)–(9.22) can be written with an equivalent inverse form of the dominance constraint:

$$\text{maximize} \quad \sum_{t=1}^{T} p_t \sum_{j=1}^{n} x_j r_{jt} \tag{9.28}$$

$$\text{subject to} \quad F_{(-2)}\big(R(x); p\big) \geq F_{(-2)}(Y; p), \quad p \in [0, 1], \tag{9.29}$$

$$x \in X. \tag{9.30}$$

The function $x \mapsto F_{(-2)}\big(R(x); p\big)$ is concave, positively homogeneous, and polyhedral, as shown in Ogryczak and Ruszczyński (2002). Therefore, (9.29) is a infinite system of convex polyhedral constraints. If the set X is a convex polyhedron, problem (9.28)–(9.30) has an equivalent semi-infinite linear programming formulation.

Under an additional assumption that $m = T$ and that the probabilities p_t of all elementary events are equal, we can express (9.29) with finitely many inequalities (see Dentcheva and Ruszczyński 2006a,b).

We use the symbol $R_{[t]}(x)$ to denote the ordered realizations of $R(x)$, that is,

$$R_{[1]}(x) \leq R_{[2]}(x) \leq \cdots \leq R_{[T]}(x).$$

Since $R(x)$ has a discrete distribution, the functions $F_2(R(x); \cdot)$ and $F_{(-2)}(R(x); \cdot)$ are piecewise linear. Owing to the fact that all probabilities p_t are equal, the break points of $F_{(-2)}(R(x); \cdot)$ occur at t/T, for $t = 0, 1, \ldots, m$. The same applies to $F_{(-2)}(Y; \cdot)$. It follows from (9.14) that the stochastic dominance constraint (9.21) can be expressed as

$$F_{(-2)}\left(R(x); \frac{t}{T}\right) \geq F_{(-2)}\left(Y; \frac{t}{T}\right), \quad t = 1, \ldots, T.$$

Note that $F_{(-2)}(R(x); 0) = F_{(-2)}(Y; 0) = 0$. We have

$$F_{(-2)}\left(R(x); \frac{t}{T}\right) = \frac{1}{T}\sum_{k=1}^{t} R_{[k]}(x), \quad t = 1, \ldots, T.$$

Therefore problem (9.28)–(9.30) can be simplified as follows:

$$\text{maximize} \quad \sum_{t=1}^{T} p_t \sum_{j=1}^{n} x_j r_{jt} \tag{9.31}$$

$$\text{subject to} \quad \sum_{k=1}^{t} R_{[k]}(x) \geq \sum_{k=1}^{t} y_{[k]}, \quad t = 1, \ldots, T, \tag{9.32}$$

$$x \in X. \tag{9.33}$$

These transformations are possible due to the assumption that the probabilities of all elementary events are equal. If they are not equal, the break points of the function $F_{(-2)}(R(x); \cdot)$ depend on x, and therefore inequality (9.14) cannot be reduced to finitely many convex inequalities. This is in contrast to the direct formulation, where the discreteness of Y alone was sufficient to reduce the stochastic dominance constraint to finitely many convex inequalities.

We are not making the assumption of equal probabilities here and it still possible to have a finite procedure for solving problem (9.28)–(9.30). This method, described in Dentcheva and Ruszczyński (2010), is even more efficient than methods using the direct formulation.

9.4.3 Floating Benchmark

In model (9.20)–(9.22), the benchmark return rate Y may also represent our aspirations, or demands of a decision maker. It may not be realistic, and it may be difficult or even impossible to dominate using the available basket of assets. On the other hand, the benchmark Y may be insufficiently ambitious.

For these reasons, we modify the problem to allow shifting $R(x)$ by a deterministic quantity σ, while introducing a penalty for the shift to the objective function. We obtain the following optimization problem:

$$\begin{aligned} \text{maximize} \quad & \mathbb{E}\big[R(x)\big] - \kappa\sigma && (9.34)\\ \text{subject to} \quad & R(x) + \sigma \succeq_{(2)} Y, && (9.35)\\ & x \in X. && (9.36) \end{aligned}$$

The scalar variable σ facilitates coping with two conflicting objectives. On the one hand, we want to have the expected return as high as possible; on the other hand, we want to stochastically dominate the benchmark Y. The parameter $\kappa > 0$ is a trade-off between the mean return and the error in dominating Y.

Let us assume now that aspiration return Y and the return rates of the assets have discrete distributions as in Section 9.4.1. Denote

$$v_i = F_2(Y; y_i) = \sum_{k=1}^{m} \pi_k (y_i - y_k)_+, \quad i = 1, \dots, m.$$

We obtain an equivalent formulation to (9.34)–(9.36):

$$\begin{aligned} \text{maximize} \quad & \sum_{t=1}^{T} p_t \sum_{j=1}^{n} x_j r_{jt} - \kappa\sigma && (9.37)\\ \text{subject to} \quad & \sum_{t=1}^{T} p_t \Big(y_i - \sum_{j=1}^{n} x_j r_{jt} - \sigma \Big)_+ \le v_i, \quad i = 1, \dots, m, && (9.38)\\ & x \in X. && (9.39) \end{aligned}$$

Problem (9.37)–(9.39) is a convex programming problem with piecewise-linear nonsmooth constraints. It can be re-formulated as a linear programming problem, similar to (9.23)–(9.27).

In a similar way, we can amend problem (9.28)–(9.30) to allow constraints with floating benchmarks.

9.5 Optimality Conditions and Duality Relations

9.5.1 Direct Form. Implied Utility Function.

From now on we assume that the probability distributions of the return rates are discrete with finitely many realizations r_{jt}, $t = 1, \dots, T$, $j = 1, \dots, n$, attained with probabilities p_t, $t = 1, 2, \dots, T$. We also assume that there are finitely many ordered realizations of the benchmark outcome Y: $y_1 < y_2 < \cdots < y_m$. The probabilities of these realizations are denoted by π_i, $i = 1, \dots, m$.

We define the set $\mathcal{U}$ of functions $u : \mathbb{R} \to \mathbb{R}$ satisfying the following conditions:

$u(\cdot)$ is concave and nondecreasing;
$u(\cdot)$ is piecewise linear with break points y_i, $i = 1, \dots, m$;
$u(t) = 0$ for all $t \geq y_m$.

Let us define the function $L : \mathbb{R}^n \times \mathcal{U} \to \mathbb{R}$ as follows:

$$L(x, u) = \mathbb{E}\big[R(x) + u(R(x)) - u(Y)\big]. \tag{9.40}$$

It will play a similar role to that of a Lagrangian. The following theorem has been proved in a more general version in Dentcheva and Ruszczyński (2003) (see also Dentcheva and Ruszczyński 2006b).

Theorem 9.1 *If $\hat{x}$ is an optimal solution of (9.20)–(9.22), then there exists a function $\hat{u} \in \mathcal{U}$ such that*

$$L(\hat{x}, \hat{u}) = \max_{x \in X} L(x, \hat{u}), \tag{9.41}$$

$$\mathbb{E}\big[\hat{u}(R(\hat{x}))\big] = \mathbb{E}\big[\hat{u}(Y)\big]. \tag{9.42}$$

Conversely, if for some function $\hat{u} \in \mathcal{U}$ an optimal solution $\hat{x}$ of (9.41) satisfies (9.21) and (9.42), then $\hat{x}$ is an optimal solution of (9.20)–(9.22).

The "Lagrange multiplier" $u(\cdot)$ is directly related to the expected utility theory of von Neumann and Morgenstern. We have established earlier that the second-order stochastic dominance relation is equivalent to (9.8) for *all* utility functions in $\mathcal{U}$. Our result shows that one of them, $\hat{u}(\cdot)$, assumes the role of a Lagrange multiplier associated with (9.21). A point $\hat{x}$ is a solution to (9.20)–(9.22) if there exists a utility function $\hat{u}(\cdot)$ such that $\hat{x}$ maximizes over X the objective function $\mathbb{E}[R(x)]$ augmented with this dual utility. We see that the optimization problem in (9.41) is equivalent to

$$\operatorname*{maximize}_{x \in X} \mathbb{E}\big[v(R(x))\big], \tag{9.43}$$

where $v(\eta) = \eta + u(\eta)$. At the optimal solution the function $\hat{v}(\eta) = \eta + \hat{u}(\eta)$ is the It attaches higher penalty to smaller realizations of $R(x)$ (bigger realizations of $L(x)$). By maximizing $L(R(x), u)$ we look for x such that the left tail of the distribution of $R(x)$ is thin.

We can also develop duality relations for our problem. With the function (9.40) we can associate the dual function

$$D(u) = \max_{x \in X} L(x, u).$$

As the set X is compact and the function $L(\cdot, u)$ is continuous, the dual function is well defined.

The dual problem has the form

$$\underset{u \in \mathcal{U}}{\text{minimize}}\, D(u). \tag{9.44}$$

The set $\mathcal{U}$ is a closed convex cone and $D(\cdot)$ is a convex function, so (9.44) is a convex optimization problem.

Theorem 9.2 *Assume that (9.20)–(9.22) has an optimal solution. Then problem (9.44) has an optimal solution and the optimal values of both problems coincide. Furthermore, the set of optimal solutions of (9.44) is the set of functions $\hat{u} \in \mathcal{U}$ satisfying (9.41)–(9.42) for an optimal solution $\hat{x}$ of (9.20)–(9.22).*

Note that all constraints of our problem are linear or convex polyhedral, and therefore we do not need any constraint qualification conditions here.

It is important to stress that the optimal function $\hat{u}(\cdot)$ is piecewise linear, with break points at the realizations $y_1, \ldots, y_m$ of the benchmark Y. Therefore, the dual problem has also an equivalent linear programming formulation.

9.5.2 Inverse Form. Implied Rank-Dependent Utility Function

We introduce the set $\mathcal{W}$ of concave nondecreasing rank-dependent utility functions (distortions) $w : [0, 1] \to \mathbb{R}$.

Recall the identity

$$\mathbb{E}[R(x)] = \int_0^1 F_{(-1)}(R(x); p)\, dp.$$

Let us define the function $\Phi : X \times \mathcal{W} \to \mathbb{R}$, as follows:

$$\Phi(x, w) = \int_0^1 F_{(-1)}(R(x); p)\, dp + \int_0^1 F_{(-1)}(R(x); p)\, dw(p) - \int_0^1 F_{(-1)}(Y; p)\, dw(p). \tag{9.45}$$

It plays a role similar to that of a Lagrangian of (9.28)–(9.30).

Theorem 9.3 *If $\hat{x}$ is an optimal solution of (9.28)–(9.30) then there exists a function $\hat{w} \in \mathcal{W}$ such that*

$$\Phi(\hat{x}, \hat{w}) = \max_{x \in X} \Phi(x, \hat{w}), \tag{9.46}$$

$$\int_0^1 F_{(-1)}(R(\hat{x}); p)\, d\hat{w}(p) = \int_0^1 F_{(-1)}(Y; p)\, d\hat{w}(p). \tag{9.47}$$

Conversely, if for some function $\hat{w} \in \mathcal{W}$ an optimal solution $\hat{x}$ of (9.46) satisfies (9.29) and (9.47), then $\hat{x}$ is an optimal solution of (9.28)–(9.30).

We can also develop a duality theory based on Lagrangian (9.45). For every function $w \in \mathcal{W}$ the problem

$$\underset{x \in X}{\text{maximize}}\ \Phi(x, w) \tag{9.48}$$

is a Lagrangian relaxation of problem (9.28)–(9.30). Its optimal value, $\Psi(w)$, is always greater than or equal to the optimal value of (9.28)–(9.30).

We define the dual problem as

$$\underset{w \in \mathcal{W}}{\text{minimize}}\ \Psi(w). \tag{9.49}$$

The set $\mathcal{W}$ is a closed convex cone and $\Psi(\cdot)$ is a convex function, so problem (9.49) is a convex optimization problem. Duality relations in convex programming yield the following result.

Theorem 9.4 *Assume that problem (9.28)–(9.30) has an optimal solution. Then problem (9.49) has an optimal solution and the optimal values of both problems coincide. Furthermore, the set of optimal solutions of (9.49) is the set of functions $\hat{w} \in \mathcal{W}$ satisfying (9.46)–(9.47) for an optimal solution $\hat{x}$ of (9.28)–(9.30).*

The "Lagrange multiplier" $w(\cdot)$ in this case is related to rank-dependent expected utility theory. We have established earlier that the second-order stochastic dominance relation is equivalent to (9.12) for *all* dual utility functions in $\mathcal{W}$. Our result shows that one of them, $\hat{w}(\cdot)$, assumes the role of a Lagrange multiplier associated with (9.29). A point $\hat{x}$ is a solution to (9.28)–(9.30) if there exists a dual utility function $\hat{w}(\cdot)$ such that $\hat{x}$ maximizes over X the objective function $\mathbb{E}[R(x)]$ augmented with this dual utility. We can transform the Lagrangian (9.45) in the following way:

$$\begin{aligned}\Phi(X, w) &= \int_0^1 F_{(-1)}(R(x); p)\, dp + \int_0^1 F_{(-1)}(R(x); p)\, dw(p) - \int_0^1 F_{(-1)}(Y; p)\, dw(p)\\ &= \int_0^1 F_{(-1)}(R(x); p)\, dv(p) - \int_0^1 F_{(-1)}(Y; p)\, dw(p),\end{aligned}$$

where $v(p) = p + w(p)$. At the optimal solution the function $\hat{v}(p) = p + \hat{w}(p)$ is the quantile utility function implied by the benchmark Y. Since $\int_0^1 F_{(-1)}(Y; p)\, dw(p)$ is fixed, the problem on the right-hand side of (9.46) becomes a problem of maximizing the implied rank-dependent expected utility in X. It attaches higher weights to quantiles corresponding to smaller probabilities p. By maximizing $\Phi(R(x), w)$ we look for x such that the left tail of the distribution of $R(x)$ is thin.

Similarly to von Neumann–Morgenstern utility function, it is very difficult to elicit the dual utility function in advance. Our model derives it from the random benchmark.

If $m = T$ and all probabilities are equal, then the optimal function $\hat{w}(\cdot)$ is piecewise linear, with break points at t/T, $t = 1, \ldots, T$. In this case, the dual problem has an equivalent linear programming formulation. However, the numerical method of Dentcheva and Ruszczyński (2010) is not dependent on this assumption.

9.6 Dominance Constraints and Coherent Measures of Risk

9.6.1 Coherent Measures of Risk

A *measure of risk* ϱ assigns to a random variable V a real value $\varrho(V)$ and can take values in the extended real line $\mathbb{R} \cup \{+\infty\}$. A *coherent measure of risk* is defined by the following axioms:

Convexity : $\varrho(\gamma V + (1-\gamma)W) \le \gamma\varrho(V) + (1-\gamma)\varrho(W)$ for all V, W and $\gamma \in [0, 1]$.
Monotonicity : If V, W and $V \ge W$ a.s., then $\varrho(V) \le \varrho(W)$.
Translation equivariance : If $a \in \mathbb{R}$, then $\varrho(V + a) = \varrho(V) - a$.
Positive homogeneity : If $\tau > 0$, then $\varrho(\tau V) = \tau\varrho(V)$.

The concept of a coherent measure of risk was introduced by Artzner et al. (1999) for measures of risk ϱ defined only for bounded random variables. For further developments and more detail, we refer to Rockafellar et al. (2006), Ruszczyński and Shapiro (2006) and the references therein. A measure of risk ϱ is called *law invariant* if it depends only on the distribution of the random variables, that is, $\varrho(V) = \varrho(W)$ if V and W have the same distribution.

Recall that the conditional value at risk at level p is defined as

$$\mathrm{CVaR}_p(-R(x)) = -\frac{1}{p} F_{(-2)}(R(x); p) = \frac{1}{p} \int_0^p \mathrm{VaR}_\alpha(-R(x))\, d\alpha.$$

A measure of risk ϱ is called *spectral* (see Acerbi 2002) if a probability measure μ on the interval $(0, 1]$ exists such that for all random variables V, $\varrho(V)$ can be calculated as follows:

$$\varrho(V) = \int_0^1 \mathrm{CVaR}_p(-V)\,\mu(dp).$$

Here the convention is that larger values of V are preferred, for example, V represents the return rate. We say that a measure of risk ϱ is a *Kusuoka measure*, if there exists a convex set $\mathcal{M}$ of probability measures on $(0, 1]$ such that for all V we have

$$\varrho(V) = \sup_{\mu \in \mathcal{M}} \int_0^1 \mathrm{CVaR}_p(-V)\,\mu(dp).$$

A fundamental result in the theory of coherent measures of risk is the Kusuoka theorem (Kusuoka 2001): *Every law invariant, finite-valued coherent measure of risk that is defined on the set of bounded random variables on a nonatomic probability space is a Kusuoka measure.*

9.6.2 The Implied Measure of Risk

In Dentcheva and Ruszczyński (2008a), we have established a correspondence between rank-dependent utility functions (distortions) and spectral measures of risk. For every function $w \in \mathcal{W}$, a spectral risk measure ϱ and a constant $\lambda \geq 0$ exist such that for all random variables X with finite expectation the following relation holds:

$$\int_0^1 F_{(-1)}(X; p)\,dw(p) = -\lambda\hat{\varrho}(X). \tag{9.50}$$

Conversely, for every spectral measure of risk ϱ there exists a function $w \in \mathcal{W}$ such that (9.50) holds with $\lambda = 1$.

We may formulate necessary conditions of optimality for problem (9.28)–(9.30) in terms of measures of risk. The Lagrangian-like function involves a spectral risk measure and represents a mean–risk model.

Theorem 9.5 *If $\hat{x}$ is an optimal solution of (9.28)–(9.30) then a spectral risk measure $\hat{\varrho}$ and a constant $\lambda \geq 0$ exist such the following conditions are satisfied:*

$$\max_{x \in X} \big\{\mathbb{E}\big[R(x)\big] - \lambda\hat{\varrho}\big[R(x)\big]\big\}, \tag{9.51}$$

$$\lambda\hat{\varrho}(\big[R(\hat{x})\big]) = \lambda\hat{\varrho}(Y). \tag{9.52}$$

Condition (9.51) means that $\hat{x}$ is also an optimal solution of an implied mean–risk problem, which can be viewed as a Lagrangian relaxation of problem (9.28)–(9.30). Furthermore, if the dominance constraint (9.29) is active (i.e., its removal would change the optimal value), then we can show that $\lambda > 0$ and the complementarity condition (9.52) take on the form

$$\hat{\varrho}([R(\hat{x})]) = \hat{\varrho}(Y).$$

In the special of discrete distributions with $m = T$ and equal probabilities of all realizations, the implied spectral measure of risk has the form:

$$\hat{\varrho}(R(x)) = -\sum_{i=1}^{m} \nu_i \sum_{k=1}^{i} [R(x)]_{[k]},$$

where $[R(x)]_{[k]}$ refers to the kth-worst (smallest) realization of the return rate of portfolio x, and $\nu_i \geq 0$, $i = 1, \ldots, m$.

Theorem 9.5 is valid for a general objective function of problem (9.28)–(9.30). In the special case of $f(x) = \mathbb{E}[R(x)]$, we can reformulate the conditions as follows.

Corollary 9.6 *If $\hat{x}$ is an optimal solution of (9.28)–(9.30), then a spectral risk measure $\bar{\varrho}$ exists such that $\hat{x}$ is also an optimal solution of the mean–risk problem*

$$\min_{x \in X} \bar{\varrho}(R(x)); \quad \textit{and} \tag{9.53}$$

$$\mathbb{E}[R(\hat{x})] + \bar{\lambda}\bar{\varrho}(R(\hat{x})) = \mathbb{E}[Y] + \bar{\lambda}\bar{\varrho}(Y), \tag{9.54}$$

with some $\bar{\lambda} \geq 1$.

The sufficient conditions of optimality for optimization problems with dominance constraints based on the Lagrangian involving spectral risk measures have the following form:

Theorem 9.7 *If for some spectral risk measure $\hat{\varrho}$ and some $\lambda \geq 0$ an optimal solution $\hat{x}$ of (9.51) satisfies (9.29) and (9.52), then $\hat{x}$ is an optimal solution of (9.28)–(9.30).*

9.7 Numerical Illustration

We have tested our approach on a basket of 719 real-world assets, using 616 possible realizations of their joint return rates (Ruszczyński and Vanderbei 2003). Historical data on weekly return rates in the 12 years from Spring 1990 to Spring 2002 were used as equally likely realizations.

We have used four benchmark return rates Y. Each of them was constructed as a return rate of a certain index composed of our assets. Since we actually know the past return rates, for the purpose of comparison we have selected equally weighted indexes composed of the N assets having the highest average return rates in this period. Benchmark 1 corresponds to $N = 26$, Benchmark 2 corresponds to $N = 54$, Benchmark 3 corresponds to $N = 82$, and Benchmark 4 corresponds to $N = 200$. Our problem was to maximize the expected return rate, under the condition that the return rate of the benchmark portfolio is dominated. Since the benchmark point was a return rate of a portfolio composed from the same basket, we have $m = T = 616$ in this case.

We have solved the problem by our method of minimizing the dual problem which was presented in Dentcheva and Ruszczyński (2006b).

The implied utility functions from (9.43) obtained by solving the optimization problem (9.41) in the optimality conditions are illustrated in Fig. 9.5. We see that for Benchmark Portfolio 1, which contains only a small number of fast growing assets, the utility function is linear on almost the entire range of return rates. Only very negative return rates are penalized.

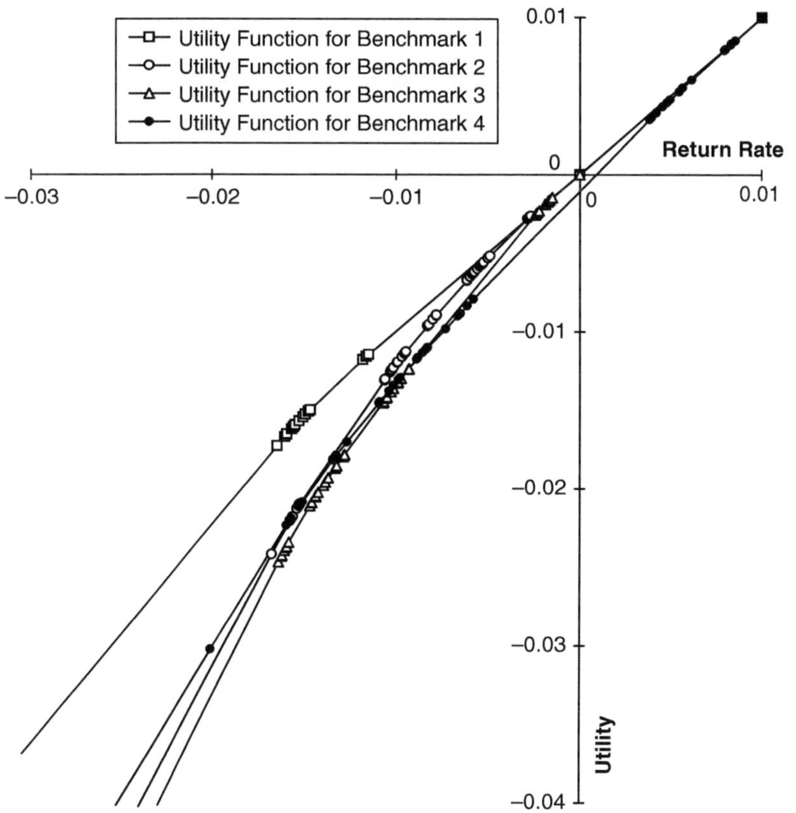

Fig. 9.5 Implied utility functions corresponding to dominance constraints for four benchmark portfolios

If the benchmark portfolio contains more assets, and is therefore more diversified and less risky, in order to dominate it, we have to use a utility function that introduces penalty for a broader range of return rates and is steeper. For the broadly based index in Benchmark Portfolio 4, the optimal utility function is smoother and is nonlinear even for positive return rates. It is worth mentioning that all these utility functions, although nondecreasing and concave, have rather complicated shapes. It would be a very hard task to guess the utility function that should be used to obtain a solution which dominates our benchmark portfolio.

Obviously, the shape of the utility function is determined by the benchmark within the context of the optimization problem considered. If we change the optimization problem, the utility function will change.

Finally, we may remark that our model (9.20)–(9.22) can be used for testing the statistical hypothesis that the return rate Y of the benchmark portfolio is non-dominated.

References

Acerbi, C.: Spectral measures of risk: A coherent representation of subjective risk aversion. J. Bank. Finance **26**, 1505–1518 (2002)

Arnold, B.C.: Majorization and the Lorenz Order: A Brief Introduction. Lecture Notes in Statistics, vol. 43. Springer, Berlin (1980)

Artzner, P., Delbaen, F., Eber, J.-M., Heath, D.: Coherent measures of risk. Math. Finance **9**(3), 203–228 (1999)

Dentcheva, D., Ruszczyński, A.: Optimization with stochastic dominance constraints. SIAM J. Optim. **14**(2), 548–566 (2003)

Dentcheva, D., Ruszczyński, A.: Optimality and duality theory for stochastic optimization problems with nonlinear dominance constraints. Math. Program. **99**(2), 329–350 (2004a)

Dentcheva, D., Ruszczyński, A.: Semi-infinite probabilistic optimization: First order stochastic dominance constraints. Optimization **53**, 583–601 (2004b)

Dentcheva, D., Ruszczyński, A.: Convexification of stochastic ordering constraints. C. R. Acad. Bulgare Sci. **57**, 5–10 (2004c)

Dentcheva, D., Ruszczyński, A.: Inverse stochastic dominance constraints and rank dependent expected utility theory. Math. Program. **108**(2–3), 297–311 (2006a)

Dentcheva, D., Ruszczyński, A.: Portfolio optimization with stochastic dominance constraints. J. Bank. Finance **30**(2), 433–451 (2006b)

Dentcheva, D., Ruszczyński, A.: Duality between coherent risk measures and stochastic dominance constraints in risk-averse optimization. Pac. J. Optim. **4**, 433–446 (2008a)

Dentcheva, D., Ruszczyński, A.: Inverse cutting plane methods for stochastic dominance constraints. Optimization **59**, 323–338 (2010)

Fishburn, C.: Decision and Value Theory. Wiley, New York, NY (1964)

Fishburn, C.: Utility Theory for Decision Making. Wiley, New York, NY (1970)

Gastwirth, J.L.: A general definition of the Lorenz curve. Econometrica **39**(6), 1037–1039 (1971)

Hadar, J., Russell, W.: Rules for ordering uncertain prospects. Am. Econ. Rev. **59**(1), 25–34 (1969)

Hanoch, G., Levy, H.: The efficiency analysis of choices involving risk. Rev. Econ. Stud. **36**(3), 335–346 (1969)

Hardy, H., Littlewood, J.E., Pólya, G.: Inequalities. Cambridge University Press, Cambridge, MA (1934)

Konno, H., Yamazaki, H.: Mean–absolute deviation portfolio optimization model and its application to Tokyo stock market. Manage. Sci. **37**(5), 519–531 (1991)

Kusuoka, S.: On law invariant coherent risk measures. Adv. Math. Econ. **3**, 83–95 (2001)

Lehmann, E.: Ordered families of distributions. Ann. Math. Stat. **26**(3), 399–419 (1955)

Lorenz, M.O.: Methods of measuring concentration of wealth. J. Am. Stat. Assoc. **9**, 209–219 (1905)

Markowitz, H.M.: Portfolio selection. J. Finance **7**(1), 77–91 (1952)

Markowitz, H.M.: Portfolio Selection. Wiley, New York, NY (1959)

Markowitz, H.M.: Mean–Variance Analysis in Portfolio Choice and Capital Markets. Blackwell, Oxford (1987)

Marshall, W., Olkin, I.: Inequalities: Theory of Majorization and Its Applications. Academic, San Diego, CA (1979)

Mosler, K., Scarsini, M., (eds.) Stochastic Orders and Decision Under Risk. Institute of Mathematical Statistics, Hayward, CA (1991)

Muliere, P., Scarsini, M.: A note on stochastic dominance and inequality measures. J. Econ. Theory **49**(2), 314–323 (1989)

Müller, A., Stoyan, D.: Comparison Methods for Stochastic Models and Risks. Wiley, Chichester (2002)

Ogryczak, W., Ruszczyński, A.: From stochastic dominance to mean–risk models: Semideviations as risk measures. Eur. J. Oper. Res. **116**, 33–50 (1999)

Ogryczak, W., Ruszczyński, A.: On consistency of stochastic dominance and mean–semideviation models. Math. Program. **89**(2), 217–232 (2001)

Ogryczak, W., Ruszczyński, A.: Dual stochastic dominance and related mean–risk models. SIAM J. Optim. **13**, 60–78 (2002)

Prékopa, A.: Probabilistic programming. In: Ruszczyński, A., Shapiro, A. (eds.) Stochastic Programming. Elsevier, Amsterdam, 267–351 (2003)

Quiggin, J.: A theory of anticipated utility. J. Econ. Behav. Organ. **3**, 225–243 (1982)

Quiggin, J.: Generalized Expected Utility Theory – The Rank-Dependent Expected Utility Model. Kluwer, Dordrecht (1993)

Quirk, J.P., Saposnik, R.: Admissibility and measurable utility functions. Rev. Econ. Stud. **29**(2), 140–146 (1962)

Rockafellar, R.T., Uryasev, S.: Optimization of conditional value-at-risk. J. Risk **2**, 21–41 (2000)

Rockafellar, R.T., Uryasev, S., Zabarankin, M.: Generalized deviations in risk analysis. Finance Stochast. **10**(1), 51–74 (2006)

Rothschild, M., Stiglitz, J.E.: Increasing risk: I. a definition. J. Econ. Theory **2**(3), 225–243 (1970)

Rudolf, G., Ruszczyński, A.: Optimization problems with second order stochastic dominance constraints: duality, compact formulations, and cut generation methods. SIAM J. Optim. **19**, 1326–1343 (2008)

Ruszczyński, A., Shapiro, A.: Optimization of convex risk functions. Math. Oper. Res. **31**, 433–452 (2006)

Ruszczyński, A., Vanderbei, R.J.: Frontiers of stochastically nondominated portfolios. Econometrica **71**(4), 1287–1297 (2003)

Shapiro, A., Dentcheva, D., Ruszczyński, A.: Lectures on Stochastic Programming: Modeling and Theory. SIAM, Philadelphia, PA (2009)

Von Neumann, J., Morgenstern, O.: Theory of Games and Economic Behavior. Princeton University Press, Princeton, NJ (1947)

Wang, S.S., Yong, V.R.: Ordering risks: expected utility versus yaari's dual theory of risk. Insur. Math. Econ. **22**, 145–161 (1998)

Wang, S.S., Yong, V.R., Panjer, H.H.: Axiomatic characterization of insurance prices. Insur. Math. Econ. **21**, 173–183 (1997)

Whitmore, G.A., Findlay, M.C. (eds.) Stochastic Dominance: An Approach to Decision–Making Under Risk. D. C. Heath, Lexington, MA (1978)

Yaari, M.E.: The dual theory of choice under risk. Econometrica **55**(1), 95–115 (1987)

Chapter 10
Single-Period Mean–Variance Analysis in a Changing World

Harry M. Markowitz and Erik L. van Dijk

10.1 Introduction

Consider a model of financial decision making along the following lines: an investor acts at discrete points in time that are a fixed interval apart (e.g., a year, quarter, month, day, or millisecond). At each point in time the investor chooses a portfolio from a universe of many securities (e.g., a dozen asset classes or a thousand individual securities). Changes in portfolio from one point in time to the next incur costs. The probability distribution of holding period security returns may change over time. Perhaps the investor (or investment management team) is uncertain as to which hypothesis about securities returns is correct—the investor's beliefs shifting in a Bayesian fashion as evidence accumulates. The investment portfolio is subject to deposits and withdrawals and occasionally distributes dividends. The investor seeks to maximize the expected value of some total utility function of the stream of present and future dividends, e.g., the sum of the present values of the single-period utility of each future dividend. A "strategy" for this model, in the sense of Von Neumann and Morgenstern (1944), is a rule that specifies, for all circumstances that can arise during the course of this investment "game," the action to be taken should that circumstance arise. An optimum strategy maximizes the expected value of total utility.

It is well beyond foreseeable computing capabilities to calculate the optimal strategy for such a game. The game could be formulated as a dynamic programming problem (Bellman 1957) but has too many state variables to solve in relevant time even with the fastest parallel computers. If the model were restricted to a finite number of periods, such as the number of months in a human lifetime, then it could be formulated as a stochastic programming problem (Dantzig 1955, Dantzig and Infanger 1993). But the set of possible paths would fan out so rapidly that the problem would be beyond foreseeable storage as well as computational capabilities. The

H.M. Markowitz (✉)
Harry Markowitz Company, San Diego, CA 92109, USA
e-mail: harryhmm@aol.com

This article originally appeared in *Financial Analyst Journal*, 2003, Vol 59, No. 2, pp. 30–44, published by The CFA Institute.

G. Infanger (ed.), *Stochastic Programming*, International Series in Operations Research & Management Science 150, DOI 10.1007/978-1-4419-1642-6_10,

most one can use, in general, for the analysis of such a "realistic" investment model is a detailed Monte Carlo simulation (see Levy et al. 2000). Such a simulation model could be used to evaluate alternate heuristic strategies.

The present chapter seeks to form some idea as to how well one particular type of heuristic strategy might perform as compared to an optimum strategy or other heuristics. The heuristic of interest is the "mean–variance surrogate for the 'derived' utility function" or, "MV surrogate heuristic" for short. We know from dynamic programming that a many-period (or unending) investment game may be solved by solving a sequence of one-period games. For example, number the points in time when decisions can be made as $t = 0, 1, \ldots$ and the intervals between them as $1, 2, \ldots$. Thus, time interval t lies between time-points $t - 1$ and t. Suppose that total utility is the sum of present values of one-period utilities:

$$U = \sum_{t=1}^{\infty} d^{t-1} u(D_t), \tag{10.1}$$

where D_t is a "dividend" paid during time interval t and $d < 1$. In this case the optimum action at point-in-time $t - 1$, given that the system is in state S_{t-1}, is the one which maximizes the expected value of

$$u(D_t) + dW(S_t), \tag{10.2}$$

where $W(S_t)$, the "derived utility function," is the expected value of

$$\sum_{i=1}^{\infty} d^{i-1} u(D_{t+i}) \tag{10.3}$$

given that the investor is in state S_t at time-point t and plays the game optimally then and thereafter. Bellman shows how to compute W and hence solve such games when the dimensionality of S is not too great. An MV surrogate heuristic replaces W by a function of the mean and variance of the investor's portfolio. We will restrict ourselves to a simple, linear surrogate function

$$W^S = E_{\mathrm{wt}} \cdot E_{\mathrm{p}} + V_{\mathrm{wt}} \cdot V_{\mathrm{p}}, \tag{10.4}$$

where E_{p} and V_{p} are the mean and variance of the portfolio, respectively, and E_{wt} and V_{wt} are their respective weights in the linear surrogate function. Definition (10.4) omits a constant term as irrelevant, since its inclusion would not affect the optimal choice of action if W^S is substituted for W in (10.2).[1] We assume that the investor (presumably a large institution investor) has a detailed simulation model with which to explore questions such as what E_{p} and V_{p} should be used. For

[1] We will see that, because of transaction costs, we cannot further simplify (10.4) at this stage by dividing through by E_{wt} as is often justified.

example, if the portfolio is reviewed monthly, should E_p be the expected return on the chosen portfolio for its first month? quarter? year? assuming that the portfolio is held at least that long without further changes? Other questions which the investing institution can explore with its simulation model include what E_{wt}, V_{wt} should be used in (10.4) and how well does the MV surrogate heuristic compare with other heuristics? The important question that the simulator cannot solve is given a good choice of the E_p, V_p definitions, and good E_{wt}, V_{wt} weights, how well does the MV surrogate do as compared to an optimal strategy?

Our general approach will be to define a simpler dynamic investment model—one for which an optimal (or nearly optimal) strategy can be computed—and evaluate how well MV surrogate heuristics do in this model as compared to other heuristics and the optimal strategy. It is practical to scale the MV surrogate and other heuristics to larger size problems for which optimum solutions cannot be computed. The object of the experiment is to give us a reading on how well the MV surrogate may do, for such more realistic problems, as compared to the optimum solution and other heuristics.

Specifically, we consider an investment game with two assets: stock and cash. The investor's portfolio can be in one of 11 states: 0, 10, 20, . . ., 100% stock. Transaction costs are incurred when the investor changes portfolio state. The investor has a stock return forecasting model which can be in one of five states: (1) very optimistic, (2) optimistic, (3) neutral, (4) pessimistic, and (5) very pessimistic. Thus the system-as-a-whole can be in any of 55 states, depending on the states of the portfolio and the forecasting model. The game is unending with a utility function as in (10.1). This simplifies the game, since time remaining is not a state variable. Specific assumptions are made concerning the ability of the forecasting model to predict and concerning the probability of transition from one prediction state to another. In particular, we assume that the forecasting model has enough predictive ability that—if there were no transaction costs—the optimum portfolio to hold would differ considerably from one forecast state to another.

The optimum strategy for this game may be written as an 11×5 *action matrix* $\tilde{A}$ that specifies choice of next portfolio as a function of the current portfolio and prediction state. For example, if the portfolio is in portfolio state 6, i.e., 50% stock, and the predictive model is in prediction state 1, very optimistic, the (6, 1) entry of the action matrix specifies which portfolio should be selected in this circumstance. Associated with action matrix $\tilde{A}$ is an 11×5 expected discounted utility matrix $\tilde{W}$. $\tilde{W}(i, j)$ is the expected value of discounted utility (10.1) if the game starts in (portfolio, prediction) states (i, j) and follows action matrix $\tilde{A}$ henceforth.

We will seek a (nearly) optimal action matrix for two levels of transaction costs (50 and 200 bp). For a given model (e.g., that with cost $c = 0.005$) the action matrix A^* we obtain using dynamic programming will not necessarily be the optimum $\tilde{A}$, but its associated discounted utility matrix W^* will be within ϵ of $\tilde{W}$, i.e.,

$$W^*(i, j) \geq \tilde{W}(i, j) - \epsilon \quad \text{for all } i, j, \tag{10.5}$$

where ϵ will be chosen so that the difference between $\tilde{W}$ and W^* is negligible.

If we are given an arbitrary action matrix A, optimal or other, we can solve for its W_A matrix—namely the expected utility of the game starting from each of the 55 portfolio and prediction states, following the given action matrix then and thereafter—by solving a certain system of 55 simultaneous equations. This system of 55 equations—with action matrix A as input and expected total utilities W as output—is used to evaluate various heuristics including MV surrogate heuristics for various choices of E_{wt}, V_{wt}, and definitions of E and V, as well as other heuristics. This allows us to say—at least for this simplified world—how well a good MV surrogate heuristic would do as compared to other heuristics and as compared to an optimum strategy. In more complex situations we would need to use simulations to estimate W_A and would not have an optimum W^* with which to compare it.

Many-period and continuous-time portfolio selection with transaction costs have been studied by Zabel (1973), Kamin (1975), Magill and Constantinides (1976), Constantinides (1979), and Davis and Norman (1990). Constantinides (1979) is closest to the model presented here.[2] The general conclusions of these papers for the two-asset case are, first, that there is an interval in which no action is taken; outside this interval the investor buys or sells to bring its position to the closest boundary of the no-action interval. Second, it is usually not easy to compute the location of the no-action interval. The many-security portfolio selection problem with transaction costs also has a no-action region which is practically impossible to compute.

We approach the problem from the other direction. We start with heuristics that are easy to compute and ask what is lost by using one or another such heuristic rather than doing the estimation and optimization required to seek an optimum. The answer, at least for the cases considered, is that surprisingly little is lost by using the MV heuristic.

Balduzzi and Lynch (1999) and Lynch and Balduzzi (1999) compute optimal consumption and investment for a two-asset dynamic model with transaction costs, considering "the empirically documented predictability of asset returns." Our objective is different. We do not claim that the changing forecasts in our model reflect the true "predictably of asset returns." Rather, they are part of an experiment used to compare the performance of portfolio choice heuristics, which can be easily scaled to larger size models, with optimization procedures that cannot. The forecasting model within the experiment is such that, if there were no transaction costs, there would be considerable shifting of portfolio. This provides a substantial challenge for the heuristics tested.

The mean–variance *surrogate heuristics* discussed here should be distinguished from mean–variance *approximations* discussed in Markowitz (1959), Chapters 6 and 13, Young and Trent (1969), Levy and Markowitz (1979), Dexter et al. (1980),

---

[2] As explained in footnote 4, although Constantinides (1979) presents a remarkably general discrete-time analysis, the model presented here is not a special case. In particular, an "inexpensive" (in that it did not increase the dimensionality of the problem) bit of realism in our model made it no longer true that if the current equity position X is, e.g., less than some level $\underline{X}$ (which depends on the state of the system) then the optimum action is to buy $\underline{X} - X$, no matter the value of $X < \underline{X}$.

Pulley (1981), Pulley (1983), Kroll et al. (1984), Grauer (1986), Ederington (1986), Simaan (1987), and Markowitz et al. (1994). In this literature a single-period $U(R)$ function is given and a function $f(E, V)$ of mean and variance is sought to approximate expected $U(R)$. In the type of many-period situation addressed in the present chapter we know that a single-period-"derived" utility function exists, albeit a complex function of many state variables. We do not seek to approximate it, since we do not know what it is. Rather, we seek a mean–variance surrogate to stand in its place. In general our proposal for complex, real-world dynamic portfolio selection problems with illiquidities is to use simulation analysis to seek a mean–variance surrogate that does as well as one can find in terms of the game-as-a-whole. The purpose of the experiments reported here was to get an initial reading on how good this might be.

Section 10.2 of this chapter defines the model; Sections 10.3 and 10.4 describe the computation of W_A and of A^* and W^*, respectively; in Section 10.5 the (nearly) optimum action matrices for $c = 0.005$ and 0.02 are examined; in Section 10.6 the MV surrogate heuristic is defined and exhibited for the two values of c. In Section 10.7 the expected utilities of various heuristics and the (nearly) optimum action are compared. In Section 10.8 further applications of the method of MV surrogates are proposed. Section 10.9 concludes the chapter.

10.2 The Model

Exhibit A summarizes the model. As noted in the first two items of the exhibit, we assume a 1-month interval between portfolio reviews and a constant risk-free rate of 0.4 of 1%, i.e., 40 basis points, per month. The table in Item 3 shows, for example, that if the predictive model is in state 1—most optimistic—then the expected return for stock is 64 bp for the forthcoming month, with a standard deviation of $\sqrt{0.000592} \approx 0.024$. The following four columns include the mean and variance of return for the forthcoming month given that the predictive model is in one of states 2 through 5. Below we discuss the suitability of these numbers for the present experiment and the meaning of other numbers in this table.

The entries in the table in Item 4 show assumed probabilities $P(j, h)$ that the prediction state h listed at the top of the table will occur at time $t + 1$ given that the prediction state j listed on the left of the table is true at time t. Given this Markov matrix one can solve for the long run, "ergodic" steady-state probabilities of the various states. These are listed in the last column of this table. Returning to Item 3, the last column shows the mean and variance of a random number that is drawn by first drawing a prediction state according to the steady-state distribution and then drawing a stock return given that prediction state.

Item 5 states that the investor seeks to maximize the expected value of discounted future utility functions, with discount factor $d = 0.99$ (or discount rate $= 0.01$ per month). Thus, e.g., utility 10 years from now counts only 30% as much as utility next month, but still counts.

Let p_{t-1} be the fraction of the investor's portfolio held in stocks at time $t-1$ and p_t be the fraction that the investor decides at time $t-1$ to accumulate up to, or sell down to, or continue to hold, for time t. What fraction should we assume that the investor holds during the interval between these two points in time when the random return on equities is determined: p_{t-1}? p_t? or something between? Item 6 defines an "effective fraction" assumed to be held during the interval as determined by a parameter θ_p, assumed to be 0.5 in these experiments. Item 6 notes the consequent portfolio expected return and variance of return for the month, at time $t-1$, given that the prediction state is j, the current stock fraction is p_{t-1}, and the next is p_t. The fraction invested is related to the integer "portfolio state" i by $p = (i-1)/10$.

As noted in Item 7 we will report the results of two experiments, one with transaction costs, including market impact and bid-asked spread, equal to $c = 0.005$ and the other with $c = 0.02$. Thus we speak of Exhibit A as defining two specific models. If we abstract from the particular parameters given, then Exhibit A defines a "general" model.

As noted in Item 8 we assume that $u(D_t)$ can be adequately defined by a mean–variance approximation. In the runs reported we let $k = 0.5$. This makes $Eu(D_t)$ approximately $E(\log(1 + \text{return}))$. As noted in the Introduction, this approximation has proved quite good in the case of historical portfolio returns (see especially Young and Trent 1969 regarding mean–variance approximations to expected log or, equivalently, geometric mean return). Larger values of k approximate more risk-averse utility functions. The assumption that the monthly contribution to total expected utility, $Eu(D_t)$, may be approximated well by a function of mean and variance does not obviously imply that a mean–variance surrogate will serve well in place of the derived utility function W. This remains to be seen from the experiment. The assumption in Item 8 does conveniently imply that we need only specify the mean and variance of the stock return distributions given each prediction state j.

Returning to the table in Item 3, the last row shows the investment which would be optimal in each state if there were no transaction costs, if θ_p were 0, if $k = 0.5$ as assumed, and if stock shorting and leverage were permitted. Specifically,

$$\begin{aligned} \hat{X} &= (E^j - r_f)/(2kV^j) \\ &= (E^j - r_f)/V^j. \end{aligned} \tag{10.6}$$

We see, e.g., that if the predictive model was very optimistic and there were no transaction costs, then it would be optimal to borrow roughly \$3 for every dollar of equity and invest the \$4 in stock. Conversely, if the model were very pessimistic it would be optimal to go short 229%. These differences in optimal position would be less extreme if risk aversion k were larger, the differences between E^j and r_f were smaller or V^j were larger. The present work is the result of a collaboration of one author especially interested in predictive models and the other author with optimization. But we do not claim here that we can offer the reader a predictive model that can actually predict 24 bp more than the risk-free rate in some circumstances and 13 bp less in others with the errors of estimation given in the table. Rather, the question considered here is how should one proceed *if* one had such a predictive model given greater or less transaction costs. We chose the parameters in Item 3 based on a judgment that a predictive model which would imply much greater leverage when optimistic, or much greater short position when pessimistic, would not be plausible, whereas a predictive model which implied much smaller moves, if there were no transaction costs, would not much challenge the alternate heuristics.

Similarly, we chose the parameters of transition probabilities in Item 4 on the assumption that good months would tend to be followed by good months; bad months by bad months; and the steady-state distribution would imply roughly that the familiar 60–40 or 70–30 stock-cash mix was optimum. As it is, as shown in the last column of Item 3, the parameters chosen imply that a 72–28 mix is optimum if faced with the steady-state distribution. Thus one might say that, with the parameters chosen, the "strategic" optimum portfolio is 72% in stock. We will see whether this strategic solution is related to the dynamic optimum in some way.

Item 9 confesses that we assume that all returns, net of cost, are "distributed" each month. In particular, we assume that losses are collected from the investors as well as gains distributed to them. This assumption expresses itself in two ways: the calculation of the contribution to expected utility for the period as shown in Item 9 and implicitly, in the assumption that if at time $t-1$ the investor selects fraction p_t then, in fact, p_t will be the fraction invested in stocks when t time occurs. The assumption that returns are distributed helps keep the state space small and thus, in particular, greatly simplifies the computation of the optimum policy. While unrealistic, this assumption seems innocuous for present purposes, since this method of scoring does not particularly favor the MV heuristic as compared to other heuristics or the optimum solution.

Item 10 assumes that security returns are bounded by some huge number.

As noted above, the model in Exhibit A, including parameter settings, was selected as a simple, easily computed, but challenging setting within which to test heuristics. Once the model was selected and testing began, no further changes in the model were made.

The following two sections discuss how results were computed. Sections 10.5 through 10.7 present the results. Readers not interested in checking our computational procedures may wish to skip to Section 10.5.

10.3 Computation of Expected Discounted Utility for a Given Action Matrix

Below we consider action matrices that represent heuristics such as an MV surrogate or a heuristic that immediately moves to a 60–40 mix of stocks and cash and stays at this mix forever. We shall report EU, the expected value of U in (10.1), if the action matrix is followed starting in portfolio state i and prediction state j. In this section we describe how $E(\sum_{t=1}^{\infty} d^{t-1}u(D_t))$ is evaluated, where D_t is current return on the investor's portfolio less transaction costs.

For a given action matrix A let

$$W_A^T(i,j) = E\left(\sum_{t=1}^{T} d^{t-1}u(D_t)\right) \tag{10.7}$$

be the expected utility of a T-period game given that the game starts in portfolio state i and prediction state j and action matrix A is followed at every move. Then

$$W_A^{T+1}(i,j) = u_{ij}^A + d\sum_{g,h} \tilde{P}^A(i,j,g,h)W_A^T(g,h), \tag{10.8}$$

where

$$u_{ij}^A = E(u|i,j,A(i,j)) \tag{10.9}$$

(see Item 9 in Exhibit A), $\tilde{P}^A(i,j,g,h)$ = the probability that state (g,h) will occur at t if state (i,j) holds at time $t-1$. Specifically $\tilde{P}^A(i,j,g,h) = 0$ unless $g = A(i,j)$. With this choice of g, $\tilde{P}^A(i,j,g,h)$ equals the probability $P(j,h)$ of transition from prediction state j to prediction state h given in Item 4.

It will be convenient to write W_A^T and u^A as vectors with 55 components rather than 11×5 matrices and write $\tilde{P}^A$ as a 55×55 matrix rather than a four-dimensional object. Toward this end we write

$$W_T^A(m) = W_T^A(i,j), \tag{10.10a}$$

$$u^A(m) = u^A(i,j), \tag{10.10b}$$

$$\tilde{P}^A(m,n) = \tilde{P}^A(i,j,g,h), \tag{10.10c}$$

where

$$m = 11\cdot(j-1)+i, \tag{10.11a}$$

$$n = 11\cdot(j-1)+g. \tag{10.11b}$$

Then (10.8) becomes

$$W_A^{T+1} = u^A + d\tilde{P}^A W_A^T, \tag{10.12}$$

where $W_A^T = (W_A^T(m))$, etc. Starting with $W_A^1 = u^A$ (10.12) presents an iterative scheme for computing W_A^T expressed as a 55 vectors. Since $d\tilde{P}^A$ is a contraction mapping, the sequence in (10.12) is convergent:

$$W_A^T \to W_A \text{ as } T \to \infty, \tag{10.13a}$$

where

$$W_A = \lim_{T\to\infty} E\left(\sum_{t=1}^{T} d^{t-1}u(D_t)\right) \tag{10.13b}$$

when action matrix A is followed each move. It is also true that

$$W_A = E\left(\lim_{T\to\infty}\sum_{t=1}^{T} d^{t-1}u(D_t)\right) \tag{10.14}$$

since Item 10 of Exhibit A implies that, for every path generated and for $S > T$,

$$\left|\sum_{t=1}^{S} d^{t-1}u^A(D_t) - \sum_{t=1}^{T} d^{t-1}u^A(D_t)\right| < \frac{Md^T}{(1-d)}. \tag{10.15}$$

Thus, every sequence converges and is bounded by the integrable function $M/(1-d)$. Therefore Lesbegue's bounded convergence theorem applies and allows us to interchange the operators.

Equation (10.12) implies that the limiting vector W_A is the fixed point satisfying

$$W_A = u^A + d\tilde{P}^A W_A. \tag{10.16}$$

This system of 55 equations in 55 unknowns is rather sparse, containing at most five nonzero coefficients per row. It is readily solved by MatLab's sparse matrix solver, which is how the various W_A reported below was obtained.

10.4 Computation of a (Nearly) Optimum Strategy

In the manner explained by Bellman, we approximate the optimum solution to the unending game by a sequence of T-period games for $T = 1, 2, \ldots, S$ for some large S. In this section we present notation for the T-period game, review Bellman's "dynamic programming" procedure as it applies to our model, and establish an upper bound on

$$\|\tilde{W} - W^*\| = \max_{i,j} |\tilde{W}(i,j) - W^*(i,j)|, \tag{10.17}$$

where $\tilde{W}(i,j)$ is the expected discounted utility of the unending game starting in state (i,j) if an optimum strategy is followed, and $W^* = \tilde{W}^T$ is that for the dynamic programming solution after T iterations, for some large T.

The T-period game involves $T+1$ time-points and T time intervals (or periods) aligned as follows:

Time-point:	0	1	2 ...	$T-1$		T
Time interval:		1	2	...	T.	

A strategy for the T-period game is a rule that specifies, for each decision point $t = 0, \ldots, T-1$ and for each state (i, j), the next portfolio selected. This can be represented as an action matrix subscripted by time-point t:

$$g = A_t^T(i, j) \quad t = 0, \ldots, T-1. \tag{10.18}$$

The optimum strategy $\tilde{A}_t^T$ maximizes the expected value of

$$U = \sum_{t=1}^{T} d^{t-1} u(D(t)). \tag{10.19}$$

This optimum expected discounted utility for the T-period game as a function of starting state (i, j) is denoted $\tilde{W}^T(i, j)$.

The optimum first and only decision in a one-period game—$T = 1$—namely $g = A_0^1(i, j)$, is found by computing

$$\tilde{W}^1(i, j) = \max_g E(u|i, j, g). \tag{10.20}$$

Given that the optimum strategy $\tilde{A}_t^T$ and the expected utility it provides, $\tilde{W}^T(i, j)$, have been determined up to some T, the optimum first move for the $T+1$-period game, $g = \tilde{A}_0^{T+1}(i, j)$, is determined by finding

$$\tilde{W}^{T+1}(i, j) = \max_g \left\{ E(u|i, j, g) + d E_h(\tilde{W}^T(g, h)|j) \right\}, \tag{10.21a}$$

where

$$E(\tilde{W}^T(g, h)|j) = \sum_{h=1}^{5} P(j, h) \tilde{W}^T(g, h). \tag{10.21b}$$

For $t > 0$,

$$\tilde{A}_t^{T+1}(i, j) = \tilde{A}_{t-1}^T(i, j) \tag{10.22}$$

(i.e., the two strategies are the same when there are the same number of periods to go).

For a large value of T (actually, $T = 1200$, i.e., a 100 years worth of months) we used (10.20) and (10.21a, b) to compute $\tilde{W}^T$ and $\tilde{A}_0^T$ for each of our two specific models, $c = 0.005$ and 0.02.

For a large T, $\tilde{W}^T(i, j)$ cannot be much less than $\tilde{W}(i, j)$ as can be seen as follows. Let

$$u_{\max} = \max_{i,j,g} E(u|i, j, g). \tag{10.23}$$

Write the expected utility of the unending game as

$$\begin{aligned} EU &= E\left(\sum_{t=1}^{T} d^{t-1}u(D_t)\right) + E\left(\sum_{t=T+1}^{\infty} d^{t-1}u(D_t)\right) \\ &\leq \tilde{W}^T + \sum_{t=T+1}^{\infty} d^{t-1}u_{\max} \\ &= \tilde{W}^T + d^T u_{\max}/(1-d), \end{aligned} \tag{10.24}$$

since $\tilde{W}^T$ is the maximum expected value of

$$\sum_{t=1}^{T} d^{t-1}u(D_t) \text{ and } Eu(D_t) \leq u_{\max}. \tag{10.25}$$

For both specific models

$$\begin{aligned} u_{\max} &= E(u|11, 1, 11) \\ &= 0.0061. \end{aligned}$$

This is $Eu(D_t)$ given that $i = 11$ (fully invested), $j = 1$ (very optimistic), and $k = 11$ (no change in the portfolio). Thus, for both specific models,

$$\tilde{W}^{1200} \geq \tilde{W} - \epsilon, \tag{10.26a}$$

where

$$\begin{aligned} \epsilon &= d^{1200}u_{\max}/(1-d) \\ &= 3.53 \times 10^{-6}. \end{aligned} \tag{10.26b}$$

A strategy for the unending game may base action on t as well as i, j as expressed by a t subscripted action matrix $A_t(i, j)$. But the optimum strategy uses an unchanging action matrix, $\tilde{A}_t = \tilde{A}$ (if $\tilde{A}$ is unique; else there exists an optimum of this nature). This may be seen as follows. As of time T write EU as

$$\sum_{t=1}^{T} d^{t-1}u(D_t) + d^T \sum_{t=1}^{\infty} d^{t-1}u(D_{T+t}).$$

The first sum has happened and cannot be changed. Given the state (i, j) at time T, the maximization of the second expected discounted sum is isomorphic to the original problem and has, therefore, the same optimum initial action matrix $\tilde{A}_T(i, j) = \tilde{A}_0(i, j)$. Since T is arbitrary, $\tilde{A}_T$ does not depend on T:

$$\tilde{A}_T = \tilde{A}_0 = \tilde{A} \quad \text{for all } T. \tag{10.27}$$

For the two specific models of Exhibit A, we used $A^* = A_0^{1200}$ as an approximation to $\tilde{A}$. For reasons illustrated in the Appendix, we cannot guarantee that $A^* = \tilde{A}$. We can, however, guarantee a lower bound on how close the associated W^* is to $\tilde{W}$ for each specific model. For example, for $c = 0.005$, given the A^* action matrix we computed W^* from the 55 equations described in the previous section. It turns out that

$$W^*(i, j) > \tilde{W}^{1200}(i, j) \quad \text{for all } (i, j), \tag{10.28}$$

and we conclude from (10.26a) that

$$W^*(i, j) > W(i, j) - \epsilon \tag{10.29}$$

for the ϵ in (10.26b). This provides the bound on the norm in (10.17). The same holds when $c = 0.02$. Inequality (10.28) is consistent with the hypotheses that $W^* = \tilde{W}$ in fact, and the remainder of the game, $t > 1200$, makes a positive contribution to total expected utility.

10.5 The (Near) Optimum Action Matrices

As we illustrate in the Appendix, we cannot conclude from the preceding that A^* in Table 10.1A is the optimum action matrix for the unending game with $c = 0.005$. All we can guarantee is that, for each starting combination of portfolio state and prediction state, the expected utility of the unending game provided by A^* is within $(3.53)(10^{-6})$ of that provided by the optimum action matrix $\tilde{A}$. We would be surprised if $\tilde{A}^{1200}$ were not optimal in fact, but all we can guarantee is that it is "(near) optimal" in the sense described.

Table 10.1B presents $A^* = \tilde{A}^{1200}$ action matrix for $c = 0.02$. This action matrix became optimum at iteration 211 and remained so. This means that a finite game had to have $T \geq 211$ months to go (rather than $T \geq 17$ as in the case of $c = 0.005$) for the A^* in Table 10.1B to be the optimum first action matrix. It is plausible that it takes more dynamic programming iterations to reach an optimum solution when $c = 0.02$ than when $c = 0.005$ (assuming both A^* matrices are optimum). This says that, with the higher cost, the finite game has to be longer to justify a move to that which is optimum for the unending game.

As is to be expected, A^* for $c = 0.005$ (i.e., $A^*_{0.005}$) shows greater activity than that for $c = 0.02$. Specifically, when prediction state = 1 (very optimistic), $A^*_{0.005}$ specifies shifting to 100% stock at t whatever the portfolio state at $t - 1$. In the same prediction state $A^*_{0.02}$ specifies moving to a stock position of 60% if the portfolio has less than that at $t - 1$; otherwise it recommends not changing the portfolio. In the case of prediction state 5 (very pessimistic), $A^*_{0.005}$ converts to 100% cash no matter what the start-of-period portfolio, whereas $A^*_{0.02}$ makes no change unless the starting portfolio is 100% stock and then it only moves down to 90% stock. $A^*_{0.005}$ is

Table 10.1 Optimum action matrixes: Fraction of stock at t

(1200 iterations)

Stock fraction at $t-1$	Prediction state 1	2	3	4	5
A. Cost = 0.005; A^* optimum from iteration 17					
0.0	1.0	0.3	0.0	0.0	0.0
0.1	1.0	0.3	0.1	0.1	0.0
0.2	1.0	0.3	0.2	0.2	0.0
0.3	1.0	0.3	0.3	0.3	0.0
0.4	1.0	0.4	0.4	0.4	0.0
0.5	1.0	0.5	0.5	0.5	0.0
0.6	1.0	0.6	0.6	0.6	0.0
0.7	1.0	0.7	0.7	0.7	0.0
0.8	1.0	0.8	0.8	0.8	0.0
0.9	1.0	0.9	0.9	0.9	0.0
1.0	1.0	1.0	1.0	1.0	0.0
B. Cost = 0.02; A^* optimum from iteration 211					
0.0	0.6	0.0	0.0	0.0	0.0
0.1	0.6	0.1	0.1	0.1	0.1
0.2	0.6	0.2	0.2	0.2	0.2
0.3	0.6	0.3	0.3	0.3	0.3
0.4	0.6	0.4	0.4	0.4	0.4
0.5	0.6	0.5	0.5	0.5	0.5
0.6	0.6	0.6	0.6	0.6	0.6
0.7	0.7	0.7	0.7	0.7	0.7
0.8	0.8	0.8	0.8	0.8	0.8
0.9	0.9	0.9	0.9	0.9	0.9
1.0	1.0	1.0	1.0	1.0	1.0

also somewhat more active than $A^*_{0.02}$ in the case of prediction state 2. Both $A^*_{0.005}$ and $A^*_{0.02}$ take no action in prediction states 3 and 4.

An investor following action matrix $A^*_{0.005}$ will have the composition of its portfolio vary between 0 and 100% invested over time. If an investor following action matrix $A^*_{0.02}$ starts with less than 60% in stock, it will bring stock holdings to this level the first time that prediction state 1 occurs, then hold that level forever. If it starts with stock holdings of 100%, it will move it to 90% in prediction state 5 and hold that level forever. If it starts with between 60 and 90% stocks, it never changes. The "steady-state" optimum, $p = 0.72$, in Item 3 of Exhibit A, is within this never-change zone of $A^*_{0.02}$.

10.6 The M–V Heuristic

For either of our specific models, e.g., the one with $c = 0.005$, if we knew $\tilde{A}_{0.005}$ then we could compute the associated expected discounted utility for each starting state (i, j)—namely $\tilde{W}(i, j)$ for $c = 0.005$—by (10.16) using $A = \tilde{A}_{0.005}$.

Conversely, if we knew $\tilde{W}(i, j)$ for $c = 0.005$ we could determine $\tilde{A}_{0.005}$ since $g = \tilde{A}_{0.005}(i, j)$ is the value of g that maximizes

$$E(u|i, j, g) + dE_h(\tilde{W}(g, h)|j). \tag{10.30}$$

The M–V heuristic presented here replaces $\tilde{W}$ in expression (10.30) by a linear function of portfolio mean and variance E_{p}, V_{p} as in (10.4). Specifically $g = A^{\text{MV}}_{0.005}(i, j)$ is the value of g that maximizes

$$E(u|i, j, g) + dE_h(E_{\text{wt}}E_{\text{p}} + V_{\text{wt}}V_{\text{p}}|j) \tag{10.31}$$

for a "good" choice of E_{wt}, V_{wt}, and the measures E_{p} and V_{p}.

We experimented with the definition of E_{p} and V_{p} and found that a quite simple definition worked surprisingly well. Specifically, in the results reported here $E_{\text{p}}(i, j)$ and $V_{\text{p}}(i, j)$ are the mean and variance of portfolio return for a single month if j is the prediction state at the beginning of the month ($j = j_{t-1}$) and the i is the beginning and ending portfolio state ($i = i_{t-1} = i_t$). With $E_{\text{p}}(i, j)$, $V_{\text{p}}(i, j)$ thus defined, $A^{\text{MV}}(i, j)$ is the value of g that maximizes

$$\psi(i, j) = E(u|i, j, g) + dE_h(E_{\text{wt}}E_{\text{p}}(g, h) + V_{\text{wt}}V_{\text{p}}(g, h)|j). \tag{10.32}$$

The procedure for finding a "good" set of weights E_{wt}, V_{wt} was as follows: For a given set of weights E_{wt}, V_{wt} the above calculation determined an action matrix A^{MV}. Equation (10.16) was solved to determine the expected present value of utility, W^{MV}, of the unending game if action table A^{MV} is followed forever starting in state (i, j). A figure of merit was assigned the weights E_{wt}, V_{wt} by summing the entries in the W^{MV} matrix:

$$\text{FOM}(E_{\text{wt}}, V_{\text{wt}}) = \sum_i \sum_j W^{\text{MV}}(i, j). \tag{10.33}$$

(The W^{MV} should be labeled with the E_{wt}, V_{wt} used to generate it; these labels have been suppressed here.) E_{wt} and V_{wt} were varied (the former by steps of 0.1, the latter by steps of 0.01) to maximize FOM.

For $c = 0.005$, one set of FOM maximizing weights is

$$\varphi_{0.005} = 4.4E_{\text{p}} - 0.44V_{\text{p}}. \tag{10.34a}$$

Actually, $E_{\text{wt}} = 4.4$ and any V_{wt} from -0.35 to -0.52 provides maximum FOM. Varying E_{wt} by 0.1 in either direction reduces FOM. The $A^{\text{MV}}_{0.005}$ action table obtained using the weights in (10.34a) is presented in Table 10.2A. It is identical to the (near) optimum action table $A^*_{0.005}$ except for 0.2 rather than 0.3 at $A(2, 2)$ and $A(3, 2)$. The effect on expected discounted utility of these differences in the action matrices is examined in the next section.

Table 10.2 Action matrix for MV heuristic: Fraction of stock at t

Stock fraction at $t-1$	Prediction state 1	2	3	4	5
A. Cost = 0.005; $E_{\text{wt}} = 4.40$; and $V_{\text{wt}} = -0.44$[a]					
0.0	1.0	0.3	0.0	0.0	0.0
0.1	1.0	0.2	0.1	0.1	0.0
0.2	1.0	0.2	0.2	0.2	0.0
0.3	1.0	0.3	0.3	0.3	0.0
0.4	1.0	0.4	0.4	0.4	0.0
0.5	1.0	0.5	0.5	0.5	0.0
0.6	1.0	0.6	0.6	0.6	0.0
0.7	1.0	0.7	0.7	0.7	0.0
0.8	1.0	0.8	0.8	0.8	0.0
0.9	1.0	0.9	0.9	0.9	0.0
1.0	1.0	1.0	1.0	1.0	0.0
B. Cost = 0.02; $E_{\text{wt}} = 10.0$; and $V_{\text{wt}} = -1.0$[b]					
0.0	0.6	0.0	0.0	0.0	0.0
0.1	0.6	0.1	0.1	0.1	0.1
0.2	0.6	0.2	0.2	0.2	0.2
0.3	0.6	0.3	0.3	0.3	0.3
0.4	0.6	0.4	0.4	0.4	0.4
0.5	0.6	0.5	0.5	0.5	0.5
0.6	0.6	0.6	0.6	0.6	0.6
0.7	0.7	0.7	0.7	0.7	0.7
0.8	0.8	0.8	0.8	0.8	0.8
0.9	0.9	0.9	0.9	0.9	0.9
1.0	1.0	1.0	1.0	1.0	1.0

[a] $E_{\text{wt}} = 4.40$ and $V_{\text{wt}} \in (-0.52, -0.35)$ produces the same result
[b] $E_{\text{wt}} = 10.0$ and $V_{\text{wt}} \in (-1.06, -0.99)$ produces the same result

The same exercise performed for $c = 0.02$ produced best E_{wt}, V_{wt} of

$$\varphi_{0.02} = 10.0E_p - 1.0V_{\text{p}}. \tag{10.34b}$$

Again, maximum FOM was reached for $E_{\text{wt}} = 10.0$ and a range of V_{wt}, namely $V_{\text{wt}} \in [-1.06, -0.99]$. The $A^{\text{MV}}_{0.02}$ is presented in Table 10.2B. This is the same as $A^*_{0.02}$ except for 1.0 rather than 0.9 in $A(11, 5)$. The expected discounted utility for A^{MV}, A^*, and certain other heuristics is discussed in the next section.

Note that for both $c = 0.005$ and 0.02, the rate of substitution between E and V in W^{MV} is 10:−1 whereas the rate of substitution in the underlying single-period utility is 2:−1. This contrasts with the model without transaction costs, analyzed by Mossin (1968) and Samuelson (1969), in which the investor seeks to maximize the expected value of a function U of terminal wealth, w_{T}. They conclude that if the utility of final wealth is the logarithm or a power function, then the derived utility functions $J_t(w_t)$ will be the same function (ignoring inconsequential location and scale constants). This implies that the mean–variance approximation to the derived utility functions J_t will have the same trade-off between mean and variance as does the given final utility function U. Since the results of our current model are quite different and the efficacy of the mean–variance heuristic as reported in Section 10.7 is quite remarkable, we have tried to minimize the probability that these are due to some bug somewhere in our programs by computing each major result by at least two distinct methods.[3]

When we substitute the definitions for $E(u|i, j, g)$ and $E_h(\ldots|j)$ from Item 9 in Exhibit A and in (10.21b) and use p_{t-1} and p_t in place of the corresponding j and g, then $\psi(i, j)$ in (10.32) can be spelled out as

$$\begin{aligned} \psi(i,j) &= \max_{p_t} \Bigg\{ (E^j_{t-1} p^e + r_f(1-p^e)) - k(p^e)^2 V^j_{t-1} \\ &- c|p_t - p_{t-1}| + (0.99) \Bigg[E_{\text{wt}} \Bigg(p_t \Bigg(\sum_{h=1}^{5} P(j,h) E^h \Bigg) \\ &+ (1-p_t) r_f \Bigg) + V_{\text{wt}} p_t^2 \sum_{h=1}^{5} P(j,h) V^h \Bigg] \Bigg\}, \end{aligned} \tag{10.35a}$$

where

$$p^e \equiv \theta_p p_{t-1} + (1-\theta_p) p_t. \tag{10.35b}$$

For computation, we substituted expression (10.35b) for p^e in (10.35a) to express $\psi(i, j)$ as a function of one variable p_t and constants p_{t-1}, E_{wt}, V_{wt}, E^h, and so on. Equation (10.35a) can be rewritten so that the expression to be maximized is a weighted sum of E' and V', where E' is a weighted average of current-period and potential next-period expected values minus a transaction cost term and V' is a weighted average of present-period and potential next-period variances.[4]

---

[3] In particular, the action matrices for the MV heuristic for various E_{wt}, V_{wt} were computed using MatLab. The action matrices for the winning weights shown in Table 10.2 were recomputed using Excel. The Excel and MatLab answers were the same except for the $(i, j) = (2, 2)$ entry for $c = 0.005$. The Excel computation gave almost identical scores to $g = A(2, 2) = 2$ and 3, but slightly favored the latter. For $c = 0.005$ and 0.02, $\tilde{W}^{1200}$ was computed using dynamic programming and W^* computed using 55 simultaneous equations. $\tilde{W}^{1200} < W^* < \tilde{W}^{1200} + \epsilon$ for ϵ in (10.26b) lending confidence to both programs. The similarity between W^* and W^{MV} is to be expected because of the similarity of their action matrices. Finally, we will note later that W_A for certain heuristic action matrices, or certain relationships among the W_A, can be determined with little or no calculation. In every instance these results confirm the simultaneous equation calculations.

[4] Table 10.2A, prediction state 2, violates the condition noted in footnote 1 that there is an $\underline{X}$ (which varies here with prediction state) such that if $X_{t-1} < \underline{X}$ then $X_t = \underline{X}$. The reason this can happen in the present model is that, substituting its definition for p^e, (10.32) includes a term $-k\theta_p(1-\theta_p)p_{t-1}p_t$. Thus the marginal cost of, e.g., increasing p_t from a contemplated $p_t = 2$ to a contemplated $p_t = 3$ depends on p_{t-1}. This is true in the exact optimization calculation as well as with the M–V approximation. The above term appears to be of "second order" since

10.7 Comparison of Expected Utilities

Figure 10.1 presents the expected discounted utility for the unending game with cost $c = 0.005$, for the near optimum $A^*_{0.005}$ action matrix and for various heuristics. The horizontal axis represents the 55 possible initial conditions for the game. From left to right, the first 11 points on the horizontal axis represent starting conditions with prediction $j = 1$ and portfolio state $i = 1, 2, \ldots, 11$, i.e., stock fraction $p = 0.0, 0.1, \ldots, 1.0$, respectively. The next 11 points on the horizontal axis represent initial conditions with prediction state $j = 2$, and portfolio state $i = 1, \ldots, 11$, etc., through the final 11 points for prediction state $j = 5$ and the 11 portfolio states.

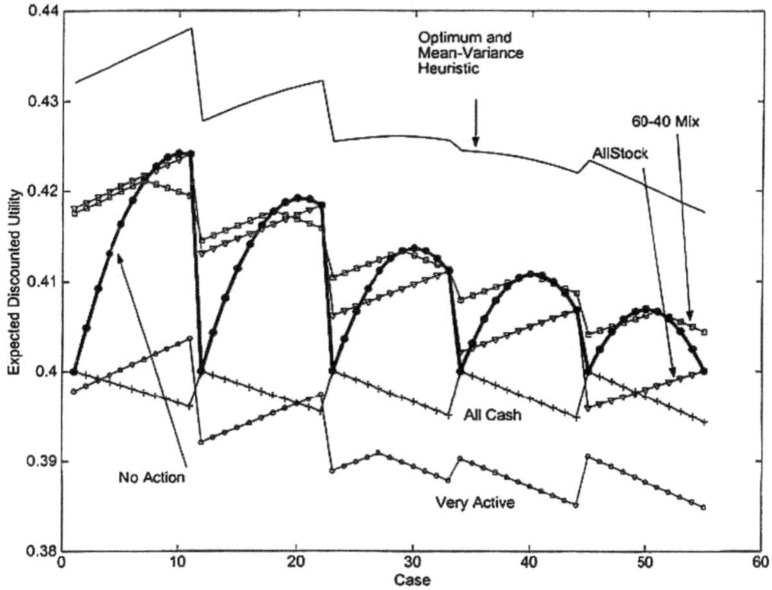

Fig. 10.1 Expected discounted utility for various strategies, unending game, cost = 0.005

The curves labeled "All Cash," "All Stock," and "60–40 Mix" show the expected discounted utility for action matrices that move immediately to the indicated mix and stay there. For example, the "60–40 mix" action matrix has $A(i, j) = 7 (p = 0.6)$ for all i, j. An unending game played using this matrix would move to a 60–40 mix as its first action and never move again.

The curve labeled "No Action" never changes its allocation. Note, for example, that the expected discounted utility of "No Action" is the same as that as "All Cash" at points 1, 12, 23, 34, 45 where the portfolio state has $p = 0.0$ and is the same as "All Stock" in cases where the initial portfolio state has = 1.0. The "Very Active"

---

the violation of the no-action principle does not occur in $\tilde{A}$ and is inconsequential in A^{MV} (see preceding footnote). The term disappears if either $\theta_p = 0$ or 1; that is, if the new position is attained instantly either at the beginning or at the end of the month.

curve shows the expected discounted utility of a heuristic which, given the current portfolio and prediction states (i, j), chooses the next portfolio state g which maximizes the expected utility in time period t—assuming zero transaction cost (and $\theta_p = 0$).

The expected discounted utility curves for the optimum and mean–variance heuristic strategies are virtually indistinguishable and dominate that of the other heuristics.

Figure 10.2 shows the same information for $c = 0.02$. In this case, the very active heuristic performs far below the others. The inactive strategy is actually optimum if initial portfolio state has $p \in [0.6, 0.9]$, as we saw at the end of Section 10.5, and virtually so for $p \in [0.5, 1.0]$. As in the case with $c = 0.005$, the expected discounted utility provided by the MV heuristic is almost indistinguishable from $\tilde{W}$ for all i, j.

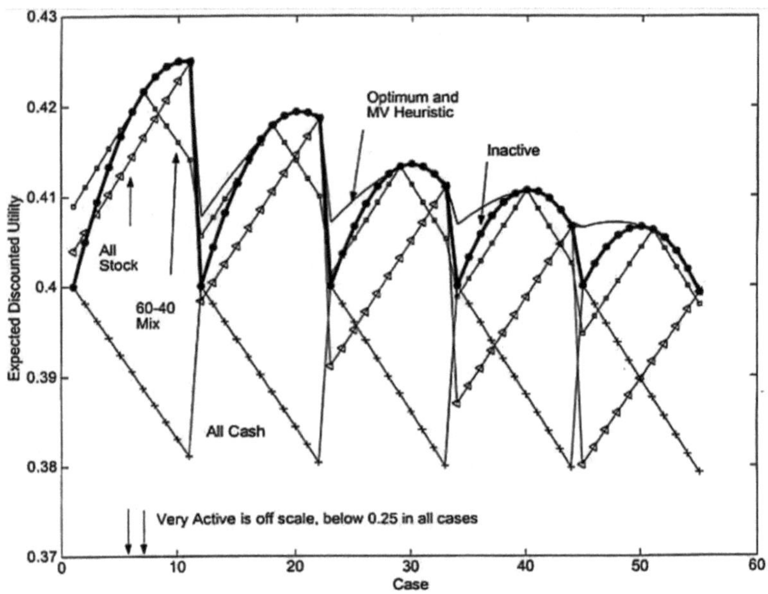

Fig. 10.2 Expected discounted utility for various strategies, unending game, cost = 0.02

10.8 Where Next?

As it stands, the model presented here is as useless for practice as the airplane that the Wright brothers flew at Kitty Hawk. The importance of the latter and the importance of the former, if it is to have any, result from what happens afterward. In the present section we sketch possible future uses of the method of mean–variance surrogates.

The general version of the model in Exhibit A could be used, with other parameter settings, to represent investment situations with

(a) large numbers of securities and
(b) other forecast models, such as ARCH-type models.

ARCH- and GARCH-type models would imply greater differences in V between prediction states and little or none in expected return E in Exhibit A, Item 3. This would imply smaller differences in optimal investment. We conjecture that this would make it easier for appropriate heuristics to approximate optimum performance. For a toy model with two securities, the methodology of the preceding sections could be used to test the conjecture.

For larger numbers of securities or asset classes, Monte Carlo runs could be used to select ("tune") E_{wt} and V_{wt} and evaluate the discounted expected utility of the mean–variance surrogate versus other heuristics. Generally, an optimum solution cannot be computed for higher dimensional problems. The simulation runs can only compare alternate heuristics. As long as optimization is out of the question, a more realistic consumption assumption, than in Exhibit A, Item 9, should not be a problem for the heuristics including mean–variance surrogate heuristics.

Moving beyond the model in Exhibit A, an important area where illiquidities are paramount is portfolio management for taxable entities. Unrealized capital gains is a much greater source of illiquidity than the transaction costs considered above. Some aspects of the problem of choosing the correct action in the presence of unrealized capital gains are well within current computing capabilities. It does not take long to trace out a mean–variance efficient frontier for a problem with thousands of decision variables, where the variables may or may not have lower and/or upper bounds and may be subject perhaps to hundreds of other linear equality or inequality constraints (see Perold 1984 or Markowitz and Todd 1959). In particular, it would not strain current computing capacity to include sales from individual lots as decision variables and have the budget equation recognize that the proceeds of such sales depend on the amount of unrealized capital gains or losses in the lot.

The hard part has to do with the objective of the analysis. For example, we could seek efficiency in terms of the mean and variance of end period of NAV. This would assume that a dollar's worth of unrealized capital gains is worth a dollar's worth of cash. An alternate approach would be to assume that a dollar's worth of unrealized capital gain is worth some fraction ϕ times as much as a dollar's worth of cash. Thus, the objective of the analysis could be efficiency in terms of the mean and variance of a weighted sum of holdings, weighting unrealized gains, and losses differently than after tax funds. A weighted average of the mean and variance of a portfolio computed in this manner could be the mean–variance surrogate, standing in place of the complex, unknown derived utility function of this problem.

10.9 Conclusion

Consider again the model in the first paragraph of this chapter, with transaction costs and many securities whose joint distribution of returns changes over time. We have examined two examples of a simple version of such a model—examples in which the optimum solutions can be calculated—and found that in these particular cases, the discounted expected utility supplied by a mean–variance surrogate heuristic is practically indistinguishable from the optimum values. This makes it plausible to conjecture that the same may be true in more complex models. At least it justifies an effort to test such rules in more realistic investment simulations.

Most or all real-world portfolio management situations involve some illiquidities, changing beliefs, and more securities or asset classes than permit optimization. Thus the portfolio manager, or management team, must necessarily use some not-know-to-be-optimal procedure for changing their portfolio. This procedure could be something they make up as they go along or they can follow some formal rule. The formal rule can be adopted because it seems plausible or because it does better than other heuristics when tested in a Monte Carlo simulation or backtest. We argue that the MV surrogate heuristics should be among those tried.[5]

Appendix Why A^* May Not Equal $\tilde{A}$

In this appendix we use a simpler example of a model of the Exhibit A structure to illustrate why we cannot conclude that A^* equals $\tilde{A}$ even when $A^* = A_t^T$ for $t = 0, \ldots, S$ for large $S \leq T$.

Let the number of prediction states equal the number of portfolio states and be 2; let the risk-free rate, r_f, be 0; d still equals 0.99, and let the prediction state transition matrix be

$$P = \begin{pmatrix} 1/2 & 1/2 \\ 1/2 & 1/2 \end{pmatrix}.$$

Label the two portfolio states "0" and "1." The 0 state is all cash, and the 1 state is all stock. The two prediction states will also be labeled "0" and "1." Prediction state 0 is pessimistic and prediction state 1 is optimistic. Assume that $\theta_p = 1$ (see Item 6 of Exhibit A); thus, $Eu(D_t)$ depends only on p_{t-1} (and the prediction state), not on p_{t-2}. Assume that if the portfolio state is 1 at $t - 1$, then

[5] The idea of using single-period mean–variance analysis in place of the derived utility function of dynamic programming appears in Markowitz (1959), Chapter 13. In particular, Markowitz asserts that "The value of the dynamic programming analysis [for us] does not lie in its immediate application to the selection of portfolios. It lies, rather, in the insight it supplies concerning optimum strategies and how, in principle, they could be computed. It emphasizes the use of the single-period utility function." p. 279. See also p. 299 on *Changing Distributions*. This is in contrast to the usual textbook characterization of mean–variance analysis as being static. However, Markowitz (1959) does not attempt to test the efficacy of this approach, as we do in the present chapter.

$$Eu(D_t) = \begin{cases} 0 & \text{if prediction state} = 0 \\ 0.0001 & \text{if prediction state} = 1. \end{cases}$$

Because stocks do as well as cash in this model in prediction state 0 and stocks do better than cash in prediction state 1, switching from all stocks to all cash is never optimum. Whether it is optimum to switch from cash to stocks depends on transaction cost c but, because of the assumed P and θ_p, does not depend on the current prediction state. Thus, the two possible optimum action matrixes are

$$A_N = \begin{pmatrix} 0 & 0 \\ 1 & 1 \end{pmatrix} \tag{10.36}$$

and

$$A_Y = \begin{pmatrix} 1 & 1 \\ 1 & 1 \end{pmatrix}. \tag{10.37}$$

For both A_N and A_Y, if the game starts in portfolio state 1, expected value of the game is

$$\frac{0.0001}{1-d} = 0.01.$$

If the unending game starts in portfolio state 0, the values of the game for A_N and A_Y are, respectively,

$$E_N = 0$$

and

$$E_Y = 0.0001\frac{d}{1-d} - c.$$

Thus, whether the initial prediction state is 0 or 1, A_Y is better than A_N if and only if

$$c < 0.0001\frac{d}{1-d} = 0.0099.$$

For a finite game of length T, as in the infinite game, if the initial portfolio state is 0, it is optimum to switch immediately or never switch. The expected utilities of the two alternatives are

$$E_N^T = 0$$

and

$$E_Y^T = 0.0001\frac{d - d^{T+1}}{1 - d} - c.$$

Thus, for a game of length T, A_Y is better than A_N if and only if

$$\begin{aligned} c &< 0.0001\frac{d - d^{T+1}}{1 - d} \\ &< 0.0001\frac{d}{1 - d}. \end{aligned}$$

To construct an example in which

$$A_t^T = A_N \quad \text{for } t = 1, \ldots, S = 10^7$$

but

$$A_N \neq \tilde{A}$$

let $u = 10^8$ and $c = 0.0001\frac{d-d^{u+1}}{1-d}$. In this case, A_N will be the optimum choice at every move for a game of, e.g., length 1,000,000, but will not be optimum for the unending game.

Exhibit A: Summary of Investment Model

1. Interval between portfolio reviews: 1 month.
2. Risk-free rate per month, assumed constant: $r_f = 0.004$.
3. Monthly stock return means $E_1, \ldots, E_5$ and variances $V_1, \ldots, V_5$ for various prediction states:

State:	1	2	3	4	5	Steady state
E:	0.0064	0.0050	0.0042	0.0038	0.0027	0.0044
V:	0.000592	0.000556	0.000538	0.000539	0.000567	0.000559
OptX:	4.05	1.80	0.37	–0.37	–2.29	0.72

4. From–to transition probabilities P between predictive states:

State:	1	2	3	4	5	Steady state
1	0.702	0.298	0	0	0	0.1608
2	0.173	0.643	0.133	0.051	0	0.2771
3	0	0.260	0.370	0.348	0.022	0.1363
4	0	0.065	0.179	0.615	0.141	0.2393
5	0	0	0.033	0.164	0.803	0.1865

5. Total utility:

$$U = \sum_{t=1}^{\infty} d^{t-1} u(D_t).$$

The investor's objective is to maximize EU. For the cases reported we use a monthly discount factor of $d = 0.99$

6. We usually refer to "portfolio state"

$$i = 1, \ldots, 11$$

representing fraction invested

$$p = 0.0, 0.1, \ldots, 1.0.$$

Note that

$$p = (i - 1)/10.$$

Thus "$p = 0.2$" in the present discussion corresponds to $i = 3$ elsewhere. In computing utility for a period, we define the "effective fraction," p_t^e, invested in stocks during time interval t, as

$$p_t^e = \theta_p p_{t-1} + (1 - \theta_p) p_t,$$

where p_{t-1} and p_t are the fraction invested in stock at the beginning and at the end of the time interval, respectively. In particular, the conditional expected return and variance of return (at time point $t - 1$) on the portfolio during time interval t given state j, current stock fraction p_{t-1}, and chosen stock fraction p_t are

$$\begin{aligned} E_t^P &= p_t^e E_{j-1}^j + (1 - p_t^e) r_f, \\ V_t^P &= (p_t^e)^2 V_{t-1}^j, \end{aligned}$$

where E_{t-1}^j and V_{t-1}^j are the mean and variance of stock return for time interval t given the prediction state at $t - 1$. Runs reported use $\theta_p = 1/2$.

7. Transaction cost incurred during time interval t is

$$c|p_t - p_{t-1}|.$$

We report cases with $c = 0.005$ and 0.020.

8. We assume that

$$Eu(D_t) = E(D_t) - kV(D_t).$$

For the cases reported $k = 0.5$.

9. We assume that the entire return on the portfolio for the month, net of costs, is "distributed." Thus the conditional expected utility $u(D_t)$ for time interval t at time point $t-1$ given prediction state j, current portfolio state i, and selected portfolio state g is

$$E(u|i, j, g) = E_t^P - c|p_t - p_{t-1}| - kV_t^P,$$

where $p_{t-1} = 0.1(i-1)$, $p_t = 0.1(g-1)$, and E_t^P, V_t^P are as defined in Item 6. Note there that E_t^P and V_t^P for time interval t depend on prediction state at $t-1$.
10. We assume that the return on equities is bounded by some (very large) number M.

References

Balduzzi, P., Lynch, A.W.: Transaction costs and predictability: Some utility cost calculations. J. Finan. Econ. **52**(1), 47–78 (1999)

Bellman, R.E.: Dynamic Programming. Princeton University Press, Princeton, NJ (1957)

Constantinides, G.M.: Multiperiod consumption and investment behavior with convex transactions costs. Manage. Sci. **25**(11), 1127–1137 (1979)

Dantzig, G.B.: Linear programming under uncertainty. Manage. Sci. **1**, 197–206 (1955)

Dantzig, G.B., Infanger, G.: Multi-stage stochastic linear programs for portfolio optimization. Ann. Oper. Res. **45**, 59–76 (1993)

Davis, M.H.A., Norman, A.R.: Portfolio selection with transaction costs. Math. Oper. Res. **15**(4), 676–713 (1990)

Dexter, A.S., Yu, J.N.W., Ziemba, W.T.: Portfolio selection in a lognormal market when the investor has a power utility function: Computational results. In: Dempster, M.A.H., (edr.) Stochastic Program, pp. 507–523. Academic, New York, NY, (1980)

Ederington, L.H.: Mean-variance as an approximation to expected utility maximization. Working Paper 86-5, School of Business Administration, Washington University, St Louis, MO (1986)

Grauer, R.: Normality, solvency, and portfolio choice. J. Finan. Quant. Anal. **21**(3), 265–278, September (1986)

Kamin, J.H.: Optimal portfolio revision with a proportional transaction cost. Manage. Sci. **21**, 1263–1271 (1975)

Kroll, Y., Levy, H., Markowitz, H.M.: Mean variance versus direct utility maximization. J. Finan. **39**(1), 47–61, March (1984)

Levy, H., Markowitz, H.M.: Approximating expected utility by function of mean and variance. Am. Econ. Rev. **69**(3), 308–317 (1979)

Levy, M., Levy, H., Solomon, S.: Microscopic Simulation of Financial Markets. Academic, New York, NY (2000)

Lynch, A.W., Balduzzi, P.: Predictability and transaction costs: The impact on rebalancing rules and behavior. J. Finance **55**(5), 2285–2310 (1999)

Magill, M.J.P., Constantinides, G.M.: Portfolio selection with transactions costs. J. Econ. Theory **13**, 245–263 (1976)

Markowitz, H.: Portfolio Selection: Efficient Diversification of Investments. Wiley, New York, NY. 1991, 2nd ed. Basil Blackwell, Cambridge, MA (1959)

Markowitz, H.M., Todd, P.: Mean-Variance Analysis in Portfolio Choice and Capital Markets. Frank, J. Fabozzi Associates, New Hope, PA (1959)

Markowitz, H.M., Reid, D.W., Tew, B.V.: The value of a blank check. J. Portfolio Manage. **20**(4), 82–91 (1994)
Mossin, J.: Optimal multiperiod portfolio policies. J. Bus. **41**(2), 215–229 (1968)
Perold, A.F.: Large scale portfolio optimization. Manage. Sci. **30**(10), 1143–1160 (1984)
Pulley, L.M.: A general mean-variance approximation to expected utility for short holding periods. J. Finan. Quant. Anal. **16**(3), 361–73 (1981)
Pulley, L.M.: Mean-variance approximations to expected logarithmic utility. Oper. Res. **31**(4), 685–696 (1983)
Samuelson, P.A.: Lifetime portfolio selection. Rev. Econ. Stat. **51**(3), 239–246 (1969)
Simaan, Y.: Portfolio Selection and Capital Asset Pricing for a Class of Non-spherical Distributions of Assets Returns. PhD thesis, Baruch College, The City University of New York (1987)
Von Neumann, J., Morgenstern, O.: Theory of Games and Economic Behavior. Princeton University Press, Princeton NJ. 1944, 3rd edn (1953)
Young, W., Trent, R.: Geometric mean approximation of individual security and portfolio performance. J. Finan. Quant. Anal. **4**(2), 179–199 (1969)
Zabel, E.: Consumer choice, portfolio decisions, and transaction costs. Econometrica **41**(2), 321–335 (1973)

Chapter 11
Mean–Absolute Deviation Model

Hiroshi Konno

11.1 Introduction

Portfolio optimization is one of the most actively studied areas in stochastic optimization. The objective here is to control the distribution of the return of the portfolio (a bundle of assets) so as to maximize the utility of risk-averse investors. It is widely known that those distributions that are not dominated by any distributions in the sense of second-degree stochastic dominance, i.e., those distributions called SSD efficient distributions, are best in terms of the principle of "maximization of expected utility" (see, e.g., Ogryczak and Ruszczyński 1999). Unfortunately, however, it is not easy to obtain SSD distributions unless the underlying assets follow some nice distribution such as the multivariate normal distribution.

The mean–variance model of Markowitz (1959) was the first practical approach for calculating a good, not necessarily the best distribution.

He formulated the problem as a two-objective optimization problem where one objective was the expected value and the other was the variance of the rate of return of the portfolio. Unfortunately, however, the mean–variance model of practical size with more than 1000 assets was not solvable until the mid-1980s, since the resulting problem became a convex quadratic programming problem with a completely dense covariance matrix. Later in 1984, Perold reformulated the problem in a compact form by introducing a factor model (Perold 1984).

However, it was not possible then to solve a problem with over a few thousands. Also, he could not solve problems with market conditions such as nonconvex commission fee, minimal transaction unit constraints, cardinality constraints, market impact effect.

To overcome these computational difficulties, the author proposed the mean–absolute deviation (MAD) model. This model can be reduced to a linear programming problem and hence can be solved fast. Also, market conditions above can be treated much easier than in the corresponding MV model.

H. Konno (✉)
Department of Industrial and Systems Engineering, Chuo University, Tokyo, Japan
e-mail: konno@indsys.chuo-u.ac.jp

G. Infanger (ed.), *Stochastic Programming*, International Series in Operations Research & Management Science 150, DOI 10.1007/978-1-4419-1642-6_11,

Computational advantages of the MAD model over the MV model are well established (Konno and Wijayanayake 1999, 2001a,b, 2002, Konno and Yamamoto 2003, 2005a,b). Unfortunately, however, these advantages are not well known since these results are scattered in various journals, some of which are of limited circulation. One of the purposes of this survey is therefore to explain the computational advantages of the MAD model over the MV model.

The second and perhaps more important purpose is to review theoretical properties of the MAD model, which was initially considered as a "quick and dirty" computational scheme without solid economic foundation. However, it was later proved that we can derive CAPM-type equilibrium relations in spite of the fact that the absolute deviation is not differentiable (Konno and Shirakawa 1994, 1995). Further, it was recently proved that the MAD model is more consistent with von Neumann's principle of "maximization of expected utility (MEU)" (Ogryczak and Ruszczyński 1999, 2001). This means that the MAD model is a "quick and clean" economic model.

Let us note here that the MAD model was proposed by Hazell (1971), an agricultural economist to overcome the computational difficulty associated with the mean–variance model as applied to the problem of best allocation of the farmyard to various crops.

Unfortunately, however, this result was criticized by statisticians and it has been forgotten for 17 years. Had any financial engineer recognized this result in the early 1970s, the history of portfolio management should have been much different since almost everybody believed then that the asset return follows a multivariate normal distribution, when the MAD model generates the same portfolio as the associated MV model.

In the next section, we will present the definition of the MAD model and discuss its relationship to the MV model and other mean–risk models. Section 11.3 will be devoted to CAPM-type results for the MAD model. We will show that all results known for the MV model apply to the MAD model as well with minor modifications. Also, readers may be surprised to know that the MAD model is more consistent with the fundamental principle of financial economics. Finally, in Section 11.4, we will discuss computational advantages of the MAD model over the MV model.

11.2 Mean–Absolute Deviation Model

Let R_j $(j = 1, 2, \ldots, n)$ be the random variable representing the rate of return of jth asset during the planning horizon. Also, let x_j $(j = 1, 2, \ldots, n)$ be the proportion of the fund to be invested into jth asset.

Then the rate of return $R(\boldsymbol{x})$ of the portfolio $\boldsymbol{x} = (x_1, x_2, \ldots, x_n)$ is represented as follows:

$$R(\boldsymbol{x}) = \sum_{j=1}^{n} R_j x_j. \tag{11.1}$$

Let $E[R(\boldsymbol{x})]$ and $V[R(\boldsymbol{x})]$ be, respectively, the expected value and variance of $R(\boldsymbol{x})$. Then the mean–variance (MV) model is formulated as follows:

$$\begin{aligned} &\text{minimize } V(\boldsymbol{x}) \\ &\text{subject to } E[R(\boldsymbol{x})] = \rho \\ &\qquad\qquad \boldsymbol{x} \in X, \end{aligned} \tag{11.2}$$

where ρ is a constant specified by an investor and X is an investable set usually defined as follows:

$$X = \left\{ \boldsymbol{x} \in R^n \mid \sum_{j=1}^{n} x_j = 1,\ 0 \le x_j \le \alpha,\ j = 1, 2, \ldots, n; \right. \\ \left. \sum_{j=1}^{n} a_{ij} x_j \ge b_i,\ i = 1, 2, \ldots, m \right\}. \tag{11.3}$$

Alternatively, one may formulate the same portfolio as follows:

$$\begin{aligned} &\text{maximize } E(\boldsymbol{x}) \\ &\text{subject to } V[R(\boldsymbol{x})] = \sigma^2 \\ &\qquad\qquad \boldsymbol{x} \in X. \end{aligned} \tag{11.4}$$

The mean–absolute deviation (MAD) model is a variant of the mean–variance (MV) model in which the measure of risk is replaced by the absolute deviation (Fig. 11.1):

$$W(\boldsymbol{x}) = E[|R(\boldsymbol{x}) - E[R(\boldsymbol{x})]|]. \tag{11.5}$$

The MAD model is formally defined as follows:

$$\begin{aligned} &\text{minimize } W(\boldsymbol{x}) \\ &\text{subject to } E[R(\boldsymbol{x})] = \rho \\ &\qquad\qquad \boldsymbol{x} \in X. \end{aligned} \tag{11.6}$$

Proposition 1 *If* $R(\boldsymbol{x})$ *follows normal distribution with mean* $r(\boldsymbol{x})$ *and variance* $\sigma^2(\boldsymbol{x})$, *then*

$$W(\boldsymbol{x}) = \sqrt{2/\pi}\,\sigma(\boldsymbol{x}).$$

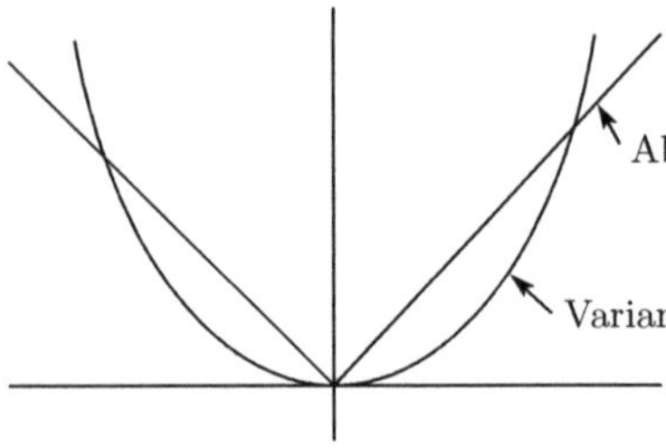

Fig. 11.1 Variance and absolute deviation

Proof

$$\begin{aligned} W(\boldsymbol{x}) &= \frac{1}{\sqrt{2\pi}\sigma(\boldsymbol{x})}\int_{-\infty}^{\infty}|z-r(\boldsymbol{x})|\exp\left\{-\frac{(z-r(\boldsymbol{x}))^2}{2\sigma^2(\boldsymbol{x})}\right\}dz \\ &= \frac{2}{\sqrt{2\pi}}\frac{1}{\sigma(\boldsymbol{x})}\int_0^{\infty} t\exp\left\{-\frac{t^2}{2\sigma^2(\boldsymbol{x})}\right\}dt \\ &= \sqrt{2/\pi}\sigma(\boldsymbol{x}). \end{aligned}$$

□

As a result, we have the following important theorem.

Theorem 2 *The MAD model* (11.6) *generates the same optimal portfolio as the corresponding MV model when* $\boldsymbol{R} = (R_1, R_2, \ldots, R_n)$ *follows multivariate normal distribution.*

When $\boldsymbol{R}$ is not normally distributed, the MAD model is not equivalent to the MV model. However, the difference should not be very significant, since absolute deviation and standard deviation are L_1 and L_2, respectively, measures of the deviation around the mean and there exists a constant $a_1 \geq a_2 > 0$ such that

$$a_2 W(\boldsymbol{x}) \leq \sigma(\boldsymbol{x}) \leq a_1 W(\boldsymbol{x})$$

for all $\boldsymbol{x}$.

Let us next define the lower semi-absolute deviation and the lower semi-variance as follows:

$$W_-(\boldsymbol{x}) = E[|R(\boldsymbol{x}) - r(\boldsymbol{x})|_-], \tag{11.7}$$

$$V_-(\boldsymbol{x}) = E[|R(\boldsymbol{x}) - r(\boldsymbol{x})|_-^2], \tag{11.8}$$

where $r(\boldsymbol{x}) = E[R(\boldsymbol{x})]$ and $|a|_- = \max\{0, -a\}$. These are typical measures of downside (lower partial) risk. Markowitz (1959) suggested that lower semi-variance or lower semi-standard deviation $\sigma_-(\boldsymbol{x}) = \sqrt{V_-(\boldsymbol{x})}$ would be a more adequate measure of risk than variance or standard deviation when the distribution of $R(\boldsymbol{x})$ was skewed, i.e., when $\sigma_-(\boldsymbol{x}) \neq \sigma(\boldsymbol{x})/2$.

Theorem 3 *The following relation*

$$W_-(\boldsymbol{x}) = (1/2)W(\boldsymbol{x})$$

holds for any distribution of $R(\boldsymbol{x})$.

Proof Let $f(\boldsymbol{x})$ be the density function of $R(\boldsymbol{x})$.

$$\begin{aligned}
W(\boldsymbol{x}) - 2W_-(\boldsymbol{x}) &= E[|R(\boldsymbol{x}) - r(\boldsymbol{x})|] - 2E[|R(\boldsymbol{x}) - r(\boldsymbol{x})|_-] \\
&= \int_{r(\boldsymbol{x})}^{\infty} (\eta - r(\boldsymbol{x}))f(\eta)d\eta + \int_{-\infty}^{r(\boldsymbol{x})} (r(\boldsymbol{x}) - \eta)f(\eta)d\eta \\
&\quad -2\int_{-\infty}^{r(\boldsymbol{x})} (r(\boldsymbol{x}) - \eta)f(\eta)d\eta \\
&= \int_{r(\boldsymbol{x})}^{\infty} (\eta - r(\boldsymbol{x}))f(\eta)d\eta - \int_{-\infty}^{r(\boldsymbol{x})} (r(\boldsymbol{x}) - \eta)f(\eta)d\eta \\
&= \int_{-\infty}^{\infty} \eta f(\eta)d\eta - r(\boldsymbol{x})\int_{-\infty}^{\infty} f(\eta)d\eta \\
&= r(\boldsymbol{x}) - r(\boldsymbol{x}) = 0.
\end{aligned}$$

□

This means that the mean–absolute deviation model is equivalent to the mean–lower semi-absolute deviation model which is a typical mean–lower partial risk model.

Let

$$f(\rho) = \min\{W(\boldsymbol{x}) \mid E[R(\boldsymbol{x})] = \rho,\ \boldsymbol{x} \in X\}. \tag{11.9}$$

Then the graph of $f(\rho)$ is called the efficient frontier associated with the MAD model. Since $W(\boldsymbol{x})$ is a convex function, $f(\rho)$ is also a convex function of ρ.

It is well known (Markowitz 1959) that those portfolios on the efficient frontier of the MV model are efficient in the sense of second-degree stochastic dominance and hence are consistent with the principle of MEU (maximization of expected utility) under either one of the following conditions:

(i) $(R_1, R_2, \ldots, R_n)$ follows multivariate normal distribution or elliptical distribution,
(ii) utility function of an investor is an increasing concave quadratic function.

Surprisingly enough, Ogryczak and Ruszczyński (1999) proved the following theorem.

Theorem 4 *Portfolios on the MAD efficient frontier to the right of point A (of* Fig. 11.2), *where* $f'(\rho) = 1$ *are efficient in the sense of second-degree stochastic dominance if the optimal solution of the problem* (11.6) *is unique.*

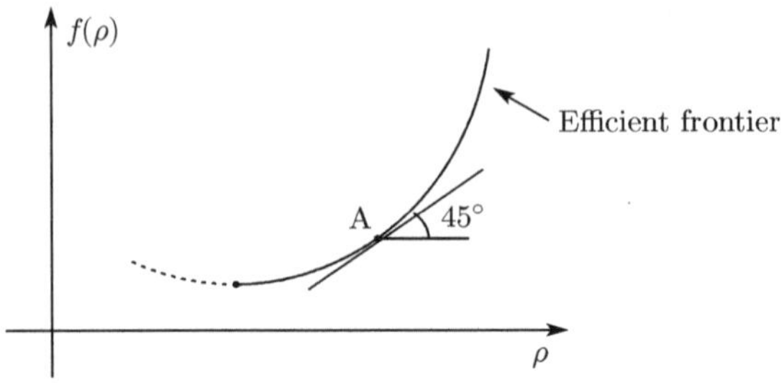

Fig. 11.2 MAD efficient frontier

Proof See Ogryczak and Ruszczyński (1999).

Let us note that this theorem holds regardless of the distribution of $R(\boldsymbol{x})$, contrary to the case of the MV model. This means that the MAD model is more consistent with the MEU principle than the MV model. Whether those portfolios on the efficient frontier to the left of the point A are efficient in the sense of second-degree stochastic dominance for any distribution of $R(\boldsymbol{x})$ remains an open problem.

Further, Ogryczak and Ruszczyński (2001) showed that Theorem 4 is valid for the model with more general lower partial risk measures:

$$\sigma_k(\boldsymbol{x}) \equiv E[|R(\boldsymbol{x}) - E[R(\boldsymbol{x})]|_-^k]^{1/k},\ k = 1, 2, \ldots.$$

11.3 Equilibrium Relations: MAD Version CAPM

Let us introduce here the riskless asset whose rate of return is r_0 and let us consider the following simplified version of the MAD model:

$$\begin{array}{ll} \text{minimize} & W(\boldsymbol{x}) \\ \text{subject to} & E[R(\boldsymbol{x})] = \rho \\ & \boldsymbol{x} \in X_0, \end{array} \tag{11.10}$$

where

$$X_0 = \left\{ (x_0, x_1, \ldots, x_n) \mid \sum_{j=0}^{n} x_j = 1,\ x_j \geq 0,\ j = 1, 2, \ldots, n \right\}. \tag{11.11}$$

Note here that x_0 can be negative. Then the following theorems hold under some standard assumptions on the capital market and investors.

Theorem 5 (Two-fund separation theorem) *Every investor invests into the market portfolio P_{M} and riskless asset and nothing else.*

Here the market portfolio P_{M} is defined as a portfolio on the tangent line to the efficient frontier $f(\rho)$ which passes the riskless point $(r_0, 0)$ (see Fig. 11.3).

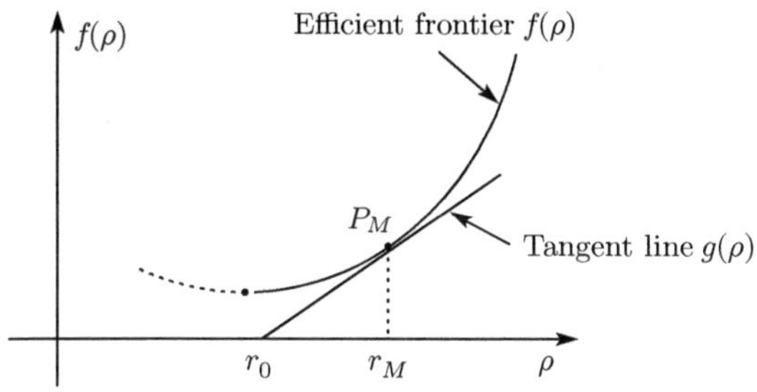

Fig. 11.3 Market portfolio

Proof of this theorem is straightforward by noting that $g(\rho)$ is convex function (Konno et al. 2002).

Theorem 6 *Let* $r_j = E[R_j]$ $(j = 1, \ldots, n)$ *and let* r_{M} *be the expected rate of return of the (mean–absolute deviation) market portfolio. Then*

$$r_j - r_0 = \theta_j (r_{\mathrm{M}} - r_0), \ j = 1, 2, \ldots, n, \tag{11.12}$$

where

$$\theta_j = \frac{\mathrm{cov}[(R_j - r_j), \ \mathrm{sign}(R_{\mathrm{M}} - r_{\mathrm{M}})]}{E[|R_{\mathrm{M}} - r_{\mathrm{M}}|]}, \ j = 1, 2, \ldots, n, \tag{11.13}$$

where

$$\mathrm{sign}(a) = 1(a > 0)\,;\ 0(a = 0)\,;\ -1(a < 0).$$

The proof of this theorem is not so easy as its counterpart Theorem 6. We will reproduce the proof in the Appendix.

Theorem 7 parallels the similar result for the MV market where θ_j is replaced by familiar beta defined by

$$\beta_j = \mathrm{cov}(R_j, R_{\mathrm{M}})/V[R_{\mathrm{M}}], \ j = 1, 2, \ldots, n. \tag{11.14}$$

Figure 11.3 shows the value of β_j's and θ_j's of a representative stock calculated by using the same set of data. We observe a remarkable similarity, as expected.

Theorems 5 and 6 have been extended to models with more general risk measures:

$$\sigma_k(\boldsymbol{x}) = E[|R(\boldsymbol{x}) - E[R(\boldsymbol{x})]|_-^k]^{1/k}, \ k = 1, 2, \ldots.$$

Also, they hold as well for the mixed market with various types of investors who employ different risk measures (see Konno et al. 2002 for these results). Let us give one important result about the equilibrium price of the asset in the MAD capital market.

Let there be m investors who own x_{ij}^0 units of jth assets before the transaction. Also, let ρ_i be the requested rate of return of investor i. Also let z_j^* $(j = 1, 2, \ldots, n)$ be an optimal solution of the following optimization problem (see the Appendix for the existence of an optimal solution):

$$\begin{aligned} &\text{minimize } E\left[\left|\sum_{j=1}^{n}(R_j - r_j)z_j\right|\right] \\ &\text{subject to } \sum_{j=1}^{n}(r_j - r_0)z_j = 1 \\ &\qquad\qquad z_j \geq 0,\ j = 1, 2, \ldots, n. \end{aligned} \tag{11.15}$$

Theorem 7 *The equilibrium price p_j of the asset j is given by*

$$p_j = \sum_{j=1}^{m}(\rho_i - r_0)x_{i0}^0 z_j^* / (1 - m_0)\sum_{i=1}^{m} x_{ij}^0,\ j = 1, 2, \ldots, n, \tag{11.16}$$

where

$$m_0 = \sum_{i=1}^{m}\sum_{j=1}^{n}(\rho_i - r_0)z_j^* x_{ij}^0 \bigg/ \sum_{i=1}^{m} x_{ij}^0. \tag{11.17}$$

Corollary 8 *The necessary and sufficient condition for the existence of nonnegative equilibrium price vector is $m_0 < 1$.*

In particular, when the condition

$$x_{ij}^0 = \alpha_i \sum_{i=1}^{m} x_{ij}^0,\ j = 0, 1, 2, \ldots, n, \tag{11.18}$$

is satisfied, then the following result holds.

Theorem 9 *The necessary and sufficient condition for the existence of the unique nonnegative equilibrium price under* (11.17) *is $\rho_M < r_M$, where $\rho_M = \sum_{i=1}^{m} \alpha_i \rho_i$. The equilibrium price is given by*

$$p_j = \frac{\rho_M - r_0}{r_M - \rho_M}(r_M - r_0)\frac{\sum_{i=1}^{m} x_{i0}^0}{\sum_{i=1}^{m} x_{ij}^0} z_j^*,\ j = 1, 2, \ldots, n. \tag{11.19}$$

Figure 11.4 shows the ratio of total value of risky assets $V_R \equiv \sum_{i=1}^{m}\sum_{j=1}^{n} p_j x_{ij}^0$ and total amount of riskless asset $V_0 \equiv \sum_{i=1}^{m} x_{i0}^0$ as a function of ρ_M. We see that the market explodes as ρ_M approaches r_M. Also, it collapses as ρ_M approaches r_0.

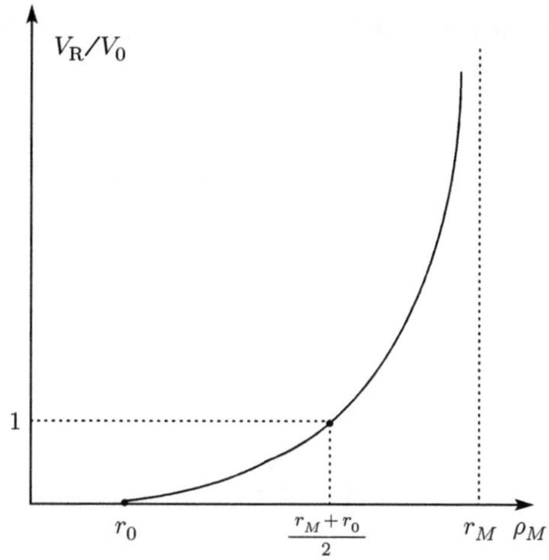

Fig. 11.4 Value of stock

Note that these results parallel those that have been proved for MV model in Konno and Suzuki (1996), where z_j^* $(j = 1, 2, \ldots, n)$ is the optimal solution of the quadratic programming problems:

$$\begin{aligned} &\text{minimize} \sum_{j=1}^{n} \sum_{k=1}^{m} \sigma_{jk} z_j z_k \\ &\text{subject to} \sum_{j=1}^{n} (r_j - r_0) z_j = 1 \\ &\qquad z_j \geq 0, \; j = 1, 2, \ldots, n, \end{aligned}$$

instead of (15).

11.4 Computational Aspects

Let us now turn to computational aspects of the MAD model. We will assume here that $(R_1, R_2, \ldots, R_n)$ is distributed over a finite set of points $(r_{1t}, r_{2t}, \ldots, r_{nt})$, $t = 1, 2, \ldots, T$. Also let

$$f_t = Pr\{(R_1, R_2, \ldots, R_n) = (r_{1t}, r_{2t}, \ldots, r_{nt})\}, \; t = 1, 2, \ldots, T,$$

be known. These data can be

(i) collected from the historical data where $(r_{1t}, r_{2t}, \ldots, r_{nt})$ is the achieved rate of return during period t

or

(ii) generated by a scenario model where $(r_{1t}, r_{2t}, \ldots, r_{nt})$ is the rate of return under scenario t.

We can calculate mean and absolute deviation as follows:

$$r_j = \sum_{t=1}^{T} f_t r_{jt}, \tag{11.20}$$

$$\begin{aligned} W(\boldsymbol{x}) &= E[|R(\boldsymbol{x}) - E[R(\boldsymbol{x})]|] \\ &= \sum_{t=1}^{T} f_t \left| \sum_{j=1}^{n} (r_{jt} - r_j) x_j \right|. \end{aligned} \tag{11.21}$$

Therefore the MAD model can be represented as follows:

$$\begin{aligned} &\text{minimize } \sum_{t=1}^{T} f_t |z_t| \\ &\text{subject to } z_t = \sum_{j=1}^{n} (r_{jt} - r_j) x_j,\ t = 1, 2, \ldots, T, \\ &\sum_{j=1}^{n} r_j x_j = \rho \\ &\boldsymbol{x} \in X. \end{aligned} \tag{11.22}$$

It is well known (Dantzig 1963) that this problem can be reduced to

$$\begin{aligned} &\text{minimize } \sum_{t=1}^{T} f_t (u_t + v_t) \\ &\text{subject to } u_t - v_t = \sum_{j=1}^{n} (r_{jt} - r_j) x_j,\ t = 1, 2, \ldots, T, \\ &u_t \geq 0,\ v_t \geq 0,\ t = 1, 2, \ldots, T, \\ &\sum_{j=1}^{n} r_j x_j = \rho \\ &\boldsymbol{x} \in X. \end{aligned} \tag{11.23}$$

By noting Theorem 3, the objective function of (11.22) can be replaced by the lower semi-absolute deviation. Thus, the problem (11.22) is equivalent to

$$
\begin{aligned}
&\text{minimize} \sum_{t=1}^{T} f_t|z_t|_- \\
&\text{subject to } z_t = \sum_{j=1}^{n}(r_{jt} - r_j)x_j,\ t = 1, 2, \ldots, T, \\
&\qquad \sum_{j=1}^{n} r_j x_j = \rho \\
&\qquad \boldsymbol{x} \in X
\end{aligned}
\tag{11.24}
$$

or

$$
\begin{aligned}
&\text{minimize} \sum_{t=1}^{T} f_t u_t \\
&\text{subject to } u_t \geq -\sum_{j=1}^{n}(r_{jt} - r_j)x_j,\ t = 1, 2, \ldots, T, \\
&\qquad u_t \geq 0,\ t = 1, 2, \ldots, T \\
&\qquad \sum_{j=1}^{n} r_j x_j = \rho \\
&\qquad \boldsymbol{x} \in X.
\end{aligned}
\tag{11.25}
$$

Computation time for solving (11.23) and (11.25) is more or less the same when T and n are small. When n and T are larger, then (11.25) is more suitable for computation since it contains fewer variables. According to our numerical experiments (Konno and Wijayanayake 2001a), the problem with $(n, T) = (10^4, 10^5)$ can be successfully solved in less than 1 h on a personal computer (Pentium IV, 1.5 GHz, 2 Gbyte memory) using the state-of-the art linear programming software CPLEX.

In addition, the MAD formulation can be adapted to solve more difficult problems under nonconvex transaction cost and minimal transaction unit constraints and cardinality constraints.

(a) *Concave and D.C. Transaction Cost*

Associated with real transaction in the market, we need to pay various costs including commission fee and tax. Let $c_j(x_j)$ be the amount of cost associated with x_j. The mean–absolute deviation model under transaction cost is then formulated as follows:

$$
\begin{aligned}
&\text{minimize } W(\boldsymbol{x}) \\
&\text{subject to} \sum_{j=1}^{n}(r_j x_j - c_j(x_j)) = \rho \\
&\qquad \boldsymbol{x} \in X.
\end{aligned}
\tag{11.26}
$$

When $c_j(\cdot)$ is linear for all j, the problem can again be reduced to a linear programming problem. When $c_j(\cdot)$ is nonlinear, nonlinear equality constraint causes computational difficulty.

In this case, we interchange the risk and the rate of return and consider the following problem:

$$\begin{aligned} &\text{maximize } \sum_{j=1}^{n}(r_j x_j - c_j(x_j)) \\ &\text{subject to } W(\boldsymbol{x}) \leq w \\ &\qquad\qquad\ \boldsymbol{x} \in X, \end{aligned} \tag{11.27}$$

where w is the level of risk determined by an investor. The efficient frontier generated by (11.26) and (11.27) are slightly different, but it is essentially the same from the practical point of view.

One advantage of this formulation is that the problem (11.27) can be solved relatively efficiently when $c_j(\cdot)$ is a concave increasing function or a d.c. function or piecewise constant depicted in Fig. 11.6.

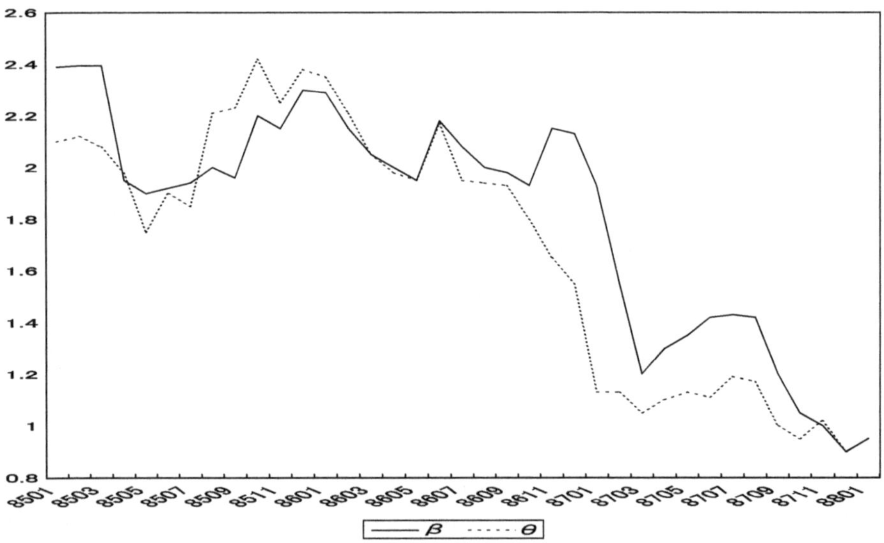

Fig. 11.5 β and θ of a typical stock

In fact it has been demonstrated in Konno and Wijayanayake (2001b, 2002), Konno and Yamamoto 2003, 2005a,b, and Konno and Yamazaki (1991) that the problem (27) can be solved in an efficient manner when the universe n is less than a few hundreds. An important fact is that the feasible region is polyhedral, so that a well-designed branch and bound algorithm can solve the resulting concave and d.c. maximization problems under linear constraints. Further, we showed in a recent

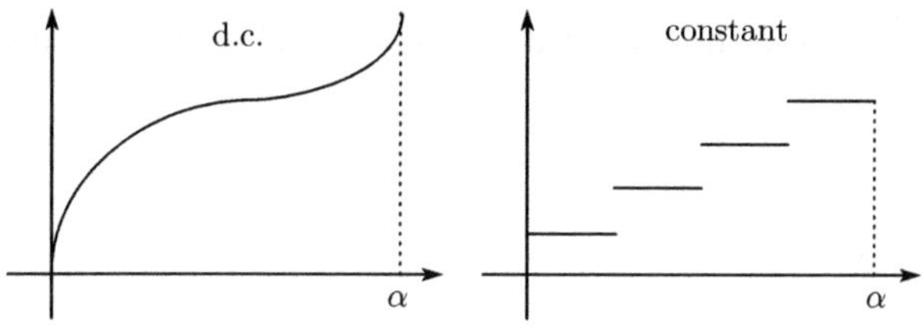

Fig. 11.6 Transaction cost

series of articles (Konno and Yamamoto 2005a,b) that these problems can be solved even faster by using a classical 0-1 integer programming approach.

If instead, the measure of risk is variance or standard deviation, it would be much more difficult to solve the corresponding problem:

$$\begin{aligned} \text{maximize } & \sum_{j=1}^{n}(r_j x_j - c_j(x_j)) \\ \text{subject to } & V(\boldsymbol{x}) \le v \\ & \boldsymbol{x} \in X, \end{aligned}$$

since this is a nonlinearly constrained convex (or d.c.) maximization problem, for which there is no efficient algorithm.

(b) *Minimal Transaction Unit Constraints*

When we apply MV or MAD model to the management of small fund, say less than 1 million dollars, we need to consider minimal transaction unit constraints. Therefore we need to replace the investable set as follows:

$$X_I = X \cap \{x_j = \text{integer multiple of minimal transaction unit}\}.$$

The resulting mean–variance model

$$\begin{aligned} \text{maximize } & \sum_{j=1}^{n}(r_j x_j - c_j(x_j)), \\ \text{subject to } & V(\boldsymbol{x}) \le v \\ & \boldsymbol{x} \in X_I, \end{aligned}$$

is intractably difficult from the viewpoint of the state-of-the-art mathematical programming technology.

If we use $W[R(\vec{x})]$ instead of $V[R(\vec{x})]$, then the problem

$$\begin{aligned} \text{maximize } & \sum_{j=1}^{n}(r_j x_j - c_j(x_j)), \\ \text{subject to } & W(\boldsymbol{x}) \le w \\ & \boldsymbol{x} \in X_I, \end{aligned} \tag{11.28}$$

can be solved in an efficient manner if $n \leq 900$ and $T \leq 60$, by using a 0-1 integer programming approach.

11.5 Concluding Remarks

In this chapter, we summarized important properties of the MAD model. We showed that the MAD model is superior to the MV model both theoretically and computationally. Also, we showed that the MAD model belongs to a class of mean–lower partial risk model that is more adequate to handle nonsymmetric return distribution.

We hope that this chapter helps people use the MAD model for formulating and solving large-scale portfolio optimization problems which may not be properly handled by the corresponding MV model. Such problems include, among others,

(i) large-scale portfolio optimization under complicated market condition (Konno and Wijayanayake, 2001a, 2002, Konno and Yamamoto 2003, 2005a,b),
(ii) very large-scale portfolio optimization associated with internationally diversified investment (Komuro and Konno 2005, Konno and Li 1998),
(iii) long-term asset management problems (Carino and Ziemba 1998, Zenios 1995, Zenios and Kang 1993),
(iv) integrated portfolio optimization using various financial instruments in addition to stocks and bonds (Komuro and Konno 2005).

Appendix. Proof of Theorem 6

Let us consider the following problem:

$$\begin{aligned} \text{maximize } & f(\vec{z}) = E\left[\left|\sum_{j=1}^{n}(R_j - r_j)z_j\right|\right] \\ \text{subject to } & \sum_{j=1}^{n}(r_j - r_0)z_j = 1 \\ & z_j \geq 0,\ j = 1, 2, \ldots, n. \end{aligned} \tag{11.29}$$

We will assume for simplicity that $r_j > r_0$, $\forall j$. Then (11.29) has an optimal solution z_j^*, $j = 1, 2, \ldots, n$, since the feasible region is compact and the objective function is continuous. We will assume here that $(R_1, R_2, \ldots, R_n)$ is distributed on R^n and that the probability measure P is absolutely continuous with respect to the n-dimensional Lebesgue measure. This means that

$$Pr\left\{\sum_{j=1}^{n} R_j z_j = r\right\} = 0 \quad \text{for} \quad \vec{z} \in R^n \setminus \{\vec{0}\},\ r \in R. \tag{11.30}$$

First we show that f is continuously differentiable on $R^n \setminus \{\vec{0}\}$. Let $f'(\vec{z}; \vec{d})$ be the one-sided directional derivative of f at $\vec{z} \in R^n \setminus \{\vec{0}\}$ with respect to the direction $\vec{d} \in R^n$ defined by

$$f'(\vec{z}; \vec{d}) = \lim_{\alpha \downarrow 0} \frac{f(\vec{z} + \alpha \vec{d}) - f(\vec{z})}{\alpha}.$$

Let $R'_j = R_j - r_j$. Then we have by Lebesgue's dominated convergence theorem,

$$\begin{aligned} f'(\vec{z}; \vec{d}) &= E\left[\lim_{\alpha \downarrow 0} \frac{|\sum_{j=1}^n R'_j(z_j + \alpha d_j)| - |\sum_{j=1}^n R'_j z_j|}{\alpha}\right] \\ &= E\left[\lim_{\alpha \downarrow 0} \frac{\alpha \sum_{j=1}^n R'_j d_j \operatorname{sign}\left\{\sum_{j=1}^n R'_j z_j\right\}}{\alpha}\right. \\ &\qquad \left. + \lim_{\alpha \downarrow 0} \frac{\alpha \cdot \vec{1}\left\{\sum_{j=1}^n R'_j z_j = 0\right\} \left|\sum_{j=1}^n R'_j d_j\right|}{\alpha}\right] \\ &= \sum_{j=1}^n E\left[R'_j \operatorname{sign}\left\{\sum_{j=1}^n R'_j z_j\right\}\right] d_j \\ &\qquad + E\left[|\sum_{j=1}^n R'_j d_j| \,\middle|\, \sum_{j=1}^n R'_j z_j = 0\right] Pr\left\{\sum_{j=1}^n R'_j z_j = 0\right\}. \end{aligned}$$

From (11.30), the second term of the above is 0.

Then we have

$$f'(\vec{z}; \vec{d}) = \sum_{j=1}^n E\left[R'_j \operatorname{sign}\left\{\sum_{j=1}^n R'_j z_j\right\}\right] d_j, \ \forall \vec{d} \in R^n,$$

and hence

$$\frac{\partial f(\vec{z})}{\partial z_j} = E\left[R'_j \operatorname{sign}\left\{\sum_{j=1}^n R'_j z_j\right\}\right]. \tag{11.31}$$

Thus the objective function f is convex, continuous, and differentiable. On the other hand, constraints in (11.29) are linear with nonempty feasible region A such that $\vec{0} \notin A$. Then the Karush–Kuhn–Tucker conditions is the necessary and sufficient conditions for the optimality of (11.29). Therefore

$$\frac{\partial f(\vec{z}^*)}{\partial z_j} - \mu(r_j - r_0) - \lambda_j = 0,\ j = 1, 2, \ldots, n, \tag{11.32}$$

$$(r_j - r_0)z_j^* = 1, \tag{11.33}$$

$$\lambda_j z_j^* = 0,\ \lambda_j^* \geq 0,\ z_j^* \geq 0,\ j = 1, 2, \ldots, n. \tag{11.34}$$

Let us note that we can assume without loss of generality that $z_j^* \geq 0,\ \forall j$, so that $\lambda_j = 0,\ \forall j$. From (11.31) we have

$$\sum_{j=1}^{n} E\left[R'_j \operatorname{sign}\left\{\sum_{j=1}^{n} R'_j z_j^*\right\}\right] z_j^* - \mu \sum_{j=1}^{n} (r_j - r_0) z_j^* = 0.$$

Thus we have

$$\mu = \frac{E\left[R'_j z_j^* \operatorname{sign}\left\{\sum_{j=1}^{n} R'_j z_j^*\right\}\right]}{\sum_{j=1}^{n}(r_j - r_0)z_j^*} = \frac{E[|R_{\mathrm{M}} - r_0|]}{r_{\mathrm{M}} - r_0}, \tag{11.35}$$

where $R_{\mathrm{M}} = \sum_{j=1}^{n} R_j z_j^* \Big/ \sum_{j=1}^{n} z_j^*$ and $r_{\mathrm{M}} = E[R_{\mathrm{M}}]$.

Therefore from (11.31) and (11.35) we have

$$r_j - r_0 = \frac{E[(R_j - r_j)\operatorname{sign}\{R_{\mathrm{M}} - r_{\mathrm{M}}\}]}{E[|R_{\mathrm{M}} - r_{\mathrm{M}}|]}(r_{\mathrm{M}} - r_0),$$

which completes the proof. □

References

Carino, D.R., Ziemba, W.T.: Formulation of the Russell-Yasuda Kasai financial planning model. Oper. Res. **46**(4), 433–449 (1998)

Dantzig, G.: Linear Programming and Extensions. Princeton University Press, Princeton, NJ (1963)

Hazell, P.B.R.: A linear alternative to quadratic and semi variance to farm planning under uncertainty. Am. J. Agric. Econ. **53**(4), 664–665 (1971)

Komuro, S., Konno, H.: Internationally diversified investment by stock-bond integrated model. J. Ind. Manage. Optim. **1**(4), 433–442 (2005)

Konno, H., Li, J.: An internationally diversified investment using a stock-bond integrated portfolio model. Int. J. Theor. Appl. Finance **1**(1), 145–160 (1998)

Konno, H., Shirakawa, H.: Equilibrium relations in the mean–absolute deviation capital market. Asia-Pacific Financ. Markets **1**(1), 21–35 (1994)

Konno, H., Shirakawa, H.: Existence of a nonnegative equilibrium price vector in the mean–variance capital market. Math. Finance **5**(3), 233–246 (1995)

Konno, H., Suzuki, K.: Equilibria in the capital market with non-homogeneous investors. Jpn. J. Ind. Appl. Math. **13**(3), 369–383 (1996)

Konno, H., Wijayanayake, A.: Mean–absolute deviation portfolio optimization model under transaction costs. J. Oper. Res. Soc. Jpn **42**(4), 422–435 (1999)

Konno, H., Wijayanayake, A.: Optimal rebalancing under concave transaction costs and minimal transaction units constraints. Math. Program. **89**(2), 233–250 (2001a)

Konno, H., Wijayanayake, A.: Minimal cost index tracking under concave transaction costs. Int. J. Theor. Appl. Finance **4**(6), 939–957 (2001b)

Konno, H., Wijayanayake, A.: Portfolio optimization under d. c. transaction costs and minimal transaction unit constraints. J. Global Optim. **22**(1–4), 137–154 (2002)

Konno, H., Yamamoto, R.: Minimal concave cost rebalance of a portfolio to the efficient frontier. Math. Program. **97**(3), 571–585 (2003)

Konno, H., Yamamoto, R.: Global optimization versus integer programming in portfolio optimization under nonconvex transaction costs. J. Global Optim. **32**(2), 207–219 (2005a)

Konno, H., Yamamoto, R.: Integer programming approaches in mean–risk models. J. Comput. Manage. Sci. **2**(4), 339–351 (2005b)

Konno, H., Yamazaki, H.: Mean–absolute deviation portfolio optimization model and its applications to tokyo stock market. Manage. Sci. **37**(5), 519–531 (1991)

Konno, H., Waki, H., Yuuki, A.: Portfolio optimization under lower partial risk measures. Asia-Pac. Financ. Mark. **9**(2), 127–140 (2002)

Markowitz, H.: Portfolio Selection: Efficient Diversification of Investments. Wiley, New York, NY (1959)

Ogryczak, O., Ruszczyński, A.: From stochastic dominance mean–risk model. Eur. J. Oper. Res. **116**(1), 33–50 (1999)

Ogryczak, O., Ruszczyński, A.: On consistency of stochastic dominance and mean–semideviation models. Math. Program. **89**(2), 217–232 (2001)

Perold, A.F.: Large scale portfolio optimization. Manage. Sci. **30**(10), 1143–1160 (1984)

Zenios, S.A.: Asset liability management under uncertainty for fixed income securities. Ann. Oper. Res. **59**(1), 77–97 (1995)

Zenios, S.A., Kang, P.: Mean–absolute deviation portfolio optimization for mortgage-backed securities. Ann. Oper. Res. **45**(1), 433–450 (1993)

Chapter 12
Multistage Financial Planning Models: Integrating Stochastic Programs and Policy Simulators

John M. Mulvey and Woo Chang Kim

12.1 Introduction

Multistage financial planning models offer significant advantages over the classical single-stage (static) portfolio approaches. First, importantly, investment performance is enhanced by capturing rebalancing gains when a portfolio is modified. Rebalancing a portfolio can be considered as an option—to be exercised when adding value to the investors performance. Second, a multistage model can address many real-world issues in considerable detail, such as transaction–market impact costs, dynamic correlation structures, integrating assets with liability and goals over time, and decisions that are conditioned on the state of the economy. Many of these issues are difficult to model within a single-period framework. In a similar fashion, a multistage model can readily depict temporal goals and objectives. For example, we can model stopping rules that activate when a target wealth value, for example, is reached. See Dantzig and Infanger (1993), Consigli and Dempster (1998), Mulvey (2000), Mulvey and Ziemba (1998), and Mulvey et al. (2003a) for examples. A theoretical discussion can also be found in Fernholz and Shay (1982) and Luenberger (1998).

Unfortunately, multistage financial models possess two significant interrelated drawbacks: (1) they give rise to complex models and require extensive computational resources, and (2) the modeling recommendations are difficult to understand by decision makers. There is considerable amount of literature that solves these problems in the academic field (Infanger 1994), but many investors are simply unable to grasp the complexity of a multistage stochastic program, and accordingly, are often reluctant to "trust" the model's recommendations for assigning their capital.

To overcome these barriers, we advocate a dual strategy illustrated in Fig. 12.1. In the first stage of the dual approach, investors apply a simplified (stylized) stochastic program to generate an optimal set of decisions defined over a scenario tree. Since

---

J.M. Mulvey (✉)
Department of Operations Research and Financial Engineering, Princeton University, Princeton, NJ 08544, USA
e-mail: mulvey@princeton.edu

G. Infanger (ed.), *Stochastic Programming*, International Series in Operations Research & Management Science 150, DOI 10.1007/978-1-4419-1642-6_12,

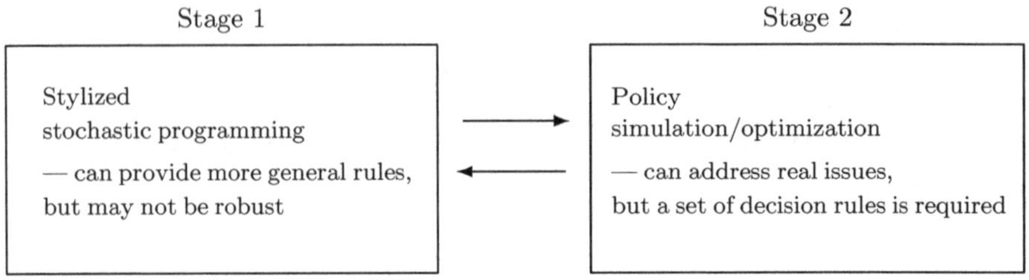

Fig. 12.1 Illustration of the dual strategy

decisions can occur anywhere in the feasible region as defined by the constraints, the stochastic program will offer general recommendations. However, by its nature, a stochastic program must present a simplified view of the real world, which may lead to a nonrobust optimal solution.

Therefore, in the second stage, a more detailed policy simulation–optimization approach is adopted. Unlike the stochastic program with generic decisions, the decisions in a policy simulator must be a function of a formal policy/decision rule. Also, the policy rules cannot employ information that is unavailable to the investor at a given time period; the model must enforce nonanticipativity. Since these approaches are complementary to each other, the dual strategy can not only help create a set of new policy rules but also evaluate the rules with respect to robustness.

To illustrate, we apply the dual strategy to an important problem in the area of asset and liability management—U.S. defined benefit (DB) pension trusts. The major goals of the integrated ALM model for DB pension trusts are to satisfy the interests of several entities—shareholders, sponsoring company, retirees, and regulators (public)—which may conflict each other. For instance, for the shareholders, the goal should be to maximize value for the company and minimize present value (and volatility) of the contribution for the pension trust. Providing an adequate level of safety for the pension plan and future beneficiaries would be the main objectives for the retirees. On the other hand, the regulators would minimize the possibility of the bankruptcy and plan termination. Therefore, there goals are best satisfied by constructing a multiobjective model whose output consists of a set of efficient frontiers.

There are numerous examples of stochastic programs implemented in different countries. A noteworthy early application is the Russell system for the Yasuda Insurance Company in Japan (Cariño et al. 1994). Mulvey et al. (2000) depict the details of the Towers Perrin CAP:Link system. Geyer et al. (2005) describe the financial planning model InnoALM developed by Innovest for Austrian pension funds. Hilli et al. (2003) develop a stochastic program for a Finnish pension company; their empirical study indicates that the results under the stochastic program outperform those under traditional fixed-mix strategies. Pflug et al. [2000] developed a modular decision support tool for asset–liability management—the AURORA Financial Management System at the University of Vienna. Pflug and Swietanowski (2000) also show that the nonconvexity of the ALM model can be resolved by approximating it to a convex problem. Dutch researchers have also achieved success in implementing ALM models for pension planning. Boender et al. (1998) describe the ORTEC model. The doctoral dissertation of Dert (1995) presents a scenario-based model for analyzing investment/funding policies of Dutch defined-benefit pension plans. Kouwenberg and Zenios (2001), Ziemba and Mulvey (1998), Arbeleche et al. (2003), Mulvey et al. (2004b), Dempster et al. (2006), and Zenios and Ziemba (2006, 2007) provide examples of related implementations. Also, Ziemba and Mulvey (1998) and Ziemba (2003) include examples of successful asset and liability management (ALM) systems within a multiperiod setting. Fabozzi et al. (2004) discuss applications of ALM in European pension funds.

Several researchers have studied ALM problems in a continuous time setting. Chapter 5 of Campbell and Viceira (2002) provides an introduction to strategic asset allocation in continuous time. A group of ALM models extend the early approach of Merton (1973). Rudolf and Ziemba (2004) present such a continuous time model for pension fund management. Although stochastic control models have difficulty in incorporating practical and legal constraints of pension plans, they possess the advantage of producing intuitive closed-form solutions or solving the system of equations satisfying optimality conditions numerically in many cases.

The remainder of the chapter is organized as follows. The next section illustrates the advantage of multiperiod models for the asset-only investors to improve asset allocation decision with several practical examples. In Section 12.3, we apply the dual strategy to the DB pension problems in the United States. The final section provides conclusions and future research topics.

12.2 Multiperiod Models for Asset-Only Allocation

This section illustrates the advantages of applying multiperiod portfolio approaches. We begin with the well-known fixed-mix investment rule due to its simplicity and profitability. Also, we provide several prominent examples that employ the fixed-mix approach for asset-only investors. This policy rule serves as a benchmark both for other types of rules and for the recommendations of a stochastic programming model.

12.2.1 Fixed-Mix Policy Rules

First, we describe the performance advantages of fixed-mix over a static, buy-and-hold perspective. This rule generates greater return than the static model by means of rebalancing. The topic of rebalancing gains (also called excess growth or volatility pumping) as derived from the fixed-mix decision rule is well understood for a theoretical perspective. The fundamental solutions were developed by Merton (1969) and Samuelson (1969) for long-term investors. Further work was done by Fernholz and Shay (1982) and Fernholz (2002). Luenberger (1998) presents a clear discussion. We illustrate how rebalancing the portfolio to a fixed-mix creates excess growth. Suppose that a stock price process P_t is lognormal so it can be represented by the equation

$$dP_t = \alpha P_t + \sigma P_t dz_t, \tag{12.1}$$

where α is the rate of return of P_t and σ^2 is its variance and z_t is Brownian motion with mean 0 and variance t.

The risk-free asset follows the same price process with rate of return equal to r and standard deviation equal to 0. We represent the price process of risk-free asset by B_t:

$$dB_t = r B_t dt. \tag{12.2}$$

When we integrate (12.1), the resulting stock price process is

$$P_t = P_0 e^{(\alpha-\sigma^2/2)t+\sigma z_t}. \tag{12.3}$$

It is well documented that the growth rate $\gamma = \alpha - \sigma^2/2$ is the most relevant measure for long-run performance. For simplicity, we assume equality of growth rates across all assets. This assumption is not required for generating excess growth, but it makes the illustration easier to understand.

Next, let us assume that the market consists of n stocks, each with stock price processes $P_{1t}, \ldots, P_{nt}$ following the lognormal process. A fixed-mix portfolio has a wealth process W_t that can be represented by

$$dW_t/W_t = \eta_1 dP_{1,t}/P_{1,t} + \cdots + \eta_n dP_{n,t}/P_{n,t}, \tag{12.4}$$

where $\eta_1, \ldots, \eta_n$ are the fixed weights given to each stock (proportion of capital allocated to each stock). In this case, the weights sum up to one

$$\sum_{i=1}^{n} \eta_i = 1. \tag{12.5}$$

The fixed-mix strategy in continuous time always applies the same weights to stocks over time. The instantaneous rate of return of the fixed-mix portfolio at anytime is the weighted average of the instantaneous rates of returns of the stocks in the portfolio.

In contrast, a buy-and-hold portfolio is one where there is no rebalancing and therefore the number of shares for each stock does not change over time. This portfolio can be represented by the wealth process W_t:

$$dW_t = m_1 dP_{1,t} + \cdots + m_n dP_{n,t}, \tag{12.6}$$

where $m_1, \ldots, m_n$ depict the number of shares for each stock.

Again for simplicity, let us assume that there is one stock and a risk-free instrument in the market. This case is sufficient to demonstrate the concept of excess growth in a fixed-mix portfolio as originally presented in Fernholz and Shay (1982). Assume that we invest η portion of our wealth in the stock and the rest $(1-\eta)$ in the risk-free asset. Then the wealth process W_t with these constant weights over time can be expressed as

$$dW_t/W_t = \eta dP_t/P_t + (1-\eta)dB_t/B_t, \tag{12.7}$$

where P_t is the stock price process and B_t is the risk-free asset.

When we substitute the dynamic equations for P_t and B_t, we get

$$dW_t/W_t = (r + \eta(\alpha - r))dt + \eta\sigma dz_t. \tag{12.8}$$

Assuming the growth rate of all assets in the ideal market should be the same over long-time periods, the growth rate of the stock and the risk-free asset should be equal. Hence

$$\alpha - \sigma^2/2 = r. \tag{12.9}$$

From (12.8), we can see that the rate of return of the portfolio, α_W, is

$$\alpha_W = r + \eta(\alpha - r). \tag{12.10}$$

By using (12.9), this rate of return is equal to

$$\alpha_W = r + \eta\sigma^2/2. \tag{12.11}$$

The variance of the resulting portfolio is

$$\sigma_W^2 = \eta^2\sigma^2. \tag{12.12}$$

Hence the growth rate of the fixed-mix portfolio becomes

$$\gamma_W = \alpha_W - \sigma_W/2 = r + (\eta - \eta^2)\sigma^2/2. \tag{12.13}$$

This quantity is greater than r for $0 < \eta < 1$. As it is greater than r, which is the growth rate of individual assets, the portfolio growth rate has an excess component, which is $(\eta - \eta^2)\sigma^2/2$. Excess growth is due to rebalancing the portfolio constantly to a fixed mix. The strategy moves capital out of stock when it performs well and moves capital into stock when it performs poorly. By moving capital between the two assets in the portfolio, a higher growth rate than each individual asset is achievable. It can be shown that the buy-and-hold investor with equal returning assets lacks the excess growth component. Therefore, buy-and-hold portfolios under-perform fixed-mix portfolios in various cases. We can easily see that the excess growth component is larger when σ takes a higher value.

12.2.2 Examples of Fixed-Mix Policy Rules

Many investors have applied versions of the fixed-mix rules with practical successes. For example, the famous 60–40 norm (60% equity and 40% bonds) falls under this policy. Here, at each period, we rebalance the portfolio to 60% equity and 40% bond. Another good example is the S&P 500 equal-weighted index (S&P EWI) by Rydex Investments (Mulvey 2005). As opposed to traditional cap-weighted S&P 500 index, stocks have the same weight (1/500) and the index is rebalanced semi-annually to maintain the weights over time. To illustrate the benefits of applying the fixed-mix policy rule, during 1994–2005, S&P EWI achieved 2% excess return with only 0.6% extra volatility compared to the S&P 500 index. Figure 12.2 illustrates log-prices of S&P 500 and S&P EWI for last 4 years.[1]

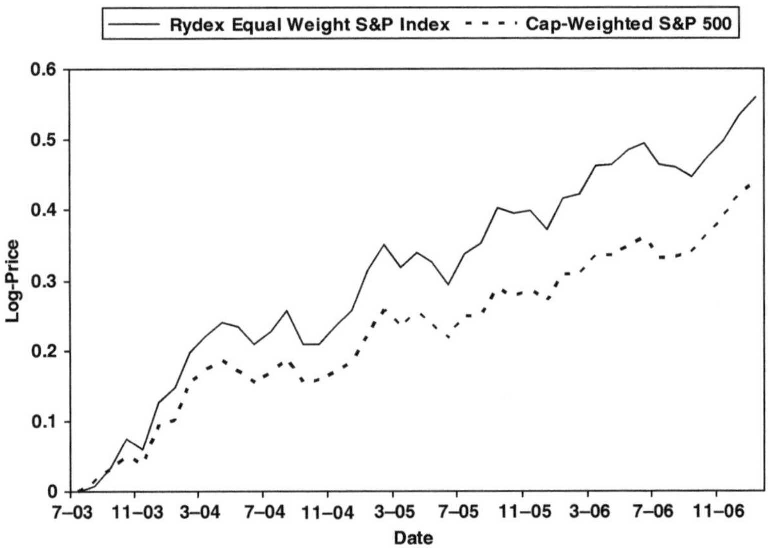

Fig. 12.2 Log-prices of S&P 500 index and S&P EWI during July 2003–December 2006. Each index is scaled to have a log-price of 0 at the beginning of the sample period. In terms of the total return, the S&P EWI outperformed the S&P 500 index for the last 4 years and this performance difference between the two assets can be interpreted as a rebalancing gain due to the fixed mix policy rule

[1] The advantages of the equal weighted S&P 500 index is partially due to rebalancing gains and partially due to the higher performance of midsize companies over the discussed period.

Another significant example involves the Mount Lucas Management (MLM) Index (Mulvey et al. 2004a). It is an equally weighted, monthly rebalanced investment in 25 futures contracts in commodity, fixed income, and currency markets. Briefly, the monthly positions (long or short) are determined by trend-following strategies. The total return of the MLM index can be decomposed into three parts. The first one is the T-bill return gained from the capital allocated for margin requirements. The second component is generated by trend following the futures prices. The third component, rebalancing gains, is earned when all markets are invested with equal weights at the beginning of each month. If trend following strategy had been applied to all the markets without reweighing at each month, then there would be no rebalancing gains. Figure 12.3 shows how those three components affected the total return of the MLM index for selected time periods. Trend following has underperformed its long-term averages for the last several years. Still, rebalancing provided positive returns over the recent period. These empirical results show how periodic reallocation of capital among diverse markets boosts the performance of a long-term investment strategy with the contribution of rebalancing gains.

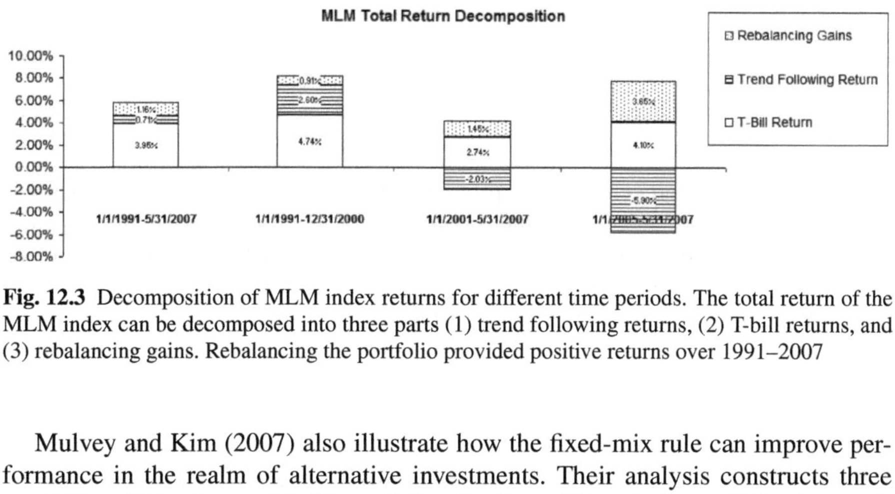

Fig. 12.3 Decomposition of MLM index returns for different time periods. The total return of the MLM index can be decomposed into three parts (1) trend following returns, (2) T-bill returns, and (3) rebalancing gains. Rebalancing the portfolio provided positive returns over 1991–2007

Mulvey and Kim (2007) also illustrate how the fixed-mix rule can improve performance in the realm of alternative investments. Their analysis constructs three portfolios: (P1) a buy-and-hold portfolio of only traditional assets, (P2) a buy-and-hold portfolio of traditional and alternative assets, (P3) a fixed-mix portfolio of both traditional and alternative assets. In this regard, they employ the fixed-mix rule at two levels. First, at the stock selection level, they substitute an equal-weighted S&P 500 index for the capital-weighted S&P 500 fund. The equal weighted index has generated better performance over the standard S&P 500 index, as would be expected due to the additional returns gotten from rebalancing the mix. Then, the portfolio is rebalanced monthly to fulfill the fixed mix policy rule at the asset selection level. Figure 12.4 illustrates the investment performance of each of three portfolios. Note that several degrees of leverage—at several values: 20, 50, and 100%—are applied to each portfolio. From the figure, it is evident that applying fixed-mix rules can improve investment performance.

Portfolio	Description	Constituents
P1	Traditional Assets Only (Buy-and-Hold)	Traditional assets: SP500, LB Bond, EAFE, NAREIT, GSCI, STRIPS
P2	With Alternative Assets (Buy-and-Hold)	Traditional assets: SP500, LB Bond, EAFE, NAREIT, GSCI, STRIPS Alternative assets: Man Futures, Hedge Fund Ind., L/S Ind., Currency
P3	With Alternative Assets (Fixed Mix)	Traditional assets: *SP EWI*, *LB* Bond, EAFE, NAREIT, GSCI, STRIPS Alternative assets: Man Futures, Hedge Fund Ind., L/S Ind., Currency

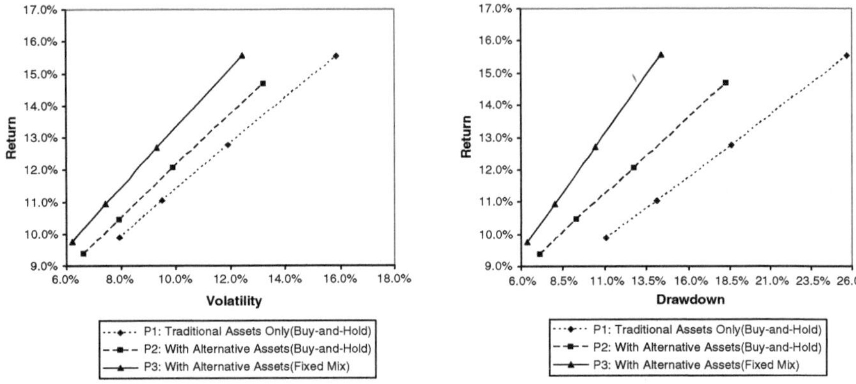

Fig. 12.4 Efficient frontiers of the portfolios with/without alternative assets. The *left figure* illustrates efficient frontiers in the volatility-return plane, while the *right one* is drawn in the maximum drawdown-return plane. The efficient frontier of P3 contains those of P1 and P2 in both cases, which clearly exhibit the role of alternative assets in portfolio construction

The fixed-mix approach can also provide solid investment performance when it is applied to another set of investment strategies. A good example is the dynamic diversification strategy, which is a fixed-mix portfolio consisting of various momentum strategies in several stock markets, suggested by Mulvey and Kim (2008).

Next, we develop a new decision rule based on momentum in equity markets. This rule is applied in conjunction with fixed-mix within a portfolio context. There is a considerable amount of literature on the predictability of stock returns based on its past performance. De Bondt and Thaler (1985, 1987) argue that the "loser" portfolio shows superior performance to the "winner" portfolio over the following 3–5 years. Lehmann (1990) and Jegadeesh (1990) illustrate short-term return reversals of "winner" and "loser" stocks. Also, Jegadeesh and Titman (1993) report the momentum of stock returns for the intermediate term (3–12 months). Rouwenhorst (1998), Kang et al. (2002), Demir et al. (2004), and Cleary and Inglis (1998) provide the evidence of the profitability of the momentum strategy in overseas stock markets. Among many variants of the momentum strategies such as the 52-week high momentum portfolio (George and Hwang 2004) and the market index momentum portfolio (Chan et al. 1996), we examine the industry-level momentum strategy.

To construct a "dynamic diversification" portfolio of equity positions, we first generate industry-level momentum strategies that follow the rules of the relative strength portfolio of Jegadeesh and Titman (1993). The implemented algorithm applies the following rules: At the beginning of the sample period, all industries defined as in Datastream service with a return history of at least t_2-month are ranked based on their past t_1- and t_2-month returns ($t_1 < t_2$). Then industries in the first decile (four industries) from each t_1 and t_2 ranking lists are chosen to be put into the portfolio with the equal weight (eight industries total). If an industry is shown in the top decile for both past t_1 and t_2 months, we place a double-weight on it. The equal-weighted portfolio is held for the given holding period (H) without any rebalancing and we repeat the industry-selecting process after H-month. We construct the momentum strategies with six different settings for (t_1, t_2) observation length pair—(3, 6), (3, 9), (3, 12), (6, 12), (6, 18), and (12, 18)—and four different holding periods—3, 6, 9, 12. In our empirical study, we apply a scoring approach based on recent historical returns. These rules are applied in five nonoverlapping regions—the USA, EU, Europe except EU, Japan and Asia except Japan—to potentially reduce the correlations over 1980–2007. Then, for each of four holding periods, we construct the dynamic diversification portfolio by applying the fixed-mix policy rule to thirty strategies.

Here, we focus on industry-level data for several reasons: (1) the globalization of trade and money flows leads many firms to broaden their exposure to international markets. These firms conduct their own diversification by a variety of approaches, such as moving production/manufacturing out of the home country, or international mergers. An example is the purchase of Chrysler by Mercedes-Benz. The combined firm fits an industrial designation, but is difficult to categorize within a single country. (Germany or the USA.?) (2) A second reason to focus on industries (rather than individual stocks) is to broaden the scope of momentum for institutional investors; there is greater capacity if the core assets depict industries as compared with individual stocks. (3) We believe that future asset allocation studies (as well as asset-liability management) will need to include industry exposure and concentration in order to provide improved risk analytics. (4) With introduction of alternative asset classes, such as ETFs, which enable us to trade industries easily, industries chosen by the momentum strategy have potential to be vehicles for international diversification.

Figure 12.5 illustrates the investment performance of the dynamic diversification portfolio at several leverage levels. Not only the market indices but also the buy-and-hold portfolio is located on the right-bottom region compared to the dynamic diversification portfolio. In fact, among 30 industry-level momentum strategies that are utilized for the dynamic diversification portfolio, only one setting is located on the left-top part, which clearly indicates the robustness of its investment performance (see Fig. 12.7).

12.2.3 Practical Issues Regarding Fixed-Mix Rules

From the derivation of the rebalancing gain, the desirable properties of assets in order to achieve a rebalancing gain can be summarized as follows. First, suppose two assets in the derivation above are perfectly correlated. Then, it can be easily shown that the rebalancing gain is zero. From this, it is evident that diversification

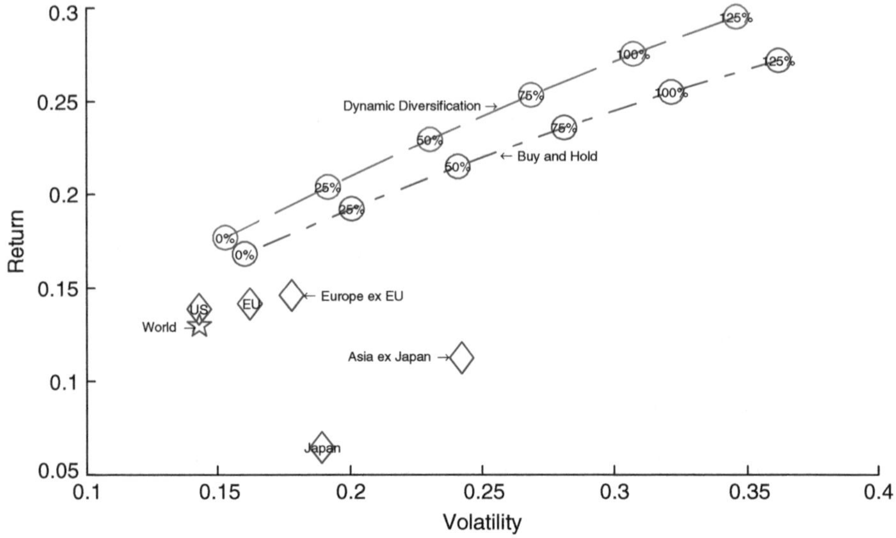

Fig. 12.5 Performance of the dynamic diversification portfolio with leverage. The dynamic diversification portfolio is an equally weighted fixed-mix portfolio of 30 momentum strategies—five regions, six settings. Each number next to a point on the line represents leverage. The 3-month US T-bill is used. The sample period is 1980–2006

among assets plays a major role to achieve an excess growth rate. This observation suggests that dynamic diversification is essential in order to produce extra gains via multiperiod approaches. Also, as always, diversification provides a source of reducing portfolio risk. Second, given a set of independent assets, the rebalancing gain ($(\eta - \eta^2)\sigma^2/2$) increases as the volatilities of assets increase. To benefit from rebalancing gain, the volatility of each asset should be reasonably high. In this context, the traditional Sharpe ratio might not be a good measure for individual assets in terms of multiperiod portfolio management, even though it is still valid at the portfolio level. Additionally, low transaction costs (fees, taxes, etc.) are desirable, since applying the fixed-mix policy rule requires portfolio rebalancing. In summary, the properties of the best ingredients for the fixed-mix rule are (1) relatively good performance (positive expected return); (2) relatively low correlations among assets; (3) reasonably high volatility; and (4) low transaction costs.

However, potential obstacles related to the fixed-mix approach should be addressed. First, it is becoming harder to locate independent or low correlated assets. For instance, oil and corn, which were once relatively independent, are now highly correlated, because ethanol is manufactured from corn. Also, due to globalization, the correlation of assets across countries is becoming higher. Second, even if we are successful in finding a set of independent assets with positive returns and high volatilities, independence is likely to disappear under extreme conditions; there is considerable evidence that stock correlations dramatically increase when the market crashes. Furthermore, it is well known that stock returns and volatilities are negatively correlated. Third, since the fixed-mix model requires portfolio rebalancing, one must consider transaction costs, such as capital gain taxes. Such costs not only deteriorate the investment performance but also make it harder to implement the model. These circumstances suggest that other policy rules will outperform the fixed-mix rule. In the next section, we take up a systematic approach for discovering and evaluating new policy rules.

12.3 Application of the Dual Strategy to ALM

In this section, we suggest that the dual strategy provides an ideal framework for assisting investors who must address future liabilities and goals. A prominent example involves the pension arena. Both the traditional defined-benefit (DB) and the growing defined-contribution (DC) plans have been successfully posed as multiperiod models. The generic domain takes several names, including asset and liability management, dynamic financial analysis (DFA) for insurance companies, and enterprise risk management (ERM) in the corporate finance setting. As an additional advantage, a multiperiod model helps investors understand the pros/cons of combining investment and savings (contribution) strategies.

A number of successful pension planning systems are based on elements of the dual strategy, including the Towers Perrin—Tillinghast system (Mulvey et al. 2000). Also, Mulvey and Simsek (2002) present a stochastic program for the asset allocation of a DB pension plan. This model takes the cash contributions by the sponsoring entity (company or nonprofit agency) as given. Contributions arise from two major sources: plan sponsors own capital and/or possible borrowing. Mulvey et al. (2003b) extend the model by linking the pension plan with its sponsoring company. In addition to the asset allocation decisions, the integrated model addresses: (1) the sponsors contribution policy and (2) the corporate borrowing policy for deciding whether to borrow funds and what proportion of the borrowing should be put into the pension plan. Zhang (2006) extends the integrated model by adding the company's dividend policy as the third corporate decision variable and enhances the corporate model by decomposing the company into a headquarters and several divisions. Earlier, Peskin (1997) argued that the economic cost of a defined-benefit pension plan is the present value of future contributions. In order to achieve savings, he showed that it is necessary for corporate sponsors (1) to integrate asset allocation and contribution policies and (2) to make significant changes in investment policies which include, among others, adopting an appropriate rebalancing rule.

Herein, we provide an example of the defined-benefit pensions for the telecommunication and the automobile sectors to illustrate the idea of the dual strategy. Since the passage of the Employee Retirement Income Security Act (ERISA) in 1974, the DB pension system in the USA has grown from a modest level of assets (less than $100 billion) to over $5 trillion in 2004. Unfortunately, numerous DB pension trusts in the USA (and elsewhere) have seen their adequate surpluses disappear during the period 2000–2003. Simply put, a pension plan is judged to be healthy if the plan's funding ratio (market value of assets to discounted value of liabilities) is greater than 100%. A plan with a low funding ratio, less than 80–90%, is called underfunded, leading to several required steps by the sponsoring company, including typically a contribution to the pension trust. A fully funded trust has a ratio equal to 100%.

The telecommunication sector in the USA possesses a relatively large pension system relative to market capitalization. Moreover, it has been and will likely continue to encounter relatively slow growth. Fortunately, the current overall funding ratio is reasonable roughly 90% in 2005. As an initial step, we set up an anticipatory multistage stochastic program. To simplify the analysis, we treat the sector as a single aggregate company. We employ a scenario generator that has been in use by one of the major actuarial firms for over a decade. The stochastic program consists of 1000 scenarios over a 9-year horizon (several hundred thousand nonlinear variables and constraints). We have found that a stochastic program consisting of 1000 scenarios is a reasonable compromise between model realism and computational costs for pension planning problems. In most cases, we advocate that the planning model be rerun on a recurring basis, at least once per year or more often if large changes take place in the markets. The scenario generator consists of a set of cascading stochastic equations (Mulvey et al. 2000), starting with interest rates in several developed countries.

After studying the solutions of the stochastic program for the telecommunication sector, we discovered a particular policy rule (called conditional ratios) that could be implemented in a Monte Carlo simulation with similar results as the stochastic program across the major objectives. A policy that shifted investments to a more conservative allocation when certain triggers occur had a beneficial impact on the overall condition of the sector and the pension plans. The trigger consists of a combination of funding ratio and the ratio of pension assets to market capitalization. To simplify the example, we set the default investment strategy to the 70/30 fixed mix (70% equity and 30% bonds). The conditional strategy keeps assets in the 70/30 growth mode until a potential problem arises as evaluated by the triggers. In fact, over the 9-year horizon, the conditional strategy not only protects the pension surplus but also reduces the NPV of contributions and maximizes the company's value at the end of the 9-year planning horizon (Table 12.1). In this example, the model develops a sound compromise solution for the diverse stakeholders. However, in general, setting priorities of the multi objectives for the integrated system presents a complex and potentially controversial issue and provides a direction for future research. Yet, it is important to evaluate the problem in its full capability so that the stakeholders will be able to understand the tradeoffs among the objectives.

The second example case study depicts a more desperate situation than the telecommunication industry. Here, we model a hypothetical automobile company. Not only is the company in a slow growth domain with a large pension obligation (relative to market cap), but also is the funding ratio much lower than the telecommunication case, in fact about 70%. We attempt to increase the performance of this company by solving an anticipatory 9-year stochastic program. Unfortunately, the recommended solution does not improve the fortunes of the company and pension system, relative to benchmark strategies such as the 70/30 fixed mix. In many scenarios, the company must make large contributions, weakening the sponsoring company, or it reduces contributions, weakening the pension trust.

Table 12.1 Means and ranges of objective function values for the telecommunication services sector under two investment strategies

	Expected final sector value			Exp. excess contribution penalty function			Probability of insolvency		
	Max	Min	Mean	Max	Min	Mean	Max	Min	Mean
Conditional	372.91	335.34	354.23	7.49	0.75	3.48	0.0590	0.0152	0.0381
Benchmark	372.97	335.47	354.38	10.56	1.09	4.93	0.0672	0.0184	0.0433
	Expected NPV of contributions			Variability of contributions			Downside risk on final funding ratio		
	Max	Min	Mean	Max	Min	Mean	Max	Min	Mean
Conditional	31.76	10.27	20.73	18.34	11.19	15.22	0.0800	0.0570	0.0698
Benchmark	31.69	10.23	20.64	19.02	11.38	15.60	0.0823	0.0580	0.0714

Assisting this company requires enhancing investment performance, relative to that obtainable from traditional assets. We employ the concept of an overlay strategy. In this case, the overlay strategies employ the futures markets and trend-following rules, discussed in Section 12.2. An overlay strategy can be implemented in several ways. First, we might borrow capital as margin requirements for the futures trades. We call this strategy overlays with borrowing (as needed for leveraging the portfolio). Second, we might employ the core assets as margin requirements for the futures markets. We implemented both strategies in the anticipatory planning model. The calibration of return parameters was carried out as before via maximum likelihood estimates based on historical performance. Rather than conducting a full optimization model, we deploy the overlay strategies as fixed-mix additions to the core policy simulations.

The results appear in Fig. 12.6 and in Table 12.2. The first approach (overlays with borrowing) has modest or no impact on overall performance. However, the second strategy improves the long-term objectives for the pension planning problem, such as reducing the net present value and volatility of contribution, while increasing

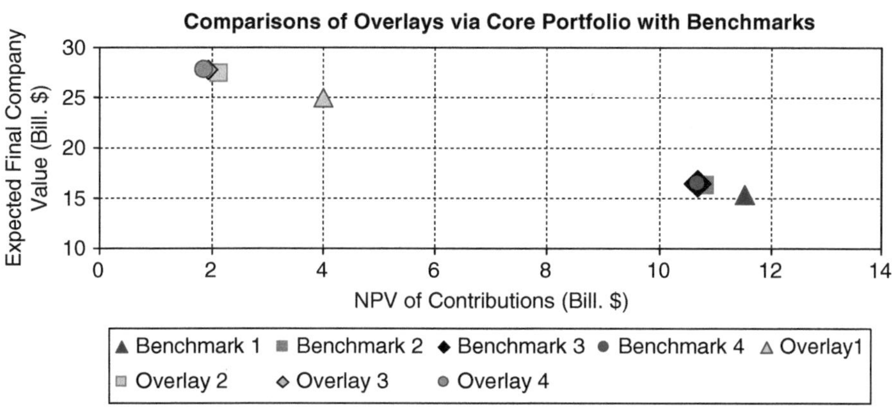

Fig. 12.6 Improvements by employing overlay strategies for an illustrative auto company. For detailed explanation of each strategy, see Table 12.2

Table 12.2 Projections of illustrative auto company under various benchmarks and under two overlay strategies

	Benchmark 1	Benchmark 2	Benchmark 3	Benchmark 4
S&P500	0.4	0.6	0.7	0.8
25-Yr strips	0.6	0.4	0.3	0.2
Commodity	0	0	0	0
Currency	0	0	0	0
Fixed income	0	0	0	0
Expected final company value (Bill. $)	15.331	16.227	16.426	16.435
Expected final plan surpus (Bill. $)	−8280	−8.188	−7.749	−7.027
Expected final funding ratio (%)	92.15	92.38	82.87	93.64
Semi-standard deviation of final funding ratio (%)	6.80	6.75	7.5	8.68
Standard deviation of final funding ratio (%)	10.39	10.01	11.31	13.43
Standard deviation of funding ratio across all periods (%)	12.86	12.75	13.45	16.67
NPV of contributions (Bill. $)	11.528	10.81	10.668	10.678
Volatility of contributions (Bill. $)	3.609	3.533	3.610	3.754
Probability of any excess contribution (%)	49.16	49.68	50.36	51.22
Excess of contribution penalty function (Bill. $)	6.332	6.212	6.418	6.801
Probability of insolvency (%)	45.74	45.88	46.32	46.64
	Overlay 1	Overlay 2	Overlay 3	Overlay 4
S&P500	0.7	0.7	0.7	0.7
25-Yr strips	0.3	0.3	0.3	0.3
Commodity	0.5	0.75	1	1
Currency	0.5	1.25	2	1
Fixed income	0.5	1.75	1	1
	Overlays via core portfolio			
Expected final company value (Bill. $)	24.930	27.419	27.746	27.791
Expected final plan surpus (Bill. $)	1.877	30.287	45.316	28.193
Expected final funding ratio (%)	101.42	123.02	133.48	121.96
Semi-standard deviation of final funding ratio (%)	12.76	33.22	45.69	31.53
Standard deviation of final funding ratio (%)	22.07	58.37	82.21	55.56
Standard deviation of funding ratio across all periods (%)	18.60	40.74	55.04	38.79
NPV of contributions (Bill. $)	3.992	2.126	1.932	1.880
Volatility of contributions (Bill. $)	2.034	1.586	1.523	1.416
Probability of any excess contribution (%)	25.86	13.64	11.46	12.44
Excess of contribution penalty function (Bill. $)	3.484	2.800	2.770	2.524
Probability of insolvency (%)	36.46	33.40	32.98	33.14

Table 12.2 (continued)

	Overlays via borrowing			
Expected final company value (Bill. $)	17.977	14.052	15.158	17.773
Expected final plan surpus (Bill. $)	−6.210	−3.688	0.343	−1.330
Expected final funding ratio (%)	94.14	96.07	99.73	98.37
Semi-standard deviation of final funding ratio (%)	9.87	17.13	21.10	16.69
Standard deviation of final funding ratio (%)	15.76	29.40	39.90	30.43
Standard deviation of funding ratio across all periods (%)	16.09	24.85	30.46	24.52
NPV of contributions (Bill. $)	9.340	12.330	11.58	9.503
Volatility of contributions (Bill. $)	3.506	4.807	4.723	3.937
Probability of any excess contribution (%)	44.24	50.22	45.94	41.36
Excess of contribution penalty function (Bill. $)	6.326	9.697	9.602	7.564
Probability of insolvency (%)	44.40	49.20	47.76	44.64

This table displays a list of summary projection results of the illustrative auto company (Section 12.3) under three set of strategies: (1) four fixed-mix strategies as benchmarks with varying degrees of risk aversion; (2) four overlay strategies via the core portfolio (70% in S&P 500 index and 30% in STRIPS); and (3) four overlay strategies via borrowing at the risk-free treasury bill rates. The overlay strategy via the core portfolio outperforms the others on most dimensions

the value of the company at the end of the 9-year horizon. There is a modest increase in volatility over shorter horizons. In summary, the wider diversification and "cheap" leverage available with the second type of overlays improves both the pension trust and the sponsoring company over the long term. The Appendix provides additional evidence of the advantages of the overlay strategies (Mulvey et al. 2006). Also, overlay strategies based on trend following rules have been successfully implemented by multistrategy hedge funds, especially the Mt. Lucas Management Company. For a complete discussion regarding this subject, see Mulvey et al. (2006).

12.4 Conclusions and Future Research

The proposed dual strategy provides benefits over both a standalone stochastic program and a policy simulator. To this point, as a significant advantage, a policy simulation model can include many real-world considerations, such as complex regulations, tax laws, and company specific guidelines. These issues can be difficult to embed within a continuous stochastic program due to nondifferentiability, jump functions, and other complications. On the other hand, a policy simulator requires a predetermined decision/policy rule. Generally, the selection of a policy rule is accomplished by means of long experience and, perhaps, tradition. How can you tell if such a policy rule is the best, relative to other possible rules? The stochastic program (or dynamic program) provides a benchmark for this evaluation. We say that a policy rule that closely approximates the optimal solution values of the stochastic program is "optimal" in so far as it will perform well under the developed restrictive conditions. The dual strategy draws advantages from each of the competing frameworks.

What are promising directions for future research? First, there is much interest in developing a systematic approach for discovering robust policy rules, coming out of the stochastic programming (or stochastic control) solution. We showed that a specific investment rule (conditional ratios) performs quite well, in comparison to alternative rules and to the stochastic program, for the telecommunication industry. As mentioned, we discovered this rule by carefully studying the solution structure of the 5000-scenario tree. However, there are strong advantages, especially as the size of solvable stochastic program grows, to automating the discovery process. Further research is needed on this topic.

A recurring research topic involves the design of efficient algorithms for solving stochastic program and for optimizing policy simulations. The latter is particularly complex due to the presence of noisy data (with sampling and parameter errors). Remember that an objective function value generated from a policy simulator depicts an output from a statistical experiment and thereby possesses sampling errors. An optimizing algorithm must address sampling errors within the search process. Likewise, policy rules by their nature can give rise to nonconvex optimization models. Again, efficient search algorithms are needed.

Last, there has been much recent research involving approximate dynamic programs. Some recent references are White and Sofge (1992), Bertsekas and Tsitsiklis (1996), Sutton and Barto (1998), and Powell (2006). This research may promote further bridges between competing frameworks. The multistage financial planning models are among the most difficult in the realm of scientific computing. Therefore, it seems likely that a synthesis of approaches will be needed to find robust solutions to these important societal problems.

Appendix

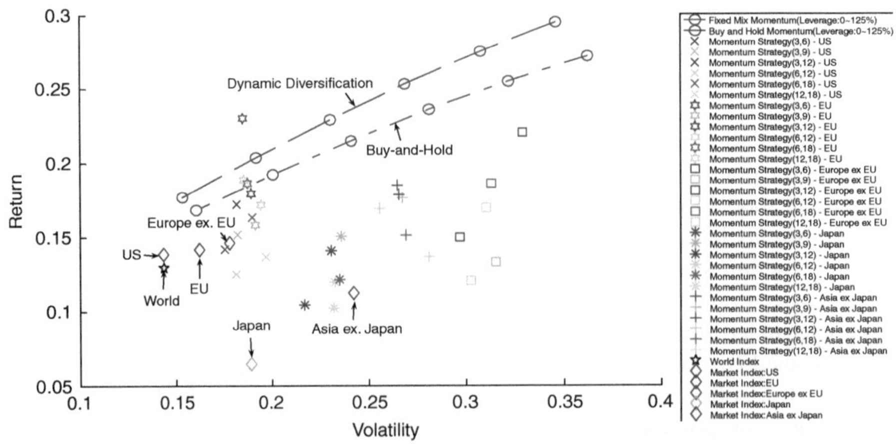

Fig. 12.7 Performance of dynamic diversification portfolio. Dynamic diversification portfolio is an equally weighted fixed mix portfolio of 30 momentum strategies—five regions, six settings. Each number next to a point on the line represents leverage; 3-month US T-bill is used. The sample period is 1980 through 2006

References

Arbeleche, S., Dempster, M.A.H., Medova, E.A., Thompson, G.W.P., Villaverde, M.: Portfolio Management for Pension Funds, vol. 2690. Springer, Berlin (2003)

Bertsekas, D., Tsitsiklis, J.: Neuro-Dynamic Programming. Athena Scientific, Belmont, MA (1996)

Boender, G.C.E., van Aalst, P.C., Heemskerk F.: Modelling and management of assets and liabilities of pension plans in the Netherlands, in w. In: Ziemba, T., Mulvey, J.M. (eds.) Worldwide Asset and Liability Modeling, pp. 561–580. Cambridge University Press, UK (1998)

Campbell, J.Y., Viceira, L.M.: Strategic Asset Allocation. Oxford University Press, New York, NY (2002)

Cariño, D.R., Kent, T., Myers, D.H., Stacy, C., Sylvanus, M., Turner, A., Watanabe, K., Ziemba, W.: The Russell-Yasuda Kasai model: An asset/liability model for a Japanese insurance company using multistage stochastic programming. Interfaces **24**(29), 29–49 (1994)

Chan, L.K.C., Jegadeesh, N., Lakonishok, J.: Momentum strategies. J. Finance **51**(5), 1681–1713 (1996)

Cleary, S., Inglis, M.: Momentum in Canadian stock returns. Revue Canadienne Des Sciences De L'Administration **15**(3), 279–291 (1998)

Consigli, G., Dempster, M.A.H.: Dynamic stochastic programming for asset-liability management. Ann. Oper. Res. **81**, 131–162 (1998)

Dantzig, G.B., Infanger, G.: Multi-stage stochastic linear programs for portfolio optimization. Ann. Oper. Res. **45**, 59–76 (1993)

De Bondt, W.F.M., Thaler, R.: Does stock market overreact? J. Finance **40**(3), 793–805 (1985)

De Bondt, W.F.M., Thaler, R.: Further evidence on investor overreaction and stock market seasonality. J. Finance **42**(3), 557–581 (1987)

Demir, I., Muthuswamy, J., Walter, T.: Momentum returns in Australian equities: The influences of size, risk, liquidity and return computation. Pac.-Basin Finance J. **12**, 143–158 (2004)

Dempster, M.A.H., Germano, M., Medova, E.A., Rietbergen, M.I., Sandrini, F., Scrowston, M.: Managing guarantees. J. Portfolio Manage. **32**, 51–61 (2006)

Dert, C.L.: Asset liability management for pension funds. PhD thesis, Erasmus University, Rotterdam, Netherlands (1995)

Fabozzi, F.J., Focardi, S., Jonas, C.: Can modeling help deal with the pension funding crisis? Working paper, The Intertek Group (2004)

Fernholz, R.: Stochastic Portfolio Theory. Springer, New York, NY (2002)

Fernholz, R., Shay, B.: Stochastic portfolio theory and stock market equilibrium. J. Finance **37**(2), 615–624 (1982)

George, T.J., Hwang, C.: The 52-week high and momentum investing. J. Finance **59**(5), 2145–2176 (2004)

Geyer, A., Herold, W., Kontriner, K., Ziemba, W.T.: The innovest Austrian pension fund financial planning model innoALM. Working paper, University of British Columbia, Vancouver, BC, Canada (2005)

Hilli, P., Koivu, M., Pennanen T., Ranne, A.: A stochastic programming model for asset liability management of a Finnish pension company. Annals of Operations Research **152**(1), 115–139 (2007)

Infanger, G.: Planning Under Uncertainty - Solving Large-Scale Stochastic Linear Programs. The Scientific Press Series, Boyd & Fraser San Francisco, CA (1994)

Jegadeesh, N.: Evidence of predictable behavior of security returns. J. Finance **45**(3), 881–898 (1990)

Jegadeesh, N., Titman, S.: Returns to buying winners and selling losers: Implications for stock market efficiency. J. Finance **48**(1), 65–91 (1993)

Kang, J., Liu, M., Ni, S.X.: Contrarian and momentum strategies in the china stock market: 1993–2000. Pac.-Basin Finance J. **10**, 243–265 (2002)

Kouwenberg, R., Zenios, S.A.: Stochastic programming models for asset liability management. Working Paper 01-01, HERMES Center on Computational Finance, & Economics, University of Cyprus, Nicosia, Cyprus (2001)

Lehmann, B.N.: Fads, martingales, and market efficiency. Quart. J. Econ. **105**(1), 1–28 (1990)

Luenberger, D.: Investment Science. Oxford University Press, New York, NY (1998)

Merton, R.C.: Lifetime portfolio selection under uncertainty: The continuous-time case. Rev. Econ. Stat. **51**(3), 247–257 (1969)

Merton, R.C.: An intertemporal capital asset pricing model. Econometrica **41**(5), 867–887 (1973)

Mulvey, J.M.: Multi-period stochastic optimization models for long-term investors. In: Avellaneda, M., (ed.) Quantitative Analysis in Financial Markets, vol. 3. World Scientific, Singapore (2000)

Mulvey, J.M.: Essential portfolio theory. A rydex investment white paper, Princeton University, Princeton, NJ (2005)

Mulvey, J.M., Kim, W.C.: Constructing a portfolio of industry-level momentum strategies across global equity markets. Princeton university report, Department of OR and Financial Engineering, Princeton University, Princeton, NJ (2007)

Mulvey, J.M., Kim, W.C.: The role of alternative assets for optimal portfolio construction. Wiley, New York, NY (2008) referenced 2007 but appeared in 2008

Mulvey, J.M., Simsek, K.D.: Rebalancing strategies for long-term investors. In: Rustem, B., Kontoghiorghes, E.J., Siokos, S. (eds.) Computational Methods in Decision Making, Economics and Finance: Optimization Models. Kluwer, Netherlands (2002)

Mulvey, J.M., Ziemba, W.T.: Asset and liability management systems for long-term investors. In: Ziemba, W.T., Mulvey, J.M. (eds.) Worldwide Asset and Liability Modeling. Cambridge University Press, Cambridge (1998)

Mulvey, J.M., Gould, G., Morgan, C.: An asset and liability management system for towers perrin-tillinghast. Interfaces **30**, 96–114 (2000)

Mulvey, J.M., Pauling, B., Madey, R.E.: Advantages of multiperiod portfolio models. J. Portfolio Manage. **29**, 35–45 (2003a)

Mulvey, J.M., Simsek, K.D., Pauling, B.: A stochastic network approach for integrated pension and corporate financial planning. In: Nagurney, A. (ed.) Innovations in Financial and Economic Networks. Edward Elgar, UK (2003b)

Mulvey, J.M., Kaul, S.S.N., Simsek, K.D.: Evaluating a trend-following commodity index for multi-period asset allocation. J. Alter. Invest. **7**(1), 54–69 (2004a)

Mulvey, J.M., Simsek, K.D., Zhang, Z., Holmer, M.: Preliminary analysis of defined-benefit pension plans. Princeton university report, Department of OR and Financial Engineering, Princeton University, Princeton, NJ (2004b)

Mulvey, J.M., Ural, C., Zhang, Z.: Optimizing performance for long-term investors: Wide diversification and overlay strategies. Princeton university report, Department of Operations Research and Financial Engineering, Princeton University, Princeton, NJ (2006)

Peskin, M.W.: Asset allocation and funding policy for corporate-sponsored defined-benefit pension plans. J. Portfolio Manage. **23**(2), 66–73 (1997)

Pflug, G.C., Swietanowski, A.: Asset-Liability Optimization for Pension Fund Management. Operations Research Proceedings, Springer Verlag pp. 124–135 (2000)

Pflug, G.C., Dockner, E., Swietanowski, A., Moritsch, H.: The aurora financial management system: Model and parallel implementation design. Ann. Oper. Res. **99**, 189–206 (2000)

Powell, W.B.: Approximate Dynamic Programming: Solving the Curses of Dimensionality. Wiley, Hoboken, NJ (2006)

Rouwenhorst, K.G.: International momentum strategies. J. Finance **53**(1), 267–284 (1998)

Rudolf, M., Ziemba, W.T.: Intertemporal asset-liability management. J. Econ. Dynam. Control **28**(4), 975–990 (2004)

Samuelson, P.A.: Lifetime portfolio selection by dynamic stochastic programming. Rev. Econ. Stat. **51**(3), 239–246 (1969)

Sutton, R., Barto, A.: Reinforcement Learning. The MIT Press, Cambridge, MA (1998)

White, D.A., Sofge, D.A. (eds.) Handbook of Intelligent Control. Von Nostrand Reinhold, New York, NY (1992)

Zenios, S.A., Ziemba, W.T. (eds.) Handbook of Asset and Liability Management. Handbooks in Finance vol. 1. Elsevier, Amsterdam (2006)

Zenios, S.A., Ziemba, W.T. (eds.) Handbook of Asset and Liability Management. Handbooks in Finance vol. 2. Elsevier, Amsterdam (2007)

Zhang, Z.: Stochastic optimization for enterprise risk management. PhD thesis, Princeton, NJ (2006)

Ziemba, W.T., Mulvey, J.M. (eds.) Worldwide Asset and Liability Modeling. Cambridge University Press, Cambridge (1998)

Ziemba, W.T.: The Stochastic Programming Approach to Asset-Liability and Wealth Management. AIMR-Blackwell Charlottesville, VA (2003)

Chapter 13
Growth–Security Models and Stochastic Dominance

Leonard C. MacLean, Yonggan Zhao, and William T. Ziemba

13.1 Introduction

In dynamic stochastic systems, the key component is the process that changes the state of the system between time points. If the transition process can be controlled, the state of the system at each time can be manipulated, and various control settings can be compared based on the associated state distributions. The distribution may be at a cross-section in time, i.e., at a planning horizon, or the distribution may be over trajectories across time. Whether it is from the perspective of the initial state and the terminal state or the path from initial to terminal state, the primary feature of the process is the total change in state. That change can be represented by the rate of growth/decay.

The study of growth is a central theme of economic theory. The conversion of inputs (capital, labor) into outputs (products) is the basis of economic growth theory. In financial economics the emphasis is on capital. It is assumed that the prices on financial instruments for raising capital, such as securities and bonds, reflect the value of economic activity. Alternatively economic activity is the force changing the prices on financial instruments over time. In the financial market comprised of various instruments and their trading prices, the individual investor buys and sells assets to create a stream of wealth over time. So parallel with economic growth is individual capital growth. Although the theory of economic growth has been studied and developed extensively, the related theory of capital growth has received limited attention. In this chapter some aspects of capital growth theory are presented.

In capital growth under uncertainty, an investor must determine how much capital to invest in riskless and risky instruments at each point in time, with a focus on the trajectory of accumulated capital to a planning horizon. Assuming prices are not affected by individual investments but rather aggregate investments, indi-

L.C. MacLean (✉)
Herbert Lamb Chair School of Business Administration, Dalhousie University,
Halifax, NS, Canada B3H 3J5
e-mail: l.c.maclean@dal.ca

G. Infanger (ed.), *Stochastic Programming*, International Series in Operations Research & Management Science 150, DOI 10.1007/978-1-4419-1642-6_13,

vidual decisions are made based on the projected price process given the history of prices to date. An investment strategy which has generated considerable interest is the growth optimal or Kelly strategy, where the expected logarithm of wealth is maximized (Kelly 1956). Researchers such as Thorp (1975), Hausch et al. (1981), Grauer and Hakansson (1986, 1987), and Mulvey and Vladimirou (1992) have used the optimal growth strategy to compute optimal portfolio weights in multi-asset and worldwide asset allocation problems. The wealth distribution of this strategy has many attractive characteristics (see, e.g., Hakansson 1970, 1971, Markowitz 1976, Ziemba and Ziemba 2007). Some properties, both good and bad, of the Kelly strategy are presented in Table 13.1, see also Ziemba and Ziemba (2007) and MacLean, Thorp and Ziemba (2010) on this.

The optimal growth strategy has many advantages, particularly in the long run. The holdings in risky assets are proportional to the market portfolio, so that the strategy is efficient. However, the literature points to a number of problems with such an aggressive strategy; see MacLean et al. (2011)

Table 13.1 Some properties of optimal growth strategy

Feature	Property	References
	Maximizes rate of growth	Breiman (1961), Algoet and Cover (1988)
	Maximizes median log-wealth	Ethier (2004)
	Minimizes expected time to large goals	Breiman (1961), Algoet and Cover (1988)
	Never risks ruin	Hakansson and Miller (1975)
	Absolute amount invested is monotone in wealth	MacLean et al. (2005)
Good	On average never behind any other investor	Finkelstein and Whitley (1981)
	The chance of being ahead of any other investor at least 0.5	Bell and Cover (1980)
	Wealth pulls way ahead of wealth for other strategies	MacLean et al. (1992)
	Growth optimal policy is myopic	Hakansson (1971)
	Can trade growth for security with power utility or fractional Kelly	MacLean et al. (2005)
	It takes a long time to outperform with high probability	Thorp (1975), Aucamp (1993), Browne (1997)
	The investments are extremely large if risk is low	Ziemba and Hausch (1986)
	The total amount invested swamps the gains	Ethier and Tavare (1983), Griffin (1984)
Bad	There is overinvesting when returns are estimated with error	Rogers (2000), MacLean et al. (2007)
	The average return converges to half the return from optimal expected wealth	Ethier and Tavare (1983)
	Levered strategies (e.g., double Kelly) can reduce growth rate to zero	Stutzer (2000), Ziemba (2003)
	The chances of losing considerable wealth in the short term (drawdown) can be high	Ziemba and Hausch (1986)

(1) The fallback in capital in the short to medium term may be unsustainable, i.e., capital could drop below an established operating minimum.
(2) The mean estimates (forecasts) for asset prices may be poor. This could result from new market dynamics or simply the volatility of markets. Errors in forecasts will translate into misdirected strategies which are suboptimal and inefficient. The risk of poor returns from price estimation errors can be controlled, but the decision model needs to be modified.

When the investor prefers less risk than presented by the wealth trajectories of the optimal growth strategy, then this can be reflected in the choice of utility function. If the utility function has an Arrow Pratt risk aversion parameter, e.g., the constant relative risk aversion (CRRA) utility, then the value of this parameter captures risk tolerance, see Kallberg and Ziemba (1983). The risk parameter moderates the volatility risk as well as the estimation risk (MacLean et al. 2004).

Another approach is to use a measure of risk that depends on the investment decision. A standard measure of risk is volatility as defined by the variance of wealth at a point in time or the variance of the passage time to a wealth target. Mean–variance analysis of wealth has been widely used to determine investment strategies, see Markowitz (1952, 1987). An alternative to variance is to use a downside risk measure (Breitmeyer et al. (1999). MacLean et al. (1992) considered, as risk measures, quantiles for wealth, log-wealth, and first passage time in identifying investment strategies that achieve capital growth with a required level of security. Security is defined as controlling downside risk. Growth is traded for security with fractional Kelly strategies. In discrete-time models with general return distributions this strategy is generally suboptimal, but it has attractive wealth/time distribution properties, see MacLean and Ziemba (1999) for extensions of this research.

The most common downside risk measure is value at risk (Jorion 1997). VaR has been studied extensively (Artzner et al. 1999). Basak and Shapiro (2001) consider VaR in a model with CRRA utility, where there is a probabilistic constraint imposed. Although VaR is an industry standard it has weaknesses in controlling risk. The most serious shortcoming is the insensitivity to very large losses which have small probability—the essence of risk. Measures based on lower partial moments (incomplete means) such as CvaR (Rockafellar and Uryasev 2000) and convex risk measures based on target violations (Carino and Ziemba 1998) attempt to deal with this problem.

There is, however, another issue complicating the use of a constraint on a risk measure. The value of the risk measure in practice is an estimate of the true risk and the error of that estimate is very sensitive to errors in the estimation of parameters in the returns distribution. So the objective and the constraint are both affected by the estimation error. Of course, this problem is also a factor in the expected utility setup, with the expectation defined by the estimated distribution for prices. A traditional approach to controlling a system which is deviating from expectations is process control, where control limits on the actual trajectory of the state of a system define boundaries requiring corrective action. MacLean et al. (2006) implemented a process control system in an application to the fundamental problem of investment in stocks, bonds, and cash over time.

In this chapter the traditional capital growth model and modifications to control risk are developed. Parameter estimation and risk control are considered in a Bayesian dynamic model where the filtration and control processes are separate. The model is a generalization to the multi-asset case of the random coefficients model of Browne and Whitt (1996). Given the estimated price dynamics, an investment decision is made to control the path of future wealth. The various approaches to risk control are analyzed in the context of stochastic dominance (Hanoch and Levy 1969b). In practice, an investment portfolio cannot be continuously rebalanced and a realistic approach is to reconsider the investment decision at discrete points in time (Rogers 2000). The time points can be at fixed intervals or random times determined by the wealth trajectory. At each rebalance time, with additional data and a change in wealth, price parameters are re-estimated and a new investment strategy is developed. This dynamic process is illustrated in Fig. 13.1.

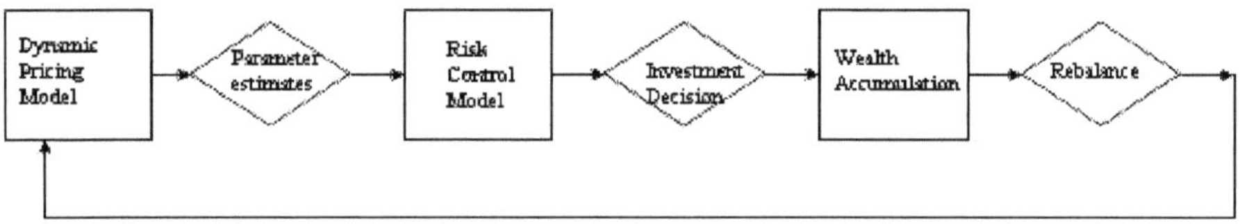

Fig. 13.1 Dynamic investment process

13.2 Capital Accumulation Models

In the dynamic investment process presented in Fig. 13.1, the returns on risky assets, as well as investor preferences for wealth and risk, are the basis for capital allocation decisions. At the time of rebalancing a portfolio, the returns distributions on assets are updated using the history of prices and beliefs/theories about the price process. The concept of considering the future prices conditional on the past and prior beliefs fits naturally into a Bayesian model. The pricing of assets is defined in continuous time, with a discrete-time version following from the continuous-time equations.

13.2.1 Asset Prices

There are m risky assets, with $P_i(t)$ the trading price of asset i, $i = 1, \ldots, m$, at time t, and a riskless asset with rate of return r at time t. The distribution for the stochastic dynamic process $P(t) = (P_1(t), \ldots, P_m(t))'$ is modeled as geometric Brownian motion.

Let $Y_i(t) = \ln P_i(t), i = 0, \ldots, m$. The price dynamics are defined by the stochastic differential equations

$$\begin{aligned} dY_0(t) &= r dt \\ dY_i(t) &= \alpha_i(t) + \delta_i dZ_i, \quad i = 1, \ldots, m, \end{aligned} \tag{13.1}$$

where dZ_i, $i = 1, \ldots, m$, are standard Brownian motions. There are a variety of specifications for the drift $\alpha(t) = (\alpha_1(t), \ldots, \alpha_m(t))'$, the volatility $\Delta = \text{diag}(\delta_1, \ldots, \delta_m)$, and the covariance of log prices $\Sigma(t) = (\sigma_{ij}(t))$. The relevant considerations in the context of capital growth are as follows:

1. At a fixed point in time the past prices are known, and the distribution over future prices is conditioned by the history. It is important, then, to have updating of distributions built into the model.
2. The model is revised at rebalance points, so the current specification is accepted until the next rebalance point.
3. Prices on individual assets are affected by market forces, but the prices also react to forces specific to that asset.

The following assumptions are made:

A1: The $dZ_i, i = 1, \ldots, m$, are independent and represent the variation in price specific to each security.

A2: The instantaneous mean rate of return $\alpha(t) = (\alpha_1(t), \ldots, \alpha_m(t))'$ is a random variable with distribution

$$\alpha_i(t) = \mu_i + \sum_{j=1}^{K} \gamma_{ij} F_j. \tag{13.2}$$

The $F_j, j = 1, \ldots, K$, are independent Gaussian variables, representing the market factors affecting all asset prices. So the rates of return are correlated, with the covariance between α_i and α_j, $\rho_{ij} = \sum_{k=1}^{K} \gamma_{ik}\gamma_{jk}$.

The random rates of return assumption generates a prior distribution over the rates. Let $\Lambda = (\lambda_{ij})$, $\mu = (\mu_1, \ldots, \mu_m)'$, and $\Gamma = \Lambda\Lambda'$. Then the prior for α is $\alpha \propto N(\mu, \Gamma)$. The conditional distribution for the price changes, given α and Δ, is integrated to obtain the conditional distribution of log prices at time t, assuming an initial value y_0. That is, $(Y(t)|\alpha, \Delta) \propto N(y_0 + \alpha t, t\Delta)$. If $\mu_t = y_0 + \mu t$ and $\Sigma_t = t^2\Gamma + t\Delta = \Gamma_t + \Delta_t$, the marginal distribution of log prices at time t is normally distributed as $Y(t) \propto N(\mu_t, \Sigma_t)$.

At a point in time when an investment decision is made, the past prices are assumed to be known and the posterior distribution is the basis of decisions. Consider the data available at time, $\{Y_s, 0 \leq s \leq t\}$, and the corresponding filtration $\Im_t^Y = \sigma(Y_s, 0 \leq s \leq t)$, the σ-algebra generated by the process up to time t. Conditioned on the data, the distribution for the rate of return can be determined from Bayes' theorem. That is,

$$\begin{aligned}
\tilde{\alpha} &= (\alpha|\Im_t^Y) \propto N(\hat{\alpha}_t, \hat{\Gamma}_t), \quad \text{where} \\
\hat{\alpha}_t &= \mu + (I - \Delta_t \Sigma_t)(\bar{Y}_t - \mu), \\
\bar{Y}_t &= \frac{1}{t} Y_t, \\
\hat{\Gamma}_t &= \frac{1}{t^2}(I - \Delta_t \Sigma_t^{-1})\Delta_t.
\end{aligned} \tag{13.3}$$

The Bayes estimate for the rate of return $\hat{\alpha}_t$ is the conditional expectation. This estimate is the minimum mean squared error forecast for the rate given the data. In the context of rebalancing, $\hat{\alpha}_t$ is the planning value for the mean rate of return until the next decision time. Since no information is added in the hold interval, it is also the best estimate throughout the interval.

The posterior distribution for the rate of return demonstrates the procedure for updating. The difficulty with the Bayes estimate is the dependence on the unknown parameters. However, the data can be used to identify values for the parameters. The estimation for the parameters follows a sequence $\hat{\mu} \to \hat{\Gamma} \to \hat{\Delta}$, following from the linear factor model representation. This is demonstrated in the applications.

This continuous-time price model is a natural framework for theoretical analysis of investment decisions. In practice, the information on prices is received at discrete times, so developing the asset prices at regular intervals in time from the stochastic differential equations provides the discrete-time context. Consider the asset prices $P'(k) = (P_1(k), \dots, P_m(k))$, for the times $t + k\nu$, $k = 0, \dots, n$, where ν is a fixed interval width. If $Y(k) = \ln P(k)$ and $E(\nu)$ is the accumulated (integrated) change in log prices over an interval of width ν, then

$$E(\nu) = \nu\alpha + \sqrt{\nu}\Delta Z, \tag{13.4}$$

where $\alpha \propto N(\mu, \Gamma)$ and $Z \propto N(0, I)$.

If $e_k(\nu)$, $k = 0, \dots, n$, are independent observations on $E(\nu)$, then a discrete-time trajectory is $y_t + \sum_{j=0}^{k} e_j(\nu)$, $k = 0, \dots, n$. So the same underlying process generates discrete-time asset prices. This is the basis for estimating the parameters in the continuous-time model from observations at discrete points in time.

13.2.2 Investment and Wealth

Assuming that at a rebalance point the data has been filtered to obtain estimates for the parameters, the decision about how much capital to allocate to the various assets is now considered. The objective is to control the path of accumulated capital generated by the decision and the unfolding asset prices. The decision is developed from the estimated price parameters, whereas capital will accumulate from the true process. Of course, if the estimates used in computing a strategy are substantially in error, then the trajectory of wealth will not proceed as anticipated, but the trajectory may still be under control.

At the decision time t, with $\hat{\alpha}_t$ the forecast for rates of return at time t, and the volatility matrix $\hat{\Delta}$, the forecast dynamics for the conditional price process are

$$dY(t) = \hat{\alpha}_t dt + \hat{\Delta}^{\frac{1}{2}} d\hat{Z}, \tag{13.5}$$

where the innovations process $d\hat{Z}$ is standard Brownian motion.

An investment strategy is a vector process,

$$\{(x_0(t), X(t)), t \geq 0\}, \tag{13.6}$$

where $X(t) = (x_1(t), \ldots, x_m(t))$ and $\sum_{i=0}^{m} x_i(t) = 1$ for any t, with $x_0(t)$ the investment in the risk-free asset. The proportions of wealth invested in the risky assets are unconstrained since can always be chosen, with borrowing or lending, to satisfy the budget constraint.

The forecast change in wealth from an investment decision is determined by the conditional price process dynamics. Suppose the investor at time t has wealth $W(t)$, with allocations of that wealth to assets $X(t)$. Then the forecast for the instantaneous change in wealth is determined by the change in prices as

$$dW(t) = W(t)\left\{\left[X(t)'(\hat{\phi}(t) - re) + r\right]dt + X(t)'\Delta^{\frac{1}{2}}dZ\right\}, \tag{13.7}$$

where $\hat{\phi}_i(t) = \hat{\alpha}_i(t) + \frac{1}{2}\hat{\delta}_i^2$, $i = 1, \ldots, m$.

There is a growth condition required of any feasible investment strategy:

$$\chi_t = \left\{X(t) \middle| \left[X(t)'(\hat{\phi}(t) - re) + r - \frac{1}{2}X(t)'\Delta X(t)\right] \geq 0\right\}. \tag{13.8}$$

With a fixed strategy over the hold interval satisfying the growth condition, the forecast wealth process defined by (13.7) follows geometric Brownian motion such that $W(\tau) \geq 0, \tau \geq t$. If $W(t) = w_t$, then integration of the dynamic equation for wealth from time t to time τ gives the forecast wealth at time τ:

$$\hat{W}(\tau) = w_t \exp\left\{\left[X(t)'(\hat{\phi}(t) - re) + r - X(t)'\Delta X(t)\right](\tau - t) + (\tau - t)^{\frac{1}{2}} X(t)'\Delta Z\right\}. \tag{13.9}$$

In discrete time, the wealth equation is based on the rates of return over the time interval of width ν represented by

$$E = (E_1(\nu), \ldots, E_m(\nu))'(\nu).$$

With $R(s) = (R_1(s), \ldots, R_m(s))'$, where $R_i(s) = \exp(E_{si}(\nu))$ = the rate of return on asset i, and $X(s)$ the investment decision in period s, where period s is the time interval $(t + (s-1)\nu, t + s\nu)$, the wealth after k periods is

$$W(k) = w_t \prod_{s=1}^{k} R'(s)X(s). \tag{13.10}$$

As $\nu \to 0$ the discrete-time wealth converges to the continuous-time formula.

13.3 Growth and Security

The accumulated wealth process from time t, $W(\tau)$, $\tau \geq t$, is defined by the wealth distribution and the dynamics. The decision at time t controls the stochastic dynamic wealth process until a decision is made to revise the strategy. A standard approach in portfolio problems is to set a planning horizon at time T and specify performance criteria for $W(T)$ as the basis of decision. The decision may be reconsidered between time t and T, either at regular intervals or at random times determined by the trajectory of wealth, but the emphasis is on the horizon. However, the path to the horizon is important, and paths with the same terminal wealth can be very different from a risk/survival perspective. Conditions which define acceptable paths can be imposed. For example, drawdown constraints are used to avoid paths with a substantial falloff in wealth (MacLean et al. 2004, Grossman and Zhou 1993).

After the investment decision is taken, it is necessary to monitor the performance of the portfolio in terms of accumulated capital. This is particularly true in the case where the dynamics of prices have changed since the time of decision. The use of control limits to detect unacceptable or out-of-control paths is an additional component in the management of the capital accumulation process. If the capital accumulation process is unacceptable, then corrective action is required. Using information on prices collected since the last decision, new forecasts are developed, an updated investment strategy is calculated (based on the new forecasts and possible financial constraints), and new control limits are proposed.

13.3.1 Stochastic Dominance Measures

The traditional performance criterion for wealth at the planning horizon is expected utility. Wealth $W_1(T)$ dominates $W_2(T)$ if and only if $Eu(W_1(T))$ is greater than or equal to $Eu(W_2(T))$ for every $u \in U$, with strict inequality for at least one utility (Hanoch and Levy 1969a). There are alternative classes of dominance based on the sign of the kth-order derivative of the utility function. Consider

$$U_k = \{u|(-1)^{j-1}u^{(j)} \geq 0, j = 1, \ldots, k\}. \tag{13.11}$$

So U_1, the monotone nondecreasing functions, define first-order dominance-FSD, U_2, the concave, monotone nondecreasing functions, define second-order dominance-SSD, and so on. Since $U_1 \supseteq U_2 \supseteq \cdots$, the dominance orders become more restrictive. The usual class of utility functions is U_2, but certainly U_1 and first-order dominance are common. If $U_\infty = \lim_{j\to\infty} U_j$, then U_∞ contains the set of power utility functions,

$$U_* = \left\{ u \mid u(w) = \frac{1}{c}(w^c - 1),\ c < 1 \right\}, \tag{13.12}$$

with the logarithmic utility function being the limit as $c \to 0$. This special class is important since it captures features such as risk aversion, and positive skewness, and is amenable to analytic results.

Increasingly restrictive orders of dominance are defined by the classes of utilities.

Definition 13.3.1 (kth-order dominance). Wealth $W_1(T)$ kth-order dominates wealth $W_2(T)$, denoted $W_1(T) \succeq_k W_2(T)$, if $Eu(W_1(T)) \geq Eu(W_2(T))$ for all $u \in U_k$, with strict inequality for at least one u.

Another formulation for stochastic dominance follows from the distribution functions for wealth. Let $W_1(T)$ and $W_2(T)$ have densities f_1 and f_2, respectively. For a density f, consider the nested integrations

$$I_k^f(w) = \int_{-\infty}^{w} I_{k-1}^f(z)dz, \tag{13.13}$$

where $I_0 = f$. Then W_1 kth-order dominates W_2 if and only if $I_k^{f_1}(w) \leq I_k^{f_2}(w)$, for all w, with strict inequality for one w.

It is convenient to write the nested interval in terms of loss functions of the wealth distribution. Consider wealth W with distribution F_W. Let

$$\rho_1(\alpha) = \inf\{\eta | F_W(\eta) \geq \alpha\}, \tag{13.14}$$

the αth percentile of the wealth distribution. For $k \geq 2$, define

$$\rho_k^W(\alpha) = \int_{-\infty}^{\rho_{k-1}(\alpha)} (\rho_{k-1}(\alpha) - w)^{k-1} dF_W(w), \tag{13.15}$$

which is a normed measure of loss along the distribution for W. Then W_1 kth-order dominates W_2 if and only if $\rho_k^{W_1}(\alpha) \leq \rho_k^{W_2}(\alpha)$, for all α, with strict inequality for one α.

Proposition 13.1 *Consider alternative terminal wealth variables $W_1(T)$ and $W_2(T)$. The following formulations of kth-order stochastic dominance are equivalent:*

1. *$Eu(W_1) \geq Eu(W_2)$ for $u \in U_k$, with strict inequality for at least one u.*
2. *$I_k^{f_1}(w) \leq I_k^{f_2}(w)$, for all w, with strict inequality for at least one w.*

3. *For* $k \geq 2$, $\rho_k^{W_1}(\alpha) \leq \rho_k^{W_2}(\alpha)$, *for all* α, *with strict inequality for at least one* α.
   For $k = 1$, $\rho_1^{W_1}(\alpha) \geq \rho_1^{W_2}(\alpha)$, *for all* α, *with strict inequality for at least one* α.

Proof The equivalence between the expected utility and the nested integral definitions of stochastic dominance is established in Levy (1998).

For $k = 1$, it follows from definitions that

$$\{I_1^{f_1}(w) \leq I_1^{f_2}(w),\ \forall\alpha\} \Leftrightarrow \{\rho_1^{W_1}(\alpha) \geq \rho_1^{W_2}(\alpha), \forall\alpha\}.$$

For $k \geq 2$, consider the integral

$$I_k^f(w) = \int_{-\infty}^{w} I_{k-1}^f(z)\, dz = [zI_{k-1}^f(z)]_{-\infty}^{w} - \int_{-\infty}^{w} zI_{k-2}^f(z)\, dz,$$

using integration by parts. Iterating the parts operation gives

$$\begin{aligned} I_k^f(w) &= \left[\sum_{j=1}^{k}(-1)^{j-1}\frac{z^j}{j!}I_{k-j}^f(z)\right]_{-\infty}^{w} - \int_{-\infty}^{w}(-1)^{k-1}\frac{z^{k-1}}{(k-1)!}\, dF(z) \\ &= (k-1)!\int_{-\infty}^{w}(w-z)^{k-1}\, dF(z). \end{aligned}$$

Consider

$$I_k^{f_1}(w) = (k-1)!\int_{-\infty}^{w}(w-z)^{k-1}\, dF_1(z)$$

and

$$I_k^{f_2}(w) = (k-1)!\int_{-\infty}^{w}(w-z)^{k-1}\, dF_2(z).$$

Let $w = \rho_{k-1}^{W_1}(\alpha_1) = \rho_{k-1}^{W_2}(\alpha_2)$. Then

$$I_k^{f_1}(w) - I_k^{f_2}(w) = (k-1)!\left[\rho_k^{W_1}(\alpha_1) - \rho_k^{W_2}(\alpha_1)\right] + \int_{\rho_{k-1}^{W_2}(\alpha_1)}^{\rho_{k-1}^{W_2}(\alpha_2)}(w-z)^{k-1}\, dF_2(z)$$

If $I_k^{f_1}(w) - I_k^{f_2}(w) \leq 0$, then $\rho_{k-1}^{W_2}(\alpha_1) \geq \rho_{k-1}^{W_2}(\alpha_2)$. Therefore

$$\left\{I_k^{f_1}(w) - I_k^{f_2}(w) \leq 0, \quad \forall w\right\} \Rightarrow \left\{\rho_k^{W_1}(\alpha_1) - \rho_k^{W_2}(\alpha_1) \leq 0, \quad \forall\alpha_1\right\}.$$

A similar argument yields the opposite direction. □

The variations on the dominance relation provide insight into the conditions for the wealth process. The loss function definition of dominance is useful. It is clear that $\rho_1^W(\alpha), 0 \leq \alpha \leq 1$, is the cumulative distribution, and ordering based on this measure expresses a preference for wealth. Since $\int_{-\infty}^{\rho_1(\alpha)}(\rho_1(\alpha) - w)\, dF(w) = \alpha\rho_1(\alpha) - \int_{-\infty}^{\rho_1(\alpha)} w\, dF(w)$, the second-order measure, $\rho_2^W(\alpha), 0 \leq \alpha \leq 1$, is equivalent to the Lorenz curve (Lorenz 1905): $L(\alpha) = \int_{-\infty}^{\rho_1(\alpha)} w\, dF(w)$ and reflects risk aversion in that it is sensitive to the lower tail of the distribution. The measure $\rho_3^W(\alpha), 0 \leq \alpha \leq 1$, captures the aversion to variance. These characteristics are components of the preferences which investors might express with a utility function.

In general, the ordering of wealth distributions at the horizon using stochastic dominance is not practical. Rather than use the full range of α values with the moments definition, it is more realistic to identify specific values and work with a component-wise order. In particular, selecting two values of α to set up bi-criteria problems $(\rho_k(\alpha_1), \rho_k(\alpha_2))$ has some appeal.

Definition 13.3.2 Consider terminal wealth variables $W_1(T)$ and $W_2(T)$ and values $\alpha_1 < \alpha_2$. Then, for $k \geq 2$, $W_1(T)$ is kth-order (α_1, α_2)-preferred to $W_2(T)$, denoted $W_1(T) \gg_k W_2(T)$, if $(\rho_k^{W_1}(\alpha_1), \rho_k^{W_1}(\alpha_2)) > (\rho_k^{W_2}(\alpha_1), \rho_k^{W_2}(\alpha_2))$. W_1 is first-order (α_1, α_2)-preferred to $W_2(T)$, if and only if $(\rho_1^{W_1}(\alpha_1), \rho_2^{W_1}(\alpha_2)) > (\rho_1^{W_2}(\alpha_1), \rho_1^{W_2}(\alpha_2))$.

The preference ordering is isotonic with respect to the dominance ordering: $W_1(T) \succeq_k W_2(T) \Rightarrow W_1(T) \gg_k W_2(T)$ (Ogryczak and Ruszczyński 2002). The (α_1, α_2) values at which the ρ_k functions are compared have some convention. In Table 13.2, a pair of α values are used with the first-, second-, and third-order dominance relations to generate measures which can be used as a basis of comparing wealth distributions.

Table 13.2 Stochastic dominance measures

Order	α_1	α_2
First	$\rho_1(0.5)$	$\rho_1(0.05)$
Second	$\rho_2(1.0)$	$\rho_2(0.05)$
Third	$\rho_3(1.0)$	$\rho_3(0.5)$

The intention in the table is to capture important aspects of wealth preferences in the choice of α values. So, the first-order pair identifies the median wealth and the fifth percentile. The second-order pair considers the mean and the lower 5% incomplete mean. The third captures the mean squared wealth and the lower semi-mean squared wealth.

The properties of wealth at the horizon are significant, but the path of wealth to the horizon is also important. There are unsustainable wealth levels and losses, so measures that reflect the chance of survival should be included in the decision process. There are two approaches considered: (i) acceptable paths based on the change or transition in wealth; (ii) acceptable paths based on wealth levels. The chance that a path is acceptable defines a path measure. Table 13.3 gives a statement of such measures. The notation τ_w refers to the first passage time to the wealth level w.

Table 13.3 Path measures

Criterion	Specification	Path measure
Change	$b < 0$	$\delta_1(b) = \Pr[\frac{1}{W(\tau)} dW(t + \xi) > b\xi, \xi > 0]$
Levels	$(w_{\mathrm{L}}, w_{\mathrm{U}})$	$\delta_2(w_{\mathrm{L}}, w_{\mathrm{U}}) = Pr[\tau_{w_{\mathrm{U}}} < \tau_{w_{\mathrm{L}}}]$

These measures could be used in place of a moment measure or even in addition to those measures to control the path of wealth.

13.3.2 Wealth Control

The wealth measures in Tables 13.2 and 13.3 provide a bi-criteria framework for determining investment decisions that have a form of optimality referred to as growth–security efficiency. The various measure combinations determine decision models displayed in Table 13.4.

Table 13.4 Alternative decision models

Model	Criterion	Problem
M_1	Expected utility: power utility with risk index $\frac{1}{1-\beta}$	$\max E\left[\frac{1}{1-\beta} W(T)^{\beta}\right]$
M_2	First-order dominance: optimal median wealth subject to a VaR constraint	$\max \left\{\rho_1(0.5) \mid \rho_1(\alpha) \geq \rho_1^*\right\}$
M_3	Second-order dominance: optimal mean wealth subject to a CVaR constraint	$\min \left\{\rho_2(1.0) \mid \rho_2(\alpha) \geq \rho_2^*\right\}$
M_4	Drawdown: optimal median wealth subject to a drawdown constraint	$\max \{\rho_1(0.5) \mid \delta_1(b) \geq 1 - \alpha\}$
M_5	Wealth goals: optimal median wealth subject to control limits.	$\max \{\rho_1(0.5) \mid \delta_2(w_L, w_U) \geq 1 - \alpha\}$

The variety of options for determining an investment policy reflects a focus on specific criteria for performance. Optimizing the median wealth is equivalent to optimizing the geometric mean or rate of growth. This objective is usually stated in terms of the mean log-wealth, which is a special case of the utility function in M_1. For M_3, the relationship $\rho_2(\alpha) = \alpha\rho_1(\alpha) - L(\alpha)$ implies that the objective is to maximize $\rho_2(1.0)$, the mean wealth. Also, optimizing the rate of growth to the horizon is equivalent to minimizing the time to a wealth goal determined from the planning horizon. So the objective in M_5 can be written as $\min\left\{E\tau_{w_{\mathrm{U}}}\right\}$ for appropriate choice of w_{U}.

Another significant factor in these decision problems is the estimates for future returns. If those estimates are substantially in error, then the anticipated wealth and the investment decision may be misdirected. The impact of estimation errors, especially for means, is major and risk aversion dependent (Chopra and Ziemba 1993,

Rogers 2000). A natural way to deal with the uncertain direction of a trajectory of the stochastic dynamic wealth process is to set process control limits as in M_5 and to adjust (update estimates for returns and resolve the growth–security problem) when a limit is reached. The limits can be selected so that they are consistent with the specifications for the growth–security problem for wealth. One effect of the control limits is that decisions are made at random times.

13.4 Efficient Strategies

The problems in Table 13.4 are defined for continuous and discrete-time models. Information on prices is recorded at discrete points in time, and decisions on allocation of capital to investment instruments are taken at discrete points in time. However, wealth accumulates continuously and it is reasonable to analyze capital growth in continuous time. So discrete price data is used to estimate parameters in a continuous-time pricing model, but the forecast growth in capital between decision points is continuous. There are some limitations to this parametric approach. It is possible that there is not a good model for prices, and discrete-time and state scenarios are better able to capture future prospects. Also, certain pseudo-investment instruments such as lotteries and games of chance are discrete by nature. In the analysis of investment strategies in this section, it is assumed that the continuous-time Bayesian pricing model is correct, although knowledge of the model parameters is uncertain.

The class of feasible investment strategies for the single risk-free and m risky opportunities is

$$\chi_t = \left\{ X(t) | \left[X(t)'(\hat{\phi}(t) - re) + r - \frac{1}{2} X(t)' \hat{\Delta} X(t) \right] \geq 0 \right\}, \tag{13.16}$$

where $e' = (1, 1, \cdots, 1)$. There is a special subclass of strategies determined by the optimal expected growth rate strategy, referred to as the Kelly strategy (Kelly 1956). Consider, then, the expected growth rate for wealth, given an investment strategy $X(t) \in \chi_t$, defined as $G(X) = E \ln W(T)^{\frac{1}{T}}$. The Kelly strategy is $X^*(t) = \arg\max G(X)$. In the continuous-time problem, the strategy has the closed form: $X^*(t) = \Delta^{-1}(\hat{\phi}(t) - re)$, where $e' = (1, \ldots, 1)$. The subclass of fractional Kelly strategies is

$$\chi_t^* = \left\{ X(t) | X(t) = f X^*(t), 0 \leq f \leq 1 \right\}. \tag{13.17}$$

Since $X^*(t) \in \chi_t$, then $\chi_t^* \subseteq \chi_t$. The significance of the fractional Kelly strategies lies in their optimality for the problems in Table 13.4, assuming the Bayesian geometric Brownian model for prices is correct.

Proposition 13.2 *(Efficiency). Let $X_{M_j}(t)$ be the optimal solution to growth problem M_j, $j = 1, 2, 3, 5, 6$, defined in Table 13.4. Then $X_{M_j}(t) \in \chi_t^*$, that is, the solution is fractional Kelly.*

Proof The result for most models is known (M_1(Merton 1992); M_2, M_3 (MacLean et al. 2006); and M_5 (MacLean et al. 2005). Consider M_4

$$\max \{\rho_1(0.5) \mid \delta_1(b) > 1 - \alpha\}.$$

Since $W(T)$ is lognormal, the problem is

$$\max \left\{(E \ln(W(T)) \mid \Pr\left[dW(t)/W(t) > b\right] \geq 1 - \alpha)\right\}.$$

Now, $E \ln(W(T)) = \ln(w_t) + \left[X'(\phi - re) + r - 1/2\, X'\Delta X\right]$ and the constraint is satisfied if

$$X_0(\phi - re) + r + z_\alpha (X'\Delta X)^{1/2} = b.$$

Consider the Lagrangian

$$\begin{aligned} L(X, \theta) = &\ln(w_t) + \left[X'(\phi - re) + r - 1/2\, X'\Delta X\right] \\ &+ \theta \left[X'(\phi - re) + r + z_\alpha (X'\Delta X)^{1/2} - b\right]. \end{aligned}$$

The first-order conditions $\frac{\partial}{\partial x_i} L(X, \theta) = 0$ produce the equations

$$(1 + \theta)(\phi_i - r) - \left[1 + \frac{\theta}{(X'\Delta X)^{1/2}}\right] \delta_i^2 x_i = 0, \qquad i = 1, \ldots, k.$$

The solution has the fractional form. □

For the continuous-time formulation, the optimal investment strategies for the various problems have the same form, i.e., the problems have the separation property, where assets are combined in a risk-free fund and a risky fund. However, the actual fraction, which controls the allocation of capital to risky and risk-free instruments, depends on the decision model and parameters. The formulas for the fractions for different models are displayed in Table 13.5. The notations $\tilde{\mu} = (\hat{\phi} - re)' X^*(t) + r$ and $\tilde{\sigma}^2 = X^{*'}(t) \Delta X^*(t)$ are used for the mean and variance of the rate of return on the Kelly strategy. Also y^* is the minimum positive root of the equation $\gamma y^{c+1} - y + (1 - \gamma) = 0$, for $c = \frac{\ln(w_U) - \ln(w_t)}{\ln(w_t) - \ln(w_L)}$.

Although it is not obvious from the formulas, the fractions in Table 13.5 reflect the orderings from the sequence of stochastic dominance relations. So the fraction $f_1 = \frac{1}{1-\beta}$ in M_1 is the most specific, and there is a set of specifications for (ρ_1^*, α) in M_2, e.g., which yield the same fraction, that is, $f_2 = f_1$.

The solutions displayed in Table 13.5 are derived from the continuous-time wealth equation, although the strategies are calculated at discrete decision points

Table 13.5 Investment fractions

Model	Parameters	Fraction
M_1	β	$f_1 = \frac{1}{1-\beta}$
M_2	(ρ_1^*, α)	$f_2 = \frac{B_1+\sqrt{B_1^2+2A_1C_1}}{A_1}$
		$A_1 = (\hat{\phi} - re)'\Delta^{-1}(\hat{\phi} - re)$
		$B_1 = A_1 + z_\alpha\sqrt{\frac{A_1}{T-t}}$
		$C_1 = r - (T-t)^{-1}\ln\frac{\rho_1^*}{w_t}$
M_3	(ρ_2^*, α)	$f_3 = \frac{B_2+\sqrt{B_2^2+2A_2C_2}}{A_2}$
		$A_2 = \left[z_\alpha\frac{\Phi'(z_\alpha)}{\alpha} + \left(\frac{\Phi'(z_\alpha)}{\alpha}\right)^2\right]A_1$
		$B_2 = A_1 - \frac{\Phi'(z_\alpha)}{\alpha}\sqrt{\frac{A_1}{T-t}}$
		$C_2 = r - (T-t)^{-1}\ln\frac{\rho_2^{**}}{w_t}$
M_4	(b, α)	$f_4 = \frac{b}{\tilde{\mu}+Z_\alpha\tilde{\sigma}}$
M_5	(w_L, w_U)	$f_5 = h_t \cdot \tilde{H} + \sqrt{\left[h_t \cdot \tilde{H}\right]^2 + \frac{2rh_t}{\tilde{\sigma}^2}}$
		$\tilde{H} = \frac{\tilde{\mu}-r}{\tilde{\sigma}^2}$
		$h_t = \frac{\ln(w_t)-\ln(w_L)}{\ln(w_t)-\ln(y^*w_L)}$

in time. The alternative problems in Table 13.4 can be based on the discrete-time wealth equation, but the optimal solution is not necessarily fractional Kelly. It is possible to restrict the feasible strategies for the discrete-time problem to the class of fractional strategies, providing sub-optimal, but effective solutions (MacLean et al. 1992).

13.5 The Fundamental Problem

The methods of capital growth are now considered with the fundamental problem of investment in stocks, bonds, and cash over time. Portfolio performance is primarily determined by the strategic allocation to these assets over time (Hensel et al. 1991, Blake et al. 1999). The data for the assets covers the period 1985–2002, with stock returns from the S&P 500 and bond returns from the US Long Government Bond.

13.5.1 Continuous-Time Problem

In Table 13.6 are the means, variances, and covariances of daily returns over the period of the data. The risk-free rate was set at 5% per annum.

With those values as parameters in the price model, daily price trajectories for 1 year were generated. At 10-day intervals along a price trajectory, investment

Table 13.6 Return parameters

Parameter	Stocks	Bonds	Cash
Mean	0.00050	0.00031	1
Variance	0.00062	0.00035	0
Covariance	0.000046		

decisions were calculated based on the model M_2. That is, the median wealth was maximized, subject to a VaR constraint. The median wealth objective is equivalent to a mean log-wealth objective. The solutions come from the fractional Kelly formulas. The price history was used to estimate the price parameters. This price generation and investment process were repeated 1000 times. The expected wealth at the end of a 1-year planning horizon is shown in Table 13.7 for a range of values for parameters (ρ_1^*, α) in the VaR constraint.

Table 13.7 Expected terminal wealth

	ρ_1^*				
α	0.95	0.96	0.97	0.98	0.99
0.01	1.1292	1.1053	1.0881	1.0764	1.0629
0.02	1.1509	1.1189	1.0953	1.0797	1.0649
0.03	1.1703	1.1310	1.1019	1.0826	1.0666
0.04	1.1895	1.1432	1.1084	1.0854	1.0681
0.05	1.2091	1.1557	1.1154	1.0884	1.0696

The VaR constraint has a significant effect on wealth. With the value at risk level set at 95% of initial wealth, the increase in wealth is good and grows as the probability level drops. With a greater VaR level, the growth in capital is somewhat less. Of course, as the VaR requirements are tightened, the security is increased. So there is a trade-off between growth and security.

13.5.2 Discrete-Time Problem

For comparison the problem M_2 was implemented in discrete time, with years as the time units. The statistics on annual returns from the data are provided in Table 13.8.

Table 13.8 Annual return statistics

Parameter	Stocks	Bonds	Cash
Mean	0.08750	0.0375	0
Variance	0.1236	0.0597	0
Correlation	0.32		

A corresponding set of scenarios was created (sampling from a lognormal distribution for stocks and bonds), as displayed in Table 13.9. The sampling process is structured so that sample statistics are as close as possible to the statistics in Table 13.8 (MacLean et al. 2004).

Table 13.9 Return scenarios

Stocks	Bonds	Cash	Probability
0.95	1.015	1	0.25
1.065	1.100	1	0.25
1.085	0.965	1	0.25
1.250	1.070	1	0.25

The planning horizon was set at 3 years, and the same scenarios were used each year. So there were 64 scenarios, each with a probability of 1/64. With this discrete-time and discrete-scenario formulation, problem M_2 was solved with $\alpha = 0.01$ and a variety of values for ρ_1^*. Starting wealth is \$1. The results from solving the problems are shown in Table 13.10. (Details on this problem are in MacLean et al. 2004.) If the annual returns are compared to the results for the continuous-time problem with the same α, the returns in this case are slightly lower. However, the continuous-time application involved rebalancing every 10 days. In Table 13.10, it can be observed that the very strict VaR condition almost eliminates the possibility of growth. As well, the optimal strategy is not fractional, with the investment mix changing as the horizon approaches.

Table 13.10 Investment strategy and rate of return

	Year 1			Year 2			Year 3			
ρ_1^*	Stocks	Bonds	Cash	Stocks	Bonds	Cash	Stocks	Bonds	Cash	Yearly return
0.950	1	0	0	0.492	0.508	0	0.492	0.508	0	1.061
0.970	1	0	0	0.333	0.667	0	0.333	0.667	0	1.057
0.990	0.456	0.544	0	0.270	0.730	0	0.270	0.730	0	1.041
0.995	0.270	0.730	0	0.219	0.590	0.191	0.218	0.590	0.192	1.041
0.999	0.270	0.730	0	0.008	0.020	0.972	0.008	0.020	0.972	1.017

13.6 Conclusion

The accumulation of wealth through investment in a variety of assets is a fundamental problem in investments. In this chapter, the problem with the objective of capital accumulation, without reference to consumption, is discussed. There is no universal solution to the capital growth problem. Within the framework of stochastic dominance, the capital growth problem can be specified as bi-criteria problems in terms of growth and security. A sequence of such problems has been described and solved. In continuous-time models, the solutions each have the important *separation property*. The solution is in two stages: (i) find the optimal growth portfolio; (ii) determine the optimal blend between the optimal growth portfolio and the risk-free portfolio. It is the blend fraction that incorporates the risk tolerance requirements of the investor as expressed in the choice of model and specification of parameter values. The elegance of the continuous-time models lies in the explicit solutions that facilitate any analysis of risk preferences and the sensitivity of solutions to specifications.

The analogous discrete-time models contain the same principles of growth and security, but the approach is computational rather than analytic. In most cases, solving the discrete-time stochastic optimization problems is complicated. A development of a number of discrete-time problems is provided in Ziemba (2003). A compromise is to impose the separation property on the discrete-time problems. The solutions in that case will be analytic (MacLean et al. 1992), and a trade-off between growth and security as the risk specifications are varied can be achieved. This trade-off is effective in the bi-criteria problems (a decrease in one criteria matches an increase in other criteria), but it is not efficient.

References

Algoet, P H., Cover, T.M.: Asymptotic optimality and asymptotic equipartion properties of log-optimum investment. Ann. Probab. **16**, 876–898 (1988)

Artzner, P., Delbaen, F., Eber, J., Heath, D.: Coherent measures of risk. Math. Financ. **9**, 203–228 (1999)

Aucamp, D.: On the extensive number of plays to achieve superior performance with geometric mean strategy. Manage. Sci. **39**, 163–172 (1993)

Basak, S., Shapiro, A.: Value at Risk based risk management:optimal policies and asset prices. Rev. Financ. Stud. **14**, 371–405 (2001)

Bell, R.M., Cover, T.M.: Competitive optimality of logarithmic investment. Math. Oper. Res. **5**, 161–166 (1980)

Blake, D., Lehmann, B.N., Timmermann, A.: Asset allocation dynamics and pension fund performance. J. Bus. **72**, 429–461 (1999)

Breiman, L.: Optimal gambling system for favorable games. Proc. 4th Berkeley Symp. Math. Stat. Probab. **1**, 63–68 (1961)

Breitmeyer, C., Hakenes, H., Pfingsten, A., Rechtien, C.: Learning from poverty measurement: An axiomatic approach to measuring downside risk. Working paper, University of Muenster, Germany (1999)

Browne, S.: Survival and growth with a liability: Optimal portfolio strategies in continuous time. Math. Oper. Res. **22**(2), 468–493 (1997)

Browne, S., Whitt, W.: Portfolio choice and the bayesian kelly criterion. Adv. Appl. Prob. **28**, 1145–1176 (1996)

Carino, D., Ziemba, W.: Formulation of the Russell Yasuda Kasai financial planning model. Oper. Res. **46**, 450–462 (1998)

Chopra, V., Ziemba, W.: The effect of errors in the mean, variance, and covariance estimates on optimal portfolio choice. J. Portfolio Manage. **Winter**, 6–11 (1993)

Ethier, S.N.: The kelly system maximizes median wealth. J. Appl. Probab. **41**, 1230–1236 (2004)

Ethier, S.N., Tavare, S.: The proportional bettor's return on investment. J. Appl. Probab. **20**, 563–573 (1983)

Finkelstein, M., Whitley, R.: Optimal strategies for repeated games. Adv. Appl. Probab. **13**, 415–428 (1981)

Grauer, R.R., Hakansson, N.H.: A half century of returns on levered and unlevered portfolios of stocks, bonds and bills, with and without small stocks. J. Business **59**, 287–318 (1986)

Grauer, R.R., Hakansson, N.H.: Gains from international diversification: 1968-85 returns on portfolios of stocks and bonds. J. Finance **42**, 721–739 (1987)

Griffin, P.A.: Different measures of win rate for optimal proportional betting. Manage. Sci. **30**(12), 1540–1547 (1984)

Grossman, S.J., Zhou, Z.: Optimal investment strategies for controlling drawdowns. Math. Financ. **3**(3), 241–276 (1993)
Hakansson, N.H.: Optimal investment and consumption strategies under risk for a class of utility functions. Econometrica **38**, 587–607 (1970)
Hakansson, N.H.: Capital growth and the mean-variance approach to portfolio selection. J. Financ. Quant. Anal. **6**, 517–557 (1971)
Hakansson, N.H., Miller, B.L.: Compound–return mean–variance efficient portfolios never risk ruin. Manage. Sci. **24**(4), 391–400 (1975)
Hanoch, G., Levy, H.: Effciency analysis of choices involving risk. Rev. Econ. Stud. **36**, 335–346 (1969a)
Hanoch, G., Levy, H.: Efficiency analysis of choices involving risk. Rev. Econ. Stud. **36**, 335–346 (1969b)
Hausch, D.B., Ziemba, W.T., Rubinstein, M.E.: Efficiency of the market for race track betting. Manage. Sci. **27**, 1435–1452 (1981)
Hensel, C.R., Ezra, D.D., Ilkiw, J.H.: The importance of the asset allocation decision. Financ. Anal. J. **July–August**, 65–72 (1991)
Jorion, J.: Value-at-Risk: The Benchmark for Controlling Market Risk. Irwin, Chicago, IL (1997)
Kallberg, J., Ziemba, W.T.: Comparison of alternative utility functions in portfolio selection problems. Manage. Sci. **29**, 1257–1276 (1983)
Kelly, J.: A new interpretation of information rate. Bell Syst. Technol. J. **35**, 917–926 (1956)
Levy, H.: Stochastic Dominance. Kluwer, Norwell, MA (1998)
Lorenz, M.: Methods for measuring the concentration of wealth. J. Am. Stat. Assoc. **9**, 209–219 (1905)
MacLean, L.C., Ziemba, W.T.: Growth versus security tradeoffs in dynamic investment analysis. Ann. Oper. Res. **85**, 193–225 (1999)
MacLean, L.C., Ziemba, W.T., Blazenko, G.: Growth versus security in dynamic investment analysis. Manage. Sci. **38**, 1562–1585 (1992)
MacLean, L.C., Sanegre, R., Zhao, Y., Ziemba, W.: Capital growth with security. J. Econ. Dyn. Control **28**, 937–954 (2004)
MacLean, L.C., Thorp, E.O., Zhao, Y., Ziemba, W.T.: How does the *Fortune's Formula* – Kelly capital growth model perform? *Journal of Portfolio Management, spring,* (2011)
MacLean, L.C., Thorp, E.O., Ziemba, W.T.: Long term capital growth: The good and bad properties of the Kelly and fractional Kelly capital growth criterion. *Quantitative Finance, Vol.* 10, No. 7, 681–687 (2010)
MacLean, L.C., Ziemba, W.T., Li, Y.: Time to wealth goals in capital accumulation. Quant. Finance **5**, 343–357 (2005)
MacLean, L.C., Zhao, Y., Ziemba, W.T.: Dynamic portfolio selection with process control. J. Bank. Finance **30**, 317–339 (2006)
MacLean, L.C., Foster, M.E., Ziemba, W.T.: Covariance complexity and rates of return on assets. J. Bank. Finance **31**, 3503–3523 (2007)
Markowitz, H.M.: Portfolio selection. J. Finance **7**, 77–91 (1952)
Markowitz, H.M.: Investment for the long run: New evidence for an old rule. J. Finance **31**, 1273–1286 (1976)
Markowitz, H.M.: Mean–Variance Analysis in Portfolio Choice and Capital Markets. Basil Blackwell, New York, NY (1987)
Merton, R.C.: Continuous-Time Finance, 2nd ed. Blackwell, New York, NY (1992)
Mulvey, J.M., Vladimirou, H.: Stochastic network programming for financial planning problems. Manage. Sci. **38**(11), 1642–1664 (1992)
Ogryczak, W., Ruszczyński, A.: Dual stochastic dominance and related mean–risk models. SIAM J. Optim. **13**, 60–78 (2002)
Rockafellar, R.T., Uryasev, S.: Optimization of conditional value-at-risk. J. Risk **2**, 21–41 (2000)

Rogers, L.C.G.: The relaxed investor and parameter uncertainty. Financ. Stochast. **5**, 131–154 (2000)

Stutzer, M.: The capital growth-optimal risk tradeoff. University of Iowa, Mimeo (2000)

Thorp, E.O.: Portfolio choice and the kelly criterion. In: Ziemba, W.T., Vickson, R.G. (eds.) Stochastic Optimization Models in Finance, pp. 599–619. Academic, New York, NY (1975)

Ziemba, R.E.S., Ziemba, W.T.: Scenarios for Risk Management and Global Investment Strategies. Wiley, New York, NY (2007)

Ziemba, W.T.: The Stochastic Programming Approach to Asset, Liability, and Wealth Management. AMIR, Virginia (2003)

Ziemba, W.T., Hausch, D.B.: Betting at the Race Track. Dr. Z Investments Inc. San Luis Obispo, CA (1987)

Chapter 14
Production Planning Under Supply and Demand Uncertainty: A Stochastic Programming Approach

Julia L. Higle and Karl G. Kempf

14.1 Introduction

Manufacturing companies that occupy a mid-echelon position in supply–demand networks play multiple roles. From the upstream perspective, they purchase production equipment and materials from suppliers and, consequently, are buyers. From the downstream perspective, they sell products to be used either directly or as components for further production and, consequently, are suppliers. Internally, the company adds value through production.

There are a number of inescapable stochastic processes involved in the dynamics of production within the company. No piece of equipment can be available all of the time, and stochastic availability of production equipment and contention among work in progress leads to stochasticity in throughput times. Similarly, no production process is perfect either in its individual operations or overall. Losses and imperfections in the production process lead to stochasticity in throughput. Practically speaking, once material is released into a manufacturing facility it is output stochastically in time, quantity, and quality.

Perhaps more concerning are the inescapable, and in many cases, uncontrollable stochastic processes involved in downstream demand from the company's customers. Surprise moves by competitors can change a customer's mind even after orders have been placed. Changes in the tastes of the customer's customers and/or problems in the customers's production facilities can lead to requests to adjust orders in the pipeline. In extreme cases, products shipped to a customer can be returned. Practically speaking, given the dynamics of the marketplace, even firm orders are simply forecasts until shipped, received, and invoices paid.

The abstraction of the practical planning and scheduling problem faced by manufacturing firms is the management of the confluence of two stochastic streams—one from production and one due to customers. Stimulated by the early work of Hadley and Whitin (1963), one popular approach to this problem, both in the literature and

J.L. Higle (✉)
Department of Integrated Systems Engineering, The Ohio State University,
Columbus, OH 43210, USA
e-mail: higle.1@osu.edu

G. Infanger (ed.), *Stochastic Programming*, International Series in Operations Research & Management Science 150, DOI 10.1007/978-1-4419-1642-6_14,

in practice, is to focus on safety stock (see, e.g., Zipkin 2000, Porteus 2002, and Axsater 2006). Building safety stock always increases both production and storage costs. In the short term it can increase revenues by buffering variable supply and satisfying stochastic demand. However, safety stock built in excess of actual demand can turn into scrap, wasting material, production costs, and manufacturing capacity that could have been applied to demand for other products.

Much of the relevant literature associated with this approach concerns the characterization of optimal inventory policies. Demand is modeled using random variables that are independent and identically distributed from one period to the next. Bollapragada and Morton (1999) use yield to represent production uncertainty and model it as the product of the executed production level and a random multiplier that is independent of the production level. Such yield models have been reviewed in Yano and Lee (1995). Uncertainty in production is represented in Ciarallo et al. (1994) as random productive capacity, modeled with independent and identically distributed random variables in multi-period settings. Wang and Gerchak (1996) and Erdem and Ozekici (2002) utilize both yield and capacity uncertainty in their treatment of the problem. In general, results in this area hold for single-stage production systems with independent and identically distributed demands and yields.

Influenced by the early work of Arrow and Karlin (1958), another popular approach for managing uncertainty focusses on production planning models. Production planning is generally concerned with using a statement of current and future manufacturing capacity and a forecast of customer demand to generate a capacity allocation that realizes maximum profit (Vollman et al. 2004, Pinedo 2006). Such planning is done strategically outside of equipment acquisition lead time with capacity as the variable to test whether the company's equipment set is appropriate. For example, Kouvelis and Milner (2002) consider the dynamic interplay of demand and supply uncertainty with known distributions and their impact on the strategic decision to expand (contract) internal capacity versus expand (contract) outsourcing. Inside of equipment lead time, production planning focusses on timing material release into factories to provide finished products matching the demand forecast. Capacity responses to variable demand can still be made, but only if overtime or prequalified subcontractors are available. Such a situation where demand is exogenous and uncertain, and capacity during regular production is independent and identically distributed from one period to the next is studied in Duenyas et al. (1997). A review of production planning for systems with stochastic supply and demand is provided in Mula et al. (2006).

Focussing on production scheduling is another popular approach for managing uncertainty (see, e.g., Pinedo 2006, Ovacik and Uzsoy 2006). Production scheduling proceeds tactically with a fixed equipment set. Given the plan for overall material releases and finished goods due dates, as well as the current detailed state of the production equipment, the goal is to choreograph the loading of individual tools. This requires consideration of such details as batch sizes and setups, while making a number of assumptions about equipment availability and yield. When those assumptions are violated during the course of execution, reactive scheduling or schedule repair is necessary to maintain the starts plan and realize the required output (Church and Uzsoy 1992). This may include altering the priority of work in progress or reallocating particular tool capacities (Smith et al. 1992). Both Sethi et al. (2002) and Aytug et al. (2005) have recently provided reviews of production scheduling and control for systems with stochastic supply and demand.

Our work considers uncertain production and demand with a fixed equipment set and without inventory targets. We propose a model that may be viewed simultaneously as a planning model and as a scheduling model. In one sense, we are planning because material release is adjusted in response to changing demand. We respond using the latest set of confirmed orders even though we know that they will change period by period. In another sense, we are scheduling because we react period by period to production execution issues impacting throughput and throughput times. However, our response does not consider such details as batching or setups.

To accomplish this, we provide an integrated model for production planning under uncertainty that simultaneously considers the evolution of demand signals as they progress toward firm orders, as well as the evolution of production yield as lots progress through the facility toward completed product. Our primary motivation comes from semiconductor manufacturing within a supply chain. As such, the timely satisfaction of demand without excessive build-up of inventory is of paramount importance. We emphasize that our primary goal in this chapter is the development of a model that incorporates many of the intricacies that typically thwart the planning process. That is, embedded within a mathematical programming model for the specification of production lots in a multiperiod planning model is a model of a multistage production facility that incorporates line losses that are observable within the facility as well as failures that can only be detected in a test facility after production has been completed. Simultaneously, the model incorporates a stochastic representation of the evolution of demand uncertainty. That is, we incorporate uncertainty in terms of new "demand signals" as well as uncertainty in the nature of changes to signals that have already been received. To the best of our knowledge, our model (which is a multistage stochastic programming model) is the first such model to incorporate these various levels of intricacy. Throughout this chapter, we use data and operating conditions that are consistent with those observed at Intel facilities, albeit in a simplified and masked fashion.

14.2 A Preliminary Planning Model

A model for production planning under uncertainty requires the development of models of the manner in which data evolves over time. In our case, this results in a need for models related to the evolution of production-related data as well as demand-related data. Our model is one in which items progress through various stages of production, ultimately leading to a supply of finished product that can be used to satisfy demand. Progression through any given production stage is influenced by events that are subject to uncertainty. If processing in a given stage is completed in a given time period, then items will progress to the next stage in the subsequent time period—otherwise, they will be retained in the current stage until processing is completed. As a result, the throughput time will vary from one lot to the next.

Yields obtained are also subject to variation of two distinct types. As processing in one stage is completed, some items may be lost or destroyed. Additionally, items that physically survive until the final stage may ultimately prove to be nonfunctional. Flaws of the former type, which can be detected during the manufacturing process, are associated with a quantity known as the "line yield." When combined across all production stages, the line yield can be as low as 95%, indicating a 5% loss of product due to detectable losses during the manufacturing process. Flaws of the latter type are associated with a quantity known as the "die yield" and are not detectable until a final test is undertaken. The die yield can be as low as 85%, indicating a 15% loss of product.

Similar to variations in the production process, the process of realizing demand is also subject to uncertainty. Well in advance of the ultimate due date, customers provide an indication of what their orders will ultimately be. We refer to this as a "demand signal," in keeping with Kempf (2004). Customer demands are dynamic and subject to uncertainty that may best be described as volatile—in many cases even more so than supply. As time passes, customers' assessments of their needs vary so that projected order quantities as well as due dates can change dramatically. Eventually, a firm order (both quantity and due date) is obtained, which must be met from the product supply. Despite the manner in which orders may change after being signaled, customers still require that orders be met within a short period after their eventual due date, even though this date may only be known with limited advance notice. In addition to this supply and demand volatility, technical innovation places a short shelf-life on product that has not been sold within a fixed period after production. It is typically "scrapped," or marketed at a substantially discounted price. In presenting our model, we consider each of these factors in turn.

In this section, our focus is on a deterministic model that we use to motivate the basic conceptual structure of a stochastic model. In Section 14.3, we present the stochastic model, and in particular, discuss the manner in which the necessary data can be modeled.

14.2.1 Supply Progression

We begin with a very basic model of supply progression based on stage-dependent production parameters. This model presumes the availability of data that is, in general, subject to uncertainty. In the following section, we propose a method for modeling this uncertainty.

Let

$P =$ the number of production stages required,

$T =$ the number of time periods under consideration,

$p \in \{0, 1, \ldots, P\}$ denotes a production stage index. By convention, $p = 0$ represents the initiation of a production lot, and $p = P$ represents completed product,

$\alpha_p^t =$ the fraction of items in production stage p at time t that progress to stage $p + 1$ at time $t + 1$,

β_p^t = the line yield from production stage p at time t, so that $1-\beta_p^t$ is the fraction of items lost from stage p,
λ^t = the die yield at time t, so that $1-\lambda^t$ is the fraction of items that are lost from any production that is completed at time t.
x_p^t = the number of items in production stage p at time t.

With this notation, x_0^t corresponds to the size of the lot released into the production facility at time t. Once released into the facility, the lot progresses through the production stages sequentially. That is, supply that is in stage p at time t was either in stage p at time $t-1$ and did not progress or was in stage $p-1$ at time $t-1$ and progressed. Note that in progressing, items are subject to the line yield. Thus, a basic model of supply progression is

$$x_p^t = \beta_{p-1}^{t-1}\alpha_{p-1}^{t-1}x_{p-1}^{t-1} + (1-\alpha_p^{t-1})x_p^{t-1} \qquad t = 1,\ldots,T,\ p = 1,\ldots,P-1. \tag{14.1}$$

Upon completion of production, supply is subject to the die yield:

$$\frac{1}{\lambda^t}x_P^t = \beta_{P-1}^{t-1}\alpha_{P-1}^{t-1}x_{P-1}^{t-1} + (1-\alpha_P^{T-1})x_P^{T-1} \qquad t = 1,\ldots,T \tag{14.2}$$

Restrictions on the accumulation of items in the production facility are imposed through the introduction of a constant work-in-process (ConWIP) constraint, which has been shown to reduce the variability of throughput times (see, e.g., Ignizio 2003):

$$\sum_{p=1}^{P-1} x_p^t \le C \qquad t = 0,\ldots,T \tag{14.3}$$

Of course, we note that $\{x_p^t\}_{p=1}^P$ are essentially determined by factory releases in previous periods and the manner in which lots progress through the facility. Thus, let

$\rho_{p,p'}^t(j)$ = the fraction of items from stage p at time j that is in stage p' at time t,

and note that these quantities can be derived from the original production facility parameters, $\{\alpha_p^t\}$, $\{\beta_p^t\}$, $\{\lambda^t\}$, although the precise derivation can be complex. In any event,

$$x_p^t = \sum_{j=0}^{t-1} \rho_{0p}^{(t)}(j)x_0^j. \tag{14.4}$$

Using the expression (14.4) we replace (14.2) and (14.3), with

$$x_P^t - \lambda_t \sum_{j=0}^{t-1} \rho_{oP}^{(t)}(j) x_0^j = 0, \qquad t = 1, \ldots, T$$

$$\sum_{p=0}^{P-1} \sum_{j=0}^{t-1} \rho_{op}^{(t)}(j) x_0^j \le C \qquad t = 1, \ldots, T. \tag{14.5}$$

Note that constraints (14.3) (or, equivalently, (14.5)) will limit releases into the production facility, $\{x_0^t\}$, in a manner that depends on the specific manner in which supply progresses as indicated by the parameters of the process.

14.2.2 Demand Progression

Our model of "demand" is one in which the demand signal fluctuates from time to time until it is a firm order. That is, an "order" begins as a "demand signal," which we refer to as a nominal forecast. As time progresses, the quantity and due date change until the order is "due." Similar to the supply progression, we begin with a very basic model of demand progression, based on fluctuations that take place while the order is in the system.

Let

F = the number of stages in the demand forecast, a nominal time between when an order is first signaled and when it is actually due.

$f \in \{0, \ldots, F\}$ denotes a "forecast stage" index, representing a nominal time until an order is due. By convention, we will use $f = F$ to represent a newly signaled order and $f = 0$ to represent an order that is currently due.

δ^t_{fn} = denotes the fraction of demand in stage f at time t whose due date changes by n time periods.

γ_f^t = the fraction by which the quantity of demand in forecast stage f at time t changes.

d_f^t = the quantity in forecast stage f at time t.

Note that an order can arrive in forecast stage f at time t through several avenues. For example, an order in forecast stage $f+1$ at time $t-1$ that does not experience a change in due date will be in stage f at time t. Similarly, an order in stage $f+3$ at time $t-1$ whose due date is decreased by two time periods will be in stage f at time t. Of course, quantities change as well as due dates, so that a basic model of demand progression is given by

$$d_f^t = \sum_{s=-2}^{s=2} \delta^{t-1}_{f+1-n,n} d^{t-1}_{f+1-n} (1 + \gamma^{t-1}_{f+1-n}). \tag{14.6}$$

Similar to the supply progression, we note that $\{d_f^t\}$ are determined by the original forecast specifications, $\{d_f^t\}$, and the manner in which the forecast process unfolds as indicated by the parameters of the process.

14.2.3 Unmet Demand and Scrapped Inventory

Once demand has been realized (i.e., when the signal reaches stage 0), it generates a sale, if there is sufficient supply available. Recognizing the impact of customer impatience, we assume that demand that is not satisfied within L periods of its due date is lost. Similarly, once an item has been successfully produced (i.e., it reaches stage P), it can be used to satisfy demand. Recognizing the impact of technological innovation, we assume that finished goods are either sold or scrapped within R days. Under these assumptions, let

s_{tk} = the amount of product finished at t that is allocated for sale at k, $k \in \{t - L, \ldots, t + R\}$,
r_t = the amount of product finished at t that is scrapped,
u_t = the amount of product that is "due" in period t that is not met.

Using $a \wedge b = \text{Min}(a, b)$ and $a \vee b = \text{Max}(a, b)$, unmet demand can be modeled as

$$\sum_{k=(t-R)\vee 0}^{(t+L)\wedge T} s_{kt} + u_t = d_{0t}. \tag{14.7}$$

Similarly, scrapped production can be modeled as

$$\sum_{k=(t-L)\vee 0}^{(t+R)\wedge T} s_{tk} + r_t - x_P^t = 0 \tag{14.8}$$

14.2.4 Objective Function

Let

c_{tk}^s denote the profit per item sold, the objective coefficient on s_{tk},
c_t^r denote the loss per item scrapped, the objective coefficient on r_t,
c_t^u denote the penalty on lost sales, the objective coefficient on u_t.

Assuming a maximization problem, we note that smaller coefficients for c_{kt}^s when $t > k$ can be used to discourage tardiness in satisfying demand. Similarly, smaller coefficients when $t < k$ can be used to discourage "holding" an excess inventory of finished good.

14.2.5 A Deterministic Planning Model

In summary, a basic representation of the model begins as follows:

$$\text{Max} \quad \sum_{t=1}^{T} \left\{ \sum_{k=(t-L)\vee 0}^{(t+R)\wedge T} c^{s}_{tk} s_{tk} - c^{r}_{t} r_t - c^{u}_{t} u_t \right\} \tag{14.9}$$

$$\begin{aligned}
\text{s.t.} \qquad & x^t_P - \lambda_t \sum_{j=0}^{t-1} \rho_{0P}^{t-j}(j) x^j_0 = 0 && t = 1, \dots, T \\
& \sum_{p=0}^{P-1} \sum_{j=0}^{t-1} \rho_{0p}^{t-j}(j) x^j_0 \le C && \\
& \sum_{k=(t-R)\vee 0}^{(t+L)\wedge T} s_{kt} + u_t = d_{0t} && t = 1, \dots, T \\
& \sum_{k=(t-L)\vee 0}^{(t+R)\wedge T} s_{tk} + r_t - x_{Pt} = 0 && t = 1, \dots, T \\
& x^t_0 \ge 0 && t = 1, \dots, T.
\end{aligned}$$

We note that while this preliminary model is relatively simple to understand, it is inadequate for the purpose of planning production in an environment that is fraught with uncertainty. For this purpose, we require a model of the manner in which uncertainty impacts the supply and demand processes. Perhaps more importantly, we need a model that permits factory release decisions to be made on the basis of partial information. That is, as factory-release decisions are made, information on the current status of order forecasts is available, as is information regarding the current status of the production facility. While neither of these are sufficient to determine precisely when orders will be due or when product will become available, they do provide quantifiable insight into when orders might be due, or when product might become available. An extension of the preliminary model that accommodates the uncertainties in the planning environment is necessary. In Section 14.3, we accomplish this by interpreting the "fractions" discussed in Section 14.2 as probabilities and incorporating them into models of appropriate stochastic processes.

14.3 Data Modeling

A model for planning under uncertainty requires the development of models of the manner in which data evolves over time. In our case, this results in a need for models related to the evolution of production-related data as well as demand-related data. We consider each of these in turn.

14.3.1 Production-Related Data

At any point in time, production in any given stage may or may not progress to the next production stage. As a result, the throughput time (the time required to complete production) is random. Typical throughput data, obtained from an Intel production facility appears in Fig. 14.1. This figure illustrates throughput times for 1091 different production lots—the horizontal axis reflects actual throughput times, while the vertical axis indicates the frequency with which they occurred.[1] Note the extreme variability present. In nearly all cases, some delay beyond the nominal throughput of five periods is experienced. Note that in many cases, the throughput time is at least three times the nominal value.

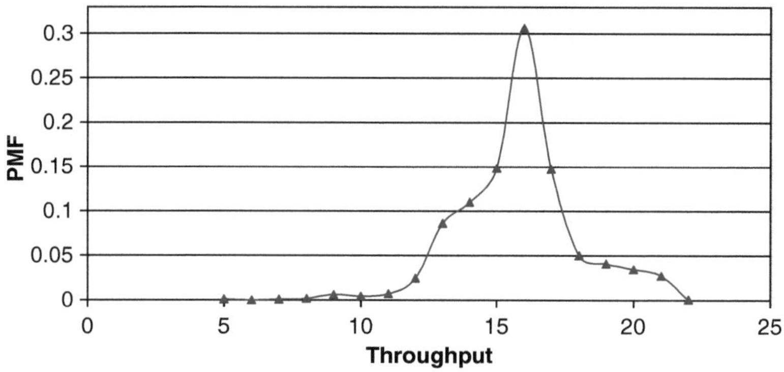

Fig. 14.1 Typical throughput distribution
This figure illustrates throughput times for 1091 different production lots. The horizontal axis reflects actual throughput times, while the vertical axis indicates the frequency with which they occurred. The data depicted has been manipulated in order to be consistent with a model of five production stages. Note the extreme variability in throughput times and the extent to which they reliably exceed five time periods

Our representation of the progression of supply through the various production stages, (14.1), suggests a discrete time Markov chain model. Let α_p represent the probability that an item progresses from stage p to stage $p+1$ in exactly one time period. Combining $\{\alpha_p\}$ with the line yields, $\{\beta_p\}$, we obtain a Markov chain model of the manner in which individual items progress through the production facility, as depicted in Fig. 14.2. Note that we model the completed production as an absorbing state. With this model, we can now compute the distribution of the throughput time associate with any particular values of $\{\alpha_p\}_{p=0}^{P}$ as the distribution of the number of transitions required to reach the absorbing state. A search among the potential

[1] This data depicted has been manipulated in order to be consistent with a model of five production stages.

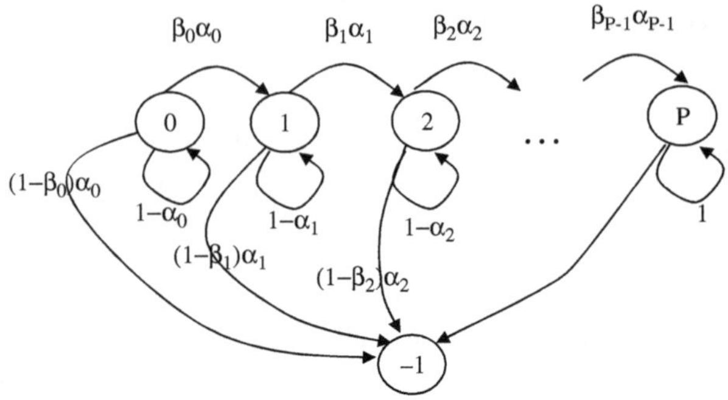

Fig. 14.2 A Markov model for throughput times
Note the existence of two absorbing states: P, corresponding to completed production and -1, corresponding to line loss. A search among the model parameters can be conducted in order to find the values that best fit the indicated throughput distribution

values of $\{\alpha_p, \beta_p\}_{p=0}^{P}$ can be conducted in order to find the values that best fit the throughput distribution indicated in Fig. 14.1.

Let $\mathcal{T}$ denote the one-step transition probability matrix associated with Fig. 14.2. That is, $\mathcal{T} = \mathcal{T}_{ij}$, where

$$\mathcal{T}_{ij} = \begin{cases} (1-\beta_i)\alpha_i & j = -1, i \in \{0, \ldots, P-1\} \\ 1-\alpha_i & j = i, i \in \{0, \ldots, P-1\} \\ \beta_i \alpha_i & j = i+1, i \in \{0, \ldots, P-1\} \\ 1 & i = j = P \;,\; i = j = -1 \\ 0 & \text{otherwise.} \end{cases}$$

Note that with a Markov chain model, we have access to a great deal of useful information regarding the manner in which supply progresses. For example, our preliminary model makes use of "$\rho_{p,p'}^{(t)}(j)$," the fraction of items from stage p in period j that are in stage p' in period t. This is related to a $t-j$-step transition probability. That is, if $\mathcal{T}^{(s)} = \mathcal{T}^s$, the s-step transition probability matrix, then

$$\rho_{p,p'}^{(t)}(j) \quad \text{corresponds to} \quad \mathcal{T}_{p,p'}^{(t-j)} \qquad \forall_j.$$

Consequently, the Markov chain model of supply progression provides access to all probabilistic descriptors for the relevant production parameters. In particular, probabilistic representations of the progression to completion from every state are available. We note that this process of modeling throughput can be applied to any production stage, not just new releases. Thus, for example, if the problem is initialized with some work already in process, complete production from this WIP can be modeled as well.

14.3.2 Demand-Related Data

Modeling the demand-related data is somewhat more involved than modeling the production-related data. An order, in terms of both the quantity and the due date, is signaled well in advance of the anticipated due date. However, customers' evaluations of their own business conditions cause fluctuations in their assessment of their actual needs. This leads to significant fluctuations in their projected due dates and order quantities, which should have a significant impact on production decisions but can be difficult to capture in a model. In this section, we illustrate a model in which due dates may change by "n" periods and order quantities may fluctuate in any time period.

As with the production-related data, we begin with an interpretation of "fractions" as probabilities as we develop a model of the manner in which demand-related data progresses. We begin with a Markov chain model to describe the due date distributions. In this case, we are modeling the evolution of the due date from the time at which the order is first intimated until it is actually due. Based on observations from Intel facilities, in any given period, the probability that the due date changes by n time periods is given by δ_n, as indicated in Table 14.1. A Markov chain model of the due date progression in which changes are possible in any given time period, independent of changes in previous periods is depicted in Fig. 14.3. The corresponding matrix of one-step transition probabilities is given by $\mathcal{D} = [\mathcal{D}_{ij}]$, where

$$\mathcal{D}_{ij} = \begin{cases} 1 & j = 0,\ i = 0 \\ \delta_{-2} + \delta_{-1} + \delta_0 & j = 0,\ i = 1 \\ \delta_{-2} + \delta_{-1} & j = 0,\ i = 2 \\ \delta_2 & j \geq 2,\ i = j - 1 \\ \delta_1 & j \geq 1,\ i = j \\ \delta_0 & j \geq 1,\ i = j + 1 \\ \delta_{-1} & j \geq 1,\ i = j + 2 \\ \delta_{-2} & j \geq 1,\ i = j + 3 \\ 0 & \text{otherwise.} \end{cases}$$

Table 14.1 Fraction of orders whose due dates change by "n" time periods

n	δ_n
−2	0.025
−1	0.075
0	0.800
1	0.075
2	0.025

Due dates are subject to change in any period and may change arbitrarily often, thereby complicating the planning process

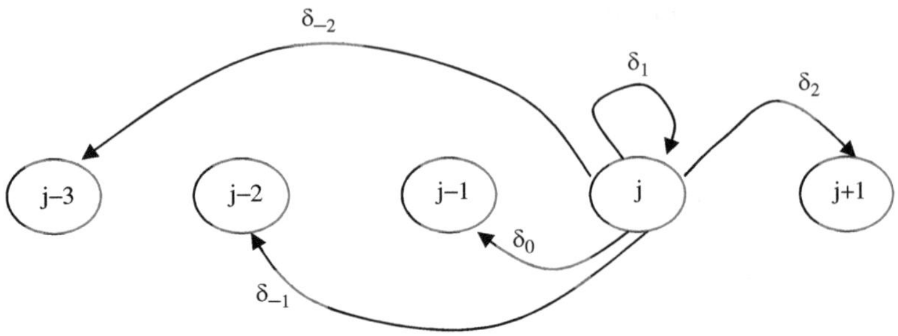

Fig. 14.3 Progression of due dates
Depiction of possible transitions from an arbitrary state j (indicating the order is due in j time periods), as well as transitions near the absorbing state "0." The absorbing state corresponds to an order that is "due"

Note that this Markov chain includes an absorbing state at "0," corresponding to an order being "due." Note also that using the one-step transition matrix $\mathcal{D}$, we can easily compute the distribution of the number of transitions (from any state) until absorption. This corresponds to the distribution of time until an order is actually due, given the current projected due date. In Fig. 14.4, we depict the resulting distribution of actual due dates associated with a "new" order based on a nominal value of $F = 10$. Note the wide variation in actual due dates.

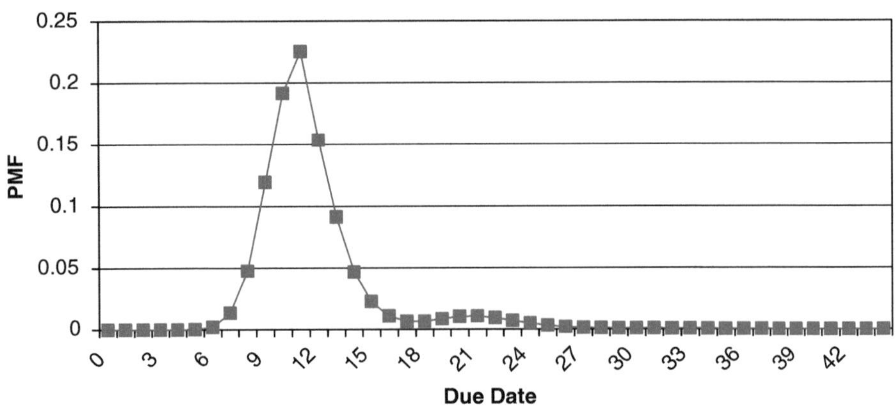

Fig. 14.4 Due date distribution

In addition to changing due dates, the order quantities might change in any period as well. Modeling the evolution of the change in order quantity is somewhat challenging as a result, since the net change will depend on the length of time that the order is in the system. Relying again on observations from Intel facilities, we have the distribution of the change in order quantity in any given period illustrated in Table 14.2.

That is, as an order progresses through the various demand stages (the actual number of which will be random), the order quantity might change several times.

Table 14.2 Fraction of orders whose order quantity changes by the indicated amount

Change in order quantity	γ_n, fraction of orders
−10%	0.025
−5%	0.075
0%	0.800
5%	0.075
10%	0.025

Order quantities are subject to change in any period and may change arbitrarily often. This also serves to complicate the planning process. The specific distribution identified here is illustrative. More elaborate models, in which demand might simply disappear (i.e., an order cancellation) are easily incorporated.

Ultimately, these changes can be accumulated in a factor, which upon multiplication by the original amount indicates the actual quantity demanded.

We again adopt a Markov chain approach to modeling the change in order quantities. However, since the changes are expressed in terms of percentages (leading to a multiplicative model), our focus is on a logarithmic model. Let

Q_t = the order quantity as it appears during the t-th period the order is in the system,
Δ_t = the fractional change in the order during the t-th period,
D = the number of time periods that the order is in the system.

Then

$$Q_t = \left\{ \prod_{k=1}^{t} \Delta_k \right\} Q_0$$

and the actual quantity that will be due is Q_D (recall that D is a random variable). To model this as a Markov chain, we focus on changes of a state that corresponds to the logarithm of the fractional change. Table 14.3 illustrates this relationship.

The natural logarithm of the fractional change on any given day is approximated by values in $\{-0.10, -0.05, 0, 0.05, 0.10\}$. Thus, we consider a Markov chain with the transition structure depicted in Fig. 14.5. The one-step transition associated with this chain is given by $\mathcal{Q} = [\mathcal{Q}_{ij}]$, where

$$\mathcal{Q}_{ij} = \begin{cases} \gamma_{j-1} & i \geq 2,\ j \in \{i-2, i-1, i, i+1, +2\} \\ 0 & \text{otherwise.} \end{cases}$$

Let $\theta_0 = 0$, and let θ_t denote the state of this process after t time transitions. Then we model the evolution of the order quantity as

$$Q_t = e^{\theta_t * 0.05} Q_0.$$

Table 14.3 Multiplicative vs. logarithmic models, changes in state values

Percentage change	Multiplier	ln (multiplier)	Approximate ln (multiplier)
−10%	0.9	−0.105	−0.10
−5%	0.95	−0.05	−0.05
0%	1.0	0	0
5%	1.05	0.0488	0.05
10%	1.1	0.095	0.10

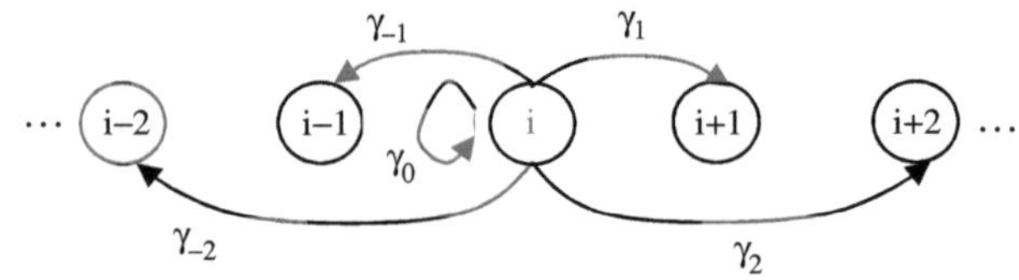

Fig. 14.5 Markov chain model for the change in order quantity
Depiction of the possible shifts in the exponent of the multiplicative change factor, from an arbitrary state i

Note that the t-step transition probability matrix, $Q^{(t)} = Q^t$, can be used to obtain the distribution of Q_t for any t.

With this combination of Markov chain models for the evolution of various data elements, we are able to obtain probabilistic representations of the ultimate due date and quantity associated with an order. That is, once an order is intimated, the distribution of the ultimate due date, D, can be described using the distribution of the time to absorption associated with the Markov chain depicted in Fig. 14.3. Conditional on D, the distribution of the ultimate order quantity can be obtained using the Markov chain model depicted in Fig. 14.5. The marginal distribution of the change in quantity obtained in this fashion is depicted in Fig. 14.6, where we note that quantities vary by as much as 30% in either direction from the nominal due date (which was set at $T = 10$).

It is important to note that this process of obtaining distributional representations of due dates and order quantities can be applied at any time, not just to new orders. If at some point we have information regarding due dates and quantities for orders that are already in the system, distributional representations of the ultimate due dates and order quantities can be obtained.

14.3.3 Decisions Based on Partial Information

In Section 14.3, we discussed probabilistic models for the problem data that is subject to uncertainty. We move now to extend the preliminary model to incorporate decisions based on partial information. That is, we now consider an extension that permits release decisions based on information regarding the state of the production facility as well as the state of orders that are currently in the system.

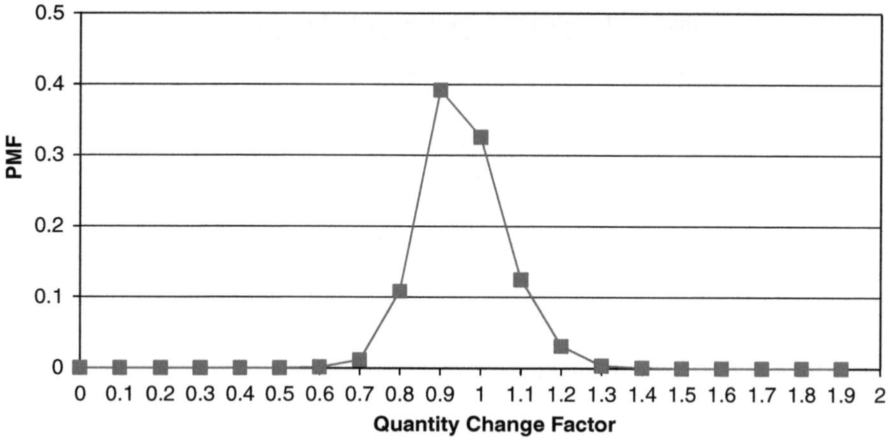

Fig. 14.6 Distribution of changes in order quantity

We begin by distinguishing between "time periods" and "decision stages." A decision stage corresponds to a time at which commitments to a production schedule must be made. For example, production schedules might be determined on a weekly basis, while production progresses through the facility daily. To accommodate this, let S denote the set of decision stages. For each $s \in S$, let $\tau(s)$ denote the time periods associated with the decision stage, and let t_s denote the time period associated with stage s. For example, if there are $T = 20$ time periods and commitments are made every five periods, then $S = \{1, 2, 3, 4\}$ and $\tau(s)$ and t_s are as indicated in Table 14.4. Thus, e.g., release decisions for periods 1–5 are made in the first stage and are based on information regarding the state of the system at time $t = 1$. Our model must be adapted to reflect the state of the system (in terms of the production facility and customer orders) at each decision stage. We begin by explicitly rewriting the formulation (14.9) using a recursion based on the decision stages as follows:

$$\text{Max} \quad \sum_{s \in S} \sum_{t \in \tau(s)} \left\{ \sum_{k=(t-L)\vee 0}^{(t+R)\wedge T} c_{tk}^s s_{tk} + c_t^r r_t + c_t^u u_t \right\} \tag{14.10}$$

$$\text{s.t.} \quad x_P^t - \lambda_t \sum_{j=0}^{t-1} \rho_{0P}^{t-j}(j) x_0^j = 0 \qquad t \in \tau(s),\ s \in S$$

$$\sum_{p=0}^{P-1} \sum_{j=0}^{t-1} \rho_{0p}^{t-j}(j) x_0^j \le C \qquad t \in \tau(s),\ s \in S$$

$$\sum_{k=(t-R)\vee 0}^{(t+L)\wedge T} s_{kt} + u_t = d_{0t} \qquad t \in \tau(s),\ s \in S$$

$$\sum_{k=(t-L)\vee 0}^{(t+R)\wedge T} s_{tk} + r_t - x_P^t = 0 \qquad t \in \tau(s),\ s \in S$$

$$x_0^t \ge 0 \qquad t \in \tau(s),\ s \in S.$$

Table 14.4 Partition of time stages into decision stages

s	$\tau(s)$	t_s
1	{1, 2, 3, 4, 5}	1
2	{6, 7, 8, 9, 10}	6
3	{11, 12, 13, 14, 15}	11
4	{16, 17, 18, 19, 20}	16

Next, we define $\bar{x}_s = \{x_0^t \mid t \in \tau(j),\ j \leq s\}$ and rewrite (14.10) as

$$\text{Max} \qquad H_2(\bar{x}_1),$$

where for $s \notin S$, $H_s(\cdot) \equiv 0$, and for $s \in S$,

$$H_s(\bar{x}_{s-1}) = \tag{14.11}$$

$$\text{Max} \quad \sum_{t\in\tau(s)} \left\{ \sum_{k=(t-L)\vee 0}^{(t+R)\wedge T} c_{tk}^s s_{tk} + c_t^r r_t + c_t^u u_t \right\} + H_{s+1}(\bar{x}_s)$$

$$\text{s.t.} \qquad x_P^t - \lambda_t \sum_{j\in\tau(s)}^{t-1} \rho_{0P}^{t-j}(j) x_0^j = \lambda_t \sum_{j<t_s}^{t-1} \rho_{0P}^{t-j}(j) x_0^j \qquad t \in \tau(s)$$

$$\sum_{p=0}^{P-1} \sum_{j\in\tau(s)} \rho_{0p}^{t-j}(j) x_0^j \leq C - \sum_{p=0}^{P-1} \sum_{j<t_s} \rho_{0p}^{t-j}(j) x_0^j \qquad t \in \tau(s)$$

$$\sum_{k=(t-R)\vee 0}^{(t+L)\wedge T} s_{kt} + u_t = d_{0t} \qquad \forall\, t \in \tau(s)$$

$$\sum_{k=(t-L)\vee 0}^{(t+R)\wedge T} s_{tk} + r_t - x_P^t = 0 \qquad \forall\, t \in \tau(s)$$

$$x_0^t \geq 0 \qquad t \in \tau(s).$$

Note that (14.11) includes $\bar{x}_{s-1}$ as a fixed quantity on the right-hand side of the constraints, reflecting a temporal decomposition in the presentation. Note also that solutions to (14.11) depend on the data elements that are incorporated $\{d_{0t}, \lambda_t, \rho_{0p}^{t-j}(j)\}$, as well as any previously determined factory-release decisions that have been made, $\bar{x}_s$. Recognizing the presence of uncertainty in this data, we note that these data are random variables, modeled in accordance with the process described in Section 14.3, and denote them as $\{\tilde{d}_{0t}, \tilde{\lambda}_t, \tilde{\rho}_{0p}^{t-j}(j)\}$. In order to reflect the state of the system at each decision epoch, it is necessary to identify the information that is available at that time, and the manner in which it influences the probabilistic representation of the future evolution of the demand and production processes.

As decisions are made at time t_s, we are able to observe the following:

- $\{x_p^{t_s}\}_{p=1}^P$, the state of the production facility at t_s.
- $\{Q_f^{t_s}\}$ (for all f), where $Q_f^{t_s}$ represents the quantity that is apparently due in f periods as of t_s.

Note that $x_P^{t_s}$ is given by

$$x_P^{t_s} = \sum_{j=0}^{t_s} \rho_{0p}^{t_s - j}(j) x_0^j.$$

Correspondingly, let $\bar{\rho}_s = \{\{\rho_{0p}^{t_s - k}\}_{k=0}^{t_s}\}_{p=0}^{P}$, reflecting the evolution of the production process through t_s. It follows that $\bar{x}_s$ and $\bar{\rho}_s$ are sufficient to describe the state of the production facility at t_s. With regard to the future evolution of the production process, note that as a result of the Markov chain model, for $t > t_s$, the state of the production process can be determined by combining $\bar{\rho}_s$ with $\{\tilde{\rho}_{pp'}^{t-t_s}\}$. That is, items in stage p at t_s will be in stage p' at time t with a probability that is easily derived from the Markov chain model.

Note that $Q_f^{t_s}$ is a function of all demand signals received by t_s and the manner in which their due dates and quantities evolve following the initial receipt of the signal. Collectively, we refer to this as $\bar{D}_s$, the state of the demand process at decision stage s. We note that as a result of the Markov chain models used to describe the demand process, $\bar{D}_s$ is sufficient to probabilistically describe the forward evolution of demand from t_s.

Incorporating these stochastic elements into the problem is accomplished through conditional expectations, as follows:

$$\text{Max} \qquad E\left[H_2(\bar{x}_1, \tilde{D}_1)\right],$$

where for $s \notin S$, $H_s(\cdot,\cdot) \equiv 0$, and for $s \in S$,

$$H_s(\bar{x}_{s-1}, D_{s-1}) = \tag{14.12}$$

$$\text{Max} \quad \sum_{t \in \tau(s)} \left\{ \sum_{k=(t-L)\vee 0}^{(t+R)\wedge T} c_{tk}^s s_{tk} + c_t^r r_t + c_t^u u_t \right\} + E\left[H_{s+1}(\bar{x}_s, \tilde{D}_s \mid D_{s-1})\right] \tag{14.13}$$

$$\text{s.t.} \qquad x_P^t - \lambda_t \sum_{j \in \tau(s)}^{t-1} \rho_{0P}^{t-j}(j) x_0^j = \lambda_t \sum_{j<t_s}^{t-1} \rho_{0P}^{t-j}(j) x_0^j \qquad t \in \tau(s) \tag{14.14}$$

$$\sum_{p=0}^{P-1} \sum_{j \in \tau(s)} \rho_{0p}^{t-j}(j) x_0^j \le C - \sum_{p=0}^{P-1} \sum_{j<t_s} \rho_{0p}^{t-j}(j) x_0^j \quad t \in \tau(s) \tag{14.15}$$

$$\sum_{k=(t-R)\vee 0}^{(t+L)\wedge T} s_{kt} + u_t = d_{0t} \qquad \forall\, t \in \tau(s)$$

$$\sum_{k=(t-L)\vee 0}^{(t+R)\wedge T} s_{tk} + r_t - x_{Pt} = 0 \qquad \forall\, t \in \tau(s)$$

$$x_0^t \ge 0 \qquad t \in \tau(s).$$

Note that although (14.12) is a single product model, it is easily extended to incorporate multiple products as well as one-way substitution, as in Hsu and Bassok (1999) and Duenyas and Tsai (2000).

14.4 Conclusions

Our primary goal in this chapter is to introduce a production planning model that incorporates multistage production and volatile demand that evolves stochastically in time. We note that the model presented in (14.12), includes opportunities to adjust the production schedule depending on the manner in which orders and yields are evolving. To the best of our knowledge, this is the first model of volatile demand to facilitate production planning.

Of course, there are numerous research issues that must be addressed. We note that the inclusion of Markov chain models for the demand and production process greatly facilitates the generation of scenarios in a sample-based solution method. However, the efficient specification of the underlying transition matrix in a manner that fits historical data regarding throughput times is an open question. For small problems, exhaustive searches can be undertaken, but that becomes impractical as the number of productions stages increases. Similarly, the solution of multistage stochastic programs can become increasingly challenging as the number of decision stages increases. Correspondingly, an investigation of the value associated with an increasing number of decision stages can provide guidance on the trade-off between model fidelity and computational tractability. These and other issues provide the motivation for continued work in this area.

Acknowledgments This work was supported by Grant No. DMS 04-00085 from The National Science Foundation.

References

Arrow, K.J., Karlin, S.: Studies in the Mathematical Theory of Inventory and Production. Stanford University Press, Stanford, CA (1958)

Axsater, S.: Inventory Control, 2nd Ed. Springer, New York, NY (2006)

Aytug, H., Lawley, M.A., McKay, K., Mohan, S., Uzoy, R.: Executing production schedules in the face of uncertainties: A review and some future directions. EJOR **161**, 86–110 (2005)

Bollapragada, S., Morton, T.E.: Myopic heuristics for the random yield problem. Oper. Res. **47**(5), 713–722 (1999)

Church, L.K., Uzsoy, R.: Analysis of periodic and event-driven rescheduling policies in dynamic shops. Int. J. Comput. Integr. Manufact. **5**, 153–163 (1992)

Ciarallo, F.W., Akella, R., Morton, T.E.: A periodic review production planning model with uncertain capacity and uncertain demand – Optimality of extended myopic policies. Manage. Sci. **40**, 320–332 (1994)

Duenyas, I., Tsai, C.-Y.: Control of a manufacturing system with random product yield and downward substitutability. IIE Trans. **32**, 785–795 (2000)

Duenyas, I., Hopp, W.J., and Bassok, Y.: Production quotas as bounds on interplant JIT contracts. Manage. Sci. **43**(10), 1372–1386 (1997)

Erdem, A.S., Ozekici, S.: Inventory models with random yield in a random environment. Intl. J. Prod. Econ. **78**, 239–253 (2002)

Hadley, G., Whitin, T.M.: Anal Invent Syst Prentice-Hall, London (1963)

Hsu, A., Bassok, Y.: Random yield and random demand in a production system with downward substitution. Oper. Res. **47**(2), 277–290 (1999)

Ignizio, J.P.: The implementation of conwip in semiconductor fabrication facilities. Future Fab. Intl. Feb. (2003)

Kempf, K.G.: Control-oriented approaches to supply chain management in semiconductor manufacturing. In: Proceeding of the 2004 American Control Conference, June 30–July 2 (2004) Boston, MA (2004)

Kouvelis, P., Milner, J.M.: Supply chain capacity and outsourcing decisions: the dynamic interplay of demand and supply uncertainty. IIE Trans. **34**, 717–728 (2002)

Mula, J., Poler, R., Garcia-Sabater, J.P., Lario, F.C.: Models for production planning under uncertainty: A review. Int. J. Produc. Econ. **103**, 271–285 (2006)

Ovacik, I.M., Uzsoy, R.: Decomposition Methods for Complex Factory Scheduling Problems. Springer, New York, NY (2006)

Pinedo, M.L.: Planning and Scheduling in Manufacturing and Services. Springer, New York, NY (2006)

Porteus, E.L.: Foundations of Stochastic Inventory Theory. Stanford University Press, Stanford, CA (2002)

Sethi, S.P., Yan, H., Zhang, H., Zhang, Q.: Optimal and hierarchical controls in dynamic stochastic manufacturing systems: A survey. Manufact. Serv. Oper. Manage. **4**(2), 133–170 (2002)

Smith, S., Keng, N., Kempf, K.: Exploiting local flexibility during execution of pre-computed schedules. In: Famili, A., Nau, D.S., Tong, S.H., (eds.) Artificial Intelligence Applications in Manufacturing, pp. 277–292. MIT Press, Cambridge, MA (1992)

Vollman, T.E., Barry, W.L., Whybark, D.C., Jacobs, F.R.: Manufacturing Planning and Control Systems for Supply Chain Management, 5th ed. McGraw-Hill, New York, NY (2004)

Wang, Y., Gerchak, Y.: Periodic review production models with variable capacity, random yield, and uncertain demand. Manage. Sci. **42**(1), 130–137 (1996)

Yano, C.A., Lee, H.L.: Lot sizing with random yields: A review. Oper. Res. **43**, 311–334 (1995)

Zipkin, P.H.: Foundations of Inventory Management. McGraw-Hill/Irwin, New York, NY (2000)

Chapter 15
Global Climate Decisions Under Uncertainty

Alan S. Manne

15.1 Introduction

No matter what one's views on global climate change, it is easy to agree that there is great uncertainty and that our models should reflect that uncertainty. The technical difficulty is that uncertainty can lead to an enormous increase in dimensionality. If, for example, we wish to consider 10 alternative states of the world, this will typically lead to an increase by something between 100 and 1000 times in the amount of computing effort that is required to solve a mathematical programming problem. This may be an acceptable increase if we add uncertainty to a small-scale model, but it may be unacceptable if we start with a technology-oriented model that is already quite large. One possibility might be to employ Benders decomposition. For an application of this technique to climate modeling, see Chang (1997).

In this chapter, we will explore an alternative approach to dealing with the problem of dimensionality in large multiregion, multiperiod models. The regions are aggregated so that we solve a "one-world" model in the later time periods. Why aggregate during the later periods? Because discounting limits the importance of distant-future uncertainties for near-future decisions.

We begin by illustrating these ideas through a small model—one with only two regions and two states of the world. We then illustrate how these ideas carry over to more complex models such as MERGE (a *m*odel for *e*valuating *r*egional and *g*lobal *e*ffects of greenhouse gas reductions). For extensive documentation on this model, see the web site authored jointly with Richard Richels: www.stanford.edu/group/MERGE.

15.2 A Small-Scale Example

It is instructive to begin with a small-scale example. This should help to clarify the basic concepts. The basic GAMS code is available upon request to the author as

A.S. Manne (✉)
Department of Management Science and Engineering, Stanford University,
Stanford, CA 94305, USA
e-mail: jackie.manne@gmail.com

G. Infanger (ed.), *Stochastic Programming*, International Series in Operations Research & Management Science 150, DOI 10.1007/978-1-4419-1642-6_15,

ATL.GMS (act, then learn). In this case, there are only two regions: North (low population, high per capita incomes, and low growth rates) and South (high population, low per capita incomes, and high growth rates). There is only one greenhouse gas (carbon dioxide). There is a *non*linear economy-wide production function—based on inputs of capital, labor, and energy. This allows us to consider price-induced conservation through a smooth rather than a bang–bang process. In turn, this means that the solution algorithm is based upon a nonlinear rather than a linear solver. In this small example, energy is provided by only two "technologies." These are perfect substitutes. They are inexhaustible, and they exhibit constant returns to scale: a low-cost fossil fuel that emits carbon and a high-cost "backstop" that is carbon-free. There are expansion constraints on the introduction of the carbon-free technology, and there are decline constraints on the use of fossil fuels.

In the debate over global climate change, perhaps the most controversial element is one's view on the consequences of greenhouse gas accumulation. This type of model may be formulated in terms of cost-effectiveness—or in terms of benefit–cost analysis. In the first case, we ask for the least cost way to comply with an arbitrary constraint on greenhouse gas accumulation. In the second, we view the accumulation as a decision variable and ask for the optimal balance between the costs of accumulation and the costs of abatement. The benefit–cost perspective is the one that we shall adopt for our small-scale example. The cost-effectiveness criterion will be employed with the larger model, MERGE.

The costs of accumulation are very difficult to estimate. One can take the position that nothing much has happened to date and that therefore nothing much is likely to happen. Or one can point out that climate change could lead to a rise in the sea level and the destruction of many species. There is also the possibility of catastrophic events such as a meltdown of the Antarctic ice sheet or a breakdown in the thermohaline circulation of the oceans. For our small model, we will illustrate these alternative perspectives with just two states of the world—either low or high damages—depending on the cumulative emissions of carbon post-2000. It will be supposed that 5% is the probability of the high-damage scenario.

Decision variables are indicated by uppercase letters and parameters by lowercase letters. The index tp defines time periods. For modeling purposes, the time periods are decades, beginning with 2000 and extending through 2200. The index sw defines alternative states of the world. Here there are only two states of the world. These imply either low or high damages—depending on the decision variable $CS_{tp,sw}$, the cumulative increase of the carbon stock post-2000. Damages are assumed to be a quadratic function of this quantity. Output is defined in terms of a numéraire good that serves to measure the economic value of the world's current GDP.

The decision variable $ELF_{tp,sw}$ defines the economic loss factor—the proportion of gross output in period tp that survives the damages in alternative states of the world. That is,

$$ELF_{tp,sw} = 1 - (CS_{tp,sw}/catcon_{sw})^2,$$

where $\text{catcon}_{\text{sw}}$ is a catastrophic concentration parameter chosen so that the value of the world's entire economic output is wiped out at this level. The two alternative values assigned to this parameter are 1800 and 600 ppmv (parts per million, by volume),

Why these two alternative levels? It is generally agreed that carbon concentrations were 275 ppmv during the pre-industrial era and that they had grown to 370 ppmv by 2000. Much of the discussion on impacts has focused on the consequences of 550 ppmv (twice the pre-industrial level). In a business-as-usual scenario, we might reach this level sometime between 2050 and 2100. Now consider the implications of a doubling of carbon concentrations above their pre-industrial level of 275 ppmv. This would imply an absolute value of 550 ppmv—that is, an increase of 180 ppmv above the carbon concentrations in 2000. The decision variable $\text{CS}_{\text{tp,sw}}$ would then be 180 ppmv. With a quadratic loss function and a $\text{catcon}_{\text{sw}}$ value of 1800, this would imply that losses are only 1% and that the economic loss factor would be 99%. (This is our "most likely" case, and it is assigned a subjective probability of 95%.) By contrast, with a $\text{catcon}_{\text{sw}}$ value of 600, the losses would be 9% and the loss factor would drop to 91%. We assign a subjective probability of only 5% to this outcome.

With decision analysis, it is important to know not only the probabilities but also the dates of resolution of these uncertainties. When will we know which of these two states of the world will occur? For illustrative purposes, it is assumed that these uncertainties will not be resolved until just after 2040. That is, during the decades prior to 2050, all decisions must be identical—regardless of which outcome eventually occurs. Prior to the resolution of uncertainty, there are "locking constraints." These provide that the identical value of each decision variable is obtained in each alternative state of the world. For example, in the case of $K_{\text{rg,tp,sw}}$ (the capital stock variable in region rg, time period tp, and state of world sw):

$$K_{\text{rg,tp,sw1}} = K_{\text{rg,tp,sw2}} \quad \text{(for tp prior to 2050)}.$$

How does this model allocate the task of carbon abatement? One can argue that North is so much richer than South that North *ought* to undertake the entire burden. One can also argue that this is an inefficient proposition and that some of the abatement *ought* to be undertaken by the South. This leads to the possibility of a conflict between equity and efficiency. The way we have attempted to resolve this conflict is to assign shares in global emissions to each individual region at each point of time and then to allow the regions to trade these carbon emission rights with each other—at prices determined by an international market. Unlike the Kyoto accord, this type of agreement places some of the burden of abatement upon the South, but it leads to economic efficiency. With unlimited international trade in emission rights, there is virtually no conflict between equity and efficiency. Why? Because the value of emission rights is just a small portion of each region's endowments. For more on this issue, see Manne and Stephan (2003).

In translating damages into their economic consequences, much of the story is centered around just two equations. One may be described as a "green output" equation. The left-hand side indicates how the economic loss factor affects total output in each region (net of climate damages). The right-hand side shows how this is allocated between the following elements: C (consumption), I (investment in building up the stock of physical capital), F and N (respectively, the costs of fossil and backstop carbon-free energy), and X (net exports of the numéraire good):

$$\mathrm{ELF}_{\mathrm{tp,sw}} Y_{\mathrm{rg,tp,sw}} = C_{\mathrm{rg,tp,sw}} + I_{\mathrm{rg,tp,sw}} + \mathrm{cfos}\, F_{\mathrm{rg,tp,sw}} + \mathrm{cbak} N_{\mathrm{rg,tp,sw}} + X_{\mathrm{rg,``num",tp,sw}}.$$

The second equation is the maximand. It shows how each region's *expected* discounted utility depends upon consumption in each time period. Each region's utility is a logarithmic function of consumption and is proportional to its "Negishi weight," $\mathrm{nw}_{\mathrm{rg}}$, These weights are revised through sequential optimization. They are chosen so that in equilibrium, they will equal each region's share of the present value of global endowments. Rutherford (1999) has shown that when production sets are convex, each agent (here each region) has homothetic preferences and each agent has fixed endowments of goods, it is possible to convert a standard economic equilibrium problem into a sequence of "joint maximization" problems. For this multiregion uncertainty problem, the Negishi maximand is written as

$$\sum_{\mathrm{tp,rg,sw}} \mathrm{nw}_{\mathrm{rg}} \mathrm{prob}_{\mathrm{sw}} \mathrm{udf}_{\mathrm{rg,tp}} \log C_{\mathrm{rg,tp,sw}},$$

where $\mathrm{prob}_{\mathrm{sw}}$ denotes the probability that state of world sw will be realized and $\mathrm{udf}_{\mathrm{rg,tp}}$ denotes the utility discount factor for the logarithm of consumption received by region rg in period tp. For the final time period, the utility discount factor is adjusted upward so as to allow for an infinite stream of post-horizon consumption. Our parameters are chosen so that this adjustment implies only a small change in the objective function. The horizon effects are minor.

To initiate the algorithm, an arbitrary Negishi weight is associated with each agent's utility. A joint maximization problem is solved—with constraints to ensure a balance between the supplies and the demands of each good. The Negishi weights are then revised so that each agent comes closer to satisfying its budget constraint. Each of the budget constraints is expressed in terms of present value prices (based on the international productivity of capital). The cumulative present value of each region's current account deficit must add up to zero.

There is no theoretical proof that Rutherford's algorithm will converge, and some numerical counterexamples have been developed. In practice, however, the procedure works well. There are not many iterations required to obtain good numerical convergence. Currently, it is routine for us to solve problems with nine Negishi agents (one for each region), 25,000 equations, 35,000 variables, and 2,000 nonlinear nonzero entries. Using GAMS in conjunction with Drud's CONOPT3 solver on a 3200 MHz machine and a "hot start," the run times seldom exceed 15 min. This may be an acceptable run time, but cannot then be extended to include many states of the world. This is our motive for aggregating regions during the later time periods.

15.3 Aggregating Regions to a "One-World" Model

Suppose that we aggregate regions during the later time periods. With regional aggregation, it is clear that the model cannot incorporate international trade during those periods. Accordingly, the meaning of the Negishi weights must be modified. During the early periods, these weights still refer to each region's share of the global present value of consumption. During the later periods, however, we cannot distinguish the regional shares. It is the global value that must be allocated between time periods.

What is this global value? We cannot provide an exact answer, but only an approximation. The approximation is based on the fact that each region's potential GDP is an *input* to the model. It is the value that GDP would take on if all future value prices remain constant over time. Another input is the international marginal productivity of capital. One of the *outputs* of the model is the value of consumption in each future period. Our approximation is based on the assumption that consumption in each region is proportional to GDP. It then becomes straightforward to calculate the total present value of the world's consumption during the later time periods. This is the *approximate* value of the Negishi weight assigned to the world during those periods. With our two-region model and a transition from two to one region after 2100, the Negishi weight assigned to the entire post-2100 world is only 10%. A 10% value also serves as the approximate Negishi weight for the much larger nine-region MERGE model.

What else must be done? Each of the intertemporal constraints that have a regional subscript during the early periods must be aggregated during the transition. For instance, consider the accumulation of physical capital. During the early periods, this is written as

$$K_{\mathrm{rg,tp+1,sw}} = 0.6\,\{K_{\mathrm{rg,tp,sw}} + 5\,I_{\mathrm{rg,tp,sw}}\} + 5\,I_{\mathrm{rg,tp+1,sw}},$$

where 60% of the capital in place survives from one decade to the next and where new capital represents an average of the gross annual investment during the first and the second decade. A similar equation is written for the world as a whole during each of the periods when there is a one-world representation.

Now suppose that the transition date is 2100. This is the last period in which there are multiple regions. Then 2110 is the first period of the one-world model. There is no regional subscript for the variables after 2100. The transitional capital accumulation equation would therefore be written as

$$K_{2110,\mathrm{sw}} = 0.6\left\{\sum_{\mathrm{rg}} K_{\mathrm{rg},2100,\mathrm{sw}} + \sum_{\mathrm{rg}} 5\,I_{\mathrm{rg},2100,\mathrm{sw}}\right\} + 5\,I_{2110,\mathrm{sw}}.$$

Optimal global carbon emissions are shown in Fig. 15.1. These rise through 2040, and they continue to rise through 2080 in the "low-damages" scenario, sw1. With sw2 (the low-probability, "high-damages" scenario), global emissions drop immediately after 2040, and they continue to drop throughout the remaining planning horizon. Note that there is very little difference between the solutions for the aggregated "one-world" model (dashed lines) and the disaggregated multiregion case (solid lines). It remains to be seen whether this invariance holds true for larger models.

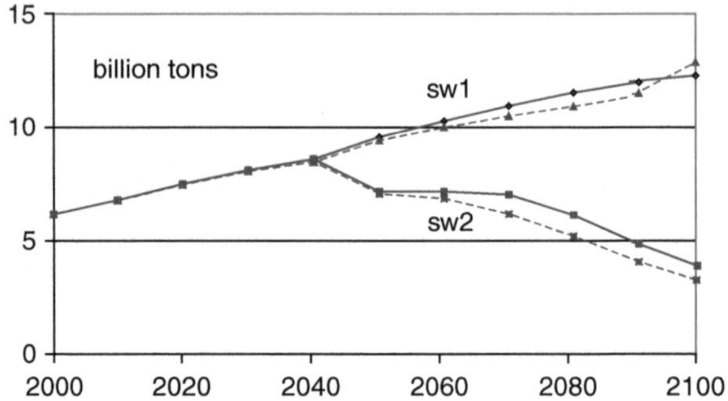

Fig. 15.1 Global carbon emissions

The international price of carbon is the dual variable associated with the cumulative carbon constraint. This is shown in Fig. 15.2—again through the transition date of 2100. With sw1, prices rise gradually from their initial level. With the high-damage and low-probability (5%) case, there is an abrupt rise immediately after the resolution of uncertainty. Again, there is very little difference between the aggregated model (dashed lines) and the multiregion model (solid lines).

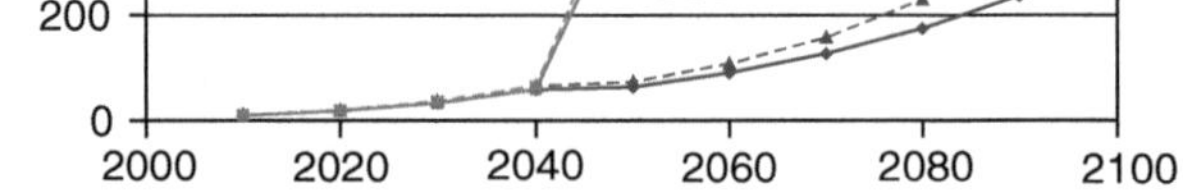

Fig. 15.2 Carbon price

15.4 MERGE: A Multiregion, Multitechnology Model

In MERGE, as in the small model, the time horizon extends by decades through 2200. During the twenty-first century, the world is described in terms of nine regions: the USA, western Europe, Japan, Canada + Australia + New Zealand,

Eastern Europe + former Soviet Union, China, India, Mexico + other oil-exporting nations, and ROW (rest of the world). Figure 15.3 shows our population projections for these regions through 2150. Thereafter, the world's population is virtually stationary at 10 billion people.

Figure 15.4 contains our productivity projections—with per capita GDP expressed at market exchange rates and prices expressed in US dollars of 2000 purchasing power. Between 2000 and 2020, the per capita GDP is taken directly from the projections of the US Energy Information Administration (2004). From 2020 onward, each region's GDP is taken from a logistic curve fitted through three points: EIA's projection for 2020, an asymptote of $200,000 per capita, and a 2100 value chosen so that there will be a smooth evolution following 2020.

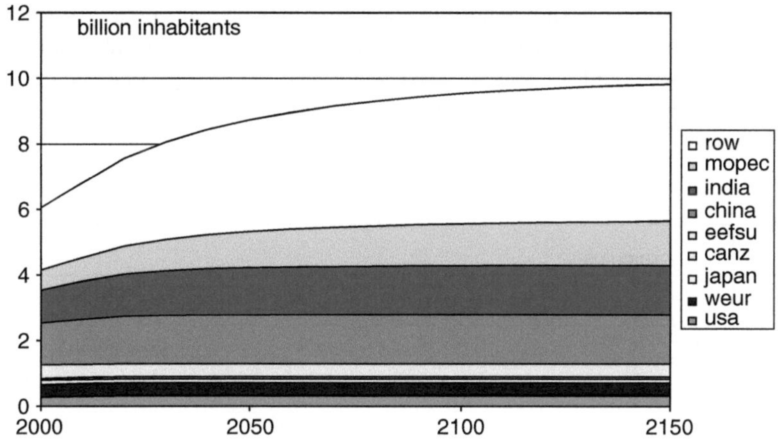

Fig. 15.3 Global and regional population

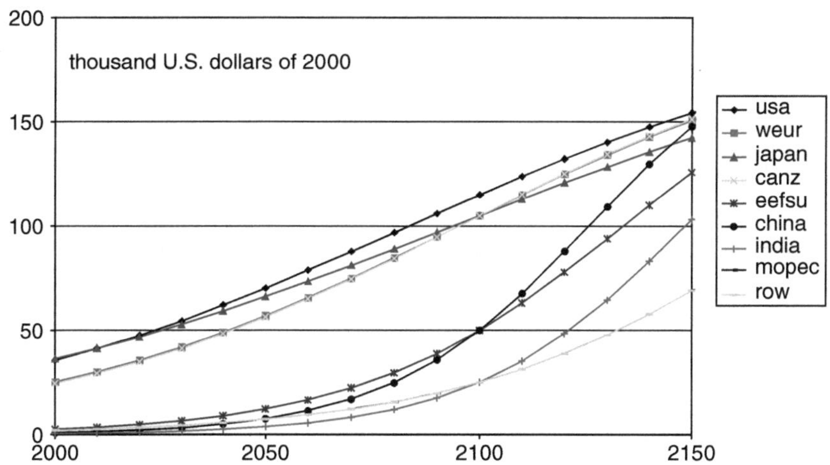

Fig. 15.4 Projections of GDP per capita at market exchange rates

Why the asymptote of $200,000? This provides a limit to growth, but does not become an effective constraint through 2200. It permits today's wealthy regions (the North) to continue growing at a roughly linear rate for the next two centuries. Why the values chosen for 2100? These are chosen so that today's poor regions (the South) will grow rapidly and converge gradually toward the North's income level. Admittedly, this is an optimistic scenario. One can imagine far more pessimistic story lines. Under these circumstances, the world's climate problems would be less severe, but the general economic outlook would become much more serious. The South would continue to be plagued by poverty, disease, and malnutrition.

How are technologies represented in MERGE? Energy supplies are handled through a bottom-up framework and energy demands through a top-down representation. Energy demands are divided into two broad categories: electric and nonelectric. Figure 15.5 shows how the world's total primary energy might originate from carbon-emitting, exhaustible resources: crude oil, natural gas, and coal. A fourth category could become increasingly important: energy that is carbon free: hydroelectric, nuclear, wind, solar, and other renewables. This category would also include carbon capture and sequestration.

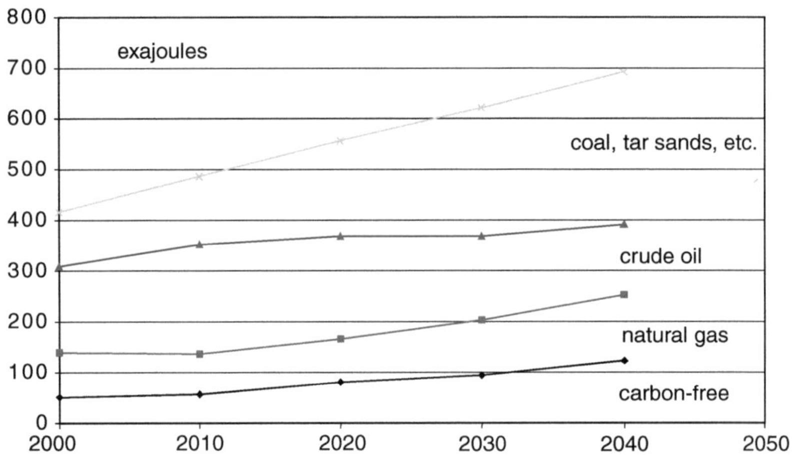

Fig. 15.5 Total primary energy

Uncertainty is introduced through the same timing and probability conventions as one of the experiments conducted by the Energy Modeling Forum (2004–2006), Study 22. That is, there is uncertainty on "climate sensitivity" through 2040, and thereafter this uncertainty is resolved. *Climate sensitivity* is defined as the equilibrium mean global temperature associated with a doubling of global carbon concentration from its pre-industrial level of 275 ppmv (parts per million, by volume). For each climate sensitivity, there is a different thermal inertia lag required for consistency with the current concentration level of 370 ppmv and a mean global temperature increase of 0.5–1.0°C.

Yohe (2005) has proposed the following discrete probability distribution to approximate the continuous distribution function estimated by Andronova and Schlesinger (2001). This distribution has a "fat tail." That is, there is a high probability of 8°, a much higher climate sensitivity than was once considered as a conventional outcome. It will be seen that this case dominates the outcome of our stochastic programming model.

Probabilities of alternative climate sensitivities:

1.5°C	25%
3°	45%
5°	15%
8°	15%

Yohe goes on to say, "We will continue to consider three temperature targets relative to pre-industrial levels: (a) 2.0° (to represent the European Union's goal), 3.0° (a more relaxed target), and 4.0° (certainly a more feasible target). The limiting temperature targets must be satisfied, if they are feasible, throughout the planning horizon (2200 at least). There is full flexibility with respect to when and where abatement is to be achieved. If you find that a temperature target is infeasible, report the lowest feasible temperature target and how it can be achieved. This means that the temperature targets are not completely limiting and so they should not dominate the hedging for high climate sensitivities that have relatively small likelihoods," see Yohe (2005) .

With the parameters employed in this version of MERGE, it turned out to be infeasible to satisfy either the 2° or the 3° temperature target. The lowest feasible target was 4°. Upon closer examination, it turned out that the lower temperature targets *could* be achieved by lowering the GDP growth rates. Alternatively, one could assume that the rate of decarbonization could exceed our standard value of 2% per year.

For our standard parameter assumptions—and with the EMF stochastic distribution of climate sensitivities—Figs. 15.6 and 15.7 show the quantities of carbon emissions and the implicit price of these emissions. Prices and quantities are identical through 2040 and thereafter diverge for each of the four climate sensitivities. Note that the global carbon price is capped at $1000 per ton. This occurs with the 8° scenario. It then becomes economical to employ a "backstop" technology—one that is high cost and unlimited in quantity.

Figure 15.8 shows how the mean global temperature might increase with these alternative sensitivities. The temperature would rise through 2040 and would drop slightly with the favorable outcome of a 1.5° sensitivity. In the other three cases, the temperature would continue to rise for the following century, but would eventually reach the 4° bound. These are the cases in which it is useful to have a horizon of 2200—even if it stretches our computational capabilities.

Figure 15.9 contains perhaps the most instructive outcome of these experiments. It shows what happens when we modify the probabilities of an 8° climate sensitivity. Instead of a 15% probability, we assume a much thinner tail of the frequency distribution—either a 1 or a 0.1% tail. To offset the reduction in the probability of 8°,

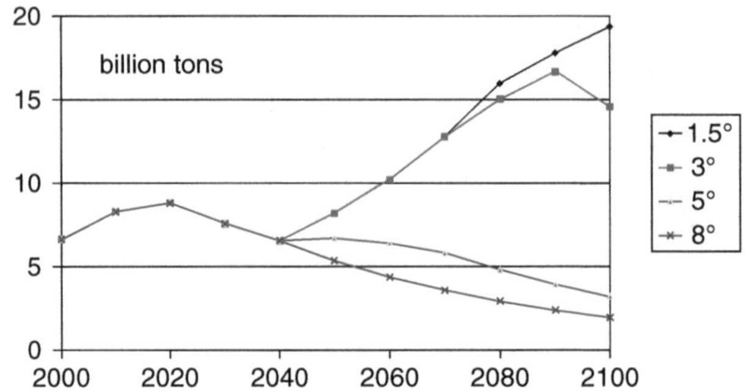

Fig. 15.6 Global carbon emissions with alternative climate sensitivities

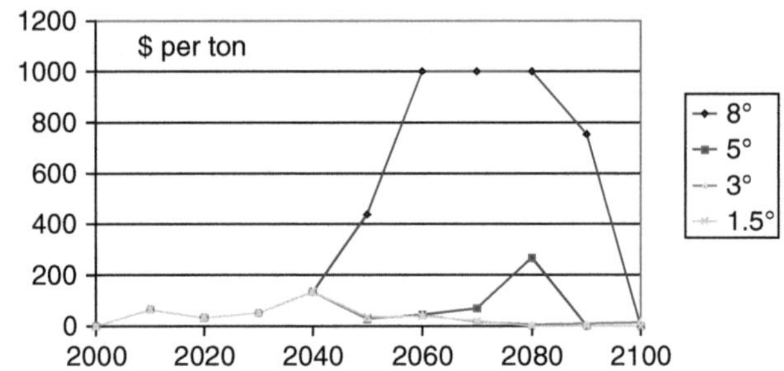

Fig. 15.7 Global carbon price with alternative climate sensitivities

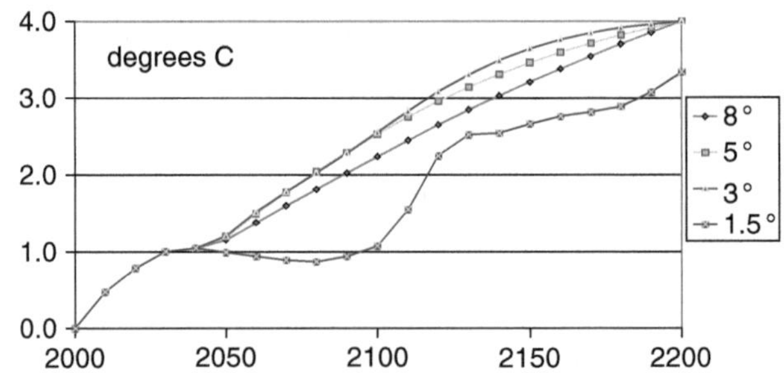

Fig. 15.8 Increase in mean global temperature with alternative climate sensitivities

the balance of the distribution is assigned to the 5° outcome. Somewhat surprisingly, these probability changes do not affect the globally optimal strategy through 2040. The optimal emission rates are virtually identical. In order to meet a 4° temperature limit, global emissions would have to be limited sharply after 2020.

With an extreme climate sensitivity of 8°, the probabilities do not affect the near-term decisions. The solution is virtually a minimax result. That is, we maximize each region's utility under the worst possible state of the world. The outcome would be more reasonable if the fixed temperature ceiling were replaced by an approach based on the expected disutility of high temperatures. This approach is similar to

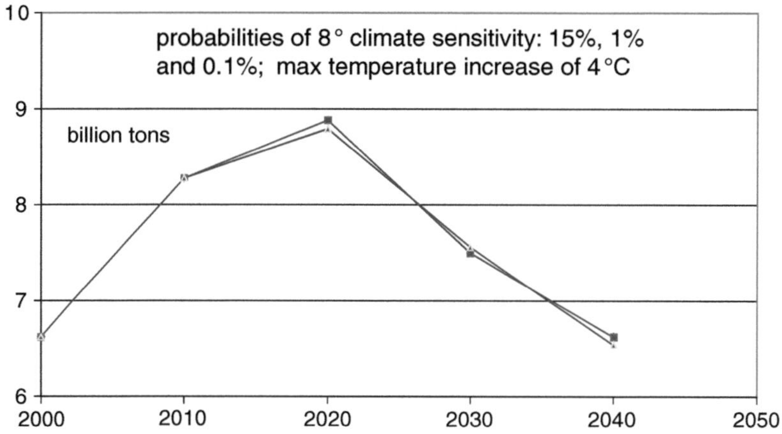

Fig. 15.9 Global carbon emissions

the one applied in the small-scale North–South model described earlier. Despite the arbitrary nature of a temperature disutility function, this leads to more reasonable results than a tightly constrained temperature limit.

Acknowledgments For helpful comments, the author is indebted to John Rowse and Thomas Rutherford.

References

Andronova, N., Schlesinger, M.E.: Objective estimation of the probability density function for climate sensitivity. J. Geophys. Res. **D19**(106), 22605–22611 (2001)

Chang, D.: Solving large-scale intertemporal CGE models with decomposition methods. PhD thesis, Department of Operations Research, Stanford University, Stanford, CA (1997)

Energy Modeling Forum. EMF 22: Climate change control scenarios, Energy Modeling Forum, Stanford University, Stanford, CA (2004–2006)

Manne, A.S., Stephan, G.: Global climate change and the equity-efficiency puzzle. Working Paper (2003)

Rutherford, T.: Sequential joint maximization. In: Weyant, J.P., (ed.) Energy and Environmental Policy Modeling. Kluwer, Norwell, MA (1999)

Yohe, G.: Second round exercise description and instructions. EMF report, Energy Modeling Forum, Stanford University, Stanford, CA (2005)

Chapter 16
Control of Diffusions via Linear Programming

Jiarui Han and Benjamin Van Roy

In this chapter we present an approach that leverages linear programming to approximate optimal policies for controlled diffusion processes, possibly with high-dimensional state and action spaces. The approach fits a linear combination of basis functions to the dynamic programming value function; the resulting approximation guides control decisions. Linear programming is used here to compute basis function weights.

What we present extends the linear programming approach to approximate dynamic programming, previously developed in the context of discrete-time stochastic control (Schweitzer and Seidmann 1985; Trick and Zin 1997; de Farias and Van Roy 2003, 2004, 2006). One might question the practical merits of such an extension relative to discretizing continuous-time models and treating them using previously developed methods. As will be made clear in this chapter, there are indeed important advantages in the simplicity and efficiency of computational methods made possible by working directly with a diffusion model.

We begin in Section 16.1 by presenting a problem formulation and a linear programming characterization of optimal solutions. The numbers of variables and constraints defining this linear program are both infinite. In Section 16.2, we describe algorithms that apply finite-dimensional linear programming to approximately solve this infinite-dimensional problem. To illustrate practical qualities, we discuss in Section 16.3 computational results from case studies in dynamic portfolio optimization. A closing section summarizes merits of the new approach.

16.1 Preliminaries

This section offers a context for algorithms to be presented in Section 16.2. To avoid the technical complexities that often arise when dealing with controlled diffusion

B. Van Roy (✉)
Department of Management Science and Engineering, Stanford University, Stanford, CA 94305, USA
e-mail: bvr@stanford.edu

G. Infanger (ed.), *Stochastic Programming*, International Series in Operations Research & Management Science 150, DOI 10.1007/978-1-4419-1642-6_16,

models, we make simplifying assumptions that can be restrictive and difficult to verify. The most notable ones concern compactness of state and action spaces and regularity of the optimal value function. These assumptions are stronger than those put forth by more thorough treatments of controlled diffusion models (Krylov 1980; Borkar 1989). Our motivation is not to develop a general model, but rather a context for explaining practical algorithms that can be applied more broadly. Indeed, the portfolio optimization models addressed in Section 16.3 fail to satisfy our assumptions.

16.1.1 Controlled Diffusion Processes

Let $B = (B^{(1)}, \ldots, B^{(D)})^\top$ be a standard Brownian motion in $\Re^D$ defined with respect to a complete probability space $(\Omega, \mathcal{F}, P)$ and let $\mathbf{F} = \{\mathcal{F}_t | t \geq 0\}$ be the augmented filtration generated by B.

We consider a controlled diffusion process with state at each time t, denoted by X_t, taking values in a state space $\mathcal{S}$ that is a compact subset of $\Re^N$. At each time t, an action $\psi \in \Psi$ is taken, where Ψ is a compact subset of $\Re^M$. The process evolves according to

$$dX_t = \mu(X_t, \psi)\, dt + \sigma(X_t, \psi)\, dB_t,$$

where μ and σ are Lipschitz continuous. The process terminates at a time $T_{\mathcal{S}}$, where for any closed set A, T_A is the first time at which X_t hits the boundary of A.

For each time t, a control function $\theta_t : \mathcal{S} \to \Psi$ maps states to actions. A policy θ is a continuum of control functions: $\theta = \{\theta_t\}_{t=0}^{\infty}$. Such a policy is said to be stationary if the same control function is used at each time. A policy is said to be admissible if its controls functions are uniformly Lipschitz continuous. Under an admissible policy θ, the state X_t evolves according to

$$dX_t = \mu(X_t, \theta_t(X_t))\, dt + \sigma(X_t, \theta_t(X_t))\, dB_t.$$

Under the assumptions we have made, it is easy to show that this stochastic differential equation admits a solution over the interval $[0, T_{\mathcal{S}}]$.

Let $\overline{\Theta}$ denote the set of all policies and Θ the set of all stationary policies. For any function f, we will use $f(x, \theta)$ as shorthand for $f(x, \theta(x))$.

A utility function $u : \mathcal{S} \times \Psi \mapsto \Re$ captures preferences among state-action pairs. The objective of our control problem takes the form

$$\sup_{\theta \in \overline{\Theta}} E_{x,\theta} \left[\int_{t=0}^{T_{\mathcal{S}}} e^{-\alpha t} u(X_t, \theta_t) dt \right] \tag{16.1}$$

for some discount rate $\alpha > 0$. The subscripts in the expectation indicate that the initial state is $X_0 = x$ and that the policy $\theta \in \overline{\Theta}$ is employed.

16.1.2 Value Functions and the HJB Equation

For each policy $\theta \in \overline{\Theta}$, the value function associated with θ is defined by

$$J_\theta(x) = E_{x,\theta}\left[\int_{t=0}^{T_{\mathcal{S}}} e^{-\alpha t} u(X_t, \theta_t) dt\right]$$

and the optimal value function is defined by

$$J^*(x) = \sup_{\theta \in \overline{\Theta}} J_\theta(x).$$

Let C^2 denote the set of twice continuously differentiable functions mapping $\mathcal{S}$ to $\mathfrak{R}$. We make the following regularity assumption:

Assumption 16.1.1 $J^* \in C^2$

We define dynamic programming operators H_θ and H by

$$\begin{aligned} H_\theta J(x) &= J_x(x)^\top \mu(x,\theta) + \frac{1}{2}\mathrm{tr}\left(J_{xx}(x)\sigma(x,\theta)\sigma(x,\theta)^\top\right) - \alpha J(x) + u(x,\theta), \\ HJ(x) &= \sup_{\theta \in \overline{\Theta}} H_\theta J(x), \end{aligned}$$

for any $J \in C^2$. With some abuse of notation, we will sometimes use H_ψ in place of H_θ when $\theta(x) = \psi$ for all x. We refer to $\theta_J \in \overline{\Theta}$ as a greedy policy with respect to J if $H_{\theta_J} J = HJ$. It is easy to see that for any $J \in C^2$, θ_J is a well-defined admissible stationary policy.

Our goal for the remainder of this section is to establish that the optimal value function satisfies the HJB equation and that policies that are greedy with respect to the optimal value function are optimal. The following lemma captures a fundamental property of H that will be used to establish our desired results.

Lemma 16.1 *For all $J \in C^2$, $\theta \in \overline{\Theta}$, $x \in \mathcal{S}$, and closed bounded sets $A \subset \mathcal{S}$,*

$$J(x) = E_{x,\theta}\left[\int_0^T e^{-\alpha t}\big(u(X_t,\theta_t) - (H_{\theta_t}J)(X_t)\big)\, dt + e^{-\alpha T} J(X_T)\right],$$

where $T = T_A \wedge T_{\mathcal{S}}$.

Proof Let X_t be a sample path generated by the policy θ. By Ito's formula,

$$d\left(e^{-\alpha t} J(X_t)\right) = e^{-\alpha t}\left((H_{\theta_t}J)(X_t) - u(X_t,\theta_t)\right) dt + e^{-\alpha t} J_x(X_t)\sigma(X_t,\theta_t)\, dB_t.$$

Therefore

$$e^{-\alpha T} J(X_T) - J(X_0) = \int_0^T e^{-\alpha t}\left((H_{\theta_t}J)(X_t) - u(X_t,\theta_t)\right) dt + \int_0^T \gamma_t\, dB_t,$$

where

$$\gamma_t = e^{-\alpha t} J_x(X_t)\sigma(X_t, \theta_t).$$

Note that γ_t is bounded for $t \le T$, since X_t is bounded and J is continuously differentiable and σ is continuous. It follows that

$$E_{x,\theta}\left[e^{-\alpha T} J(X_T) - J(X_0)\right] = E_{x,\theta}\left[\int_0^T e^{-\alpha t}\big((H_{\theta_t} J)(X_t) - u(X_t, \theta_t)\big)dt\right].$$

Rearranging terms, we obtain the desired result. □

The following theorem verifies that J^* is a solution to the HJB equation.

Theorem 16.2 *Under Assumption 16.1.1, $HJ^* = 0$.*

Proof We begin by showing that $HJ^* \ge 0$. Let x be in the interior of $\mathcal{S}$ and assume for contradiction that $(HJ^*)(x) < 0$. By Assumption 16.1.1, HJ^* is continuous and, therefore, there exists a compact set A containing x in its interior, such that $(HJ^*)(x') < (HJ^*)(x)/2$ for all $x' \in A$. Let $T = T_A \wedge T_{\mathcal{S}}$. Since the state requires time to escape,

$$\inf_{\theta \in \overline{\Theta}} E_{x,\theta}\left[\int_{t=0}^T e^{-\alpha t} dt\right] > 0.$$

Let θ^ϵ be such that $J^*(x) - J_{\theta^\epsilon}(x) \le \epsilon$. Then,

$$\begin{aligned}
0 &\le J_{\theta^\epsilon}(x) - J^*(x) + \epsilon \\
&\le E_{x,\theta^\epsilon}\left[\int_{t=0}^T e^{-\alpha t} u(X_t, \theta_t)dt + e^{-\alpha T} J^*(X_T)\right] - J^*(x) + \epsilon \\
&= E_{x,\theta^\epsilon}\left[\int_{t=0}^T e^{-\alpha t} (H_{\theta_t^\epsilon} J^*)(X_t)dt\right] + \epsilon \\
&\le E_{x,\theta^\epsilon}\left[\int_{t=0}^T e^{-\alpha t} (HJ^*)(X_t)dt\right] + \epsilon \\
&\le \frac{(HJ^*)(x)}{2} E_{x,\theta^\epsilon}\left[\int_{t=0}^T e^{-\alpha t} dt\right] + \epsilon \\
&\le \frac{(HJ^*)(x)}{2} \inf_{\theta \in \overline{\Theta}} E_{x,\theta}\left[\int_{t=0}^T e^{-\alpha t} dt\right] + \epsilon,
\end{aligned}$$

where the one equality follows from Lemma 16.1. Since ϵ can be chosen arbitrarily small, we have a contradiction. It follows that $HJ^* \ge 0$.

We will now establish that $HJ^* \leq 0$. Let x be in the interior of $\mathcal{S}$ and assume for contradiction that $(HJ^*)(x) > 0$. Note that

$$E_{x,\theta_{J^*}}\left[e^{-\alpha T}\left(J^*(X_{T_S}) - J_{\theta_{J^*}}(X_{T_S})\right)\right] = 0.$$

By Assumption 16.1.1, HJ^* is continuous and, therefore, there exists a closed bounded set A containing x in its interior, such that $(HJ^*)(x') > (HJ^*)(x)/2$ for all $x' \in A$. Since the state requires time to escape,

$$E_{x,\theta_{J^*}}\left[\int_{t=0}^{T\wedge T_A} e^{-\alpha t} dt\right] > 0.$$

By Lemma 16.1, we obtain

$$\begin{aligned}
0 &\leq J^*(x) - J_{\theta_{J^*}}(x) \\
&= E_{x,\theta_{J^*}}\left[\int_{t=0}^{T} e^{-\alpha t}(u(X_t,\theta_t) - (H_{\theta_{J^*}}J^*)(X_t))dt + e^{-\alpha T}J^*(X_T)\right] \\
&\quad - E_{x,\theta_{J^*}}\left[\int_{t=0}^{T} e^{-\alpha T}u(X_t,\theta_t) + e^{-\alpha T}J_{\theta_{J^*}}(X_T)\right] \\
&= -E_{x,\theta_{J^*}}\left[\int_{t=0}^{T} e^{-\alpha t}(HJ^*)(X_t)dt\right] \\
&\leq -E_{x,\theta_{J^*}}\left[\int_{t=0}^{T\wedge T_A} e^{-\alpha t}(HJ^*)(X_t)dt\right] \\
&\leq -\frac{(HJ^*)(x)}{2}E_{x,\theta_{J^*}}\left[\int_{t=0}^{T\wedge T_A} e^{-\alpha t}dt\right],
\end{aligned}$$

where the second-to-last inequality follows from the fact that $HJ^* \geq 0$. We have a contradiction. It follows that $HJ^* \leq 0$. □

We close this section with a characterization of optimal policies in terms of greedy policies with respect to the optimal value function.

Theorem 16.3 *Under Assumption 16.1.1, a stationary policy $\theta \in \overline{\Theta}$ is optimal if $H_\theta J^* = 0$.*

Proof Suppose $H_\theta J^* = 0$. Then θ is a greedy strategy with respect to J^*. By Lemma 16.1,

$$\begin{aligned}
J^*(x) &= E_{x,\theta}\left[\int_0^{T_S} e^{-\alpha t}(u(X_t,\theta) - (H_\theta J^*)(X_t))dt\right] \\
&= E_{x,\theta}\left[\int_0^{T_S} e^{-\alpha t}u(X_t,\theta)dt\right] \\
&= J_\theta(x).
\end{aligned}$$

Hence, θ is an optimal stationary policy.

16.1.3 A Linear Programming Characterization

The HJB equation offers a characterization of the optimal value function. In this section, we discuss an alternative though closely related characterization which motivates the approximation algorithms of the next section. In particular, we consider the following optimization problem:

$$\begin{array}{ll} \text{minimize} & \int J(x)\rho(dx) \\ \text{subject to} & HJ \leq 0 \\ & J \in C^2, \end{array} \tag{16.2}$$

where the function J is the variable to be optimized and ρ is a prespecified positive definite measure. Note that the objective functional is linear. Furthermore, for each θ, $H_\theta J$ is affine in J, so each constraint $(HJ)(x) \leq 0$ can be converted into a continuum of linear constraints, each taking the form $(H_\theta J)(x) \leq 0$. As such, this optimization problem can be viewed as a linear program.

The following theorem establishes that the unique optimal solution to this linear program is the optimal value function J^*. Here, uniqueness is in the sense that two solutions can only differ on a zero measure set under measure ρ.

Theorem 16.4 *Under Assumption 16.1.1, J^* uniquely attains the optimum in (16.2).*

Proof From Theorem 16.2 we know that $HJ^* = 0$, so J^* is a feasible solution.

Consider an arbitrary feasible solution $J \in \mathcal{J}$. By Lemma 16.1, we have

$$J(x) = E_{x,\theta_{J^*}}\left[\int_0^{T_S} e^{-\alpha t}\left(u(X_t,\theta_t) - (H_{\theta_{J^*}}J)(X_t)\right)dt\right].$$

Moreover, since $HJ \leq 0$, we also have

$$J(x) \geq E_{x,\theta_{J^*}}\left[\int_0^{T_S} e^{-\alpha t}u(X_t,\theta_t)\,dt\right] = J_{\theta_{J^*}}(x) = J^*(x).$$

Since ρ is a positive definite measure and $\int J^*(x)\rho(dx) < \infty$, only J^* can obtain the optimum. □

16.2 Approximate Dynamic Programming

It is not generally possible to solve the linear program (16.2), as there are infinite numbers of variables and constraints. In this section, we present approximation methods that extend analogous ones previously developed for discrete-time stochastic control problems (Schweitzer and Seidmann 1985; Trick and Zin 1997; de Farias and Van Roy 2003, 2004, 2006).

16.2.1 Approximation via Basis Functions

The approach we consider approximates the optimal value function J^* using a linear combination $\sum_{k=1}^{K} r_k \phi_k$ of preselected basis functions[1] $\phi_1, \ldots, \phi_K \in C^2$. Here, $r \in \Re^K$ denotes a vector of weights. The basis functions are handcrafted using intuition about the form of the value function, and the choice may be improved via trial-and-error. Given a set of basis functions, weights are computed to fit the optimal value function. We will present in this section algorithms for computing weights. The case studies of Section 16.3 will serve to illustrate, among other things, examples of basis functions.

Let us first briefly discuss how an approximation may be used to control the system. The idea is to employ a greedy policy with respect to this approximation in lieu of an optimal one, which would be greedy with respect to the optimal value function. Note that when at a state $x \in \mathcal{S}$, a greedy action with respect to $\sum_{k=1}^{K} r_k \phi_k$ can be obtained in real time by solving

$$\max_{\psi \in \Psi} \left\{ \sum_{k=1}^{K} r_k \left((\nabla_x \phi_k)^\top (x) \mu(x, \psi) + \tfrac{1}{2} \mathrm{tr}\left((\nabla_x^2 \phi_k)(x) \sigma(x, \psi) \sigma(x, \psi)^\top \right) \right) + u(x, \psi) \right\}. \tag{16.3}$$

This is a constrained nonlinear program, and one might consider solving for a local optimum using an interior point method. It is worth noting, though, that in many contexts of practical relevance, such expressions are amenable to efficient global optimization. For example, in the portfolio optimization problems to be discussed in Section 16.3, this optimization problem is a convex quadratic program.

As a framework for computing weights, we consider a variation of the linear program (16.2) that characterizes the optimal value function. We maintain the same constraints but restrict the solution space to the span of the basis functions. The resulting linear program takes the form

$$\begin{array}{ll} \text{minimize} & \int (\Phi r)(x) \rho(dx) \\ \text{subject to} & H \Phi r \leq 0, \end{array} \tag{16.4}$$

where $\Phi r = \sum_{k=1}^{K} r_k \phi_k$. Note that the C^2 constraint is not needed here because the basis functions are themselves in C^2. As established in Han (2005), this linear program is feasible if the constant function lies in the span of the basis functions and, further, under suitable conditions, the solution to this linear program converges to the solution of (16.2) as K grows.

This new linear program has a finite number of variables. However, the number of constraints remains infinite. There is one linear constraint per state-action pair. Further, the objective function involves an integral over states. In the next section,

---

[1] The term *basis functions* is commonly used in the approximate dynamic programming literature to refer to a set of functions that form a basis for the space from which an approximation is generated.

we present a method for alleviating the dependence on the number of states through randomized sampling. We will later discuss methods that alleviate the dependence on the number of actions.

It is worth mentioning that the choice of measure ρ influences the set of optimal weight vectors for our new linear program (16.4). This is in contrast with the previous linear program (16.2) for which, by Theorem 16.4, the unique optimal solution is the optimal value function J^* regardless of the choice of ρ. There is no clear understanding of how ρ should be chosen in (16.4), though results in de Farias and Van Roy (2003, 2006) suggest desirability of a discounted relative frequency measure associated with the greedy policy that will ultimately be derived from the resulting weight vector. The case studies reported in Section 16.3 make use of measures chosen in this spirit.

16.2.2 Sampling States

We consider approximating the linear program (16.4) using one that replaces the integral with a sum over a sampled set of states and retains only constraints associated with this sampled set. In particular, this method generates a set of Q independent identically distributed samples $x^{(1)}, \ldots, x^{(Q)} \in \mathcal{S}$, drawn according to the positive definite measure ρ, and then solves a linear program of the form

$$\begin{aligned} &\text{minimize } \sum_{i=1}^{Q} (\Phi r)(x^{(i)}) \\ &\text{subject to } (H\Phi r)(x^{(i)}) \leq 0, \quad \forall i = 1, \ldots, Q. \end{aligned} \tag{16.5}$$

This method is entirely analogous to one developed for discrete-time problems in de Farias and Van Roy (2004), which also offers theoretical results to motivate why sampling may not distort the solution too much. Here we only provide some less formal heuristic motivation. For any r, the sum $\sum_{i=1}^{Q}(\Phi r)(x^{(i)})$ offers an unbiased estimate of the integral $\int(\Phi r)(x)\rho(dx)$. Intuitively, as the number of samples grows the estimate should become accurate uniformly across relevant values of r and, therefore, replacing the integral with the sum seems reasonable. Regarding constraints, note that the linear program (16.4) defines a set in $\Re^K$ using an infinite number of constraints. As such, one may expect almost all of the constraints to be irrelevant. A tractable subset may lead to a reasonable approximation.

Our new linear program (16.5) alleviates the dependence on the number of states. The objective involves a sum over Q sampled states rather than the entire state space. The number of constraints $(H\Phi r)(x^{(i)}) \leq 0$ also scales with the number of sampled states. However, there are still as many linear constraints as there are actions. In particular, each constraint $(H\Phi r)(x^{(i)}) \leq 0$ is equivalent to a set of linear constraints: $(H_\psi \Phi r)(x^{(i)}) \leq 0$ for all $\psi \in \Psi$. In the next section, we explain two methods for alleviating this dependence on the number of actions.

16.2.3 Dealing with the Action Space

The two methods we will present pose different merits. The first is guaranteed to obtain an optimal solution to (16.5). However, it requires solving a nonlinear convex program which cannot always be represented in a manner amenable to efficient computation. Further, even when an appropriate representation exists, though this convex program can be solved in polynomial time, for the large instances that arise in practical applications, computational requirements can be onerous. The second method is heuristic and comes with no guarantee of finding an optimal solution to (16.5). However, it relies on solving linear rather than nonlinear convex programs, and in computational studies, this method has proved to be effective and more efficient. The method solves a sequence of linear programs, each with Q constraints. The set of constraints is adapted based on results of each iteration.

16.2.3.1 Convex Programming

Note that (16.5) is itself a convex program since the objective function is linear and the constraint associated with each sampled state is convex. However, the left-hand side of each constraint $(H\Phi r)(x^{(i)}) \leq 0$ is the result of an optimization problem:

$$\begin{aligned}(H\Phi r)(x^{(i)}) &= \max_{\psi\in\Psi}(H_\psi \Phi r)(x^{(i)})\\ &= \max_{\psi\in\Psi}\left\{\sum_{k=1}^{K} r_k\Big((\nabla_x\phi_k)^\top(x^{(i)})\mu(x^{(i)},\psi)\right.\\ &\quad+\frac{1}{2}\mathrm{tr}\left((\nabla_x^2\phi_k)(x^{(i)})\sigma(x^{(i)},\psi)\sigma(x^{(i)},\psi)^\top\right)\\ &\quad\left.-\alpha\phi_k(x^{(i)})\Big)+u(x^{(i)},\psi)\right\}.\end{aligned}$$

If this optimization problem can be solved in closed form to obtain a simple expression for the maximum as a function of r, there is hope of representing the convex constraint in a way that is amenable to efficient computation. However, this is unlikely to be the case for a constrained optimization problem. As such, we consider converting it to an unconstrained problem by pricing out the constraints.

We will assume that the action set Φ is a polytope, represented by L linear inequalities. In particular, there is a matrix A and vector b such that

$$\Psi = \left\{\psi \in \Re^M | A\psi \leq b\right\}.$$

Then, by Lagrangian duality,

$$(H\Phi r)(x^{(i)}) = \min_{\lambda\in\Re_+^L}\ \max_{\psi\in\Re^M}((H_\psi \Phi r)(x^{(i)}) - \lambda^\top(A\psi - b)).$$

Now, for fixed λ, the maximization problem is unconstrained. Our convex programming approach is applicable in cases where this unconstrained maximization problem can be solved in closed form to obtain a simple expression for the maximum as a function of r and λ. The case studies of Section 16.3, for example, fall into this category because in each case the expression being maximized is a concave quadratic.

Let

$$g_i(r, \lambda) = \max_{\psi \in \Re^M} ((H_\psi \Phi r)(x) - \lambda^\top (A\psi - b)).$$

Recall that the constraint we wish to impose is $(H\Phi r)(x^{(i)}) \leq 0$. This is equivalent to imposing $\min_{\lambda \in \Re^L_+} g_i(r, \lambda) \leq 0$, which is equivalent in turn to imposing a pair of constraints

$$\begin{aligned} g_i(r, \lambda) &\leq 0 \\ \lambda &\geq 0. \end{aligned}$$

In particular, r satisfies the constraint if there exists a $\lambda \in \Re^L_+$ such that $g_i(r, \lambda) \leq 0$. Now, if $g_i(r, \lambda)$ is a simple expression, we arrive at a convex program represented in a manner amenable to efficient computation:

$$\begin{aligned} &\text{minimize } \sum_{i=1}^{Q} (\Phi r)(x^{(i)}) \\ &\text{subject to } g_i(r, \lambda^{(i)}) \leq 0, \ \forall i = 1, \ldots, Q \\ &\qquad\qquad \lambda^{(i)} \geq 0, \qquad\quad \forall i = 1, \ldots, Q. \end{aligned} \tag{16.6}$$

16.2.3.2 Adaptive Constraint Selection

In many cases of practical interest $g_i(r, \lambda)$ can indeed be written as a simple expression. However, when this is not the case or when the resulting convex program (16.6) is computationally burdensome, an alternative method is called for. We now describe a heuristic which in computational studies has proved to be effective and much more efficient.

Our heuristic solves a sequence of linear programs, each with Q linear constraints. In each iteration, one constraint is chosen per sampled state $x^{(i)}$. For each of these states, the choice is governed by the action that is greedy with respect to the weight vector generated in the previous iteration. A more detailed description of the heuristic follows:

The linear program solved in line 5 involves K variables and Q linear constraints. Use of this heuristic requires computation of greedy actions as well. The associated optimization problem of line 3 may be an arbitrary nonlinear program, but in many contexts of practical interest, the problem is a simple convex program. In the case studies of Section 16.3, for example, this problem is a convex quadratic program.

Algorithm 1 Adaptive constraint selection.

for $t = 1$ to ∞ **do**
  for $i = 1$ to Q **do**

$$\psi^{(i)} \in \underset{\psi \in \Psi}{\operatorname{argmax}} (H_\psi \Phi r^{(t-1)})(x^{(i)})$$

  end for

$$\begin{array}{rll} r^{(t)} \in & \operatorname{argmin} & \sum_{i=1}^{Q} (\Phi r)(x^{(i)}) \\ & \text{subject to} & (H_{\psi^{(i)}} \Phi r)(x^{(i)}) \leq 0, \quad \forall i = 1, \ldots, Q. \end{array}$$

end for

The adaptive constraint selection algorithm does not necessarily generate an optimal solution to the linear program (16.5) it aims to solve. However, if the sequence $r^{(t)}$ converges, it must converge to an optimal solution. To see why, consider a limit of convergence $\overline{r}$. This limit must attain the minimum of

$$\begin{array}{ll} \text{minimize} & \sum_{i=1}^{Q} (\Phi r)(x^{(i)}) \\ \text{subject to} & (H_{\psi^{(i)}} \Phi r)(x^{(i)}) \leq 0, \quad \forall i = 1, \ldots, Q, \end{array} \tag{16.7}$$

where each action $\psi^{(i)}$ is greedy with respect to $\Phi \overline{r}$. Now assume for contradiction that $\overline{r}$ is not an optimal solution of (16.5) and let r^* be an optimal solution. Then,

$$\sum_{i=1}^{Q} (\Phi r^*)(x^{(i)}) < \sum_{i=1}^{Q} (\Phi \overline{r})(x^{(i)})$$

and

$$(H_{\psi^{(i)}} \Phi r^*)(x^{(i)}) \leq (H \Phi r^*)(x^{(i)}) \leq 0, \quad \forall i = 1, \ldots, Q.$$

It follows that r^* is a feasible solution to (16.7) with a smaller objective value than $\overline{r}$, yielding a contradiction.

16.3 Case Studies in Portfolio Optimization

In this section we will formulate a dynamic portfolio optimization model and present two case studies that apply the approximation approach of the previous section. There is significant prior work on approximate dynamic programming methods for portfolio optimization (Judd and Gaspar 1997; Brandt 1999; Munk 2000; Barberis 2000; Lynch 2001; Brandt et al. 2003; Wang 2003; Haugh et al. 2006). The primary source of difference in our work, which was originally reported in the first author's dissertation (Han 2005), is the algorithm used. In particular, it is based on linear programming and works directly with the controlled diffusion (without requiring discretization). These features enable more efficient computations and a more streamlined experimental process, involving far less tinkering.

16.3.1 Market Model

We consider a market with M risky assets and one risk-free asset. Prices follow a diffusion processes with drift and diffusion coefficients modulated by market state. Let $B = (B^{(1)}, \ldots, B^{(D)})^\top$ be a standard Brownian motion in $\Re^D$, and let $\mathbf{F} = \{\mathcal{F}_t | t \geq 0\}$ be the associated filtration.

At each time t, the state of the market is described by a vector $Z_t \in \Re^N$, which evolves according to

$$dZ_t = \mu_z(Z_t)\,dt + \sigma_z(Z_t)\,dB_t,$$

where $\mu_z : \Re^N \to \Re^N$ and $\sigma_z : \Re^N \to \Re^{N \times D}$ are Lipschitz continuous functions. We assume that Z_0 is constant.

There is an instantaneously risk-free asset, which we will refer to as the money market, whose price $S_t^{(0)}$ follows

$$dS_t^{(0)} = r(Z_t)S_t^{(0)}\,dt,$$

where $r : \Re^N \to \Re$ is a Lipschitz continuous function such that $r(z) \geq 0, \forall z \in \Re^N$, meaning that the interest rate is always nonnegative.

Further, there are M risky assets, whose prices $S_t = (S_t^{(1)}, \ldots, S_t^{(M)})^\top$ follow

$$\frac{dS_t}{S_t} = \mu_s(Z_t)\,dt + \sigma_s(Z_t)\,dB_t,$$

where $\mu_s : \Re^N \to \Re^M$, $\sigma_s : \Re^N \to \Re^{M \times D}$, and $\mathrm{tr}(\sigma_s(\cdot)\sigma_s(\cdot)^\top) : \Re^N \to \Re$ are Lipschitz continuous functions, and $\sigma_s(z)$ has full rank for all $z \in \Re^N$.

Note that Merton's classical intertemporal model is a special case of this model in which r, μ_s, and σ_s are constants. The model we consider generalizes this classical model by allowing the set of investment opportunities to vary with a stochastic market state process.

16.3.2 Portfolio Choice

We consider an investor who manages a portfolio, which he can rebalance at any time without incurring any transaction cost. The portfolio can include long and short positions in any asset, though there can be margin constraints. In mathematical terms, the portfolio is represented as a vector $\psi \in \Psi$, where Ψ is a polytope in $\Re^M$ containing the origin. Each component $\psi^{(m)}$ indicates the fraction of an investor's wealth invested in the mth risky asset. The fraction of wealth invested in the money market is denoted by

$$\psi^{(0)} = 1 - \sum_{m=1}^{M} \psi^{(m)}(x).$$

Note that these fractions can be greater than one or negative, because an investor may trade on margin or sell short.

From the perspective of an investor with wealth $W_t > 0$, the state at time t is $X_t = (Z_t, W_t) \in \Re^N \times \Re_{++}$. Let the state space be denoted by $\mathcal{S} = \Re^N \times \Re_{++}$. A *portfolio function* $\theta : \mathcal{S} \to \Psi$ maps the state to a portfolio. Let the set of portfolio functions be denoted by Θ.

A *portfolio strategy* is a continuum of portfolio functions $\theta = \{\theta_t \in \Theta | t \geq 0\}$. If an investor employs a portfolio strategy θ, his wealth process follows

$$\frac{dW_t}{W_t} = \left(r(Z_t) + \theta_t(X_t)^\top \lambda(Z_t)\right) dt + \theta_t(X_t)^\top \sigma_s(Z_t)\, dB_t, \qquad (16.8)$$

where $\lambda(z) = \mu(z) - r(z)\mathbf{1}$ is the vector of excess rates of return.

We denote the set of admissible portfolio strategies by $\overline{\Theta}$. A portfolio strategy θ is *stationary* if for all x, t, and s, $\theta_t(x) = \theta_s(x)$. When referring to a portfolio function $\theta_t \in \Theta$ associated with a stationary strategy θ, we will drop the subscript, denoting the function by $\theta \in \Theta$. Since there is a one-to-one correspondence between portfolio functions and stationary strategies, we will also denote the set of stationary strategies by Θ.

16.3.3 Utility Function

We represent investor preferences using a power utility function over his wealth w:

$$u_\beta(w) = \frac{w^{1-\beta}}{1-\beta}.$$

Here, the parameter $\beta > 0$ captures an investor's level of risk aversion. In fact, the Arrow–Pratt coefficient of relative risk aversion for this power utility function is

$$-\frac{w u_\beta''(w)}{u_\beta'(w)} = \beta,$$

which is a constant. Note that the limiting case of $\beta \to 1$ concurs with the logarithmic utility function, whose coefficient of relative risk aversion is 1. It is well known that when the investor has logarithmic utility, myopic strategies are optimal.

We aim to maximize expected utility at a random time $\tilde{\tau}$, which is independent of the evolution of the market and is exponentially distributed with mean $\tau > 0$. Hence, the optimization problem of interest is

$$\sup_{\theta \in \overline{\Theta}} E_{x,\theta} \left[u_\beta(W_{\tilde{\tau}}) \right]$$

or, equivalently,

$$\sup_{\theta \in \overline{\Theta}} E_{x,\theta} \left[\int_{t=0}^{\infty} e^{-t/\tau} u_\beta(W_t) dt \right].$$

The subscripts in the expectation indicate that the initial state is $X_0 = x = (z, w)$ and that the portfolio strategy $\theta \in \overline{\Theta}$ is employed. We will generally suppress the parameter β and use $u(w)$ to denote $u_\beta(w)$.

16.3.4 Relation to the Controlled Diffusion Framework

The correspondence between our dynamic portfolio optimization problem and the controlled diffusion framework described in Section 16.1 is straightforward. The portfolio optimization problem is a controlled diffusion for which states are comprised of market states and wealth levels and for which actions are portfolio choices. The drift and diffusion functions μ and σ can be derived from those of the market state (μ_z and σ_z) and those of the asset prices (μ_s and σ_s). Utility in the portfolio optimization problem is a function of state through wealth. The discount rate is $\alpha = 1/\tau$.

Assumptions on the problem primitives posed in Section 16.1 carry over with one notable exception that the state space in the portfolio optimization problem is not compact. When the state space is noncompact, additional technical conditions are required to support HJB equation and linear programming characterizations of the optimal value function. Such technical conditions and results for the portfolio optimization context are provided in the first author's dissertation (Han 2005). We will not discuss them here. Rather, we will focus on application of our approximation algorithms, which apply readily to problems with compact or noncompact state spaces.

16.3.5 Special Structure

The portfolio optimization problem possesses special structure that simplifies computation of greedy actions and facilitates basis function selection. First of all, $(H_\psi J)(x)$ is a concave quadratic function of ψ. Specifically,

$$
\begin{aligned}
(H_\psi J)(x) = {} & J_w(x)w\big(\psi^\top\lambda(z)+r(z)\big)+\frac{1}{2}J_{ww}(x)w^2\psi^\top\sigma_s(z)\sigma_s(z)^\top\psi \\
& +J_z(x)^\top\mu_z(z)+wJ_{wz}(x)\sigma_z(z)\sigma(z)^\top\psi \\
& +\frac{1}{2}\mathrm{tr}\big(J_{zz}(x)\sigma_z(z)\sigma_z(z)^\top\big) \\
& -J(x)/\tau+u(w),
\end{aligned}
$$

where $x=(z,w)$. Since Ψ is a polytope, the problem of computing a greedy action is a convex quadratic program.

The use of a power utility function gives rise to some useful special structure in the optimal value function, as captured by the following well-known result.

Theorem 16.5 *Under Assumption 16.1.1, there exists a twice continuously differentiable function* $V^*:\Re^N\to\Re$ *such that*

$$
J^*(z,w)=u(w)V^*(z),\quad \forall w>0, z\in\Re^N.
$$

Proof By definition, we have

$$
\begin{aligned}
J^*(z,w) &= \sup_{\theta\in\overline{\Theta}} J_\theta(z,w) \\
&= \sup_{\theta\in\overline{\Theta}} E_{(z,w),\theta}\left[\int_{t=0}^{\infty} e^{-t/\tau}u(w_t)\,dt\right] \\
&= \sup_{\theta\in\overline{\Theta}} w^{1-\beta}E_{(z,w),\theta}\left[\int_{t=0}^{\infty} e^{-t/\tau}u\left(\frac{w_t}{w}\right)\,dt\right] \\
&= w^{1-\beta}\sup_{\theta\in\overline{\Theta}} E_{(z,1),\theta}\left[\int_{t=0}^{\infty} e^{-t/\tau}u(w_t)\,dt\right].
\end{aligned}
$$

So if we define $V^*:\Re^N\to\Re$ by

$$
V^*(z)=(1-\beta)\sup_{\theta\in\overline{\Theta}} E_{(z,1),\theta}\left[\int_{t=0}^{\infty} e^{-t/\tau}u(w_t)\,dt\right],
$$

then we have

$$
J^*(z,w)=u(w)V^*(z).
$$

□

For value functions J that factor in the same way as J^*, the following theorem reflects the dependence of $H_\psi J$ on wealth.

Theorem 16.6 *For all $\psi \in \Psi$ and $J \in C^2$ for which $J(z,w) = u(w)V(z)$,*

$$\frac{(H_\psi J)(z,w)}{u(w)} = \frac{(H_\psi J)(z,\overline{w})}{u(\overline{w})},$$

for any $(z,w) \in \mathcal{S}$ and $\overline{w} \in \Re_+$,

Proof We have

$$\begin{aligned}
(H_\psi J)(x) = {} & J_w(x)w\big(\psi^\top \lambda(z) + r(z)\big) + \frac{1}{2}J_{ww}(x)w^2\psi^\top \sigma_s(z)\sigma_s(z)^\top \psi \\
& + J_z(x)^\top \mu_z(z) + wJ_{wz}(x)\sigma_z(z)\sigma(z)^\top \psi \\
& + \frac{1}{2}\mathrm{tr}\big(J_{zz}(x)\sigma_z(z)\sigma_z(z)^\top\big) \\
& - J(x)/\tau + u(w) \\
= {} & (1-\beta)u(w)V(z)\big(\psi^\top \lambda(z) + r(z)\big) + \frac{1}{2}\beta(\beta-1)u(w)V(z)\psi^\top \sigma_s(z)\sigma_s(z)^\top \psi \\
& + u(w)V_z(z)^\top \mu_z(z) + (1-\beta)u(w)V_z(z)\sigma_z(z)\sigma(z)^\top \psi \\
& + \frac{1}{2}\mathrm{tr}\big(u(w)V_{zz}(a)\sigma_z(z)\sigma_z(z)^\top\big) \\
& - u(w)V(z)/\tau + u(w) \\
= {} & u(w)\Bigg(1-\beta)V(z)\big(\psi^\top \lambda(z) + r(z)\big) + \frac{1}{2}\beta(\beta-1)V(z)\psi^\top \sigma_s(z)\sigma_s(z)^\top \psi \\
& + V_z(z)^\top \mu_z(z) + (1-\beta)V_z(z)\sigma_z(z)\sigma(z)^\top \psi \\
& + \frac{1}{2}\mathrm{tr}\big(V_{zz}(a)\sigma_z(z)\sigma_z(z)^\top\big) \\
& - u(w)V(z)/\tau + 1\Bigg).
\end{aligned}$$

The result follows. □

An immediate corollary of Theorem 16.6 is a well-known result concerning optimal portfolio strategies for investors with power utility.

Corollary 16.7 *There exists a policy $\theta \in \Theta$ that is greedy with respect to J^* for which $\theta(z,w)$ does not depend on w for $(z,w) \in \mathcal{S}$.*

16.3.6 Factorization of Basis Functions

In light of Theorem 16.5, the function $J^*(z,w)$ that we wish to approximate factors into $u(w)V^*(z)$. We know u but not V^*, which is in essence what we need to approximate. As such, it is natural to choose basis functions that factor in the same way. In particular, we will use basis functions that take the form $\phi_k(z,w) = u(w)\tilde{\phi}_k(z)$ for some functions $\tilde{\phi}_k(z)$. Then, weights r are computed with an aim of approximating J^* by $\sum_{k=1}^K r_k \phi_k$ or, equivalently, V^* by $\sum_{k=1}^K r_k \tilde{\phi}_k$.

By Theorems 16.5 and 16.6, the linear program (16.5) we aim to solve can be rewritten as

$$\begin{array}{ll} \text{minimize} & \sum_{i=1}^{Q} u(w^{(i)}) \sum_{k=1}^{K} r_k \tilde{\phi}_k(z^{(i)}) \\ \text{subject to} & (H_{\psi^{(i)}} \Phi r)(z^{(i)}, 1) \leq 0, \qquad \forall i = 1, \ldots, Q, \end{array}$$

where $x^{(i)} = (z^{(i)}, w^{(i)})$. Hence, the $w^{(i)}$ samples do not enter into the constraints and only influence the objective by weighting the values $\sum_{k=1}^K r_k \tilde{\phi}_k(z^{(i)})$ associated with $z^{(i)}$ samples.

16.3.7 Measure and Sampling

Loosely guided by the results of de Farias and Van Roy (2003, 2006), we consider a measure ρ associated with discounted relative state frequencies. First, define the discounted relative frequency measure for the market state process:

$$\tilde{\rho}(dz) = \frac{1}{\tau} E\left[\int_{t=0}^{\infty} e^{-t/\tau} \mathbf{1}(Z_t \in dz) dt\right].$$

The relative frequencies associated with X_t depend also on the evolution of wealth and therefore on the portfolio strategy in use. It is not clear how a measure should be defined here, but as a simple heuristic, we will use the measure

$$\rho(dz, dw) = \tilde{\rho}(dz) \mathbf{1}(1 \in dw).$$

Note that the conditional measure over w, conditioned on z, bears no impact on the sampling of constraints since they do not depend on the $w^{(i)}$s. There is, however, some impact on the objective through the weights $u(w^{(i)})$.

To construct the linear program, we must generate Q independent identically distributed state samples. Since wealth is constant under our measure, we need only to sample $z^{(1)}, \ldots, z^{(Q)}$. We generate each sample by simulating a discrete-time approximation to the market state dynamics, terminating the simulation at an exponentially distributed stopping time with expectation τ. We take the state at the time of termination as a sample.

16.3.8 Case Studies

In this section, we will present numerical results from two cases that make use of the adaptive constraint selection algorithm (1).[2] The first involves a problem that admits a simple closed-form solution and serves as a sanity test for our algorithm. For this problem, the algorithm appears to always deliver exact solutions. The second case study involves a 10-factor term structure model for which exact solution is likely to be intractable. Our computations make use of ILOG CPLEX to solve linear programs and quadratic programs.

16.3.8.1 Case Study 1: Constant Investment Opportunities

In Merton's classical dynamic portfolio optimization model (Merton 1971), the set of investment opportunities is constant. This enables solution of the HJB equation and derivation of an optimal portfolio strategy in closed form. We will consider such a model. Specifically, we consider constant asset price drift and diffusion functions:

$$\mu_s(z) = \mu_s, \quad \sigma_s(z) = \sigma_s, \quad r(z) = r, \quad \lambda(z) = \lambda = \mu_s - r.$$

We impose no constraints on portfolio choice, so $\Psi = \Re^M$.

In this model, the drift and diffusion functions do not depend on any market state. It is therefore natural to think of the market state being constant. However, this would lead to a trivial computational problem as there would be no function V^* to approximate. Our intention here is to test our approximation algorithm, and, as such, we model the market state as a five-dimensional Ornstein–Uhlenbeck process:

$$dz_t = -z_t\, dt + \sigma_z\, dB_t,$$

where $\sigma_z \in \Re^{5\times 5}$ is a constant and full rank matrix and B_t is a five-dimensional Brownian motion. The function V^* now maps $\Re^5$ to $\Re$, but is constant. We try approximating V^* using a polynomial to see whether our algorithm produces the desired constant function.

We first provide an analytic derivation of the optimal value function and policy. The HJB equation can be written in terms of V^* as

$$\begin{aligned}0 = \min_{\theta\in\Theta} \Bigg\{ & (1-\beta)V^*(z)\big(\theta(x)^\top\lambda + r\big) + \frac{\beta(\beta-1)}{2}V^*(z)\theta(x)^\top\sigma_s\sigma_s^\top\theta(x) \\ & + V_z^*(z)^\top\mu_z + (1-\beta)V_z^*(z)^\top\sigma_z(z)\sigma_s^\top\theta(x) \\ & + \frac{1}{2}\mathrm{tr}\big(V_{zz}^*(z)\sigma_z(z)\sigma_z(z)^\top\big) - V^*(z)/\tau + 1\Bigg\}.\end{aligned}$$

[2] We experimented with the convex programming approach, but found that not to be efficient enough to address problems of practical scale.

The first-order condition gives a candidate for the optimal strategy:

$$\theta_V^*(z) = \frac{1}{\beta}\left(\sigma_s\sigma_s^\top\right)^{-1}\lambda + \frac{1}{\beta V^*(z)}\left(\sigma_s\sigma_s^\top\right)^{-1}\sigma_s\sigma_z(z)^\top V_z^*(z).$$

Plugging this portfolio strategy into the HJB equation leads to

$$0 = \frac{1}{V^*} + ((1-\beta)r - \frac{1}{\tau}) + \frac{1-\beta}{2\beta}\lambda^\top(\sigma_s\sigma_s^\top)^{-1}\lambda + \frac{1-\beta}{\beta}\lambda^\top(\sigma_s\sigma_s^\top)^{-1}\sigma_s\sigma_z^\top\left(\frac{V_z^*}{V^*}\right)$$
$$+ \frac{1-\beta}{2\beta}\left(\frac{V_z^*}{V^*}\right)^\top\sigma_z\sigma_s^\top(\sigma_s\sigma_s^\top)^{-1}\sigma_s\sigma_z^\top\left(\frac{V_z^*}{V^*}\right) + \mu_z^\top\left(\frac{V_z^*}{V^*}\right) + \frac{1}{2}\mathrm{tr}\left(\frac{V_{zz}^*}{V^*}\sigma_z\sigma_z^\top\right).$$

It is easy to check that this equation has a constant solution:

$$V^*(z) = \left(\frac{1}{\tau} - (1-\beta)r - \frac{1-\beta}{2\beta}\lambda^\top\left(\sigma_s\sigma_s^\top\right)^{-1}\lambda\right)^{-1}.$$

Hence, the HJB equation is solved by

$$J^*(z,w) = u(w)V^*(z).$$

Using techniques from Duffie (2001, chapter 9), it can be proved that J^* and θ^* are indeed the optimal value function and an optimal policy, respectively

We now consider application of our approximation algorithm. We will employ as basis functions tensor products of Chebyshev polynomials. The Chebyshev polynomials are

$$\begin{aligned}
P_0(x) &= 1,\\
P_1(x) &= x,\\
P_2(x) &= 2x^2 - 1,\\
P_3(x) &= 4x^3 - 3x,\\
P_4(x) &= 8x^4 - 8x^2 + 1,\\
P_5(x) &= 16x^5 - 20x^3 + 5x,\\
P_6(x) &= 32x^6 - 48x^4 + 18x^2 - 1,\\
&\vdots
\end{aligned}$$

Chebyshev polynomials are orthogonal, and, in experiments, the orthogonality appears to avoid numerical instabilities that can arise in computations. The basis functions we use are the complete polynomials up to third degree:

$$\left\{P_i(z^{(a)})P_j(z^{(b)})P_k(z^{(c)}), \quad i,j,k \geq 0, i+j+k \leq 3, a \neq b, b \neq c, c \neq a\right\}.$$

We tried the adaptive constraint selection algorithm many times with different problem data and independently sampled sets each of 500, 5,000, or 50,000 states. Each time, the approximation converged to the correct function

$$V^*(z) = \left(\frac{1}{\tau} - (1-\beta)r - \frac{1-\beta}{2\beta}\lambda^\top \left(\sigma\sigma^\top\right)^{-1}\lambda\right)^{-1}$$

within two iterations.

16.3.8.2 Case Study 2: 10-Factor Term Structure Model

In this section we consider a high-dimensional problem for which an exact solution is likely to be intractable. We use a 10-factor CIR model (Cox et al. 1985), and our choice of problem data is guided by results of the empirical study of a three-factor CIR model (?, "Unpublished").

The market state $Z_t \in \Re^{10}$ follows

$$dZ_t^{(i)} = \kappa_i(\zeta_i - Z_t^{(i)})\,dt + \sigma_i\sqrt{Z_t^{(i)}}\,dB_t^{(i)}, \quad i = 1, 2, \ldots, 10,$$

where κ_i, ζ_i, and σ_i are all positive constants. So

$$\mu_z(z) = \begin{pmatrix} \kappa_1 & 0 & \cdots & 0 \\ 0 & \kappa_2 & \cdots & 0 \\ \vdots & \vdots & \ddots & \vdots \\ 0 & 0 & \cdots & \kappa_{10} \end{pmatrix} \left(\begin{pmatrix} \zeta_1 \\ \zeta_2 \\ \vdots \\ \zeta_{10} \end{pmatrix} - z\right),$$

$$\sigma_z(z) = \begin{pmatrix} \sigma_1\sqrt{z^{(1)}} & 0 & \cdots & 0 \\ 0 & \sigma_2\sqrt{z^{(2)}} & \cdots & 0 \\ \vdots & \vdots & \ddots & \vdots \\ 0 & 0 & \cdots & \sigma_{10}\sqrt{z^{(10)}} \end{pmatrix}.$$

The spot rate r as a function of market state is given by

$$r(z) = z^{(1)} + z^{(2)} + \cdots + z^{(10)}.$$

Let $P(z_t, T-t)$ be the price of a zero coupon bond with $T-t$ periods until maturity. This price is given by

$$P(z, T-t) = \prod_{i=1}^{10}\left(A_i(T-t)e^{-B_i(T-t)z^{(i)}}\right),$$

where

$$A_i(T) = \left[\frac{2\gamma_i e^{\frac{1}{2}(\kappa_i+\lambda_i-\gamma_i)T}}{2\gamma_i e^{-\gamma_i T} + (\kappa_i + \lambda_i + \gamma_i)(1 - e^{-\gamma_i T})}\right]^{\frac{2\kappa_i \zeta_i}{\sigma_i^2}},$$

$$B_i(T) = \frac{2(1 - e^{-\gamma_i T})}{2\gamma_i e^{-\gamma_i T} + (\kappa_i + \lambda_i + \gamma_i)(1 - e^{-\gamma_i T})},$$

where

$$\gamma_i = \sqrt{(\kappa_i + \lambda_i)^2 + 2\sigma_i^2}.$$

Price dynamics follow

$$\frac{dP(z_t, T-t)}{P(z_t, T-t)} = \sum_{i=1}^{10} \left(z^{(i)}[1 - \lambda_i B_i(T-t)]\,dt - B_i(T-t)\sigma_i\sqrt{z^{(i)}}\,dB_t^{(i)}\right).$$

So

$$\mu(z) = \begin{pmatrix} 1-\lambda_1 B_1(T_1-t) & 1-\lambda_2 B_2(T_1-t) & \cdots & 1-\lambda_{10} B_{10}(T_1-t) \\ 1-\lambda_1 B_1(T_2-t) & 1-\lambda_2 B_2(T_2-t) & \cdots & 1-\lambda_{10} B_{10}(T_2-t) \\ \vdots & \vdots & \ddots & \vdots \\ 1-\lambda_1 B_1(T_{10}-t) & 1-\lambda_2 B_2(T_{10}-t) & \cdots & 1-\lambda_{10} B_{10}(T_{10}-t) \end{pmatrix} z,$$

$$\sigma(z) = \begin{pmatrix} -B_1(T_1-t)\sigma_1\sqrt{z^{(1)}} & -B_2(T_1-t)\sigma_2\sqrt{z^{(2)}} & \cdots & -B_{10}(T_1-t)\sigma_{10}\sqrt{z^{(10)}} \\ -B_1(T_2-t)\sigma_1\sqrt{z^{(1)}} & -B_2(T_2-t)\sigma_2\sqrt{z^{(2)}} & \cdots & -B_{10}(T_2-t)\sigma_{10}\sqrt{z^{(10)}} \\ \vdots & \vdots & \ddots & \vdots \\ -B_1(T_{10}-t)\sigma_1\sqrt{z^{(1)}} & -B_2(T_{10}-t)\sigma_2\sqrt{z^{(2)}} & \cdots & -B_{10}(T_{10}-t)\sigma_{10}\sqrt{z^{(10)}} \end{pmatrix},$$

where T_i are the maturity of the ith zero coupon bond. In this case study, we let $T_i = i, \forall i = 1, \ldots, 10$.

An empirical study (**?**, "Unpublished") produced the following estimates for three-factor model of a real market:

$$\begin{aligned} \kappa &= (1.4298, 0.01694, 0.03510), \\ \zeta &= (0.04374, 0.002530, 0.003209), \\ \sigma &= (0.16049, 0.1054, 0.04960), \\ \lambda &= (-0.2468, -0.03411, -0.1569). \end{aligned}$$

Based on these estimates and an interest in considering a model of higher dimension, we will use the following parameter values:

$$\kappa = (1.4298, 0.01694, 0.03510, 0.03510, \ldots, 0.03510),$$
$$\zeta = (0.04374, 0.002530, 0.003209, 0.001, 0.001, \ldots, 0.001),$$
$$\sigma = (0.16049, 0.1054, 0.04960, 0.04960, \ldots, 0.04960),$$
$$\lambda = (-0.2468, -0.03411, -0.1569, -0.1569, \ldots, -0.1569).$$

Note that with this problem data, the first component of Z_t is generally an order of magnitude larger than any other component. With this in mind, we selected the following basis functions:

1. Chebyshev polynomials for $z^{(1)}$ up to sixth degree:

$$\{1, P_1(z^{(1)}), P_2(z^{(1)}), P_3(z^{(1)}), P_4(z^{(1)}), P_5(z^{(1)}), P_6(z^{(1)})\}.$$

2. Chebyshev polynomials for $z^{(i)}$ up to second degree, $\forall i = 2, \ldots, 10$:

$$\{P_1(z^{(i)}), P_2(z^{(i)}), i = 2, \ldots, 10\}.$$

3. Second-degree cross-product terms between the first component and the other components:

$$\{P_1(z^{(1)})P_1(z^{(i)}), i = 2, \ldots, 10\}.$$

This generates a total of 34 basis functions.

We sampled 10,000 market states by simulating 10,000 trajectories using a model with time discretized into steps of size 0.01. Each step here represents about half a calendar week, which seems like a reasonably small time period for fixed income portfolio rebalancing considerations. It takes about 20 min to execute the adaptive constraint selection algorithm and arrive at an approximation to the value function. We used C++ code in concert with ILOG CPLEX on a Sun Blade 2000 machine. We used as an initial state $z = (0.03, 0.00253, 0.003209, 0.001, 0.001, \ldots, 0.001)$, a horizon time $\tau = 2$, and several levels of risk aversion: $\beta \in \{1.1, 2, 3, 4, 5, 6\}$.

Since we do not know the optimal portfolio strategy, we will use simple heuristics as benchmarks for performance comparison. One heuristic we consider is the myopic strategy, which can be thought of as the greedy strategy with respect to an approximate value function $J(z, w) = u(w)$. A second heuristic we consider is the risk-free strategy, which maintains all funds in the money market, earning the instantaneous risk-free rate at each time.

Another basis for comparison is provided by the approximate value function Φr. In particular, if Φr satisfies the constraints $(H\Phi r)(x) \leq 0$ for all $x \in \mathcal{S}$ then $\Phi r \geq J^*$. In this case, $(\Phi r)(x_0)$ would provide an upper bound on performance of an optimal portfolio strategy. In our context, Φr satisfies only a sampled subset of such constraints. Regardless, one might think of $(\Phi r)(x_0)$ as an approximation to an upper bound and compare this to the performance of heuristic portfolio strategies.

One way to measure performance of a portfolio strategy θ is in terms of our objective

$$E_{x,\theta}\left[\int_{t=0}^{\infty} e^{-t/\tau} u_\beta(W_t)dt\right],$$

with $x = (z_0, 1)$. Let us denote this objective value by U. One issue with this measure of performance is that it can be difficult to interpret. We consider a more easily interpreted measure defined by the constant rate of return r^{ce} that would attain the same objective value. In particular, r^{ce} solves

$$U = \tau E\left[\int_{t=0}^{\infty} e^{-t/\tau} u_\beta(e^{r^{\text{ce}}t})dt\right]$$

and can be written as

$$r^{\text{ce}} = \frac{1}{\tau(1-\beta)} - \frac{1}{U(1-\beta)^2}. \tag{16.9}$$

We will refer to r^{ce} as the *certainty equivalent return rate* of the associated portfolio strategy.

We use Monte Carlo simulation to assess certainty equivalent return rates for portfolio strategies resulting from solving the linear program as well as myopic and risk-free strategies. This involves simulating sample paths for a discrete-time model to estimate the objective values and then converting them to estimates of certainty equivalent return rates according to (16.9). When simulating a discrete-time model, the portfolio is revised based on the strategy in use at the beginning of each time period. To obtain each objective value estimate, we simulate 40,000 sample paths for a model with time steps of duration 0.01. It takes hours to estimate certainty equivalent return rates for the three policies, and we have observed almost no difference in cases where we have compared estimates generated from 40,000 and 80,000 sample paths.

Table 16.1 presents certainty equivalent return rates from our experiments. Further, the rightmost column offers approximate upper bounds given by the approximate value function, evaluated at the initial state and converted to units of a certainty equivalent return rate. Our results indicate that strategies generated by the linear programming approach significantly outperform myopic strategies across a broad range of levels of risk aversion. Both types of strategies greatly outperform risk-

Table 16.1 Performance comparison

	Certainty equivalent return rate (%)			
β	LP strategy	Myopic strategy	Risk-free strategy	LP value
1.1	17.4218	17.2734	8.6298	17.3907
2	12.2776	12.1151	5.6263	12.2710
3	10.2067	10.0090	5.4624	10.1140
4	9.0417	8.8739	5.4563	8.9293
5	8.2736	8.0785	5.4578	8.1706
6	8.1728	7.9365	5.4653	8.1516

free strategies. Further, performance of the LP-based strategies generally exceed the approximate upper bounds.

16.4 Closing Remarks

The linear programming approach of this chapter extends one developed for discrete-time problems (Schweitzer and Seidmann 1985; Trick and Zin 1997; de Farias and Van Roy 2003, 2004, 2006). One might alternatively discretize a continuous problem, say using the techniques of Kushner and Dupuis (2001). However, there are several advantages to working directly with a continuous-time model. First, constraints in the discrete-time version of the linear program involve the discrete-time dynamic programming operator and therefore expressions with one-step expectations. Dealing with these expectations becomes computationally taxing as the dimension of the state space grows. Second, the optimization problems that must be solved to determine greedy actions are often more complex for discrete-time models. As we saw with portfolio optimization, these problems often amount to simple convex programs in continuous-time contexts. Finally, one might argue for aesthetics of working directly with a continuous-time model rather than an auxiliary model, especially when there are no practical benefits to discretization.

Many controlled diffusion problems can be addressed by the linear programming approach. In the area of dynamic portfolio optimization alone, it would be interesting to explore models involving transaction costs or taxes, either of which greatly increases model complexity. Some preliminary work that applies the approach to transaction cost optimization is reported in Yan (2005).

It should be noted, however, that the approach discussed in this chapter may need significant modification to deal either with diffusions exhibiting non-trivial boundary behavior or in settings in which one is dealing with a singular control problem.

References

Barberis, N.: Investing for the long run when returns are predictable. J. Financ. **55**(1), 225–265 (2000)

Borkar, V.: Optimal Control of Diffusion Processes. Longman, Harlow, England (1989)

Brandt, M.: Estimating portfolio and consumption choice: A conditional Euler equation approach. J. Financ. **54**(5), 1609–1645 (1999)

Brandt, M., Goyal, A., Santa-Clara, P., Stroud, J.: A simulation approach to dynamic portfolio choice with an application to learning about return predictability. Rev. Financ. Stud. 3, 831–873 (2003)

Cox, J., Ingersoll, J., Ross, S.: A theory of the term structure of interest rates. Econometrica **53**(2), 385–408, Mar (1985)

de Farias, D.P., Van Roy, B.: The linear programming approach to approximate dynamic programming. Oper. Res. **51**(6), 850–865 (2003)

de Farias, D.P., Van Roy, B.: On constraint sampling in the linear programming approach to approximate dynamic programming. Math. Oper. Res. **29**(3), 462–478 (2004) de Farias, D.P., Van Roy, B.: A cost-shaping linear program for average-cost approximate dynamic programming with performance guarantees. Math. Oper. Res. **31**(3), 597–620 (2006)

Duffie, D.: Dynamic Asset Pricing Theory 3rd edn Princeton University Press, Princeton, NJ (2001)

Han, J.: Dynamic portfolio management: An approximate linear programming approach. PhD thesis, Stanford University (2005)

Haugh, M., Kogan, L., Wu, Z.: Portfolio optimization with position constraints: An approximate dynamic programming approach. Working paper (2006)

Judd, K., Gaspar, J.: Solving large-scale rational-expectations models. Macroecon. Dyn. **1**, 45–75 (1997)

Krylov, N.V.: Controlled Diffusion Processes. Springer, New York, NY, (1980) Translated by A.B. Aries.

Kushner, H.J., Dupuis, P.G.: Numerical Methods for Stochastic Control Problems in Continuous Time. Springer, New York NY, (2001)

Lynch, A.: Portfolio choice and equity characteristics: characterizing the hedging demands induced by return predictability. J. Finan. Econ. **62**, 67–130 (2001)

Merton, R.: Optimum consumption and portfolio rules in a continuous-time model. J. Econ. Theory **3**, 373–413 (1971)

Munk, C.: Optimal consumption/investment policies with undiversifiable income risk and liquidity constraints. J. Econ. dynam. Control. **24**(9), 1315–1343 (2000)

Schweitzer, P., Seidmann, A.: Generalized polynomial approximations in Markovian decision processes. J. Math. Anal. Appl. **110**, 568–582 (1985)

Trick, M., Zin, S.: Spline approximations to value functions: A linear programming approach. Macroecon. Dyn. **1**, 255277 (1997)

Wang, A.: Approximate value iteration approaches to constrained dynamic portfolio problems. PhD thesis, MIT (2003)

Yan, X.: Transaction-cost-conscious pairs trading via approximate dynamic programming. Computer Science Honors Thesis, Stanford University (2005)

Index

A
Abatement, 318–319, 325
Action matrix, 215–216, 219–221, 219–227, 229, 233
Action space, 329–330, 337–339
Adaptive constraint selection, 338–339, 346, 348, 350
Affinely linear functions, 169
σ-Algebra, 70, 281
Algorithmically sound, 167
ALM, 258–259, 267–271
Alternative decision models, 288
Approximate distribution, 26, 68–69
Approximate dynamic programming, 329, 334, 339
Approximate solution, 67, 122, 141, 143–144, 146–147, 149, 153–154
Arrow–Pratt, 341
Arrow Pratt risk aversion parameter, 279
Asset allocation, 38, 97–135, 191, 259, 265, 267, 278
Asset–liability, 258, 265
Asset and liability management (ALM), 258–259, 267–271
Asset prices, 279–282, 340, 342, 346
Asymptotic convergence, 59, 62, 99, 135
Autoregressive moving average (ARMA), 160
Auxiliary probability measure, 74

B
Banach space, 141, 144, 146–147, 152
Barycentric approximation, 110
Barycentric bounds, 67–93
Barycentric measures, 69, 76–77, 80–81, 89–90, 92
Barycentric probability measure, 74–78
Basis functions, 329, 335–336, 342, 344–345, 347, 350
Batching, 299
Bayes, 281
Bayes estimate, 282
Bayraksan, G., 37–53
Bellman, R., 3, 10, 213–214, 221
Benchmark return, 190, 192, 200–201, 208
Benders, J.F., 14–21, 57–60, 134, 317
Bias, 31, 39–40, 44, 47–50, 52–53, 133
  reduction, 38–39, 44, 47–50
Biased estimator, 40
Bi-criteria
  optimization, 185
  problems, 287, 293–294
Bipartitioning procedure, 123
Birge, J.R., 13, 67–68, 78, 99–101, 103, 121–122, 135, 166, 183
Block
  -diagonal autoregressive processes, 72–73, 87, 92, 140
  structures, 177–180
Bootstrapping, 62
Borel probability measure, 75, 166, 175
Borel space, 69
Bound
  -based approximation, 99–103
  tightness, 101
Bounding sets, 68–69, 78, 81–82, 92
Branch and bound algorithm, 250
Brownian motion, 260, 280–281, 283, 330, 340, 346
Bundle methods, 182
Business-as-usual, 319

C
Candidate solution, 37–42, 44–45, 47, 49, 61
Capital
  accumulation, 280, 284, 293, 321
  growth theory, 277
Carbon concentrations, 319, 324
Cardinality constraint, 239, 249

G. Infanger (ed.), *Stochastic Programming*, International Series in Operations Research & Management Science 150, DOI 10.1007/978-1-4419-1642-6,

Catastrophic events, 318
Cell redefining, 123, 125–126
Certainty equivalent return rate, 351
Charnes, A., 6, 184
Chebyshev polynomials, 347, 350
Class of two-stage programs, 3
CLM property, 177
Coherent measures of risk, 206–207
Compact
  polyhedron, 103
  state space, 70, 342
Compactness, 146, 173, 176, 330
Complete local minimizing (CLM) set, 177
Complete recourse, 87, 92, 98, 143, 166
Computational aspects, 247–252
Conditional correction term, 80, 88–89, 91–92
Conditional distribution, 75, 92, 140, 281
Conditional expectation, 147, 282, 313
Conditional mean
  and covariance information, 120
  partitioning, 124–125
Conditional ratios, 268, 272
Conditional value at risk, 170, 173, 180, 196–199, 206
Confidence interval, 14, 20–21, 27–28, 38, 40–43, 45–46, 49, 52–53
CONOPT, 320
Conservative lower bound, 25–27, 31
Consistency, 38–41, 46, 61, 70, 140, 167–168, 171, 184, 324
Constrained nonlinear program, 335
Continuous time problem, 289, 291–292
Continuum of VaR constraints, 198
Control of diffusion, 329–352
Controlled diffusion processes, 329–330
Control limits, 279, 284, 288–289
Convergence, 38–39, 50–52, 58–59, 61–63, 65, 68–69, 72, 78, 80, 82, 99, 121–122, 133, 135, 140–141, 152–153, 172–173, 175, 177, 221, 253, 320, 339
  results, 38, 62, 69
Convex
  function, 4–7, 9–10, 72, 92, 98, 102, 108, 111, 127–128, 167, 204–205, 243, 245
  hull representation, 104
  program/programming, 98, 202, 205, 337–338, 346, 352
  risk measures, 279
Convexity, 7–8, 10, 75, 90, 98, 100, 104, 112, 119, 146, 152, 168, 185, 206, 258
Coordination procedure, 180
Correction term, 28, 30, 75, 78–80, 87–89, 91–92
Corrective action, 279, 284
Cost of embarrassment, 127–128
Coverage, 30–33, 41, 43, 46, 52, 86, 90–92
CPLEX, 64, 249, 346, 350
CVaR, 196, 198–199, 279, 288

D

Dantzig, G.B., 1–11, 13–34, 37–38, 43, 50, 57, 67, 78, 99, 139, 213, 248, 257
D.C. transaction cost, 249–252
Decomposition methods, 13–14, 57–58, 140, 180–185
Defined benefit, 258–259, 267
Defined contribution, 267
Degree of nonlinearity, 123
Demand
  progression, 299, 302–303
  -related data, 299, 304, 307–310
  signal, 299–300, 302, 313
  stages, 308
  uncertainty, 297–314
Dentcheva, D., 184–185, 189–210
Deterministic equivalent, 13–14, 99
Deterministic planning model, 304
Deviation measures, 170, 185
Diet, 1–2
Difference of convex functions (d.c.), 72–73, 75, 78–80, 87–88, 249–252
Diffusion, 160, 329–352
Dirac distribution, 71
Direct form, 193, 199–203
Discounted relative state frequencies, 345
Discrete probability measure, 74–75, 100, 140
Discrete random vector, 100
Discrete time problem, 291–294
Distortion, 189–190, 196, 204, 207
Distribution, 1–3, 7–9, 15, 19, 22–32, 37, 39–42, 45, 49, 51, 62–63, 68–69, 71, 74–76, 86, 88–90, 92
Distributional approximations, 99–100, 135
Dominance constrained optimization, 191
Downstream, 297
Drift, 84, 281, 340, 342, 346
Dual bounding vectors, 79
Duality gap, 182, 185
Dual utility theory, 189
Due date, 298, 300, 302–303, 307–308, 310, 313
Dynamic correlation structures, 257
Dynamic diversification, 264–266, 272
Dynamic financial analysis, 267
Dynamic investment process, 280

Dynamic programming, 3, 10, 213–215, 221, 224, 228, 232, 329, 331, 334–336, 339, 352

E
Economic consequences, 319
Economic growth theory, 277
Edge selection, 123–125
Edirisinghe, N.C.P., 68–69, 75–76, 93, 97–135, 140
Edmundson, H.P., 13, 68, 99, 103, 111, 116
Edmundson–Madansky inequality, 68, 103, 116
Effective dimension, 51
Efficient frontier, 130–135, 185, 231, 243–245, 250, 258, 264
Efficient strategies, 289–291
Electricity portfolios, 159
Electricity risk management, 154
Electric power, 28, 31, 37
Energy modeling forum (EMF), 324–325
Enterprise risk management, 267
Epi-consistency, 61
Epiconvergence, 82, 122
Equilibrium relations, 240, 244–247
Equivalent deterministic, 15
Error(s)
  estimate, 38, 51, 80, 82, 141
  in forecast, 279
  term, 20, 23–24
Euclidean space, 69, 72, 74, 142, 173
Evolution of demand uncertainty, 299
Excess probability, 170, 175, 185
Existence result, 141, 144
Expectation models, 166, 168–169, 175, 177, 179–180, 183, 185
Expected cost, 1–2, 4, 8, 10–11, 18, 127, 130
Expected discounted utility, 215, 219–222, 225–227, 229–230, 320
Expected excess, 170–171, 175, 184
Expected recourse function, 71, 97, 99, 103–104
Expected utility theory, 189–190, 203, 205
Expected value, 2, 4–5, 7–8, 10–11, 24, 75–76, 170, 184, 193, 195–196, 199, 213–215, 217, 219, 222–223, 227–228, 233, 239, 241
  of total utility, 213

F
Factorization of basis functions, 344–345
Failures, 3, 299
Fat tail, 325
Ferguson, A., 3, 10
σ-Field, 139, 144
Filtration, 147, 153, 280–281, 330, 340
Filtration distance, 147, 153
Finance, 37, 67, 69, 139, 154, 185, 190, 267
Financial risk management, 82–83
Finite first moment, 166
Finite master program, 62
First-moment upper bound, 107
First-order
  dominance, 195, 284–285, 288
  moment bound, 103–108
  moment information, 68, 104
  moments, 68, 103–108
Fixed-mix policy, 259–262, 265–266
Fixed recourse, 58, 97–98, 139
Fixed sample size procedures, 38
Fortet–Mourier distance, 141
Forward selection algorithm, 155
Forward tree construction, 141, 157, 160
Four-stage problem, 89
Fractional Kelly, 278–279, 289–292
Frauendorfer, K., 13, 67–93, 99, 110–111, 120, 123–124, 135, 140
Freight scheduling, 31, 64
Functional approximations, 99–100
Fundamental problem of investment, 280, 291
Future research, 27, 259, 268, 271–272

G
GAMS, 43, 317, 320
Gap estimator, 48, 50
Gaussian variables, 281
Generalized moment problem, 68, 99, 101
Generalized Slater condition, 73
General linear structure, 7–9
Generating scenario trees, 140–142, 154–158
Generator of the order, 190
Geometric Brownian motion, 280, 283
Global
  climate change, 317–318
  climate decisions, 317–327
Glynn, P.W., 14, 38, 43, 50, 99
Greedy policies, 333
Greenhouse gas, 317–318
Growth
  optimal, 278
  and security, 277–294

H
Hamilton–Jacobi–Bellman equation (HJB), 331–332, 334, 342, 346–347
Han, J., 329–352
Hausdorff topological space, 144
Heitsch, H., 139–163

Heteroscedasticity, 85–86, 88
Heuristics, 180, 182, 185, 214–217, 219, 227, 229–232, 350
Higher order moments, 76, 120, 135
  bound, 118–120
Higle, J.L., 14, 31, 38, 43, 50–51, 57, 59–61, 99, 140, 297–314
Huang, K., 13, 57–65, 68, 99

I
Illiquidity, 231
Implied measure of risk, 207–208
Implied rank-dependent utility function, 204–206
Implied utility function, 202–204, 209
Incomplete information, 67
Incumbent, 60–61
Infanger, G., 13–34, 38, 43, 50, 57, 64, 99, 140, 213, 257
Information constraint, 165
Integer requirements, 165
Integrated model, 267, 299
Interior-point methods, 13, 335
Intertemporal constraints, 321
Interval estimator, 38, 40–41, 43
Inverse form, 190, 195–197, 200–201, 204–206
Investment and wealth, 282–284

J
Jackknife estimator, 44, 47–50, 52
Jensen, J.L., 13, 68, 99, 102, 105, 108, 111, 116
Jensen's inequality, 68, 102, 105, 116
Joint continuity, 174–177

K
Kelly, J., 278–279, 289–292
Kempf, K., 297–314
Kim, W.C., 257–272
Konno, H., 190, 239–254
Kuhn, D., 43, 67–93, 123, 135, 140, 253
Kusuoka measure, 207
Kyoto accord, 319

L
Lagrangian heuristics, 182, 185
Lagrangian relaxation, 180, 182–185, 191, 205, 207
Latin hypercube, 43, 51–52
Law invariant, 206–207
Lead time, 298
Lebesgue measure, 74, 169, 172–173, 252
Lemke, C.E., 6
Limited moment information, 97–135
Limit points results, 41
Linear affine functions, 75
Linear programming
  characterization, 329, 334, 342
  problem, 2, 5, 7, 202, 239, 250
Line yield, 300–301, 305
Lipschitz continuous, 169, 330, 340
Lipschitz property, 148–149
Lipschitz stability, 141
Locking constraints, 319
Logarithm of wealth, 278
Lognormal process, 260
Lorenz curve, 190, 195, 287
Lorenz dominance, 195
Loser portfolio, 264
Lower bound, 8, 13–34, 40, 43–44, 46–48, 52, 59–60, 68, 88–90, 93, 99, 102, 105, 107–109, 111–114, 116–120, 122, 128, 180, 182–184, 224
  estimator, 48, 52
Lower semicontinuous, 45, 146, 169, 172–174
L-shaped decomposition, 99, 133
L-shaped method, 13, 58

M
MacLean, L.C., 277–280, 282, 284, 286, 288, 290–293
MAD, 239–249, 251–252
Madansky, A., 13, 68, 99, 103, 111, 116
Manne, A.S., 317–327
Market
  impact, 218, 239, 257
  model, 340
Markov chain, 305–310, 313–314
Markowitz, H.M., 2, 4, 168, 190–191, 213–236, 239, 242–243, 278–279
Mean–absolute deviation, 239–254
Mean–covariance upper bound, 114, 120
Mean reversion, 84, 87
Mean–risk model, 166–168, 171, 184–185, 189–190, 192, 207, 240
Mean–variance, 116, 118, 213–236, 239–241, 251, 279
  heuristics, 214–217, 219, 227, 230–232
Measurability, 71, 139, 169
  constraint, 139
MERGE, 317–318, 321–327
Mid-echelon, 297
Minimal transaction, 239, 249, 251
Mixed-integer linear programming, 170, 178–179, 182
Mixed-integer value function, 168

Momentum, 264–266, 272
Monotonicity, 48, 206
Monte Carlo, 14, 19, 26–27, 40, 43–44, 50, 53, 140, 214, 231–232, 268, 351
Monte Carlo sampling, 14, 19, 40, 53
Morgenstern, O., 189, 193, 203, 206, 213
Morton, D.P., 37–53, 68, 99, 117–120, 135, 140, 298
Mount Lucas, 263
Multi-cut, 13
Multifunction, 73, 79, 81–82, 177
Multiple optimal solutions, 39, 41, 46
Multiple replication procedure, 48
Multiregion, 317, 320, 322–327
Multistage, 7, 9, 74, 77, 135, 299, 314
  financial planning, 257–272
  stochastic program, 67, 69–71, 73, 75–76, 78, 81, 83–84, 88, 90, 92, 139–163, 185, 257, 268, 299, 314
Multitechnology, 322–327
Multivariate normal distribution, 239–240, 242–243
Multivariate stochastic processes, 139
Mulvey, J.M., 184, 257–272, 278
Myopic strategy, 350–351

N

Negishi weight, 320–321
Non-anticipative decision process, 70
Nonanticipativity, 70, 143, 146, 179–185, 258
  subspace, 143, 146
Nonanticipatory decisions, 166
Nonconvex commission, 239
Nonconvexity, 239
Nonlinear, 40, 43–44, 73, 87, 89, 91–92, 123–125, 154, 169, 180, 192, 209, 250–251, 268, 318, 320, 335, 337–338
Nonnormal errors, 39–41, 44
Nonprobabilistic error estimate, 80, 82
Nonsmooth concave maximization, 182

O

Objective function, 5, 11, 15, 30–31, 43, 60, 64–65, 70–71, 73, 75, 79, 83, 87, 89–92, 142–143, 146, 171–174, 180, 199, 202–203, 205, 208, 248, 252–253, 269, 272, 303–304, 320, 334–335, 337
One-world model, 317, 321–322
Optimal value function, 71, 330–331, 333–336, 342–343, 346–347
Option, 40, 44, 53, 257, 288
Orders of dominance, 285
Ornstein–Uhlenbeck, 346
Out of control paths, 284
Outer optimization, 99–100
Over-optimize, 40, 52

P

Pareto efficient portfolio, 190
Partani, A., 37–53
Partial information, 304, 310–314
Partitioning, 68–69, 89, 111, 120–126, 128, 130, 180
  strategies, 123
Passenger demand, 141
Path, 221, 277, 280, 282, 284, 287–288, 331
  measure, 287–288
Pentagon, 90
Piecewise linear approximation, 58–60
Policy
  rule, 37, 258–263, 265–268, 271–272
  simulators/simulation, 257–272
Polyhedral, 81, 98, 100, 102–104, 111–112, 114, 120, 139, 144, 154, 169, 185, 191, 200, 204, 250
Pompeiu–Hausdorff distance, 148
Pompeiu–Hausdorff metric, 74
Portfolio(s), 24, 159, 190–192, 198, 209, 232, 243–244, 261–264
  choice, 216, 340–342, 346
  function, 341
  optimization, 141, 168, 189–210, 239, 252, 329–330, 335, 339–340, 342, 346, 352
  problem, 189, 191–192, 199–200, 284
  selection, 168, 216–217
  strategy, 341–342, 345–347, 350–351
Positive homogeneity, 206
Posterior distribution, 281–282
Prékopa, A., 13, 166, 184, 195
Primal bounding vectors, 79
Probabilistic, 13–34, 69, 195, 279, 306, 310, 312
Probability
  measure, 74–77, 98–99, 101, 111, 166, 172, 174, 177, 206, 252
  space, 37, 58, 69, 74, 139, 154, 207, 330
Production
  planning, 297–314
  related data, 299, 304–307
  scheduling, 43, 298–299
  uncertainty, 298
Progressive hedging, 13
Pseudo cut, 15, 19–20, 31
Pseudo master problem, 20, 22, 25, 27, 30–31

Q

Quantiles, 170, 192, 195, 198, 206, 279
Quasi Monte Carlo, 44, 50, 53, 140

Quick and clean, 240
Quick and dirty, 240

R
Randomized quasi Monte Carlo (QMC), 44, 50–51, 53
Randomized sampling, 336
Random vector, 37, 42, 50–51, 69–71, 75, 86, 98, 100, 104, 114, 116, 119, 128, 135, 144, 185
Rank-dependent utility function, 192, 196, 204–207
Rebalancing, 257, 259–263, 265–267, 280, 282, 293, 350
  gains, 257, 259, 262–263, 265–266
Recourse function, 58, 65, 68–69, 71, 73, 75–78, 80–81, 88, 92, 97, 99, 103–104, 112, 116–118, 120, 122–124, 128, 135
Recourse matrix, 15, 71, 143
Rectangular domain, 111–112, 119, 124–125, 135
Rectangular partitioning, 111, 123
Refinement procedure, 89
Regularity condition, 69, 71–73, 75, 78–81, 86–89, 92
Regularized approximation, 60
Regularized decomposition, 13
Relatively complete recourse, 87, 92, 143
Reorder, 25–31, 33
Resampling, 61–65, 141, 154, 160
Risk averse optimization, 189
Risk aversion, 128, 165–185, 189, 218, 271, 279, 285, 287–288, 341, 350–351
Risk control, 189–210, 280
Römisch, W., 139–163, 175, 185
Russell system, 258
Ruszczyński, A., 13, 31, 139–140, 142, 166–167, 170–171, 173, 184–185, 189–210, 239–240, 243–244, 287
Rutherford, T., 320, 327

S
S&P 500 equal-weighted index, 262
S&P 500 index, 262–263, 271
Saddle function, 68–69, 75–77, 80, 98
Saddle property, 88, 102, 104, 112
Safety stock, 298
Sample
  average approximation, 14
  space, 7, 69
Sampling, 13–15, 19–20, 28, 38, 40, 43–44, 50–53, 57, 59, 99, 120, 140–141, 154, 272, 292
  error, 19, 38, 40, 44, 52
  scheme, 38, 140
  states, 336
Samuelson, P.A., 227, 259
Scenario(s), 13–14, 28, 31–32, 52, 85, 93, 99, 103, 120–122, 128–131, 134–135, 140–141, 154–161, 179–181, 183–184, 268, 289, 292–293, 314
  clustering, 130–131
  reduction procedure, 155
Schürle, M., 67–93
Scenario tree
  construction, 74–75
  process, 154
Schultz, R., 165–185
Scrapped inventory, 303
Second-degree stochastic dominance, 239, 243–244
Second-order, 33–34, 40, 68, 75, 104, 112, 117–118, 122–123, 190, 193–198, 203, 205, 228, 284, 287–288
  dominance, 194, 196–197, 284, 288
Semi-deviation, 170–171, 175, 184–185, 192
Semi-infinite linear programming, 200
Sen, S., 14, 31, 38, 43, 51, 57–65, 99, 135, 140
Sensitivity analysis, 130
Separation property, 290, 293–294
Sequence of linear programs, 337–338
Sequential decisions, 67
Setups, 298–299
Simple recourse, 98
Simplex, 86, 89–90, 93, 109, 114, 119, 121–126, 130
  algorithm, 67, 78
Simplicial domain, 109, 112, 116, 119–121, 123, 135
Simplicial sequential approximation, 120
Simulation-based optimality test, 37–53
Single-period, 213–236, 257
Single replication procedure, 45–46
Single-stage, 257, 298
Spectral risk measure, 207–208
Speed of convergence, 122
SSD, 193, 239, 284
Stability, 122, 133–134, 139–162, 174–175, 177
  analysis, 153, 174–175
  scenario tree, 139–162
Standard Brownian motion, 281, 283, 330, 340
Standard normal distribution, 86, 90
State-action pair, 330, 335
State space, 69–70, 72–73, 76–77, 88, 92, 154, 219, 330, 336, 341–342, 352
States of the world, 317–320

Static portfolios, 257
Stationary policy, 331, 333
Stationary strategy, 341
Statistical model, 159
Stigler, G.F., 1
Stochastically sound, 167
Stochastic decomposition (SD), 14, 57–65, 99
Stochastic dominance, 167, 171, 184, 189–210, 239, 243–244, 277–294
  constraints, 189–210
  measures, 284–288
  relation, 189, 192–199, 203, 205, 290
Stochastic integer programming, 165–185
Stochastic order, 190
Stochastic process, 69–70, 72, 77, 84, 93, 139–140, 142, 154–156, 158, 297, 304
Stochastic program, 37, 44, 50–51, 57, 67–93, 97–135, 142–143, 146, 154, 166, 169–170, 175, 184–185, 213, 239, 257–259, 267–268, 271–272, 297–314
Stochastic programming approximations, 97–135
Stochastic quasi gradients, 99
Stock price process, 260–261
Stopping rules, 61, 257
Structurally sound, 167
Successive approximations, 99
Sufficiently expensive recourse, 166
Supply
  and demand networks, 297
  progression, 300–303, 306
  uncertainty, 298
Surrogate function, 214
Surrogate heuristics, 215–216, 231–232
System of inequalities, 22, 193

T

Target, 97–98, 127–128, 130, 140, 170, 193, 257, 279, 299, 325
  shortfall, 127–128, 130
  violations, 279
  wealth, 257
Taxable entities, 231
Technology matrix, 71, 98
Temperature targets, 325
Termination, 18, 21, 65, 123, 130, 133, 258, 345
Term structure model, 84–85, 346, 348
Test problem, 28, 30–34
Tetragon, 90
Three-stage, 3
Throughput, 297, 299, 301, 305–306, 314
Tightening first moment bounds, 108–111
Tightening second-order bounds, 117–118
Towers Perrin, 258, 267
Trajectories, 277, 279–280, 282, 284, 289, 291, 350
Transaction cost, 83, 127–128, 130–134, 214–216, 218–219, 227–228, 230–233, 235, 249, 251, 266–267, 340, 352
Transition
  matrix, 15, 232, 308, 314
  probabilities, 74–75, 77–78, 219, 234, 306–307, 310
  process, 277
Translation equivariance, 206
Triangle, 86, 90, 147–148
True cut, 15–17, 19, 31
Two-fund separation theorem, 244
Two replication procedure, 44–47, 52
Two-stage, 2–7, 10–11, 13–34, 37–38, 43–44, 58, 82, 97, 99, 127, 135, 140, 142, 153, 165–185
  stochastic linear program, 14, 37, 43, 58, 97

U

Unbiased estimator, 40, 42, 51
Uncertain demand, 2–3, 10–11
Uniform random vector, 50
Unique limit, 58, 61–62, 65
Uniqueness, 37, 58, 61, 65, 72, 75, 77, 89–90, 109–111, 114, 119, 152, 156, 223, 243, 246, 334, 336
Unmet demand, 303
Upper bound, 13, 18, 24–26, 28–29, 42–43, 68, 87, 92, 101–103, 105, 107–109, 111, 114–116, 119–120, 122, 221, 350–351
Upper bounding function, 99, 102
Upstream, 297
Utility, 67, 189–190, 192–193, 196, 202–207, 209–210, 213–232, 235–236, 239–240, 243, 278–279, 284–288, 320, 326, 330, 341–344
  based approach, 189
  function, 189–190, 192, 196, 202–207, 209–210, 213–215, 217–218, 227–228, 231–232, 243, 279, 284–285, 287–288, 327, 330, 341–343

V

Value function, 71, 90, 99, 166, 168–170, 329–331, 333–336, 342–343, 346–347, 350–351

Value at risk (VaR), 24, 42, 45, 51, 154, 170, 173, 180, 185, 191, 195–199, 206, 279, 288, 292–293
Van Dijk, E.L., 213–236
Van Roy, B., 329–352
Van Slyke, R.M., 13–14, 58, 99, 133
VaR, 24, 42, 45, 51, 154, 170, 173, 180, 185, 191, 195–199, 206, 279, 288, 292–293
VaR constraint, 196, 198, 288, 292
Variable demand, 298
Variance, 2, 4, 21, 23–24, 27–28, 31, 38, 42, 45–46, 49, 51–52, 60, 99, 117–118, 167–169, 191, 213–236, 239, 241–242, 251, 260–261, 279, 287, 290, 292
  and covariance information, 114, 116–117, 119–120
  reduction, 14–15, 44, 50–52, 99
Volatility, 84, 258–259, 262, 264, 266, 269–272, 279, 281, 283, 300
  pumping, 259
Von Neumann, J., 189, 193, 203, 206, 213, 240
Von Neumann–Morgenstern, 189, 193, 203, 206

W
Wasserstein distance, 141
Weak topology, 144, 146, 151–152
Wealth
  control, 288–289
  process, 260–261, 283–284, 286, 341
Weighted difference criterion, 123
Wets, R.J., 13–14, 38, 58, 68, 73, 82, 97–101, 103, 121–122, 133, 143, 145, 148, 150, 153
Winner portfolio, 264
Work-in-process, 301
Worst-case lower bound, 24–25, 29–31

Y
Yasuda, 258
Yield curve, 83–86

Z
Zhao, Y, 277–294
Zhou, Z., 57–65, 284
Ziemba, W.T., 37, 68, 75, 98–99, 103, 108–111, 123–125, 139, 252, 257, 259, 277–294

MIX
Papier aus verantwortungsvollen Quellen
Paper from responsible sources
FSC® C105338

If you have any concerns about our products,
you can contact us on
ProductSafety@springernature.com

In case Publisher is established outside the EU,
the EU authorized representative is:
Springer Nature Customer Service Center GmbH
Europaplatz 3, 69115 Heidelberg, Germany

Printed by Libri Plureos GmbH
in Hamburg, Germany